Power Systems

More information about this series at http://www.springer.com/series/4622

Duco W.J. Pulle • Pete Darnell • André Veltman

Applied Control of Electrical Drives

Real Time Embedded and Sensorless Control using VisSim^{TM} and PLECS^{TM}

Springer

Duco W.J. Pulle
EMsynergy
Sydney
Australia

André Veltman
Piak Electronic Design
Culemborg
The Netherlands

Pete Darnell
Senior VP Altair Engineering
Westford, Mass
USA

ISSN 1612-1287 ISSN 1860-4676 (electronic)
Power Systems
ISBN 978-3-319-20042-2 ISBN 978-3-319-20043-9 (eBook)
DOI 10.1007/978-3-319-20043-9

Library of Congress Control Number: 2015942902

Springer Cham Heidelberg New York Dordrecht London

Printed on acid-free paper

Springer International Publishing AG Switzerland is part of Springer Science+Business Media (www.springer.com)

In memory of Prof. ir. J.A. Schot

1927–2015

Foreword

Control of adjustable speed drives has developed at a rapid pace ever since the introduction of digital signal processors (DSPs), first by Texas Instruments, in the early eighties. Digital control of inverter fed machines, such as induction, synchronous machines or switched reluctance machines, made possible the implementation of dynamic torque control algorithms developed in the late seventies. Combining DSPs with programmable logic devices to control and protect the power electronic PWM voltage source inverters enabled torque control with high precision and wide bandwidth. Today, as more computational power can be integrated at low cost, more sophisticated algorithms can be implemented to make drives more robust, smarter, and easier to integrate in various applications. For example, modern drives are equipped with algorithms that automatically tune field oriented (torque) control parameters of rotating field machines. Sensor-fusion for eliminating expensive position encoders is yet another development that finds its way into state-of-the-art drive systems. No doubt, when looking at modern, high-performance drives today, any novice to the field may feel overwhelmed by the complexity of the controllers developed over the past 35 years. With this book, the authors aim to provide tools for practicing engineers and researchers alike, helping them to become familiar with the control of modern electrical drive systems, in particular drives based on induction (IM) and synchronous permanent magnet (PM) machines. In contrast to many works and publications on this matter, the authors present a systematic learning-by-doing approach. Using well-known software simulation tools, such as PLECS and TI's MotorWare laboratory software, the laboratory sessions of the book invite the reader to learn, step-by-step, field oriented control for rotating field (IM and PM) machines, open and closed loop torque control and position sensorless torque control. Using VisSim, an object oriented embedded control software package, the control algorithms can be readily validated experimentally in a (low voltage) drive system that has been developed specifically for this purpose.

The book is structured in well-defined modules, taking the reader through a sequence of laboratory sessions with increasing complexity. The learning targets of each module are clearly described, so that any drive designer can step in at his or her level of experience. As such, this work can be recommended as a reference book, even to experienced drive designers. In addition, case studies illustrate how the transition of the control algorithms can be made to other (industrial) drive platforms.

The authors have been able to develop a unique learning platform to help readers quickly gain experience on how modern drive control algorithms function and how they can be developed. The book bundles state-of-the-art algorithms and software design tools that make development of inverter fed machines, not only a lot more productive, but also a fun experience, which I can recommend to anyone interested in the exciting field of electrical drive systems.

Aachen, Germany Prof. Dr. ir. Dr. h.c. Rik W. De Doncker
January 2015

Preface

Our previous books, 'Fundamentals of Electrical Drives' [14] and 'Advanced Electrical Drives' [4], provide the reader with an in-depth understanding of the theoretical aspects of a drive, complemented by a series of simulations. The novelty of this book is that it takes a 'hands-on' electrical drive approach, designed to take the reader through a series of specific laboratory examples. These start with a basic AC drive and progress to sensored field-oriented control, and ultimately to sensorless control, using Texas Instruments InstaSPIN [9] technology for induction (three and single phase) and permanent magnet machines.

For this purpose, a low cost, low power dual drive hardware platform LAUNCHXL-F28069M, developed by Texas Instruments, is introduced together with an induction/PM machine combination. An embedded control approach using VisSim is used throughout this laboratory series, which allows the reader to concentrate on understanding drive operation. The comprehensive set of InstaSPIN based sensorless control laboratories for induction and PM machines included in this book is designed to familiarize the reader with this technology. Furthermore, case study examples are discussed that demonstrate the use of InstaSPIN in industrial applications. An alternative software development approach using VisSim and PLECS is also discussed where use is made of a 'processor in the loop' approach, which allows sensorless InstaSPIN drive development using either a model of the machine/converter or the actual drive. In this context, use is made of the Texas Instruments Code Composer software environment.

This book should appeal to industry and/or university based readers who have a need to understand electrical drives starting at a basic level with minimum theoretical content and/or develop their own AC drive application using sensored or sensorless technology. To accommodate this approach, the book has been configured in such a manner that the reader can step into the chapter of interest, without having to 'back flip' through the book.

In terms of content, this work starts with a basic overview of electrical drive principles, designed to provide the reader with key information on the critical parameters, which govern speed and current control. This is followed by a brief introduction on space vector modulation and fixed point scaling, given that many applications still use 'fixed point' processors. The next three chapters deal with VisSim based 'hands-on' laboratory sessions on sensored drive control, sensorless PM and IM drive control using InstaSPIN. The final two chapters consider case studies where attention is given to industrial sensorless drive examples using VisSim and PLECS 'processor in the loop' technology. These last two chapters have been purposely added to assist the reader with making the transition from the material presented in this book to his/her own drive application. In terms of overall content, the reader is reminded of the fact that approximately 70% of the material presented in this book is associated with hardware/software. A disproportionately large amount of time, prior to publication, was used to remove hardware and software errors (bugs), but the reader is reminded to use the Springer 'extra materials' website link [11] for obtaining the latest files of the experiments that accompany this book.

It is our sincere hope and desire that this book will make electrical drives more accessible and understandable to those who have formerly been reluctant to engage with this technology.

Sydney, Australia — Duco W. J. Pulle
Westford, Mass, USA — Pete Darnell
Culemborg, The Netherlands — André Veltman

Acknowledgments

The process of writing this book has proved to be extremely challenging given the widespread use of experimental/practical examples. Furthermore, a decision was made to use the latest Texas Instruments drive hardware so that the reader would have the benefit of low cost companion hardware for this book. In addition, the development of software tools by VisSim and PLEXIM needed to be undertaken so that the reader can fully and seamlessly evaluate InstaSPIN based sensorless drive technology. Furthermore, the introduction of a 'case study' chapter where the reader can peruse the use of commercial drive platforms has led to some very intensive cooperation with the companies concerned. In all of the activities above, a significant number of individuals and companies have been involved, all of whom have been instrumental in getting this book to fruition. We would like to particularly thank the following persons (in alphabetical order) for their contributions: Beat Arnet, Robert Beekmans, Swaroop Bhushan, Chris Clearman, Wolfgang Hammer, Paul van der Hulst, Geert Kwintenberg, Dave Magee, Adam Reynolds, Eric Thomas, Gang Tian, Orhan Toker, Dave Wilson, Feitse van der Zou, Zhen Yu, and Jorge Zambada.

Last but not least, Prof. Pulle would like to acknowledge the feedback from the many attendees of my worldwide workshops that are associated with this book. No doubt that those who have attended will recognize the material presented and their feedback has been instrumental in getting this book where it is now.

Contents

Chapter 1
Introduction

Prior to discussing the control of electrical drives, it is prudent to provide an introductory overview of the concept and the hardware used in this book. In addition the notational conventions, both electrical and mechanical are introduced together with the control blocks used to build drive controllers in successive chapters. Both, so called floating and fixed point control blocks are introduced together with continuous and discrete time modules given that an 'embedded control' approach is used throughout this book. The term 'embedded control' refers to control algorithms that are located within the processor (MCU) used to control an electrical drive.

1.1 Importance of Electrical Drives and Electrical Machines

Electric motors are around us everywhere. Generators in power plants are connected to a three-phase power grid. Pumps in your heating system, refrigerator and vacuum cleaner are connected to a single phase AC grid and switched on or off by means of a simple contactor or electronic converter. In cars a direct current (DC) battery is used for example, to provide power to the starter motor. This type of motor runs on direct current and in most cases is activated by a relay switch without any control.

Many applications driven by electric motors require more or less advanced control. Lowering the speed of a fan or pump can be considered a relatively simple task. Perhaps one of the most difficult ones is the dynamic positioning of a tug in a wafer-stepper application with nanometer accuracy while accelerating at several g's. Another challenging controlled drive is an electric crane in a harbor that needs to be able to move an empty hook at high speed, navigate heavy loads up and down at moderate velocities and make a soft touchdown as close as possible to its intended final position. Other applications such as assembly robots, electric elevators, electric motor control in hybrid vehicles, trains, streetcars, or CD-players can, with regard to complexity, be situated somewhere in between.

D.W.J. Pulle et al., *Applied Control of Electrical Drives*,
Power Systems, DOI 10.1007/978-3-319-20043-9_1

Design and analysis of all electric drive systems requires not only knowledge of dynamic properties of different motor types, but also a good understanding of the way these motors interact with power-electronic converters and their loads. These power converters are used to control motor currents or voltages in various manners.

1.2 Key Components of the Electrical Drive System

The drive concept of primary interest in this book is capable of what is referred to as 'adjustable speed' operation [4, 14] which means that the machine can be made to operate over a wide speed range. A simplified structure of an adjustable speed drive is shown in Fig. 1.1, which makes use of a Texas Instruments F28069M microcontroller (MCU). In this book this MCU is used for embedded control, however the algorithms can be readily adapted to different processors and scaled to suit different drive voltage/current requirements. A brief description of the components is given below:

- Load: this component is central to any drive in that the purpose of the drive is to meet specific mechanical load requirements. It is emphasized that it is important to fully understand the nature of the load and the user requirements which must be satisfied by the drive. The load component may or may not have sensors to measure either speed, torque or shaft angle. In our case the load is a low voltage PM machine connected to an incremental encoder or an induction machine.
- Motor: the majority of motors in use are three-phase induction (asynchronous) or permanent magnet synchronous machines. Both machine types are discussed in this book. In this book either a 24 V three-phase PM or IM machine is used. Note that the term 'motor' refers to a machine which operates as a motor, i.e. energy flows from the motor to the load. When the energy flows in the opposite direction a machine is said to operate as a generator.
- Converter: this unit contains a set of power electronic (semiconductor) switches which are used to manipulate the energy transfer between power supply and motor. The use of switches is important given that no power is dissipated (in the

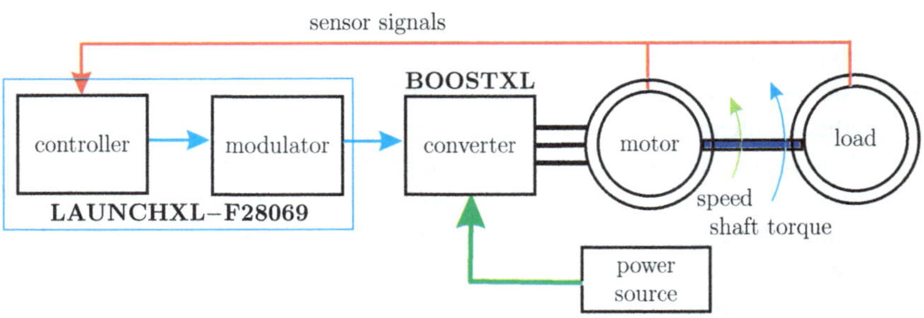

Fig. 1.1 Typical drive set-up

ideal case) when the switches are either open or closed. Hence, theoretically, the efficiency of such a converter is 100 %, which is important particularly for large converters given the fact that semiconductor devices cannot operate at high temperatures. Hence, it is not possible to absorb high losses which inevitably appear in the form of heat. In this book the basic converter used is a $\approx 6 \times 6$ cm Texas Instruments `BOOSTXL-DRV8301` 24 V/10 A module.

- Controller/Modulator: the controller, in this book is a Texas Instruments digital Micro Controller Unit `F28069M` (MCU) which houses a real-time embedded control algorithm known as InstaSPIN which implements sensorless control and protects the converter and attached Machines. The MCU is located on the Texas Instruments `LAUNCHXL-F28069M` board which can also accommodate two `BOOSTXL-DRV8301` converter boards.
- Power supply: the converter requires, in our case, a 24 V DC voltage source. The power can be obtained directly from a DC power source, e.g battery when available. However, in existing drive systems the DC voltage usually results from rectifying a single-phase voltage or three-phase AC power grid voltages (referred to as the 'grid').

Shown in Fig. 1.2 is the drive hardware predominately used in this book that consists of the `LAUNCHXL-F28069M` board which has provision for two converter `BOOSTXL-DRV8301` boards. This hardware setup is convenient for both sensored and sensorless laboratories with either a low voltage PM or IM three-phase machine. The ability for dual machine operation is particularly useful as it allows one machine to be used as a 'dynamometer' (which measures torque and speed) whilst the other is setup to accommodate a specific laboratory. Control of both machines is carried out by a single `F28069M`MCU located on the low cost `LAUNCHXL-F28069M` board.

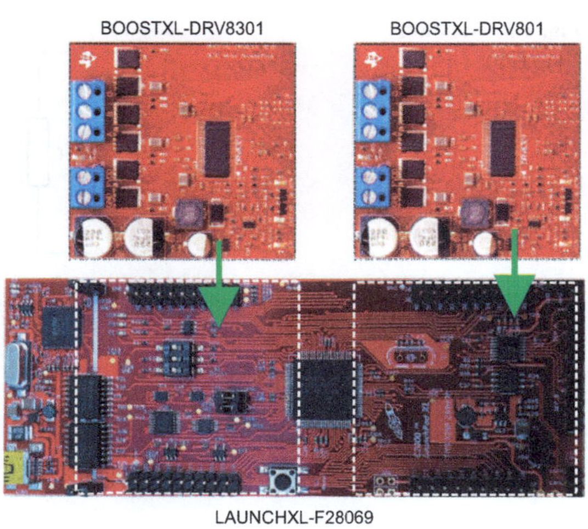

Fig. 1.2 LAUNCHXL-F28069M with dual BOOSTXL-DRV8301

1.3 Notational Conventions

1.3.1 Voltage and Current Conventions

The conventions used in this book for the voltage and current variables are shown with the aid of Fig. 1.3. The diagram shows the variables voltage u and current i, which are specifically given in 'lower case' notation, because they represent instantaneous values, i.e. are a function of time. The voltage and current 'arrows' shown in Fig. 1.3 point to the negative terminal of the respective circuit, i.e motoring arrow system, in which positive power $p = u\,i$ means power absorbed by the electrical circuit (load).

1.3.2 Mechanical Conventions

The mechanical conventions used in this book are shown with the aid of Fig. 1.4. The electro-magnetic torque T_{e} produced by the machine corresponds with a power output $p_e = T_{\mathrm{e}}\,\omega_{\mathrm{m}}$, where ω_{m} represents the rotational speed, otherwise known as the angular frequency. The load torque T_{l} is linked to the power delivered to the load $p_L = T_{\mathrm{l}}\,\omega_{\mathrm{m}}$. Ignoring bearing and windage losses in the machine, the torque difference $T_{\mathrm{e}} - T_{\mathrm{l}}$ results in an acceleration $Jd\omega_{\mathrm{m}}/dt$ of the total rotating mass, which is characterized by its inertia J. This rotating structure is represented as a lumped mass formed by the rotor of the motor, motor shaft and load. The corresponding mechanical equation which governs this system is of the form

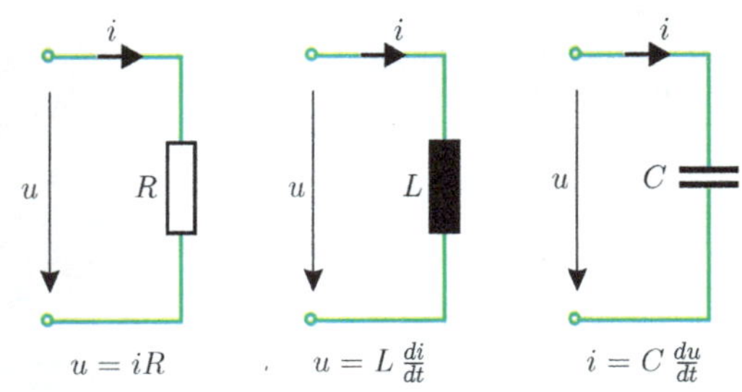

Fig. 1.3 Notation conventions used for electrical ideal components [14]

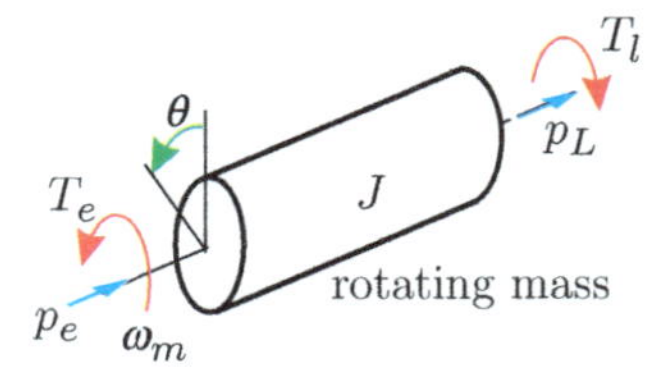

Fig. 1.4 Notation conventions used for mechanical quantities [14]

$$J\frac{d\omega_{\mathrm{m}}}{dt} = T_{\mathrm{e}} - T_{\mathrm{l}} \tag{1.1}$$

The angular frequency may also be written as $\omega_{\mathrm{m}} = d\theta/dt$ where θ represents the rotor angle. Figure 1.4 shows the machine operating as a motor, i.e. $T_{\mathrm{e}} > 0$, $\omega_{\mathrm{m}} > 0$. Note that motor operation also takes place when the conditions $T_{\mathrm{e}} < 0$, $\omega_{\mathrm{m}} < 0$ are satisfied. These motor conventions are used throughout this book.

1.4 Use of Building Blocks to Represent Equations

In this book use is made of PLECS [10] software, which is able to represent: electrical, mechanical or magnetic circuits with corresponding elements. For example the mechanical system shown in Fig. 1.4 can be readily represented by a set of 'mechanical rotational' blocks, as may be observed from Fig. 1.5. This type of approach can be extended to model complex rotation and/or linear mechanical systems, where sensors such as those shown for shaft speed and angle in the above example can be used for control purposes. When modeling electrical circuits using PLECS it is possible to also include power electronic components such as diodes and switching devices (such as IGBT's and MOSFET's). For the development of control structures use is made of so called 'control blocks' to represent equations, in which case either PLECS or VisSim [15] is deployed in this book. This approach is also useful for representing an electrical circuit with one or more control blocks. Furthermore, discrete time control blocks can be combined to develop more complex 'compound' models, which can be directly used to generate a real time controller, without requiring any knowledge of C code. This control development approach, referred to as 'embedded control' can in the case of VisSim be extended to include the actual motor and converter.

Models in this form can then be analyzed by the reader in terms of the expected transient or steady-state response. Furthermore, changes can be made to a model to observe their effect. This interactive type of learning process is particularly useful in order to become familiar with the material presented.

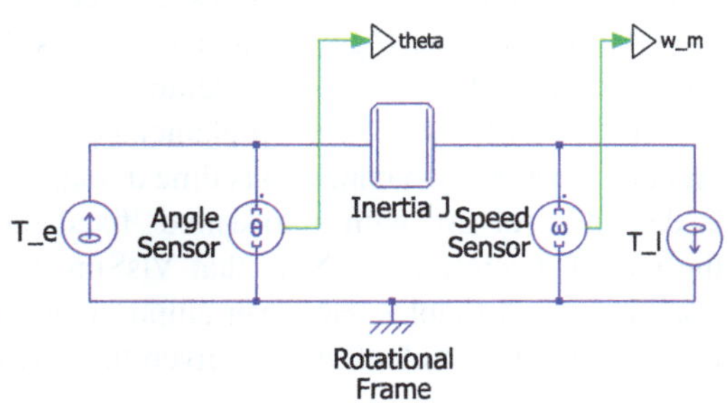

Fig. 1.5 PLECS representation of a lumped inertia system

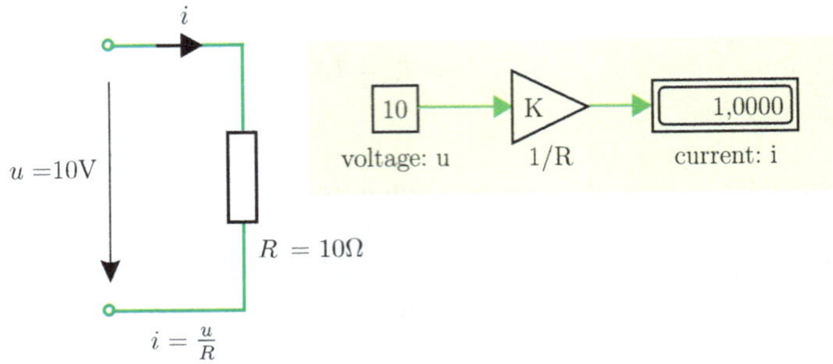

Fig. 1.6 Electrical circuit and PLECS based control block representations

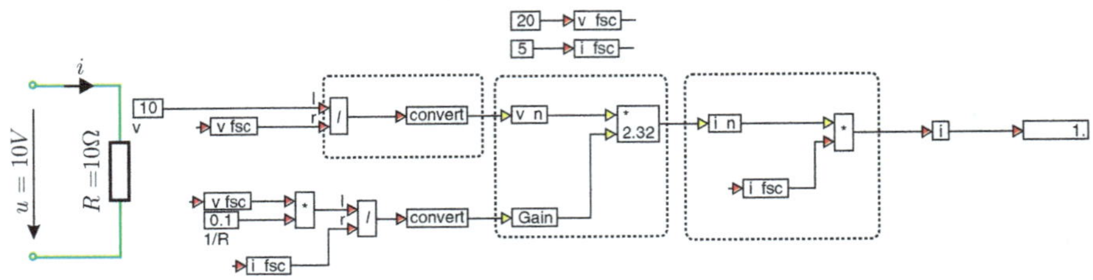

Fig. 1.7 Electrical and VisSim based control block representations

An example of moving from electrical circuit to control block representation is given in Fig. 1.6. The electrical circuit shown in Fig. 1.6 represents a resistance. This component has a known relation between voltage and current, as defined by Ohm's law: hence one can calculate current from voltage, voltage from current or resistance from both voltage and current. The control block approach assumes in this case that the voltage u, set to 10 V, is an input and the current i represents the output variable for this module known as a 'gain' module. The gain for this module must in this case be set to $1/R = 0.1$ for the chosen numerical example, in which case the current output will be equal to $i = 1.0$ A as shown on the numerical display. Note that the 'electrical circuit' model given in Fig. 1.6 can also be directly represented using PLECS software.

The same example can also be generated using the VisSim 'fixed point' library, as shown in Fig. 1.7, which is used here because the intention is to generate control structures which can be experimentally verified, i.e. implemented in real time. Observation of Figs. 1.6 and 1.7 learns that the PLECS and VisSim control block implementations are considerably different. The fundamental reason for this is that the PLECS version is implemented in so called 'floating point' and (for all practical purposes) in continuous time domain, whereas the VisSim implementation is designed for use with a so called 'fixed point' MCU, which in turns implies the use of discrete time. Note that VisSim, like PLECS is also able to represent floating point variables. However emphasis is placed here on the use of the VisSim fixed point control block library given that the latter will be used to build/develop a set of ‘embedded control’ modules in this book. Prior to discussing the VisSim implementation, it is therefore prudent to briefly discuss the underlying concept of fixed and floating point representation, as well as continuous and discrete time systems.

1.4.1 Fixed Point Versus Floating Point Representation

Typically, microprocessor controllers such as the one used in this book operate with 32 bit word length. Numerical representation can be either in fixed point or floating point format. In floating point MCU’s, the bit format as shown in Eq. (1.2) is used.

$$\mathrm{S}\,\underbrace{eeeeeeee}_{\text{8 bit exponent}}.\underbrace{fffffffffffffffffffffff}_{\text{23 bit quotient(fraction)}} \tag{1.2}$$

Floating point format works with a sign-magnitude representation. The sign of the number is the ‘S’ bit, where 0 is allocated to positive numbers and 1 to negative numbers. The remainder of the 32 bits are the ‘magnitude’. The eight ‘e’ bits represent the exponent of the floating point magnitude, between $-126\ldots127$. The quotient or fraction contains 23 ‘f’ bits which are to the right of the radix point. Using this approach any number within a very wide given range can be defined. The range and resolution (smallest definable increment) can be readily calculated, as is shown in Tables 1.1 and 1.2. For sign-magnitude representation (floating point format) the range is defined by $\pm$ maximum magnitude. In fixed point format two’s compliment representation is used. The positive and negative extreme values defer by the value of one least-significant bit (LSB).

Table 1.1 Range for 32 bit floating and fixed point representations

Format	Max decimal value	Min value
Floating point	$\pm 2^{(255-127)}\left(1-2^{-23}\right) \approx \pm 3.4028 \cdot 10^{38}$	$\pm 1.1755e \cdot 10^{-38}$
Fixed point: IQ30	$\left(2-2^{-30}\right) \approx 1.9999999991$	-2.000000000
Fixed point: IQ24	$\left(128-2^{-24}\right) \approx 127.9999999404$	-128.000000000

Table 1.2 Comparison of resolution for 32 bit floating and fixed point representations

Format	Relative resolution	Absolute resolution
Floating point	$2^{-24} \approx 5.9605 \cdot 10^{-8}$	rel. res. $\cdot 2^{(\exp-127)}$
Fixed point: IQ30	$2^{-32} \approx 2.3283 \cdot 10^{-10}$	$2^{-30} \approx 9.3132 \cdot 10^{-10}$
Fixed point: IQ24	$2^{-32} \approx 2.3283 \cdot 10^{-10}$	$2^{-24} \approx 5.9605 \cdot 10^{-8}$

$$\underbrace{SI}_{\text{2 bit integer}} \cdot \underbrace{ffffffffffffffffffffffffffffff}_{\text{30 bit quotient(fraction)}} \quad (1.3)$$

Expression (1.3) shows a so called 2Q.30 (normally simply referred to as 'IQ30') fixed point format in two's compliment form, where the 30 represents the number of 'f' bits to the right of the radix point. This implies that (for a 32 bit word length) there are two remaining bits on the left side of the radix point, of which the 'S' bit is the sign, where 0 is allocated to positive numbers and 1 to negative numbers. The 'I' bit in this example is therefore one single bit. Hence the maximum positive value of the 'integer' part of the number is decimal '1'(integer bits $S = 0$, $I = 1$) and the minimum decimal value is -2 (integer bits $S = 1$, $I = 0$). In theory, the radix point can be moved to any location within the 32 bit word, but in practice it is sufficient to work with two formats namely 2Q.30 (IQ30) and 8Q.24 (IQ24) for reasons that will become apparent shortly. In the IQ24 format the radix point is moved to realize a 24 bit quotient or fraction. Accordingly the range and resolution values will also change as may be observed from Tables 1.1 and 1.2.

Observation of Table 1.1 learns that the floating point format range is large compared to the ranges of fixed point formats shown in said table. Moreover the range is further reduced as the radix point is shifted to the left, i.e. format change from IQ24 to IQ30. However the resolution increases as the radix point is moved to the left, as may be observed from Table 1.2. The floating point format is superior in terms of available range. This advantage comes with the penalty of a less favorable relative resolution. The eight bits used for the exponent are not available for resolution any more.

Historically, floating point MCU based controllers have been more expensive then fixed point controllers. Furthermore a fixed point controller algorithm can be executed on a floating point MCU, but not vice-versa. Hence in the book, for the sake of generality we will use fixed point formatting in either IQ30 or IQ24. Note that in the VisSim models, the notation IQ30, IQ24 is replaced by 2.32, 8.32 respectively. In addition, the sensorless control algorithms used in the laboratory sessions discussed in this book are implemented in fixed point, hence it is deemed important for the reader to be familiar with scaling concepts. Generally the IQ30 format is used where high resolution is deemed mandatory, whilst IQ24 is used when a larger (than IQ30) range is required (in controllers with gain, for example).

1.4.2 Use of Scaling for Fixed Point Formatted Numbers

The use of fixed point formatted numbers, implies the need for 'scaling' variables to a per unit value as will be demonstrated in this subsection. This scaling process will be expanded, as needed for each of the laboratory sessions, as will be come apparent shortly. In this case the scaling will be applied to the resistance example show in Fig. 1.7.

Basically scaling implies the allocation of a 'full scale' value to a given variable. For example, voltage u and currents i variables may be written as

$$u = u_\text{n}\, u_\text{fs} \tag{1.4a}$$

$$i = i_\text{n}\, i_\text{fs} \tag{1.4b}$$

where the variables u_n, i_n represent the 'per unit' or scaled voltage and current values respectively. The 'fullscale' voltage and current value are defined as u_fs, i_fs respectively. In this example the expression $i = {}^{u}/_{R}$ needs to be scaled, which upon use of Eq. (1.4) gives

$$i_\text{n} = u_\text{n} \underbrace{\left(\frac{u_\text{fs}}{R\, i_\text{fs}}\right)}_{\text{gain}} \tag{1.5}$$

Shown in Eq. (1.5) is a 'gain' factor, which is precisely the value of the 'Gain' variable shown in Fig. 1.7. The input and output variable shown in the same figure represent the voltage and current respectively and accordingly these must be appropriately scaled using Eq. (1.4) in the 'pre-scaling' and 'post-scaling modules'. Pre-scaling implies that the input variable is multiplied by the inverse of the full scale value, whilst post-scaling requires multiplication of the per unit output variable by its full scale value. Note that a 'convert' module is used on the inputs to the 'fixed point' module shown in Fig. 1.7, which converts the floating point number to the designated fixed point (in this case IQ30) format.

1.4.3 Continuous and Discrete Time Operational Issues

For the purpose of explaining continuous time models it is helpful to consider an integrator circuit example as given in Fig. 1.8. The relationship between output and input may be written as

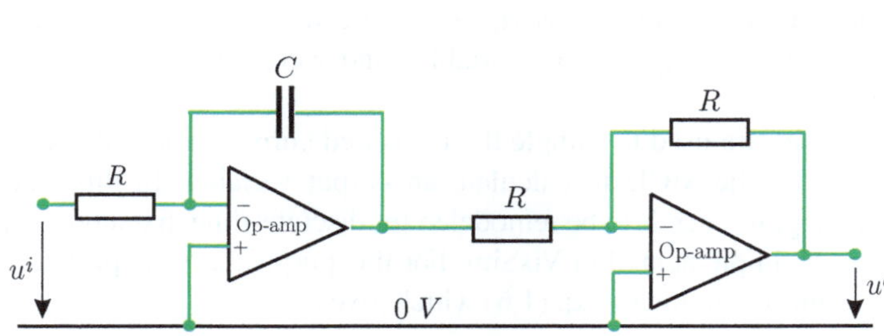

Fig. 1.8 Integrator example: continuous time domain

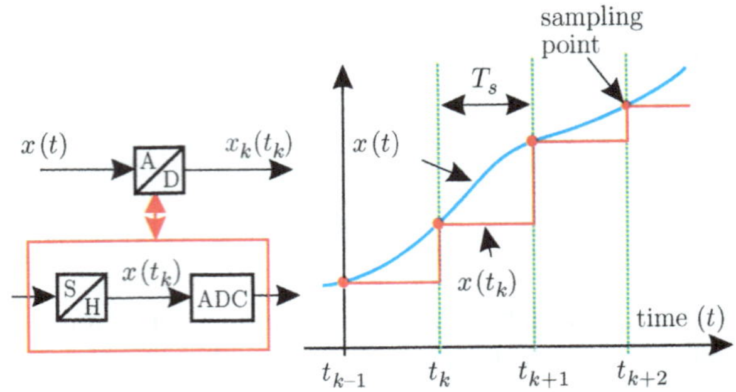

Fig. 1.9 A/D converter unit with example input/output waveforms

$$u^o = \underbrace{\left(\frac{1}{RC}\right)}_{\text{gain}} \int u^i dt \tag{1.6}$$

where the signals u^o, u^i are assumed to be continuous, i.e. have values which are defined at any instance in time.

The controller is in our case digital, which means that the analog input variable, here in the form of the integrator input voltage $u^i(t)$ must be converted to a digital form using an 'analog-digital' (A/D) converter. The function of the unit is readily shown with the aid of Fig. 1.9. Figure 1.9 shows an input function $x(t)$ to the A/D converter. In reality this module consist of two sub-modules (as indicated in Fig. 1.9) namely a sample & hold (S/H) module and an analog to digital (ADC) converter which converts the analog input variable $x(t_k)$ to a digital representation $x_k(t_k)$.

The diagram shows an example waveform together with a set of discrete time points t_{k-1}, t_k, t_{k+1} where k can be any integer value. The difference in time between any two time points is constant and equal to the 'sampling period' T_s. The output of the S/H module is such that at these time points the input is 'sampled' i.e. the output is then set to equal the input value. Hence, at these 'sampling points' the output changes to match the instantaneous value found at the input of the converter. The output is therefore held constant during the sampling time. For example, the output $x(t_k)$ represents the value of the input variable as sampled at the time mark t_k. Subsequently these variables are then converted to digital values $x_k(t_k)$. In subsequent discussions the subscript k (for the variable x) will be dropped with the implicit understanding that the variables under consideration represent digital numbers.

The A/D units are used to sample the measured currents and voltages and these are then used by the MCU to calculate an output variable. In this example the continuous integrator needs to be remodeled for discrete-time, fixed point operation, so that it can be implemented in VisSim. For this purpose it is helpful to undertake a first order approximation of Eq. (1.6) which gives

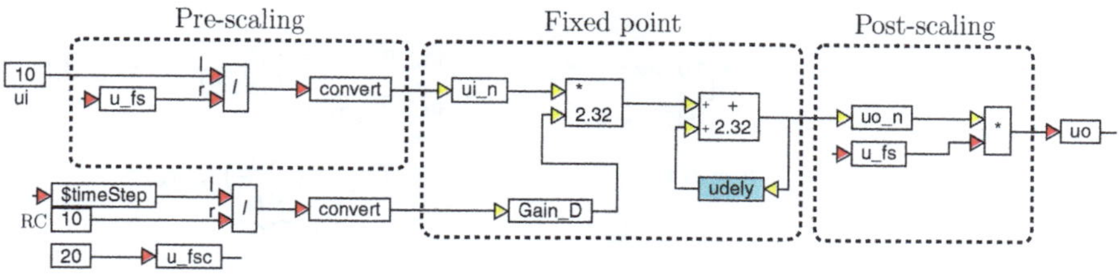

Fig. 1.10 Discrete, fixed point integrator example using VisSim

$$u^o(t_{k+1}) \cong \underbrace{\left(\frac{\mathrm{T_s}}{RC}\right)}_{\mathrm{gain_D}} u^i(t_k) + u^o(t_k) \tag{1.7}$$

The final step in realizing a fixed point implementation of Eq. (1.7) is scaling. Hence use of Eq. (1.4), which implies multiplication of expression (1.7) by the inverse of the full scale voltage which gives

$$u^o_{\mathrm{n}}(t_{k+1}) \cong \underbrace{\left(\frac{\mathrm{T_s}}{RC}\right)}_{\mathrm{gain_D}} u^i_{\mathrm{n}}(t_k) + u^o_{\mathrm{n}}(t_k) \tag{1.8}$$

Note that the only difference between expressions (1.7), (1.8) is the use of per unit values, instead of floating point variables. An example, as given in Fig. 1.10, shows an implementation of Eq. (1.8). In this example, the ‘fixed’ point module, implemented in IQ30, contains the discrete fixed point integrator, with gain module `Gain_D`, which is found using the gain value shown in Eq. (1.8). A one sample delay module `udely` is used to generate the $u^o_{\mathrm{n}}(t_k)$ signal from $u^o_{\mathrm{n}}(t_{k+1})$. The simulation example was run over a period of 1 second, with a sampling time $\mathrm{T_s} = 1$ ms, a value which is placed in the *timestep* variable. Input to the integrator is a 10 V constant voltage, hence the theoretical integrator output value according to Eq. (1.6) should be 1 V, given that product of the resistance and capacitor values `RC` was set to 10. Pre/Post scaling modules are again used for the conversion between per unit and actual (non-scaled) variables.

1.4.4 Basic Control Block Library

Throughout this book additional control blocks will be introduced as they are required. At this point, a basic set will be given which will form the basis for the first set of models to be discussed in this book. In each case the PLECS and VisSim equivalent in per unit, IQ30 format will be shown. The Pre- and Post-Scaling module are not shown, to aid readability.

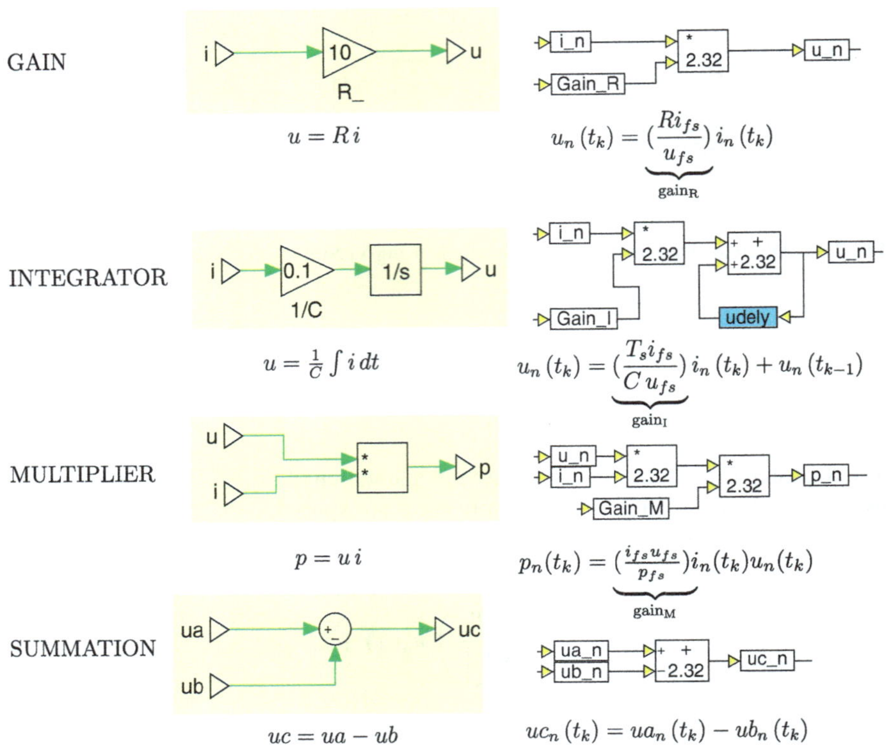

Fig. 1.11 Basic building block set

The first set of building blocks, as given in Fig. 1.11, are linked to 'example' transfer functions. For example, the GAIN module has as input the current i and as output u, the gain is set to R. In the discrete version the gain is implemented using a multiplication module and input variable `Gain_R`. The INTEGRATOR example module has as input the variable i and output u whilst the gain of the integrator is $1/c$. The corresponding discrete gain `Gain_I` for the VisSim module will be a function of the sampling time T_s used as discussed previously. When multiplying two variables in the time domain, a MULTIPLIER module is used. This module differs from the given GAIN module in that the latter is used to multiply a variable with a constant. In the discrete version an additional `gain_M` factor is present, in the event that the full scale power value p_{fs} differs from the full scale voltage and current product $u_{fs}\,i_{fs}$ Finally, an example of a SUMMATION module is given. In this case the output is a variable uc and subtracts the input variable ub from input variable ua. An example of combining some of these modules is readily given by considering the Eq. (1.9)

$$u = iR + L\frac{di}{dt} \tag{1.9}$$

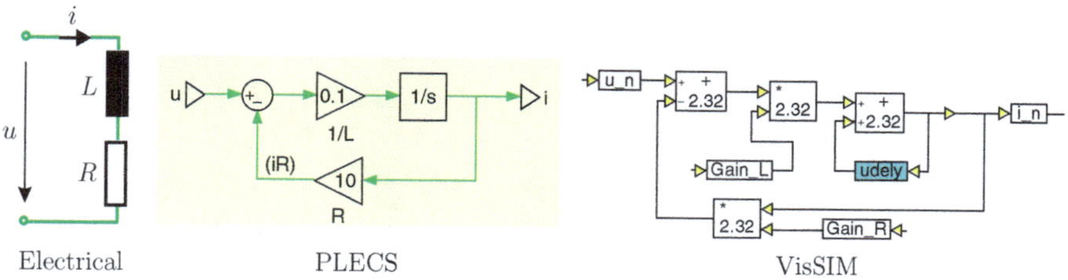

Fig. 1.12 Example of using basic building blocks

which represents the voltage across a series network in the form of an inductance L and resistance R. To build a PLECS control block representation, with the voltage as input variable and current as output variable, it is helpful to rewrite the expression in its differential equation form

$$\frac{di}{dt} = \frac{1}{L}(u - iR) \tag{1.10}$$

In this case the output of the integrator is the variable i and the input of the integrator is given as $(u - iR)$, hence

$$i = \frac{1}{L}\int (u - iR)\,dt + i(0) \tag{1.11}$$

The initial current is assumed to be zero, i.e. $i\,(0) = 0$. An observation of Eq. (1.11) learns that the integrator input is formed by the input variable u from which the term iR must be subtracted where use is made of a summation unit, as shown in the PLECS model presented in Fig. 1.12. The gain '$1/L$' present in Eq. (1.11) appears in the PLECS representation as a '$1/L$' gain as discussed previously. A discrete fixed point IQ30 format VisSim implementation requires discretization and scaling of expression (1.11) as discussed in Sect. 1.4.3. Subsequent mathematical handling of said equation gives

$$i(t_k) \cong \underbrace{\left(\frac{\mathrm{T_s} u_{\mathrm{fs}}}{L i_{\mathrm{fs}}}\right)}_{\mathrm{Gain_L}} \left(u_{\mathrm{n}}(t_k) - i_{\mathrm{n}}(t_k) \underbrace{\left(\frac{R i_{\mathrm{fs}}}{u_{\mathrm{fs}}}\right)}_{\mathrm{Gain_R}} \right) + i_{\mathrm{n}}(t_{k-1}) \tag{1.12}$$

which shows the presence of gains $\mathrm{Gain_R}$, $\mathrm{Gain_L}$, which also appear in the discrete VisSim implementation shown in Fig. 1.12 as `Gain_R` and `Gain_L` respectively. This diagram also shows the electrical and PLECS control block implementation for this example. In PLECS this example can also be directly implemented using two 'electrical' components that represent the inductance and resistance respectively.

Chapter 2
Drive Principles and Development

This chapter provides a compact overview of the underlying theoretical aspects of an electrical drive. The aim being to convey to the reader, with a minimum of mathematics, the key issues which are critical to controlling an electrical drive. In this context a fundamental understanding of AC machines is required in terms of their ability to generate torque. In the second part of this chapter the 'tool chain', otherwise referred to as the 'development approach' is discussed. Hence the steps used in this book to move from a simulation model to an actual experimental laboratory drive are shown with the aid of the hardware and software packages used in this book.

2.1 Control of Electrical Drives

This section outlines the basic control principles of any type of drive, starting from the need to achieve precise speed control. Consequently the basic control steps required to realize this task are outlined starting from the speed controller, which in turn sets the torque level in a machine. Controlling torque means controlling the current, which in turn requires the use of a current controller that makes use of a power electronic converter. This converter sets the discrete voltage levels in the connected AC machine. This logical chain of modules is discussed and the critical parameters that govern the speed and current controller gain and bandwidth are identified. Furthermore, a brief overview of the PM and induction machine are given in order to show how torque control can be achieved.

D.W.J. Pulle et al., *Applied Control of Electrical Drives*,
Power Systems, DOI 10.1007/978-3-319-20043-9_2

Fig. 2.1 Model helicopter example [17]

2.1.1 Helicopter Example

The model helicopter setup shown in Fig. 2.1, is a typical example of a modern drive system, hence it is used here by means of illustration. The helicopter utilizes a 3 pole pair, high-speed 25000 rpm, three-phase permanent magnet motor, which drives, via a 13:150 mechanical gear train, the main and tail rotor. Flight operation via a transceiver is identical to its full scale counterpart, where collective pitch control is used for the main rotor, to control lift, roll and pitch. Yaw control is achieved by controlling the pitch of the tail rotor unit. For this example an umbilical cable is connected between the motor and the DRV8301-69M-KIT high current/low voltage control board which houses the converter and the 28069M control card that contain a so called field-oriented, sensorless control algorithm known as InstaSPIN, to be discussed at a later stage. The laptop, connected to the motor kit via a USB cable, runs the VisSim software used to compile the C-code that accesses the InstaSPIN algorithm located in the ROM of the processor. This algorithm provides all the functionality to realize sensorless speed control of the motor and identification of the motor parameters. Power for the converter is provided by the 12 V car battery. A more detailed overview of the drive itself, as given in Fig. 2.2, shows the helicopter motor, which is connected to the three-phase converter. In each of the converter legs a resistance is located between the 0 V bus and the lower switches, in order to measure the phase current. In addition a 'voltage measurement unit' is used to filter and attenuate the phase voltages. Both voltages and currents are fed to an

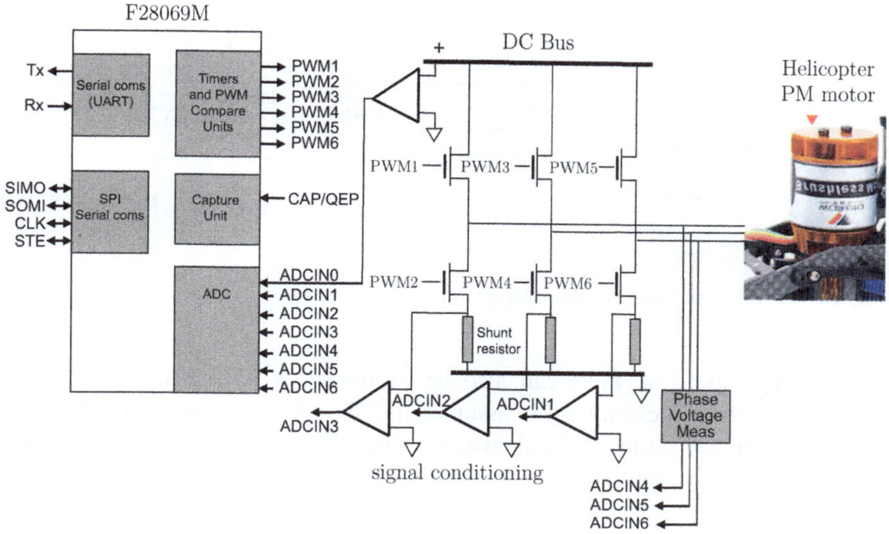

Fig. 2.2 Model helicopter drive setup [17]

ADC unit inside the F28069M processor, which is part of the Texas Instruments C2000 MCU family. Control between the MCU and laptop is done via the JTAG on the F28069M board. The hardware drive system setup, as shown for this helicopter example, reflects the type of setup need to drive any three-phase permanent magnet or induction motor under sensorless operation. Developing the required control algorithms to be executed by the 28069M card, is usually the most challenging aspect of drive development. When using the InstaSPIN-FOC technology all of the functionality needed to achieve sensorless control is provided. However it is still instructive and at times desirable to understand and execute the basic control functions, such as field-oriented current control and speed control. Hence the basic algorithms and principles, are outlined in the following subsections.

2.1.2 Use of a Speed Control Loop

Speed control for any drive is of primary concern and a typical example, as given in Fig. 2.3, shows an ideal machine that is able to generate a machine torque T_e. The machine is connected to a load T_l and is assumed to have a lumped inertia of J. The relationship between machine/load torque and inertia is represented by Eq. (2.1).

$$T_e - T_l = J\frac{d\omega_m}{dt} \tag{2.1}$$

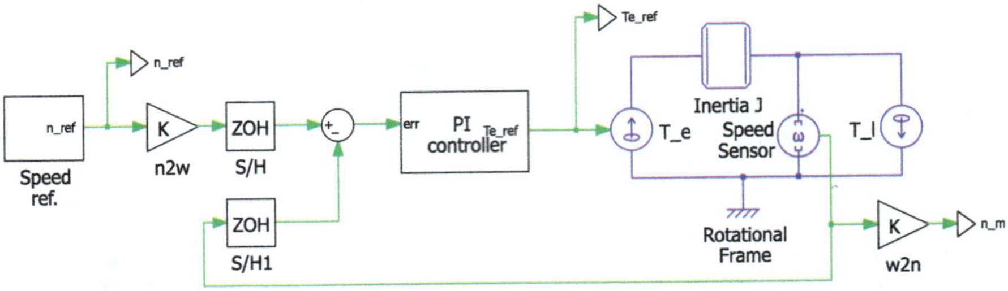

Fig. 2.3 PLECS based speed control loop example

The shaft speed ω_m rad/s can also be expressed as $n_m = \omega_m {}^{30}/_{\pi}$ rpm for convenience. In this example it is assumed that the ideal machine can deliver the required reference torque T_e^{ref} calculated by the 'PI' speed controller, hence the following condition applies

$$T_e = T_e^{ref} \tag{2.2}$$

The speed controller, is typically a Proportional-Integral (PI) concept, which may be represented by the expression

$$T_e^{ref} = K_p \Delta\omega_m + K_i \int \Delta\omega_m dt \tag{2.3}$$

where K_p and K_i represent the proportional and integral gains of the speed controller respectively. The variable $\Delta\omega_m$ is the speed error which is the difference of the user defined reference speed ω_m^{ref} and measured shaft speed (from the attached motor) ω_m. In many cases (as also considered here) the speed controller expression is used in the following format

$$T_e^{ref} = K_p \left(\Delta\omega_m + \omega_i \int \Delta\omega_m dt \right) \tag{2.4}$$

where ω_i is the PI corner frequency (rad/s), which may also be expressed as $\omega_i = {}^{K_i}/_{K_p}$. Suitable values for the Speed gain and time constant can according to [4] be expressed as

$$K_p = J\omega_b \tag{2.5}$$

$$\omega_i = \frac{\omega_b}{4} \tag{2.6}$$

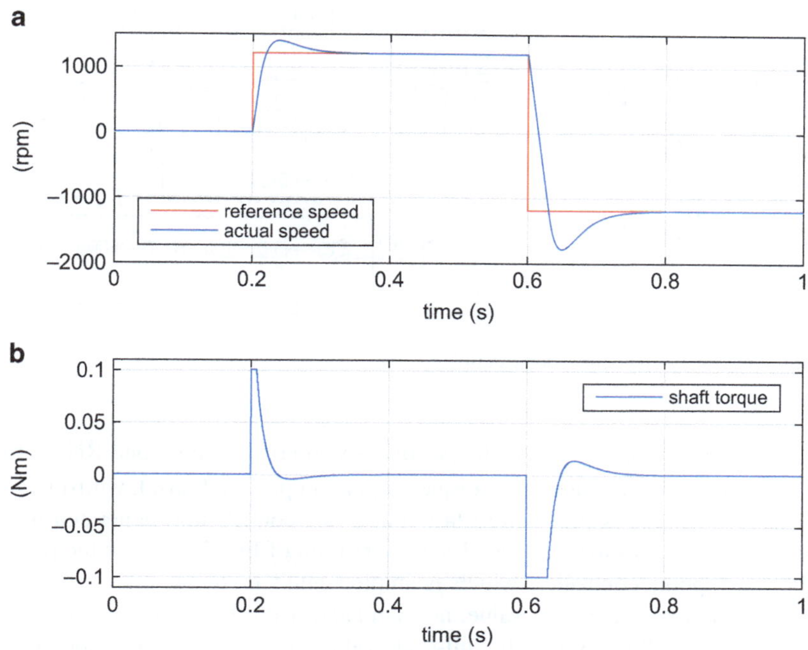

Fig. 2.4 Speed and torque time response

where ω_b is the speed controller bandwidth, which is typically in the order of 100 rad/s. A PLECS simulation, as given in Fig. 2.4, shows the speed response and torque reference output of the speed loop. This example uses an inertia value of $J = 12\,\mu\mathrm{kgm}^2$ (same order of magnitude as the PM motor used in this book), which according to Eq. (2.5) correspond to a speed gain and time constant of $\mathrm{K_p} = 1.2\,\mathrm{mNm/(rad/s)}$ and $\omega_i = 25\,\mathrm{rad/s}$, where a speed bandwidth of $\omega_b = 100\,\mathrm{rad/s}$ is assumed. The controller in use is provided with an output limit capability which in this case is set to $\pm$ 100 mNm, as may be observed from Fig. 2.4. Note that in the example shown, the load torque T_l is assumed to be zero, i.e. motor operating under no load and undergoing a speed step from $0 \rightarrow 1200$ rpm followed by a speed reversal $1200 \rightarrow -1200$ rpm at $t \approx 0.6$ s.

A discrete, fixed point implementation of the same example can also be undertaken in VisSim, which requires scaling and discretization of expression (2.3) using the approach discussed in Sects. 1.4.2 and 1.4.3. In terms of scaling it is helpful to introduce a full-scale torque value T_fs set to 150 mNm and full-scale frequency f_fs given that speed is represented as a frequency in Hz. Subsequent mathematical handling gives

$$T_\mathrm{en}(t_k) \cong \underbrace{\left(\frac{2\pi\,\mathrm{K_p} f_\mathrm{fs}}{T_\mathrm{fs}}\right)}_{\text{Gain Kp}} \Delta f_\mathrm{n} + \underbrace{\left(\frac{2\pi \mathrm{K_i} \mathrm{T_s} f_\mathrm{fs}}{T_\mathrm{fs}}\right)}_{\text{Gain } \mathrm{K_i}} \Delta f_\mathrm{n} + T_\mathrm{ein}(t_{k-1}) \tag{2.7}$$

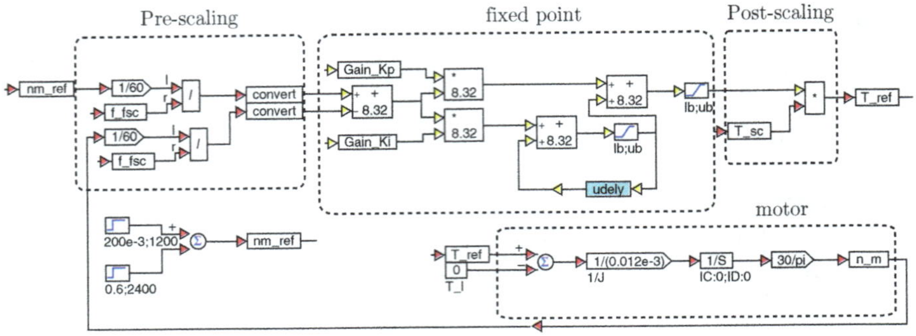

Fig. 2.5 VisSim speed control loop example

where $\Delta f_{\rm n} = \frac{1}{60 f_{\rm fs}} \left(n_{\rm m}^{\rm ref} - n_{\rm m}\right)$ is the per unit speed error. The second RHS term of Eq. (2.7) represents the integrator torque output component $T_{\rm ein}(t_k)$, whilst the third RHS term $T_{\rm ein}(t_{k-1})$ is the 'before last' integrator contribution generated using a '`udely` module' as shown in Fig. 2.5. Observation of Fig. 2.5 shows the presence of the fixed point controller, which is implemented in IQ24 in this case given that the numerical `Gain_Kp` value, as calculated using Eq. (2.7) exceeds the +2 value available in IQ30, given the full-scale values used. Two limiter modules with max/min range: `ub,lb` are added to the fixed point controller to avoid 'integrator windup'[4] and limit the per unit output torque to $\pm$ 0.1 Nm/$T_{\rm fs}$, in accordance with the values used for the PLECS example.

A pre-scaling and post-scaling unit are again used to interface between fixed point and non-scaled variables. The motor is modeled in the floating point, continuous time domain. Results achieved with this simulation, using the same reference speed versus time sequence as used for the PLECS example, are, as may be expected, identical.

2.1.3 Torque Production Mechanism in Machines

Central to the ability to achieve accurate speed control is the requirement to satisfy condition (2.2), which states that the machine must be able to deliver the torque specified by the speed controller. Achieving torque control in a machine, essentially implies controlling the orientation and magnitude of a current distribution (in either stator or rotor) relative to a given stationary or rotating magnetic field [4]. For example, in the DC machine shown in Fig. 2.6, a stationary magnetic field is present, which is horizontally oriented (relative to the figure). This field can be generated by either a field winding (connected to a DC current source) or by permanent magnets. The rotor (or armature) winding is fed by a DC current source via a brush/commutator assembly, which ensures that the current distribution in the rotor is stationary relative to the magnetic field generated by the field winding.

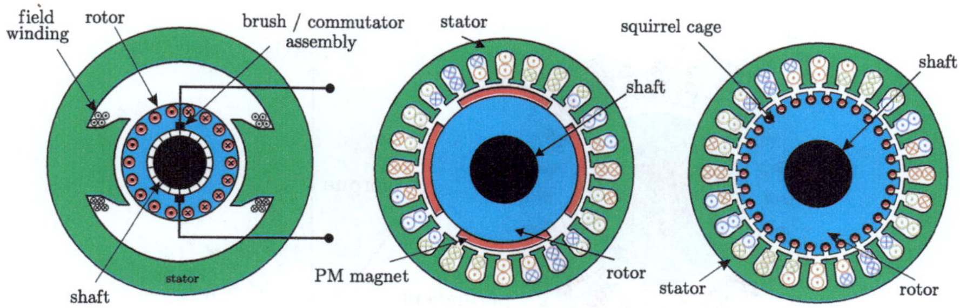

Fig. 2.6 Examples from left to right: DC, permanent magnet synchronous and squirrel cage induction machine

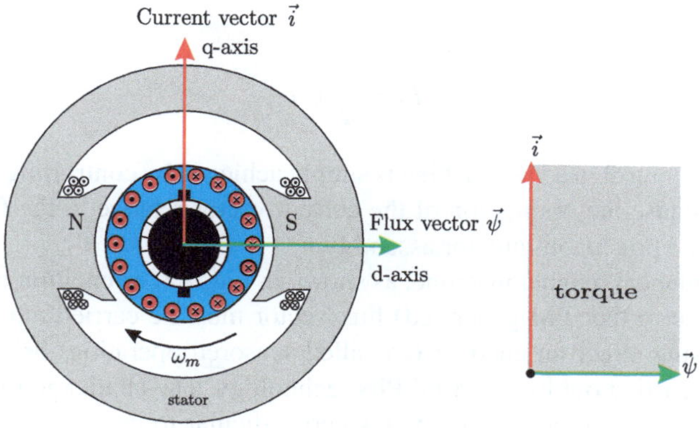

Fig. 2.7 Torque production mechanism in a DC machine

In modern drive control theory [4] it is customary and helpful to represent the current distribution and flux using so called 'spacevectors' [14]. Shown in Fig. 2.7 are the current and flux vectors in a DC machine, where the field is provided by a field winding connected to a DC current source i_f. The flux vector is aligned with the main magnetic field axis, known as the d-axis, while the current vector is aligned with the magnetic (armature) field that is caused by the current distribution in the rotor (armature), referred to as the q-axis. The torque of the machine is proportional to the 'cross product' of the flux and current vectors and may be written as

$$T_e = \frac{3}{2}\left(\vec{\psi} \times \vec{i}\right) \tag{2.8}$$

where the factor 3/2 is introduced because we are using throughout this book so called 'amplitude-invariant' vectors [14]. The 'cross product' of two vectors is equal to the 'grey' surface times 3/2 area shown in Fig. 2.7 and for a DC machine Eq. (2.8) simply reduces to

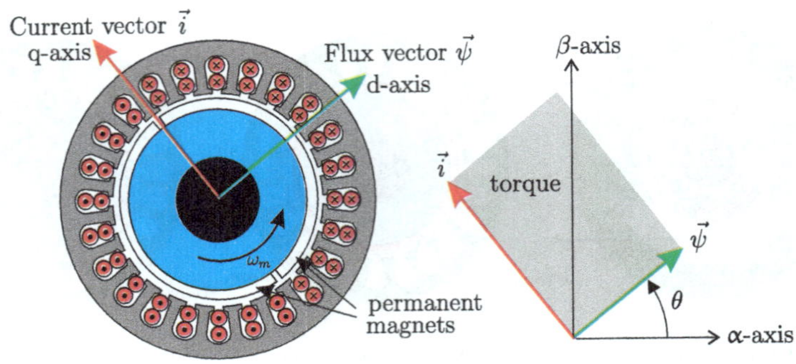

Fig. 2.8 Torque production mechanism in a PM synchronous machine

$$T_e = \frac{3}{2}\,\psi\,i \tag{2.9}$$

hence torque control in a DC machine is simply achieved by controlling the armature current i, because the orientation of the current vector relative to the flux vector is handled by the brush/commutator assembly.

For a permanent magnet machine, as shown in Fig. 2.8, orientation of the current vector relative to the (PM generated) flux vector must be carried out by a control algorithm using either an encoder (so called sensored operation) or sensorless, as for example carried out by TI InstaSPIN technology. The challenge to be met is to generate a current vector which remains perpendicular to the magnetic flux vector for all speeds, using a three-phase winding. Hence the need to control the current in each of the three phases with an attached power electronic converter, as to realize a resultant current distribution in the stator windings as shown in Fig. 2.8 for the rotor position shown. Under normal conditions the current vector is chosen perpendicular to the flux vector because this gives the best torque/current ratio (lowest current value which yields the highest torque). The torque under these circumstances is equal to

$$T_e = \frac{3}{2}\,\psi\,i_q \tag{2.10}$$

where i_q is the quadrature axis component of the current vector $\vec{i}$, hence in this case $\vec{i} = ji_q$. In some cases, particularly at higher operating speeds, so-called 'field weakening' is normally used, which implies the presence of a direct axis current component i_d which is purposely chosen to oppose the field produced by the magnets.

Torque production in an induction motor, otherwise known as 'asynchronous' machine is slightly more complicated when compared to the DC or PM machines discussed above. The primary reason for this is that the controller must generate a stator current distribution which will produce a magnetic field (given that there are

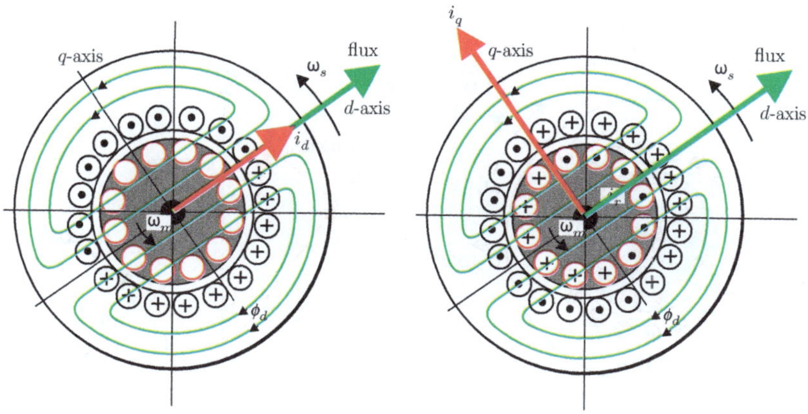

Fig. 2.9 Torque and flux generating mechanisms in a IM (asynchronous) machine

no magnets or field winding) and generate a constant torque. Given this dual task it is convenient to represent and discuss both components of the current distribution separately. The two figures given in Fig. 2.9 show the stator current distributions in question. In reality both stator current distributions are superimposed to form one single current distribution as will be discussed at a later stage. The stator current distribution in the 'left' figure is responsible for the generation of the magnetic flux and correspondingly the current vector and flux vector are both oriented along the d-axis. Torque production is achieved by orientating a second current distribution perpendicular to the flux, hence along the q-axis as shown in the 'right' figure. However in an induction machine this q-axis current distribution must be mirrored in the rotor because there is no resultant flux in the q direction. To maintain the current distribution in the rotor a voltage must be induced in said rotor by the magnetic flux, which means that the later must rotate relative to the stator, hence the following relationship must hold to achieve field oriented control

$$\omega_{\mathrm{s}} - \omega_{\mathrm{m}} = \frac{i_{\mathrm{q}} R_{\mathrm{r}}}{\psi} \tag{2.11}$$

where R_{r} is the rotor resistance of the squirrel case (to be discussed further in the next section). The resultant stator current distribution in the machine is now formed by the superposition of the two stator current distributions which are represented by the direct and quadrature current vectors, as shown in Fig. 2.10. The torque of the machine is again given by the cross product of the flux and current vector times $^3/_2$, which leads to

$$T_{\mathrm{e}} = \frac{3}{2}\left(\vec{\psi} \times \vec{i}\right) \tag{2.12}$$

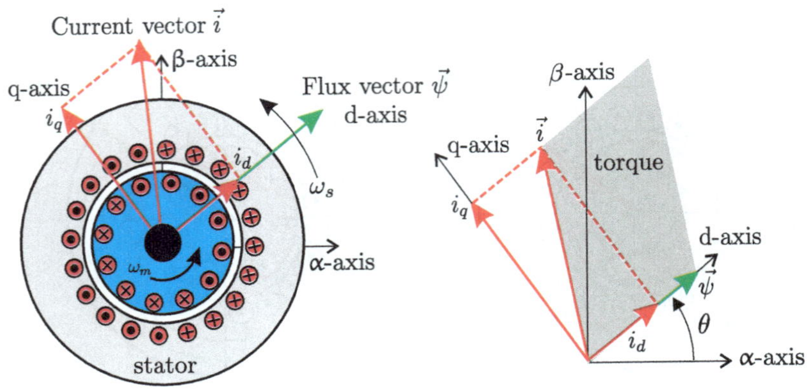

Fig. 2.10 Torque production mechanism in a IM (asynchronous) machine

The 'cross product' of the two vectors is proportional to the 'grey' surface area shown in Fig. 2.12. To achieve a torque step in an induction machine means increasing the current i_q value and simultaneously increasing the speed ω_s in order to satisfy slip Eq. (2.11). Alternatively the flux could also be increased, but this cannot be done instantaneously as this event is tied to a time constant set by the magnetizing inductance divided by the rotor resistance. In the latter case the stator frequency must according to Eq. (2.11) be reduced, given that the q-axis current distribution remains constant. Reducing the flux (known as field weakening) is often undertaken at high operating speeds given the inherent DC bus voltage limitations of a drive.

Central to achieving precise torque control for machines shown in Fig. 2.6 is the need to correctly orientate and control the current distribution in the (rotor for DC/ stator for AC) windings relative to the flux, which implies the need for current control, as will be discussed shortly.

2.1.4 Machine Modeling

Prior to discussing current control it is helpful to have an understanding of the machine, both Permanent Magnet and Induction machine, when viewed from the three phase terminals. Both of these machines can (as will be shown) be simply represented by three phase circuits, where each phase consists of a sinusoidal EMF source e_1 in series with a stator resistance R_s and stator inductance L_s, as shown in Fig. 2.11. For a Permanent Magnet machine the parameters R_s, L_s represent the stator resistance and inductance respectively. The three EMF sources (usually sinusoidal) are equal in amplitude $\hat{e}$, but are displaced electrically by 120°. The EMF amplitude can also be written as $\hat{e} = p\,\omega_m\,\psi$, where ω_m, p are the shaft speed

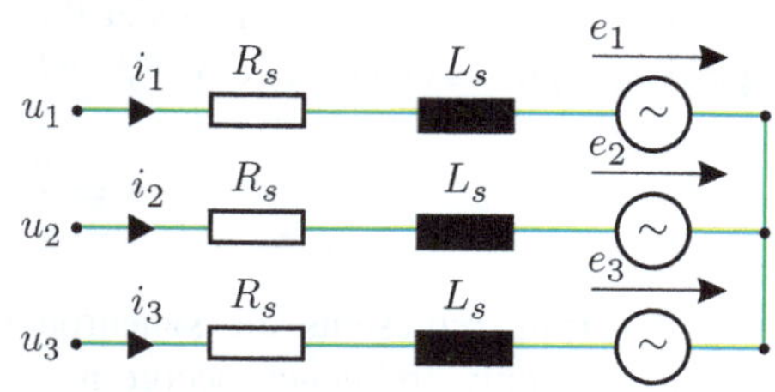

Fig. 2.11 Typical three-phase machine model

Table 2.1 Texas instruments LVSERVOMTR PM machine parameters [9]

Parameter	Value	Units
Inductance: L_s	0.181	mH
Resistance: R_s	0.34	Ω
Magnet flux: ψ_{PM}	6.46	mWb
Pole pairs: p	4	–
Inertia: J	10	μkgm^2

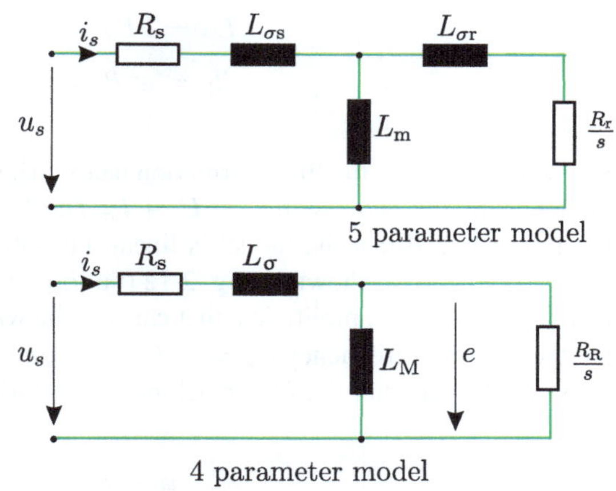

Fig. 2.12 5 and 4 parameter IM motor models

(rad/s) and pole pair number of the machine. The flux ψ in a non-salient [4] PM machine is equal to

$$\psi = \psi_{PM} \tag{2.13}$$

where ψ_{PM} represents the flux due to the magnets. The parameters of the PM machine used in this book are given in Table 2.1. Note that parameter variations relative to the values shown below will appear in this book.

The model structure according to Fig. 2.11 can also be applied to an induction machine. However, in this case it is helpful to consider the traditional steady-state single phase five parameter induction machine model given in Fig. 2.12 (top diagram) first. Shown in the top figure are magnetizing inductance L_m and leakage inductances $L_{\sigma s}$, $L_{\sigma r}$ associated with the stator and (squirrel cage) rotor respectively. In addition, the model shows the resistance R_s, R_r linked to the stator windings and squirrel cage. In this type of model the effective rotor resistance (for given shaft speed n_{m} (rpm) is determined by R_{r}/s, where s represents the 'slip' defined as

$$s = \frac{n_{\mathrm{s}} - n_{\mathrm{m}}}{n_{\mathrm{s}}} \tag{2.14}$$

where n_{s} rpm, represents the synchronous shaft speed. The latter is tied to the electrical supply frequency according to $n_{\mathrm{s}} = 60 f_{\mathrm{s}}/p$, where f_{s}, p represent the electrical frequency (Hz) and pole pair number respectively. It is shown in [4] that the five parameter model can be converted to an equivalent (as seen from the terminals) four parameter representation as also shown in Fig. 2.12 (bottom diagram) using

$$L_{\sigma} = \left(\frac{L_{\mathrm{s}}}{L_{\mathrm{m}}} - a\right) L_{\mathrm{m}} \tag{2.15a}$$

$$L_{\mathrm{M}} = a\, L_{\mathrm{m}} \tag{2.15b}$$

$$R_{\mathrm{R}} = a^2\, R_{\mathrm{r}} \tag{2.15c}$$

where a is known as the 'transformation factor [4], which is linked to the 5 parameter model using $a = L_{\mathrm{m}}/L_{\mathrm{r}}$, with $L_{\mathrm{r}} = L_{\mathrm{m}} + L_{\sigma\mathrm{r}}$ and $L_{\mathrm{s}} = L_{\mathrm{m}} + L_{\sigma\mathrm{s}}$. This transformation is valid provided that the model is linear, i.e. saturation effects are assumed to be relatively low. Also shown in Fig. 2.12 (bottom diagram) is the (usually sinusoidal) EMF, which has an amplitude $\hat{e}$ that can also be written as $\hat{e} = p\,\omega_e\,\psi$, where ω_e, p are the electrical frequency $\omega_e = 2\pi f_{\mathrm{s}}$ (rad/s) and pole pair number of the machine respectively. The flux ψ of the machine can be written as

$$\psi = L_{\mathrm{M}} i_{\mathrm{d}} \tag{2.16}$$

where i_{d} is the direct axis current component as discussed in the previous section. Hence one phase of a three-phase IM machine can according to Fig. 2.12 (bottom diagram) be represented by a series network consisting of a resistance R_{s}, leakage inductance L_{σ} and EMF source with amplitude $\hat{e}$. A three phase representation of the model is therefore given by the model shown in Fig. 2.11, provided the inductance L_{s} is replaced by L_{σ}.

The parameters of the (4 parameter) IM machine used in this book are given in Table 2.2. For motor simulation purposes, so called 'Ideal Rotating Transformer' (IRTF) space vector models, are used for representing permanent magnet and (4 parameter) squirrel case induction machines. These models are used in this book, but the reader is referred to [4] for further details, given that the emphasis of this book is on applied control of such motors.

Table 2.2 Texas instruments LVACIMTR IM machine parameters [9]

Parameter	Value	Units
Inductance: L_σ	5.39	mH
Resistance: R_s	1.84	Ω
Inductance: L_M	30	mH
Resistance: R_R	1.23	Ω
Pole pairs: p	2	–
Inertia: J	70	μkgm^2

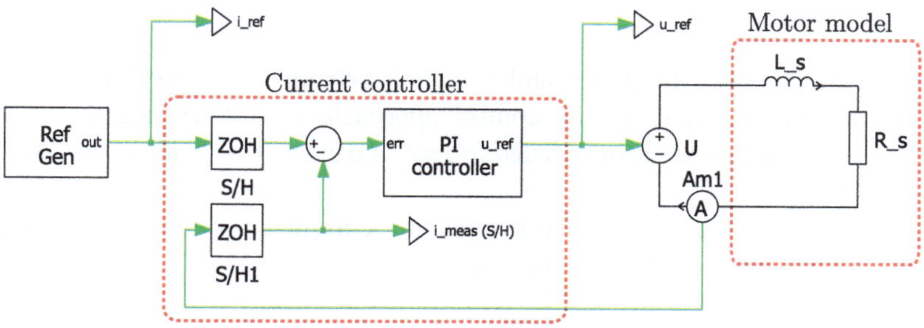

Fig. 2.13 Model based current control using an $R - L$ network to represent the motor

2.1.5 *Current Control for Electrical Drives*

In this subsection we will briefly outline the basic principles of so called 'model based' current control as applied in this book. An inductance/resistor type load is used in order to show how the controller parameters need to be chosen. Any machine can, in fact, be represented by such a network, as was discussed earlier (see Fig. 2.11) hence its use here. The use of 'model based' control in the so called synchronous reference frame is then discussed.

The current controller given in Fig. 2.13 shows a discrete PI control module, which has as inputs the reference and measured currents from a load represented by an inductance and resistance. The converter in this case is a controlled (by the PI output) voltage source. For regularly sampled control systems as considered here, we are able to derive a simple controller structure which can be used to control the current in the $R - L$ circuit. Central to our proposed controller is the use of a modulator/converter combination, represented by the voltage source, which is designed to deliver in (a given sampling interval T_s) an average voltage per sample quantity $U(t_k)$, which corresponds to the set-point (reference) $U^*(t_k)$ value provided by the current control module. The task of the current control module is thus reduced to determining this set-point (reference) average voltage per sample value on the basis of the measured discretized load current $i(t_k)$ and the set-point current value i^*. The nature of this control philosophy is shown with the aid of Fig. 2.14. Shown in this figure are the non-sampled reference current $i^*(t)$ and typical (for PWM based control) converter current $i(t)$. These currents are sampled by the controller at time marks $0,\ t_1,\ t_2, ..$ etc. At time $t = t_1$ a current error

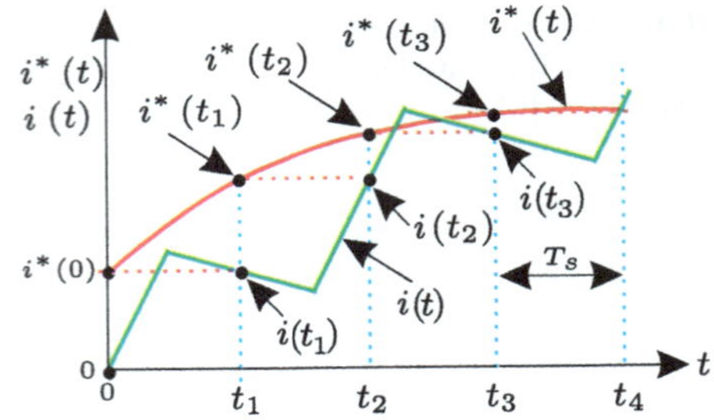

Fig. 2.14 Model based current control [4]

(between the sampled reference and sampled converter current) $i^*(t_1) - i(t_1)$ is present and the objective of the control approach is to determine the required average voltage reference value needed to zero said error. This leads to the condition $i^*(t_1) = i(t_2)$.

The control objective aimed at driving the current error to zero during each sample interval may therefore be written as

$$i(t_{k+1}) = i^*(t_k) \tag{2.17}$$

for a regularly sampled system with sampling time $\mathrm{T_s}$. The modulator/converter module will insure that the following condition is satisfied

$$U^*(t_k) = \int_{t_k}^{t_{k+1}} u(\tau)\, d\tau \tag{2.18}$$

where u represents the voltage across the load (see Fig. 2.13). An observation of Fig. 2.13 learns that the load voltage may be expressed as

$$u = Ri + L\frac{di}{dt} \tag{2.19}$$

Use of Eq. (2.19) with Eq. (2.18) allows the latter to be written as

$$U^*(t_k) = \frac{R}{\mathrm{T_s}}\int_{t_k}^{t_k+\mathrm{T_s}} i(\tau)\, d\tau + \frac{L}{\mathrm{T_s}}\int_{i(t_k)}^{i(t_k+\mathrm{T_s})} di \tag{2.20}$$

Discretization of Eq. (2.20), as discussed in Sect. 1.4.3 allows the latter to be written as:

$$U^*(t_k) \cong R\, i^*(t_k) + \frac{L}{\mathrm{T_s}}\Delta i(t_k) \tag{2.21}$$

with $\Delta i(t_k) = i^*(t_k) - i(t_k)$, which allows this expression to be written as

$$U^*(t_k) = R\,\Delta i(t_k) + \frac{L}{\mathrm{T_s}}\Delta i(t_k) + R\, i(t_k) \tag{2.22}$$

which can also be written as

$$U^*(t_k) = \mathrm{K_p}\, \Delta i(t_k)(1 + \omega_i \mathrm{T_s}) + R\, i(t_k) \tag{2.23}$$

where $\mathrm{K_p} = {}^{L}/_{\mathrm{T_s}}$ and $\omega_i = {}^{R}/_{L}$ are known as the proportional gain and bandwidth of the current controller respectively. In a drive the modulator input is the modulation index defined as $m^*(t_k) = {}^{U^*(t_k)}/_{(U_{\mathrm{DC}}/2)}$, where U_{DC} is the DC bus voltage.

$$m^*(t_k) = \mathrm{K_p} \frac{\Delta i(t_k)}{(U_{\mathrm{DC}}/2)}(1 + \omega_i \mathrm{T_s}) + \underbrace{\frac{R\, i(t_k)}{(U_{\mathrm{DC}}/2)}}_{m_i(t_k)} \tag{2.24}$$

where $m_i(t_k)$ is the modulation index value generated by the integrator alone, sampled prior to the current value. The analysis carried out to this point was done with a floating point based controller. In practice a fixed point controller is often used hence scaling is needed, as discussed for the speed controller. This implies that scaling, as discussed in Sect. 1.4.3, needs to be undertaken for Eq. (2.24) which leads to

$$m^*(t_k) = \underbrace{\left(\frac{i_{\mathrm{fs}}\, \mathrm{K_p}}{u_{\mathrm{fs}}}\right)}_{\text{Gain } \mathrm{K_p}} \frac{2\Delta i_n(t_k)}{U_{\mathrm{DCn}}} \left(1 + \underbrace{\omega_i \mathrm{T_s}}_{\text{Gain } \mathrm{w_i}}\right) + m_i(t_k) \tag{2.25}$$

The VisSim current controller example given in Fig. 2.15 shows the PI control with the gain settings as defined by Eq. (2.25). The `udely` module is used to generate the $m_i(t_k)$ term. A series network consisting of an inductance/resistance network acts as a load, which is connected to a ‘converter’ module which generates

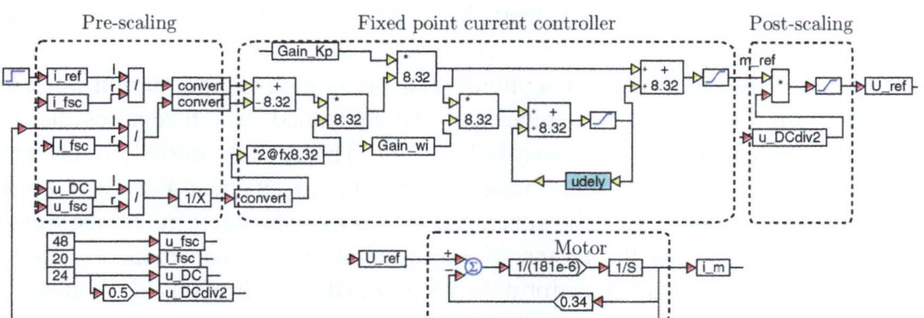

Fig. 2.15 VisSim, fixed point Model based current control example

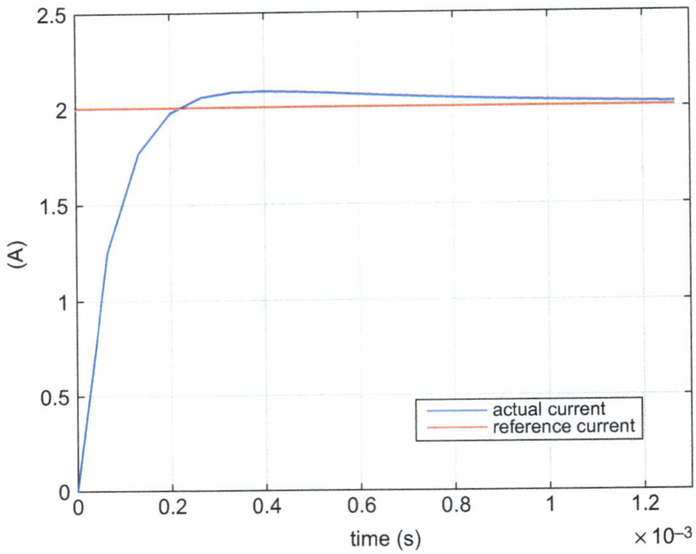

Fig. 2.16 Step response with VisSim, fixed point model based controller and LVSERVOMTR PM motor

a voltage output defined as $U^{\text{ref}} = m^{\text{ref}}.U_{\text{DC}}/2$, this module also includes a limiter which constrains the output to limits $\pm U_{\text{DC}}/2$. In practical implementations the PI controller in its current form is prone to 'windup' which occurs when the limits of a linear system are reached. Windup can occur in this case when the reference average voltage value generated by the PI controller exceeds the maximum value which can be delivered by the converter. Under such circumstances a current error occurs at the input of the controller which will cause the integrator output to ramp up or down. Practical controllers have an 'anti-windup' feature as shown in Fig. 2.15 by way of the `lb,ub` limit modules, which limits integrator action when the user defined limits are reached. The *L*, *R* parameters used in Fig. 2.15 are those of the Teknic motor shown in Table 2.1. A typical controller step response under locked rotor conditions with a DC bus voltage of 24 V and 1 kHz sample frequency is given in Fig. 2.16.

In three-phase drives both magnitude and orientation of the current vector $\vec{i}$ relative to the 'd, q' reference frame must be controlled, which requires the use of the conversion module, as shown in Fig. 2.17. This module initially transforms the measured currents into a vector with components i_α, i_β, using a 'Forward Clarke' transformation. This vector is however, referenced to a stationary reference frame, i.e. tied to the stator of the motor, hence an additional transformation, called a 'Forward Park' is needed to transform the vector $\vec{i}$ to the synchronous 'd,q' reference frame used to achieve torque control.

Input to this module are the three (in most cases, but two is also possible) measured currents and the 'flux- angle' which represents the orientation of the

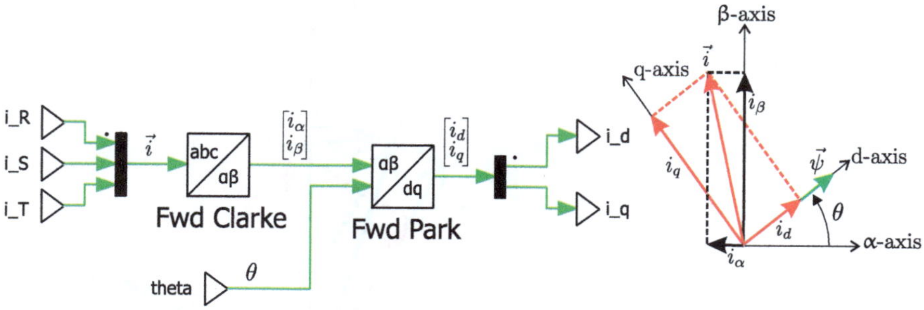

Fig. 2.17 Current conversion to 'd,q' coordinates

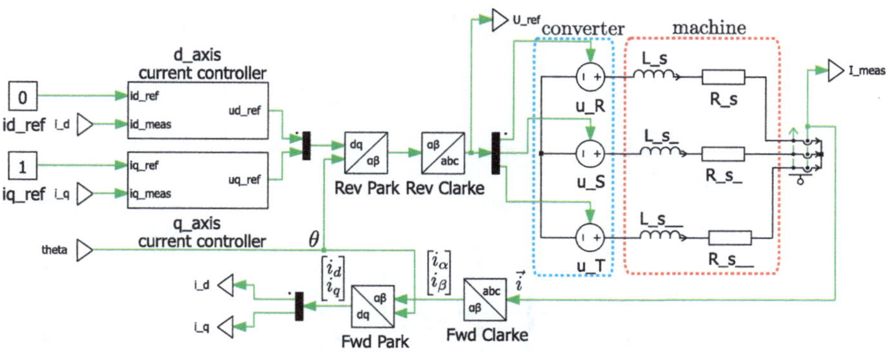

Fig. 2.18 Synchronous current control of a machine represented by an 'R-L' network

magnetic flux ψ. In a sensored drive this is provided via an encoder. In a sensorless drive this angle has to be calculated using, for example InstaSPIN technology.

Current control, in this so called synchronous reference frame, requires two model based current controllers, as shown in Fig. 2.18. The outputs of the two controllers are the $U_{\mathrm{d}}^{\mathrm{ref}}$, $U_{\mathrm{q}}^{\mathrm{ref}}$ values, which must be transformed to the stationary (motor) reference frame using a 'Reverse Park' as shown in Fig. 2.18. For electrical machines the 'L-R' load is predominantly inductive, i.e. $L/\mathrm{T_s} > R$ in which case the voltage vector can be represented in the synchronous reference frame as

$$\vec{U}^{\mathrm{ref}} \cong \frac{L}{\mathrm{T_s}}\left(\vec{i}^{\mathrm{ref}} - \vec{i}\right) \tag{2.26}$$

where $\vec{i}^{\mathrm{ref}} = i_{\mathrm{d}}^{\mathrm{ref}} + j i_{\mathrm{q}}^{\mathrm{ref}}$ is the reference current vector, set by the user, while $\vec{i}$ represents the measured current. The task of the current controller is therefore, to generate the voltage vector needed to reduce the error between the currents to zero during each sampling interval $\mathrm{T_s}$. An example which shows the relationship between current error and voltage vector is given in Fig. 2.19.

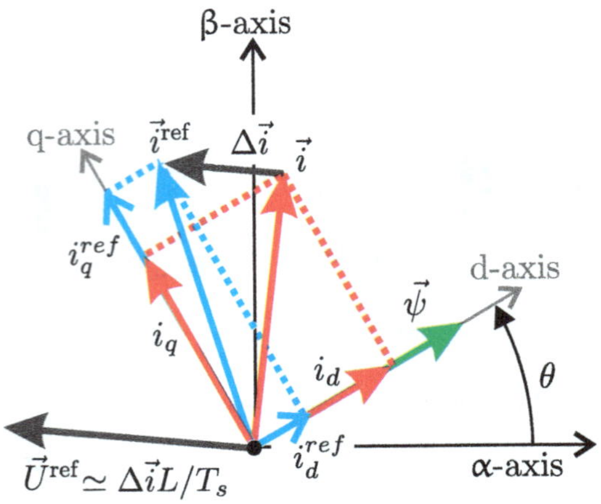

Fig. 2.19 Required voltage vector needed to minimize a current error $\Delta\vec{i}$

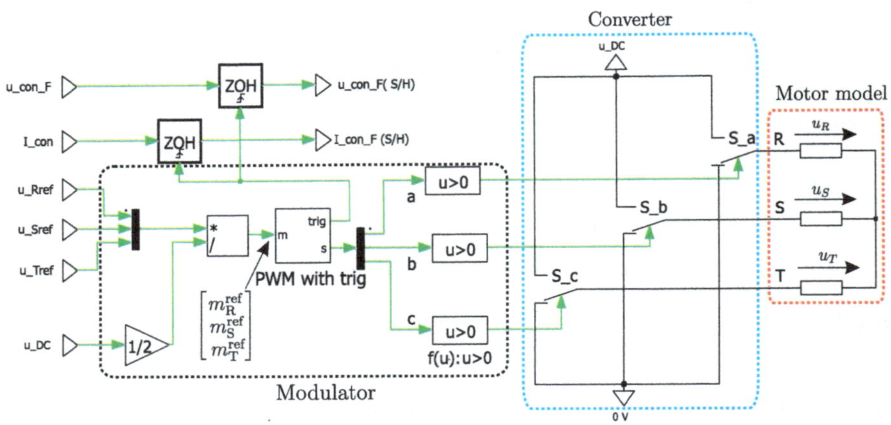

Fig. 2.20 Converter/modulator

The task of the modulator/converter is to generate a voltage vector which matches the reference vector, as will be discussed in the following subsection.

2.1.6 *Space Vector Modulation and Timing Issues*

In the example shown in Fig. 2.18 the reference voltage was converted to three phase voltages u_{Rref}, u_{Sref}, u_{Tref} which were then used as inputs to a converter represented by linear amplifiers. In reality use is made of a converter, which can be conveniently represented by three bipolar switches S_a, S_b, S_c that are controlled by logic signals a, b, c as shown in Fig. 2.20. The converter is connected to a DC supply u_{DC}, whilst the outputs R, S, T of said module are connected to a motor, which has been conveniently represented by three resistors. The voltage across the motor phases u_R, u_S, u_T will depend on the status of the switches and can be either $\pm 2u_{DC}/3$, $\pm u_{DC}/3$ or zero. In the latter case the outputs of the converter are simultaneously connected either the DC bus supply u_{DC} or the 0 V terminal. The modulator unit, also shown in Fig. 2.20, generates the desired logic signals that control the bipolar switches. The chosen logic sequence of the signals a, b, c and the duration for which these should be active is determined by the three phase reference signals u_R^{ref}, u_S^{ref}, u_T^{ref}. These input signals are then scaled by half the DC bus voltage $u_{DC}/2$, in which case they are referred to as modulation indices: m_R^{ref}, m_S^{ref}, m_T^{ref}, which are connected to the pulse width modulation unit that generates the three logic signals a, b, c for the switches. When a gate signal is '1' the corresponding bipolar switch is connected to the DC supply source. Vice-versa a logic '0' implies that the output of the converter is connected to the 0 V of the DC supply. For completeness, two 'zero order hold' `ZOH` modules are also shown which represent the sample and hold units that sample the measured voltages and currents of the converter. The triggering of the `ZOH` modules must be tied to the modulator as will become apparent shortly. Only a limited number of logic signal a, b, c combinations are possible and these can be represented by six active and two 'zero' output voltage vectors as shown in Fig. 2.21. These output vectors are generated by considering the voltages across the three motor phases for a given logic sequence a, b, c, and then applying a forward Clark transformation. Each vector, with amplitude $2u_{DC}/3$, is shown in terms of the logic signal combination a, b, c, hence $\vec{u}_{120}(010)$ implies $a = 0$, $b = 1$ and $c = 0$. A 'zero' vector denotes the conditions where the output voltage will be zero, i.e. all switches connected to the DC or zero bus voltage terminal. The task of the modulator is to identify the two adjacent vectors (as shown in Fig. 2.21) which must be used to achieve a given average voltage per sample

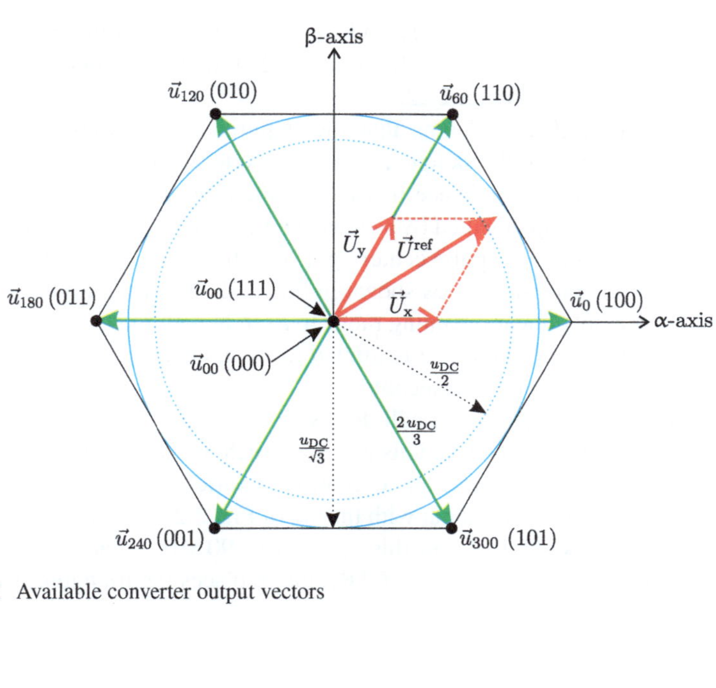

Fig. 2.21 Available converter output vectors

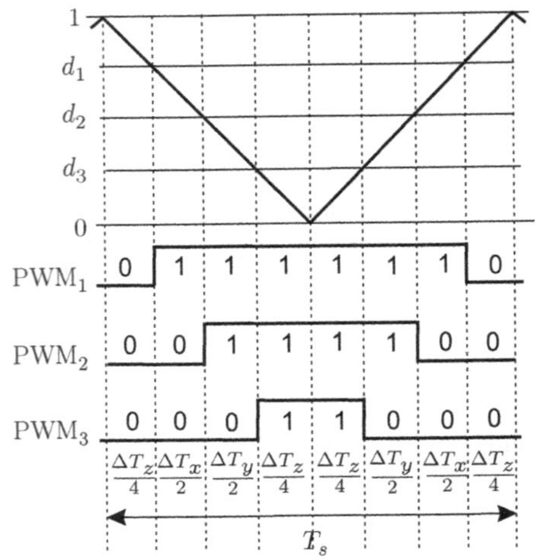

Fig. 2.22 PWM generation

$\vec{U}^{\text{ref}}(t_k)$ and the time interval (within one sample) in which either one or two active vectors must be selected. In the example shown the output voltage vector to the motor will be of the form

$$\vec{U}^{\text{ref}}(t_k) = \underbrace{\frac{\Delta T_x}{\text{T}_\text{s}}\vec{u}_0}_{\vec{U}_\text{x}} + \underbrace{\frac{\Delta T_y}{\text{T}_\text{s}}\vec{u}_{60}}_{\vec{U}_\text{y}} \tag{2.27}$$

where the time interval variables ΔT_x, ΔT_y are found by comparing the duty cycle values defined as $d_i = {}^{(m_i^*+1)}/_2$, for $i = 1, 2, 3$ with a triangular waveform (amplitude '1') as shown in Fig. 2.22.

Observation of Fig. 2.22 learns that the required time variables are evenly distributed to each **half** of the sample interval T_s. Note that the sum of ${}^{(\Delta T_x+\Delta T_y)}/_2$ must be less or equal to ${}^{\text{T}_\text{s}}/_2$, hence if it is less, the time interval ${}^{\Delta T_z}/_{/2} = {}^{\text{T}_\text{s}}/_2 - {}^{(\Delta T_x+\Delta T_y)}/_2$ must be complemented by a convenient choice of zero vectors. So called 'third harmonic' injection or 'pulse centering' [4], distributes the available time ${}^{\Delta T_z}/_2$ to either side of the interval ${}^{\text{T}_\text{s}}/_2$, as shown in Fig. 2.22. The use of pulse centering increases the available linear operating range to a circle with radius ${}^{u_\text{DC}}/_{\sqrt{3}}$. Without pulse-centering the operating range is limited to ${}^{u_\text{DC}}/_2$, hence a factor ${}^2/_{\sqrt{3}} \approx 1.15$ increase is achieved by using this technique.

In the sequel to this section we will discuss the timing aspects, which basically describe at which time interval various activities such as data acquisition, defining the active set of converter vectors and execution of various control algorithms take place. This discussion is carried out with the aid of Fig. 2.23, which (among others) shows the CPU clock (CPU_clk), in this case set to 90 MHz, that is connected to an up/down 'Time Base Counter' (TBCTR), that replaces the triangular waveform

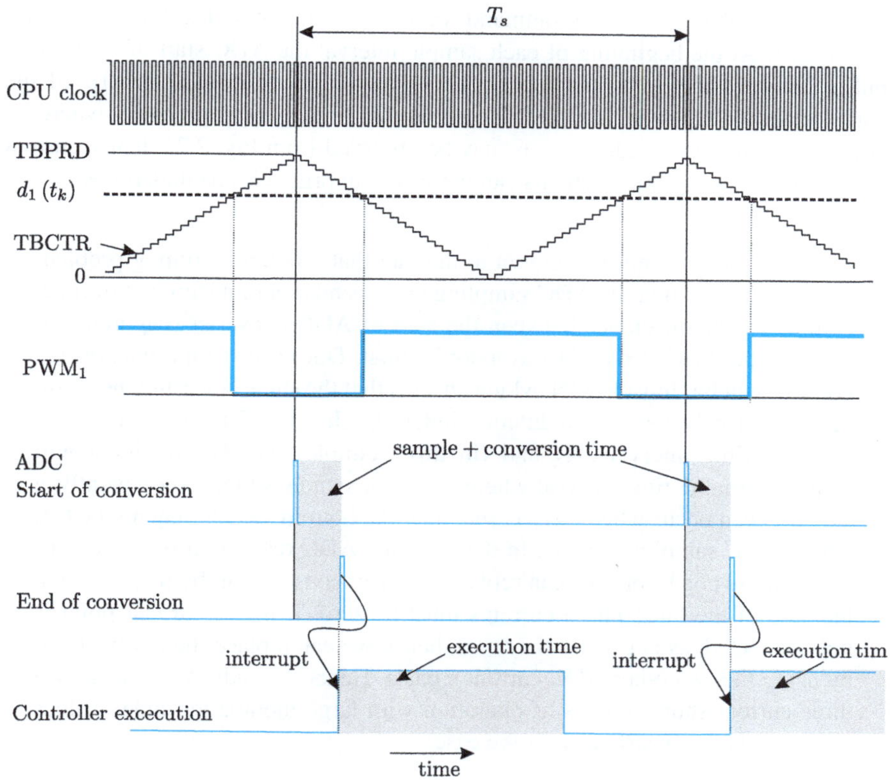

Fig. 2.23 Timing issues [9]

shown earlier in Fig. 2.22. The start of each sampling time coincides with the highest counter value (referred to as Time base Period) TBPRD, and its value is therefore determined by

$$\text{TBPRD} = \frac{\text{T}_\text{s}}{2}\text{CPU}_\text{clk} \tag{2.28}$$

hence a PWM frequency of 10 kHz, which corresponds to $\text{T}_\text{s} = 0.1$ ms, requires a maximum counter value of TBPRD=4500. The required ON/OFF times of the power switches for a given converter leg is found by comparing the up/down counter value with the discrete duty cycle value d_c which is now defined as

$$d_c = \text{TBPRD}\frac{(m_i^*(t_k) + 1)}{2} \tag{2.29}$$

where m_i^* is the reference (command value) modulation index for each phase $i = 1, 2, 3$. At the beginning of each sample interval the ADC start of conversion pulse, activates the ADC (analog Digital) data acquisition phase, where all the currents and voltages are sampled. Note that this time mark is precisely where all bottom switches are conducting, as may be observed from Fig. 2.22, hence current measurements using in line shunts can therefore (in principle) be undertaken.

The readers attention is drawn to the fact that a potential timing problem can occur, given that the ADC sampling process takes a finite amount of time, defined by the difference between the start of ADC converter (beginning of T_s) and the ADC 'End of Conversion' signal. During this time interval the bottom switches must be ON, which ensures that the shunts measure the phase currents. If however the modulation index d_1, shown in Fig. 2.22, increases then the time interval ΔT_z and (in this example) the PWM_1 signal will exhibit a smaller time interval where a bottom switch is ON. Consequently, a situation can occur where ΔT_z is such that the bottom switch may not be ON whilst ADC sampling occurs. In that case the ADC result will be erroneous. Under these conditions we can replace the erroneous output, by realizing that the sum of the three phase currents must be zero (if the motor star point is not connected, as is usually the case), hence we can replace the faulty signal by using the two other ADC current signals. Hence it is advantageous to use three current shunt sensors, if operation with large modulation index values is required (as is nearly always the case).

The 'End of Conversion' signal, triggers an 'EOC' interrupt which signifies that data is ready for processing by a control algorithm such as, for example, InstaSPIN. The control algorithm will calculate the reference modulation indices m_i^*, for $i = 1, 2, 3$ for the next time interval. Note that the EOC interrupt can be used to trigger the execution of a control algorithm during each cycle, or an integer multiple thereof. For example, the following timing setup can be used:

- each EOC pulse: current control algorithm use to calculate the modulation indices
- every second EOC pulse: A sensored, or sensor-less FOC algorithm which sets the required i_d, $i_q$values for the current controller algorithm
- every fourth EOC pulse: a speed control algorithm which calculates the required torque reference value for the FOC algorithm

This so called segmenting of various control algorithms's is needed to ensure that execution times stay within the time available. This issue becomes particularly important when very high PWM frequencies are used, which result in a very low sampling interval period T_s. Note that the sampling interval T_s refers to the sampling of the ADC unit and can be different to the time-step used by the various control algorithms, as was discussed above.

2.2 Drive Development for Real-Time Control

The principal objective of this book is to empower the reader with the ability to work with a typical, low cost, industrial drive. The drive setup shown in Fig. 2.24 is able to meet these requirements, as it contains a LVSERVOMTR PM machine connected to a LVACIMTR induction machine. In some cases a 2MTR-DYNO dual PM machine set up may be required, in which case the induction machine must be replaced by a second PM machine. In this example a Texas Instruments LAUNCHXL-F28069M with two BOOSTXL-DRV8301 booster-packs is shown, given that it is convenient for operating two machines simultaneously, i.e. one acting as a test motor, the other as a 'load' with torque measuring capability (using for example a sensored field oriented controller). For laboratories which require the use of a single machine, the 'forward' (closest to the USB connector) BOOSTXL-DRV8301 is removed. Hence only the 'aft' (location 2) BOOSTXL-DRV8301 is used for single machine operation in the ensuing laboratories, which then provides the reader with a clear view of the LEDS on the LAUNCHXL-F28069M, some of which are used for monitoring drive operation. Also shown in this figure are a DC power supply (which can be replaced by the 24V DC/AC adapter provided

Fig. 2.24 Typical experimental setup

with the kit). The supply voltage level is purposely held low for safety reasons. A DC current probe and oscilloscope, both of which are optional can be used to experimentally verify results. In addition to the above a laptop or pc is required which contains the following software packages:

- PLECS: used for detailed drive simulation with floating point controllers and electrical circuit models. Also used for so called 'processor in the loop' (PIL) control.
- VisSim: used to do real time control and simulation using fixed point controllers.
- Texas Instruments: Code Composer Studio (CCS) for code compilation and debugging operations.

With hardware and software tools in place a so called 'A-B-C' development approach is used in this book and applied to a series of 'laboratory' sessions designed to familiarize the reader with specific drive topics. The basic structure of this approach is shown in Fig. 2.25.

The following development phases are shown in Fig. 2.25 namely:

- Phase A: Drive simulation with 'test' controller, this phase is undertaken with PLECS simulation software. Its objective is to analyze the behavior of a specific controller connected to a drive, in the form of a converter and either the IM or PM machine shown in the experimental setup. This type of analysis is very detailed and includes, for example, the switching of the converter power devices.
- Phase B: Controller implementation for MCU, using VisSim software. This phase of development uses a discrete, fixed point representation of the control module, used in Phase A. Testing of this module is then undertaken with a continuous-time model of the machine (as used in phase A). However in this phase the converter switching process is not modeled. Its primary purpose is to verify fixed point controller operation. In addition, sensorless operation using so called 'Processor In the Loop' (PIL) technology will be introduced to demonstrate InstaSPIN operation.
- Phase C: MCU version of the controller, using VisSim software. In this phase the additional building blocks such as ADC, PWM and diagnostic buffers (used to examine data in the next phase) are added to the module, which is then compiled and linked with the Texas Instruments and VisSim libraries, which leads to an 'out-file' that must be down-loaded to the MCU via the JTAG connection. The file can reside in either MCU RAM, or FLASH memory.
- Phase C+: MCU Run version of the controller, using VisSim software, in this phase the user can execute the 'out-file' file created in phase C and view outputs whilst controlling the drive. This out-file is downloaded to the MCU prior to running the laboratory. If using MCU RAM this download is carried out via VisSim, otherwise (when using MCU Flash memory) Texas Instruments Codecomposer (CCS) can, for example, be used.

In the series of laboratories which will be considered in the following three chapters, not all development phases will necessarily be invoked. In some cases a detailed inverter switching analysis, may for example, not be required which means that

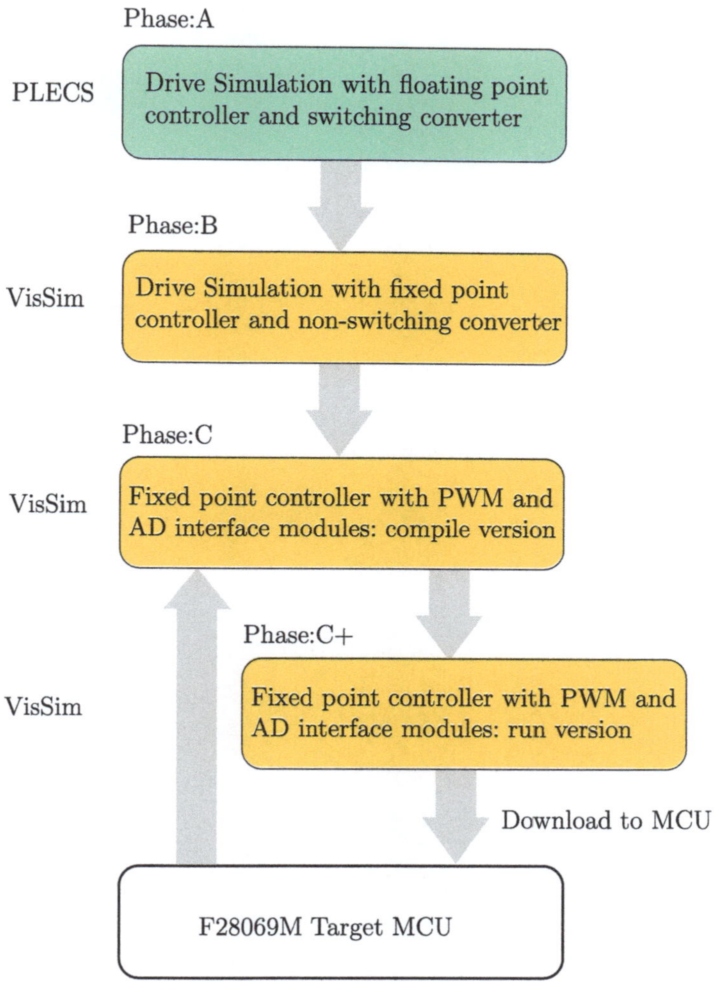

Fig. 2.25 'A-B-C' development approach used for laboratory development

phase A is then omitted. During the debugging stage it is often necessary to cycle between phases C $\rightarrow$ C+ $\rightarrow$ C where internal variables are visualized on a diagnostic scope. In some cases only phase A and/or B are used, which occurs when for example no experimental machine is available for testing. Hence a model of said machine can then be used.

Chapter 3
Laboratory Sessions: Module 1

A series of laboratory modules will be discussed in this chapter, which have been designed to lead the reader through the development phases needed to understand and work with a modern electrical drive. Hence basic voltage/frequency control, current/frequency control are introduced as a prelude to more elaborate drive structures which utilize field-oriented control. In this context a sensored field oriented PM drive and an induction machine drive are (among others) considered. Dual drive operation, where the PM sensored drive acts as a load/torque transducer (which is effectively a dynamometer) for the induction motor provides the reader with the opportunity to explore motor/generator operation.

Simultaneously the reader will be guided through the hardware and software tools needed to analyze and debug an electrical drive, knowledge which can then be readily applied to solving specific industrial drive problems. In this chapter only sensored operation (using the shaft sensor of the PM motor) is considered. Sensorless operation of drives is discussed in subsequent chapters.

3.1 Laboratory 1:1: Open-Loop Voltage Control

The purpose of this laboratory is to realize open loop voltage control of a PM motor. At the same time the lab is used to make a functional check on the current and voltage measurements and overall functioning of the drive board and software.

Electronic supplementary material: The online version of this chapter (doi: 10.1007/978-3-319-20043-9_3) contains supplementary material, which is available to authorized users.

D.W.J. Pulle et al., *Applied Control of Electrical Drives*,
Power Systems, DOI 10.1007/978-3-319-20043-9_3

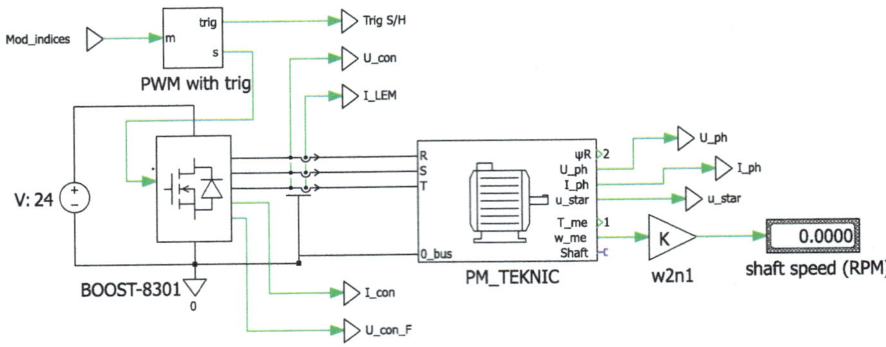

Fig. 3.1 Drive setup

3.1.1 Lab 1:1: Phase A

This module considers the simulation of the PM motor drive in use, with an open loop voltage controller. In this example the switching process of the converter is specifically modeled, to allow the reader to comprehend the waveforms that typically occur in a drive. The following information is relevant for this laboratory component:

- Reference program [11]: `Lab1-1_LaunchXL-069phaseAv2.plecs`.
- Description: Open loop voltage control of a PM motor.
- Equipment/Software: PLECS simulation program.
- Outcomes: Show the various waveforms that occurs in the drive when using a converter with discrete power switches.

The overall drive setup for this laboratory as shown in Fig. 3.1 represents the `BOOSTXL-DRV8301` converter (located on the `LAUNCHXL-F28069M` module 'aft' position) which is connected to the PM Teknic (Texas Instruments LVSERVOMTR) motor. An electrical circuit implementation of the converter is used in this part of the laboratory so that the reader can understand the voltage and current waveforms that occur in the drive. The PM motor model inputs are (among others) three electrical voltage variables which are transformed in the motor module to a set of control blocks that represent an IRTF [14] based representation of the machine. Motor module output is the vector `U_ph` which represents the three phase voltages of the star connected machine. In addition, the star-point to 0 V voltage `u_star` is made available for perusal by the reader. Furthermore, the shaft torque `T_me` (Nm), shaft speed `w_me` (rad/s) and current vector `I_ph` are provided together with the rotor flux vector `psi_R`. The motor module is also provided with a 'mechanical' output `Shaft`, which can be used to connect a set of mechanical 'control' blocks. A 24 V DC supply is connected to the converter, which is represented by the electrical circuit model shown in Fig. 3.2. The converter is represented by six ideal power electronic devices as shown in Fig. 3.2. Each of the three converter legs is provided with a shunt that measures the current in the connected motor when the

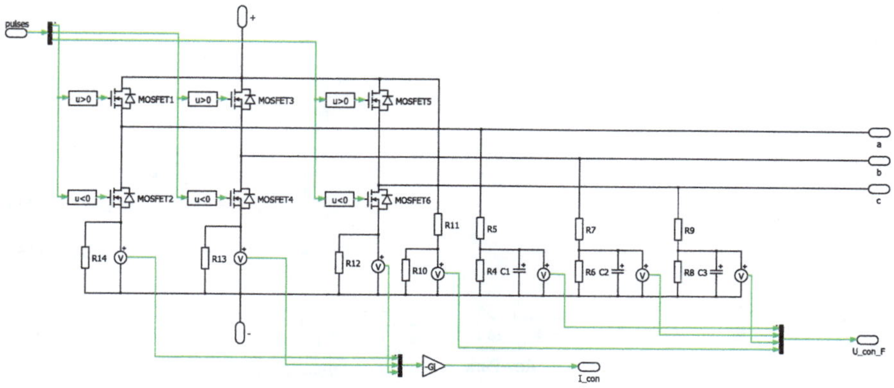

Fig. 3.2 Converter model of BOOSTXL-DRV8301

lower device is turned on. A pulse vector, which consists of three gate control signals is used to switch on the upper or lower device of a converter leg. Connected to the output of each converter terminal is a resistance/capacitor network that is used to measure and filter the converter output voltage. The three low pass filtered converter voltages and the voltages across the shunt resistors are amplified by gain `Gi` and represented by two (column matrices referred to as) vectors `U_con_F` and `I_con` respectively. Also present are the converter to 0V voltage vector `U_con` and the phase current vector `I_LEM`, both of which are measured directly between motor and converter (see Fig. 3.1). The latter two vectors are provided in order to allow a comparison between the phase voltages and current measured with the aid of shunt resistors.

Operation of the PWM modulator/converter is according to the theory given in Sect. 2.1.6. Accordingly, the input to this module is the vector `mod_indices`, which contains the three phase reference modulation indices that must be provided by a control module.

Central to these laboratory sessions is the controller structure. In this case the objective is to create a controller that can generate a rotating voltage space vector as shown in Fig. 3.3. Inputs to the controller are the `u_d_ref`, `u_q_ref` components of a synchronous d, q reference frame. A Reverse Park transformation is used to create a rotating voltage vector with a reference speed set by the (user defined) frequency `f_ref`. The controller structure, as given in Fig. 3.4, shows the integrator and Reverse Park module needed to generate the voltage vector $\vec{u}^{\mathrm{ref}}$. This vector is initially converted to a modulation reference vector $\vec{m}^{\mathrm{ref}} = \vec{u}^{\mathrm{ref}}/(u_{\mathrm{DC}}/2) = m_{\alpha}^{\mathrm{ref}} + jm_{\beta}^{\mathrm{ref}}$, which is then converted to three reference modulation indices: $m_{\mathrm{A}}^{\mathrm{ref}}, m_{\mathrm{B}}^{\mathrm{ref}}, m_{\mathrm{C}}^{\mathrm{ref}}$ via the pulse-centering (SVM) unit. These variables are then combined to a single column vector `mod_indices`. An additional output vector `U_ref` is also provided, which represents the three reference phase voltages, that are found by multiplying the modulation index vector by half the DC bus voltage as shown in Fig. 3.4.

Central to sensorless operation is the ability to accurately measure the phase voltages and currents. The experimental drive board(s) used in this book are so

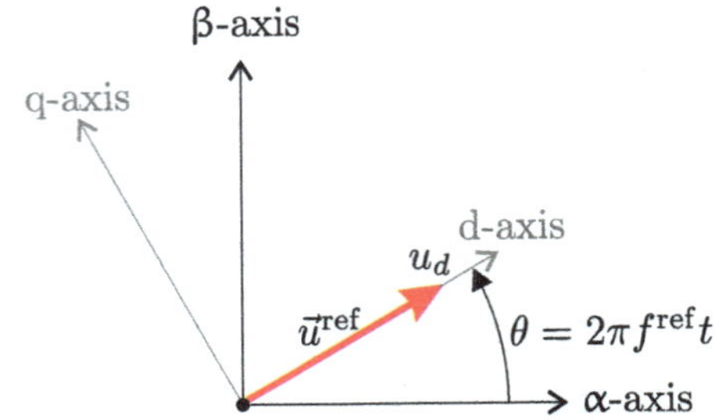

Fig. 3.3 Reference voltage vector

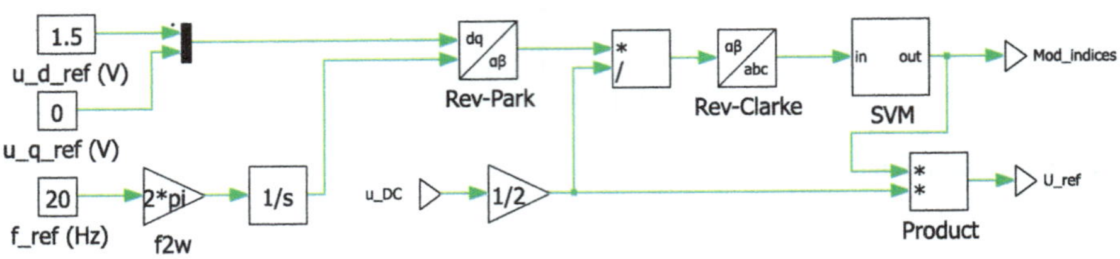

Fig. 3.4 Voltage controller module

called 'FAST prepared', which implies that they have been purposely designed with a data acquisition module (ADC) that undertakes the signal conditioning of the voltage/currents as needed for the Texas Instruments InstaSPIN algorithm. The converter model discussed previously (see Fig. 3.2) is 'FAST prepared' and at this stage it is helpful to examine the data signal condition of the circuit in more detail, where use is made of Fig. 3.5. This is required because a number of critical hardware based parameters must be provided to the controller to ensure correct operation of the drive. Readily apparent in Fig. 3.5 is the resistor/capacitor network used to attenuate and low-pass filter the converter output voltage. The gain $\mathrm{G_R}$ and time constant τ_{R} of this circuits is equal to

$$\mathrm{G_R} = \frac{R_4}{R_4 + R_5} \tag{3.1a}$$

$$\tau_{\mathrm{R}} = \left(\frac{R_4 R_5}{R_4 + R_5}\right) C_1 \tag{3.1b}$$

The time constant $\tau_{\mathrm{R}}(\mathrm{s})$ or corner frequency $\omega_c = {}^{1}/{\tau_{\mathrm{R}}}$ (rad/s) and gain of this filter is board dependent, hence their value must be identified by inspection of the components (or provided). The example given in Fig. 3.5 shows the voltage filter components as present in the `BOOSTXL-DRV8301` module. The low-pass filtered converter voltage `u_LPF` is then buffered prior to sampling by the ADC converter of the microprocessor unit (MCU). Measurement of the converter DC bus voltage is similar to that undertaken for the converter voltages, however no filtering is undertaken.

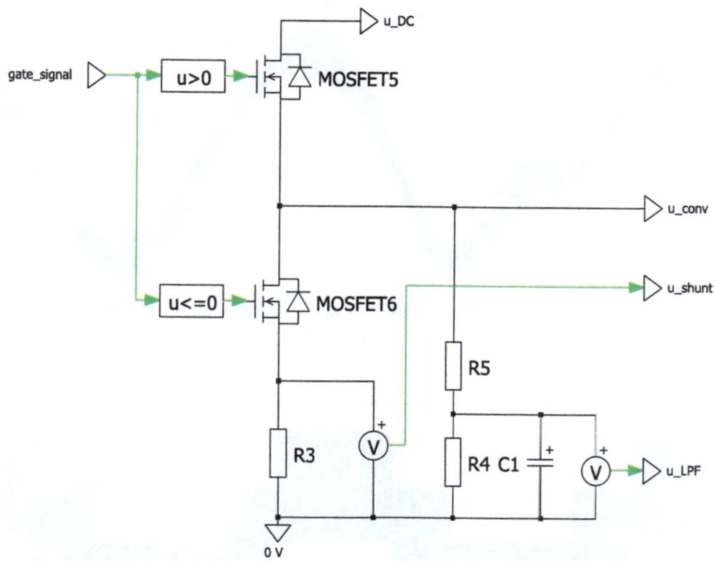

Fig. 3.5 Signal condition circuit for one converter phase

The final and critical part of the data acquisition process is the measurement of the phase currents using the shunt resistors in the converter leg (see Fig. 3.5). When a converter bottom switch is active the voltage across the shunt resistor is given as

$$u_{\mathrm{shunt}} = -i_{\mathrm{conv}} R_3 \tag{3.2}$$

where i_{conv} and R_3 are the phase current and shunt resistance values respectively. The shunt voltage `u_shunt` is then amplified with a gain G_i, prior to sampling by the ADC unit of the MCU. Note that the gain G_i is board dependent. For example, the `BOOSTXL-DRV8301` converter has a gain of $G_i = -10.0$ which is negative, hence the current inversion required to compensate the sign in Eq. (3.2) (in order to ensure that the outgoing converter current is positive) is done in hardware.

The reader is encouraged to undertake a detailed examination of this simulation using the PLECS Simulation platform and this specific file. Only by understanding each aspect of operation can one hope to understand the modeling process of the drive as a whole. An example of drive operation is shown in Fig. 3.6. This example corresponds to user control settings: $u_{\mathrm{d}} = 1.5\,\mathrm{V}$ $u_{\mathrm{q}} = 0\,\mathrm{V}$, $f^{\mathrm{ref}} = 20\,\mathrm{Hz}$, hence a voltage vector of 1.5 V, rotating at a speed of 1200 rpm is applied to the motor. This implies that the shaft speed of the motor will be 300 rpm (motor in use has four pole pairs). The following observation of the results can be made:

- The PWM switching frequency of 2500 Hz is purposely set low in order to better visualize the switching process. A typical PWM frequency would be in the order of 10–20 kHz, pending the switching devices used.

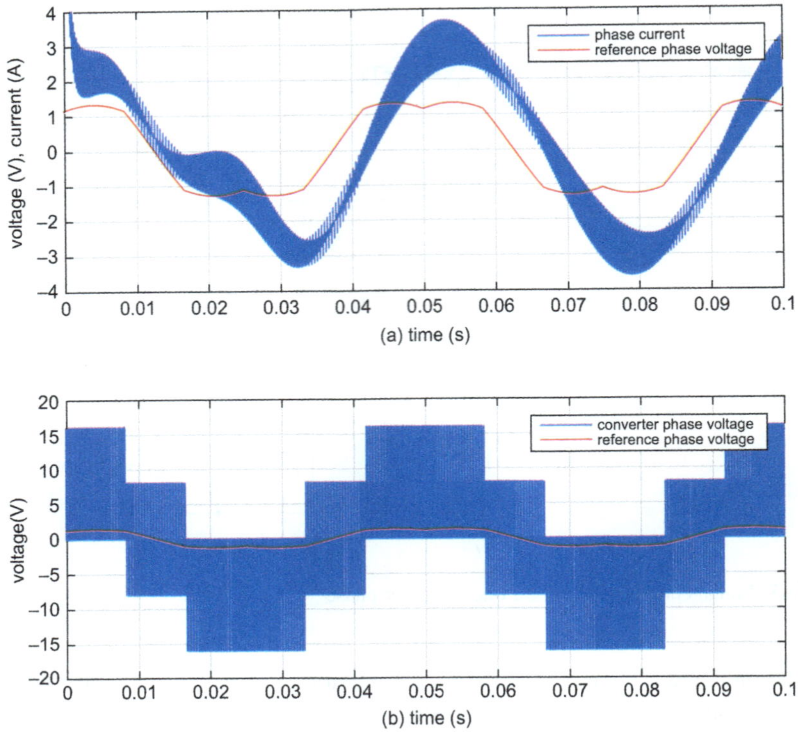

Fig. 3.6 Phase A: Typical waveforms obtained using a voltage controller

- A 100 ms operation sequence is shown with a controller reference voltage u_A^{ref} that is calculated using $u_A^{ref} = m_A^{ref} \cdot u_{DC}/2$ as discussed earlier. Note that the peak value of the phase voltage waveform is NOT 1.5 V, because its value is changed by a factor $(\sqrt{3}/2)$ due to the presence of the pulse centering module. The motor phase voltage u_A is generated by the converter, using 'ideal' switches and the per PWM cycle average value must in theory match the instantaneous reference voltage value. However in practice the average converter voltage per sample T_s phase voltage generated by the converter is typically lower than the theoretical reference voltage u_A^{ref}generated by the controller in use. The reason for this is that there is a voltage drop across the power electronic devices. Furthermore, dead-time introduced to avoid 'shoot through' (a situation where two switches of one converter leg conduct simultaneously) will also cause a error between average converter voltage per sample T_s and reference phase voltage values.
- The phase current, clearly shows a current ripple component caused by the PWM of the converter. Note that the ripple component is relatively large, because the PWM frequency was purposely set to 2.5 kHz (as to better observe the switching process. For actual experimental operation a PWM frequency of 30 kHz is typically selected, which leads to a much lower current ripple component.

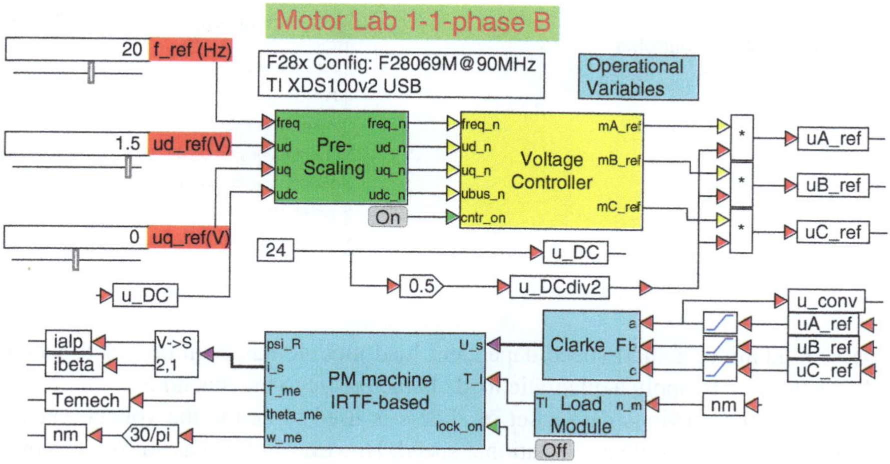

Fig. 3.7 Phase B simulation of a drive with a voltage controller

Note that the drive configuration used for this simulation closely matches the experimental drive in use, hence these results are representative of what can be expected in the actual drive, as will become apparent in subsequent development phases.

3.1.2 Lab 1:1: Phase B

The following information is relevant for this laboratory component:

- Reference program [11]: `lab1-1_LaunchXLphBv2.vsm`.
- Description: Open loop voltage control of a PM motor.
- Equipment/Software: VisSim simulation program.
- Outcomes: Show the various waveforms that occur in the drive when using a fixed point representation of the controller.

This module considers the simulation of the PM motor drive in use, with a fixed point open loop voltage controller as shown in Fig. 3.7. Scaling of the currents/voltages as discussed in Sect. 1.4.2 must therefore be undertaken. For phase B and C simulations, circuit models are not used, hence the outputs of the controller, which are the three reference modulation indices `mA_ref`, `mB_ref`, `mC_ref`, are multiplied by the term $u_{DC}/2$. The average voltage (per sample) value are then used by a controller, which in this case is simply represented by three limiters, which limit the voltage to the motor to $\pm u_{DC}/2$. A Forward Clarke transformation is used to generate the voltage vector for the machine model. The latter is modeled in floating point and is a so called IRTF based model [4]. An `on/off` switch has been added to this motor model to simulate a locked rotor situation (rotor locked, when set to

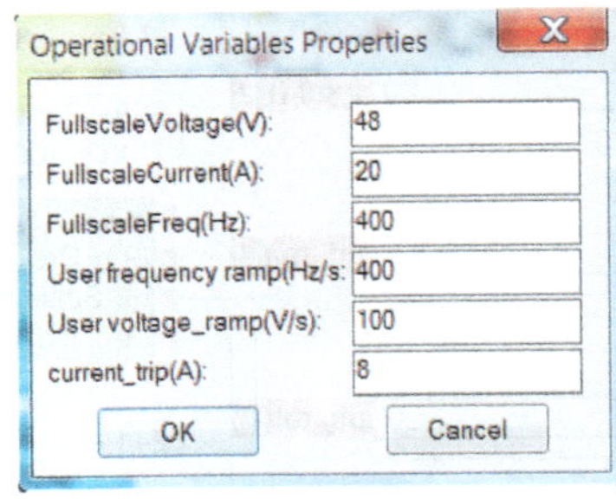

Fig. 3.8 Dialog box used for drive with a voltage controller

ON). In addition, a shaft speed dependent load module has been provided, which allows the user to apply a quadratic load characteristic to the machine.

As with the previous case a set of sliders is used to define the voltage vector, in terms of its amplitude and rotational speed. However in this case a 'pre-scaling' module (as discussed in Sect. 1.4.2) has been added to convert the floating point variables required for the controller to fixed point format.

The use of a fixed point controller implies the need to scale the variables. For this purpose a set of dialog boxes has been introduced in the simulation, where the user must assign values as shown in Fig. 3.8. An 'operational Variables' dialog box is used to assign full scale value for the voltage, current and frequency. Also assigned in this box are the rate of variable change for voltage and frequency, i.e. introduced as a safety measure to avoid erratic slider action from causing over-currents in the drive. The dialog box entry `current_trip(A)` defines a current level where operation will be terminated by the software. This entry is not used in this phase B simulation. The reason it is shown here is that this 'operational' dialog box is also used in development phases C and C+.

An example of drive operation is shown in Figs. 3.9 and 3.10. Subplot (a) given in Fig. 3.9, shows the reference voltage waveform `U_conv` (which is also the motor phase voltage in this case) together with the motor current `ialp`, which is also the motor phase current. Note that this diagram is almost identical (same user input settings) to Fig. 3.6. However, the current ripple, observable in the phase A simulation is not present because the actual converter switching process is not modeled in phase B. In this case the simulation was run over a 0.3 s time interval. Note also that in this example the time step for the simulation was set to $T_s = 66.66\,\mu s$, i.e. a 15 kHz sampling frequency is used for the discrete controller model.

Sub plot (b) given in Fig. 3.9 shows the reference and actual shaft speed during start-up, for a given frequency ramp setting and, in this case, operating frequency of 20 Hz, which corresponds to a steady-state operating shaft speed of 300 rpm, as may be observed from Fig. 3.9.

To help visualize, often complex drive operations, it is prudent to represent key vectors in a vector plot as shown in Fig. 3.10. Shown in Fig. 3.10 are the steady-state vectors which represent the reference voltage vector $\vec{u}_s$ and current vector $\vec{i}_s$. In addition, a 'd-axis' vector is introduced to show the orientation of the d,q reference frame, where the positive d-axis is displayed as a 'green' vector. In this example, the reference u_q value is set to zero, hence the resulting voltage vector

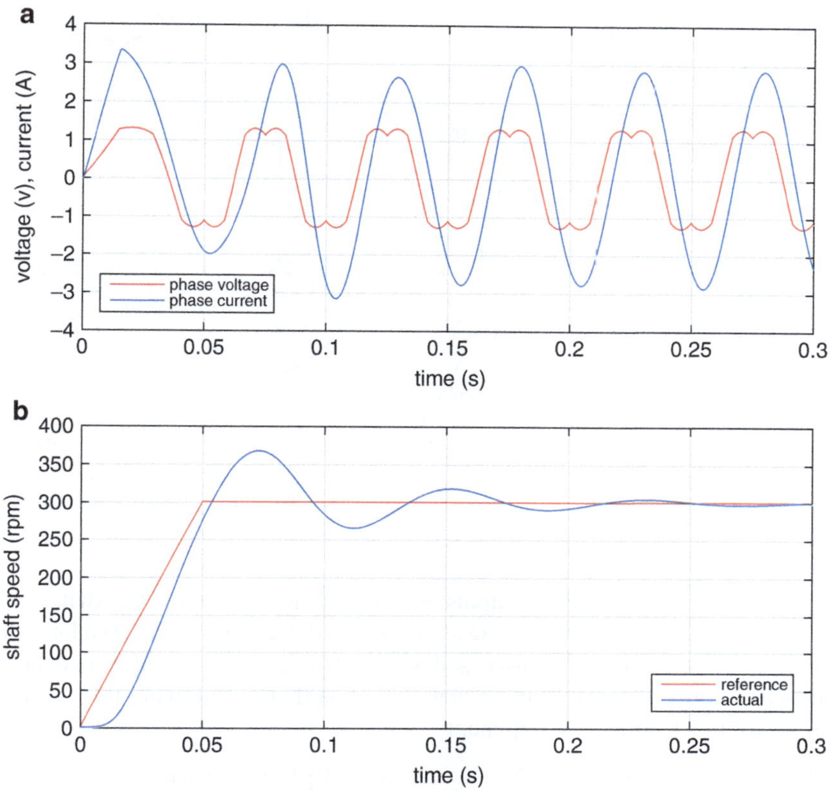

Fig. 3.9 Phase B: Typical waveforms obtained using a voltage controller

is aligned with the d-axis vector (and has an amplitude equal to 1.5), as is indeed the case. Note that part of the endpoint trajectory of the vectors are also shown in the vector plot. The reader is again encouraged to undertake a detailed examination of this simulation using the VisSim Simulation platform and this specific file. Only by becoming familiar with each aspect of operation and the program used,can one hope to understand the modeling process of the drive as a whole. Note that further information on the controller implementation is given in the next, phase C, subsection.

3.1.3 Lab 1:1: Phase C

In the previous phase, only knowledge of the control card to be used was required, in order to test the fixed point controller. In this phase the drive board itself needs to be known in considerable detail. Aspects such as allocation of data acquisition channels relative to the converter voltage/current sensors, choice of PWM outputs

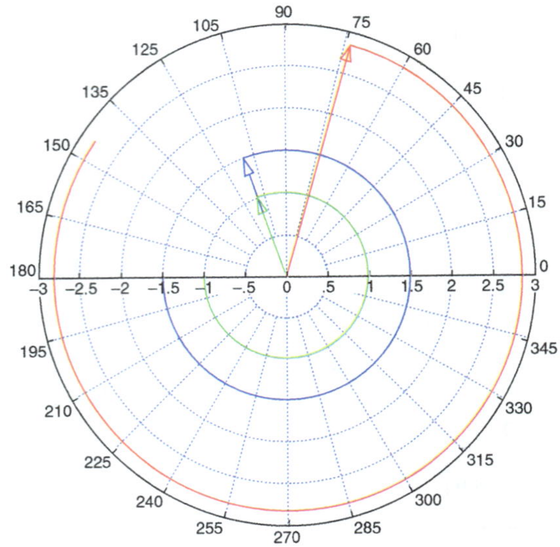

Fig. 3.10 Phase B: Vector plot obtained using a voltage controller with motor current vector $\vec{i}_s$('red'), reference motor voltage vector $\vec{u}_s$('blue') and unity d-axis vector ('green')

and allocation of gate-driver outputs which activate the converter switches (or converter module, if an integral instead of discrete devices are used). At a later stage 'commissioning of a drive' is discussed, see Sect. 4.5, which considers in detail the steps which must be taken to use a new drive board. The following information is relevant for this laboratory component:

- Reference program [11]: `lab1-1_LaunchXLphCv6.vsm`.
- Description: Open loop voltage control of a PM motor.
- Equipment/Software: Texas Instruments LAUNCHXL-F28069M, with BOOSTXL-DRV8301 module ('aft' position) and VisSim simulation program.
- Outcomes: Develop a complete drive algorithm which handles current/voltage measurement, converter PWM and setup MCU timing as required. Compile and download an `.out` file for use in phase C+.

A phase C development stage, cannot be used to run a drive, but its primary task is to assemble all the modules needed for drive operation with a given controller configuration. The drive setup given in Fig. 3.11 shows the 'open-loop voltage controller' module, which must be compiled to generate an `.out` file. As mentioned above, a single boost pack located in the 'aft' position (furthest away from the USB connector) is used for single motor operation, in which case the following jumper and dip-switch positions on the LAUNCHXL-F28069M module are required:

- Jumpers `JP1` and `JP2` OPEN
- Jumpers `JP4` and `JP5` CLOSED
- Jumpers `JP3`, `JP6` and `JP7` CLOSED
- Dip-switches `SW1` to `SW3` ON

Inputs, shown in development phase C as constants (for readability purposes) for this controller module are the user reference voltages `ud`,`uq`, which are transformed

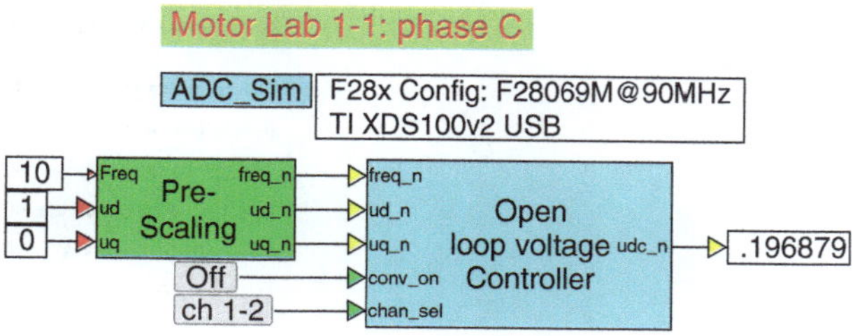

Fig. 3.11 Phase C simulation of a drive with a voltage controller

Fig. 3.12 One level down into the controller module

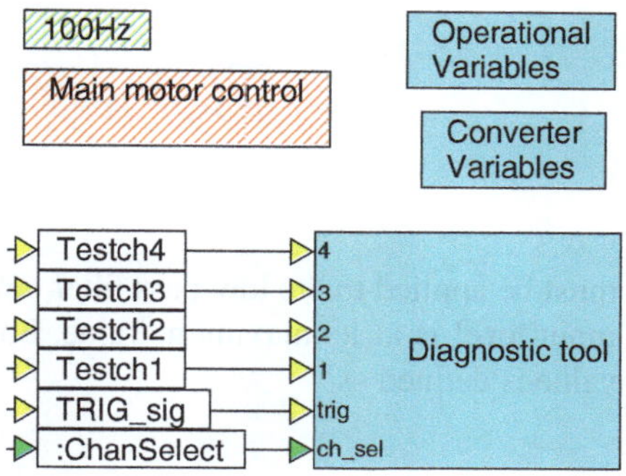

to 'per unit' values in the 'Pre-scaling' module. In addition, the electrical speed reference `Freq` is required, which is also converted to a 'per unit' value in the 'Pre-Scaling' module. Two additional logic inputs are provided, which allow the user to turn on the converter and to select which combination of diagnostic channels should be viewed. Further information on this tool will be given at a later stage.

Outputs of the Controller module is the per unit DC bus voltage `udc_n`, which must be scaled by the 'post-scaling' module (in development phase C+) to provide a reading in Volts. Also shown in Fig. 3.11 is an 'ADC-Sim' module, which is used to provide artificial inputs (as preset in this module) to the ADC channels. Henceforth these values will appear in the controller when this file is run. This approach is useful because it allows the user to check if signal levels within the module are appropriate, i.e. as anticipated.

Moving one level down into the Controller module ('right' mouse click, after selecting module) leads to the set of modules/dialog boxes shown in Fig. 3.12. In this case two dialog boxes are used, as given in Fig. 3.13, which contain the parameters that need to be assigned by the user for this laboratory. An 'Operational Variables' dialog box, as discussed in the previous section, is used to assign full scale values for the voltage, current, frequency and the ramp change for voltage as well as frequency. A current trip variable `current_trip` sets the software over-current value, i.e. the absolute trip value when the converter will be de-activated by the controller. The 'Converter Variables' dialog box defines the scaling used by the Analog to Digital converter, which has a maximum input value of 3.3 V and is directly linked to the signal processing undertaken for voltages and currents as discussed in the previous section. For example, the 'ADC-fullscale-voltage' value is the input voltage which

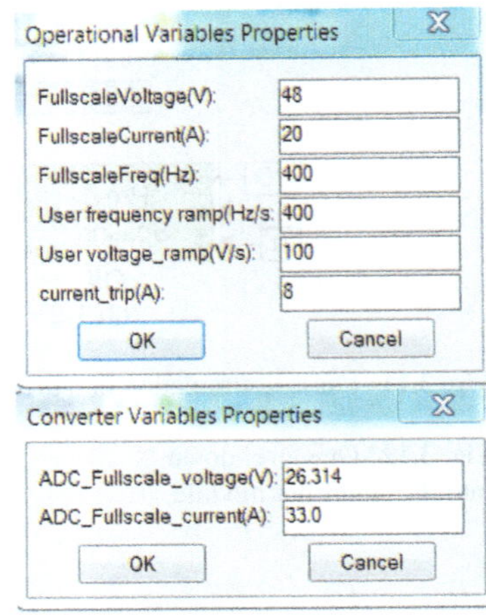

Fig. 3.13 Phase C: Dialog boxes used for drive with a voltage controller

must be applied to the low-pass filter network (see Fig. 3.5) to ensure that the ADC input level is at its maximum value. Subsequent use of Eq. (3.1a) learns that this value is defined as

$$\text{ADC-fullscalevoltage}\,(\mathrm{V_{pp}}) = \frac{3.3\,\mathrm{V}}{\mathrm{G_R}\,\mathrm{G_v}} \tag{3.3a}$$

$$\text{ADC-fullscalecurrent}\,(\mathrm{A_{pp}}) = \frac{3.3\,\mathrm{V}}{\mathrm{G_i}\,R_\mathrm{h}} \tag{3.3b}$$

where $\mathrm{G_R}$ is the low-pass filter attenuation used for measuring the converter voltages. The variable $\mathrm{G_v}$ is the signal amplification present between low-pass filter and ADC input. Similarly, the 'ADC-fullscale-current' value is the peak to peak current in the shunt resistance, which corresponds to an ADC input voltage range of 3.3 V. Hence its value is determined by the shunt resistance value R_h and any signal amplification $\mathrm{G_i}$ between said resistor and ADC input. The voltage and current measurements are unipolar and bipolar respectively. This implies that the voltage measurement is referenced to the negative converter DC bus voltage, whilst the current is referenced to the center of the ADC voltage span, i.e. $3.3/2$ V. Note that the ADC dialog values, are board dependent and need to be found by inspection of the circuit (or provided). In a subsequent chapter a laboratory on 'drive commissioning' (see Sect. 4.5) we will elaborate further on these issues.

A 'Diagnostic tool' module is used to buffer (store) two variables, which can subsequently be displayed in a graph when operating in phase C+. A two channel multiplexer is used to allow the user to either display variables `Test1`, `Test2` or `Test3`, `Test4` by using a 'channel select' button discussed above. This diagnostic tool is extremely valuable for debugging purposes.

A '100 Hz' module undertakes all the background tasks for the drive. Among these is the ADC offset control task (a state machine used to determine ADC offsets,

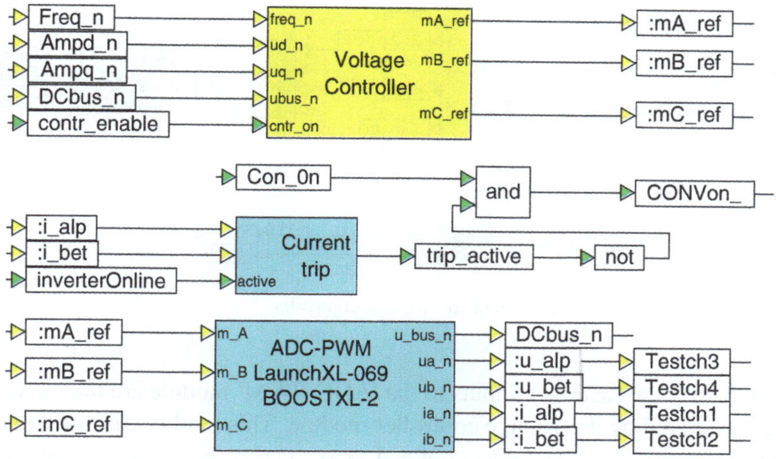

Fig. 3.14 One level down into the main motor control module

prior to drive operation). In addition, this module controls a 'heart beat' LED, which flashes on the control card. A 'blue' LED remains ON if a current trip condition occurs. Both LED's are located on the `LAUNCHXL-F28069M` board. Moving one level lower into the 'Main motor Control' module reveals a set of modules, as shown in Fig. 3.14. Of these the 'Voltage controller' has been mentioned previously, given that it is the module tested in phase B. The per unit variables: DC bus voltage, phase currents and voltages are generated by a board specific 'ADC-PWM' module. This module is configured for this lab component with the following basic settings:

- PWM frequency: 30 kHz.
- ADC sampling rate: 15 kHz.
- Converter dead-time setting: 0.11 μs.
- Board in use: `LAUNCHXL-F28069M`.

This ADC-PWM module conforms with MCU timing, as discussed in Fig. 2.23, which implies that the 'Voltage Controller' module is executed after each EOC pulse, i.e. after the ADC sampling and data conversion process has been completed. ADC Offsets are found with the aid of a low-pass filter, with an `ADC_offset_freq` corner frequency of 1 Hz (set in the dialog box of the ADC/PWM module). Estimation of the ADC offset's is controlled by the '100 Hz' module, as mentioned previously. A user `Con_On` ('converter on') signal and over-current trip signal `trip_active` control the `CONV_on` signal, which starts drive operation. Consequently, ADC offsets are determined, after which the variables `inverterOnline` (BOOST active) and `contr_enable` (control enable) are set to logic one, which implies that the controller will be activated. The 'current trip' module monitors the per unit $i_{\alpha}^{n}, i_{\beta}^{n}$ variables (defined as `i_alp,i_bet`), generated by the 'ADC-PWM' module and sets an over-current flag `trip_active` if the value set in the 'operations' dialog box is exceeded, in which case drive operation

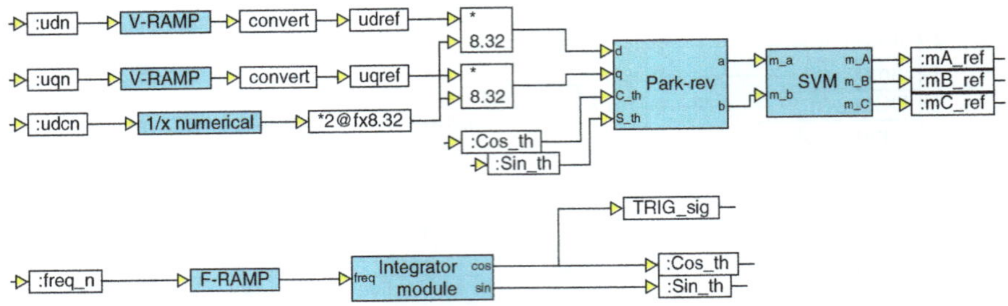

Fig. 3.15 Fixed point representation of the voltage controller

is terminated (blue LED on). Inputs to the 'ADC-PWM' module are the modulation indices generated by the voltage controller module. This module also generates the per unit voltages u_α^n, u_β^n variables, defined as `u_alp,u_bet` respectively and the DC bus voltage `DCbus_n`.

Note that for this lab, the current/voltage variables are not directly used, except to verify that the drive is working correctly and for over-current protection (used in addition to hardware protection). For diagnostic purposes the per unit current i_α^n, i_β^n and voltage variables u_α^n, u_β^n have been connected to variables `Testch1`, `Testch2` and `Testch3`, `Testch4` respectively, so that these can be displayed in phase C+.

Moving one level lower into the 'Voltage Controller' module reveals a set of modules as given in Fig. 3.15. The analogy with the floating point version given in Fig. 3.4 is hopefully apparent. The major difference is the use of scaled fixed point (per unit) variables. In contrast to the floating point version, the integrator directly generates cosine and sine signals, `Cos_th`, `Sin_th` rather then the instantaneous angle of the voltage reference vector (see Fig. 3.3). These two variables are then consistently used for subsequent Park transformations which have been adapted to allow the use of Cosine/Sine instead. The principal reason for taking this approach is to reduce computation time, given that the Cosine/Sine of the voltage vector angle only needs to be calculated once each sample (otherwise the Cos/Sin must be calculated within each Park module). A variable `TRIG_sig` is connected to the output of the integrator and is used by the diagnostic scope to trigger the buffering process. A Reverse Park transformation is used to generate the modulation indices, which are then processed by a 'SVM' (otherwise known as a pulse centering module) that centers the variables relative to the average value of the PWM triangular waveform.

3.1.4 Lab 1:1: Phase C+

Phase C+, is the operational component of the laboratory and is basically a run version of the `.out` file compiled and downloaded to the MCU in phase C (see previous subsection).

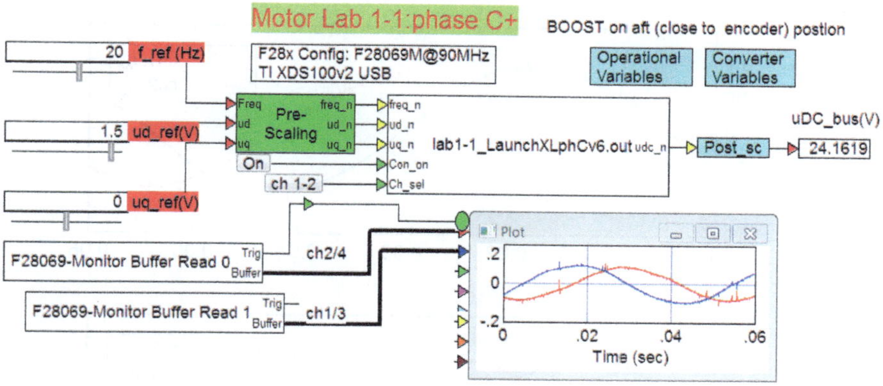

Fig. 3.16 Phase C+ operation of a drive with a voltage controller

The following information is relevant for this laboratory component:

- Reference program [11]: `lab1-1_LaunchXLphCv6_d.vsm`.
- Description: Open loop voltage control of a PM motor.
- Equipment/Software: Texas Instruments LAUNCHXL-F28069M, with BOOSTXL-DRV8301 module ('aft' position), Texas Instruments LVSERVOMTR PM motor and VisSim simulation program.
- Outcomes: achieve familiarity with using VisSim for Real-time control and test overall operation of the board, in terms of current/voltage waveforms and correct converter operation.

Note that the required jumper and dip-switch settings for the LAUNCHXL-F28069M module are given in phase C.

The run version as shown in Fig. 3.16 uses a VisSim run module which executes the `.out` file, shown in said module. Three-sliders are used to set the voltage reference amplitude and instantaneous rotational speed (in this case represented in terms of the electric frequency). A post-scaling module is used to convert the per-unit measured DC bus voltage to actual voltage, as shown with a numeric display. Two 'Monitor Buffer' modules are used to display two selected (using a button connected to the `Ch_sel` input) diagnostic signals. The plot module given in Fig. 3.16 and reproduced more clearly in Fig. 3.17, shows the scope outputs for channel select setting `ch1-2` and `ch3-4` respectively. Accordingly, subplot (a) shows the per unit α, β currents (channel select button `ch 1-2` active), while subplot (b) provides the per unit α, β voltages (channel select button `ch 3-4` active).

Note that the reference slider is set to `ud_ref=1.5` ,`uq_ref=0`, hence the per unit voltage amplitude should be $|u^{\mathrm{n}}_{\alpha}, u^{\mathrm{n}}_{\beta}| = {}^{1.5}/_{48} = 0.031$, given that a full-scale voltage of u_{fs}= 48 V is used. Observation of the scope result confirms that the amplitude is approximately equal to the anticipated value.

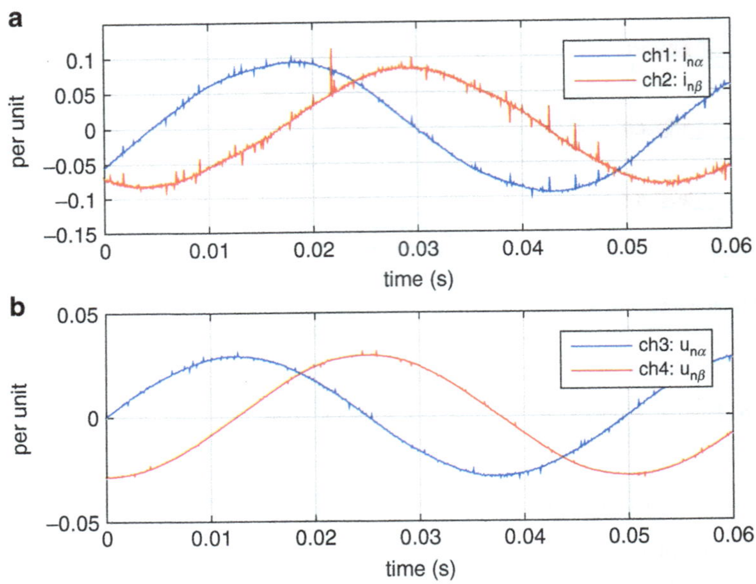

Fig. 3.17 Phase C+ operation: scope outputs, for ch1-2 and ch3-4 button values

Prior to activating this type of laboratory, the reader needs to be aware of the fact that he or she is about to active a complex electrical system, with live voltages. Hence it is prudent, **always**, to execute the following 'Pre-Drive' check list:

- Dialog boxes used in C+ mode, match those of phase C: the run version is compiled with the dialog box entries specified under phase C. The dialog boxes shown in phase C+ are used by the 'Pre/Post' scaling modules.
- Ensure that the sample time used is correct and the latest (and correct) `.out` file has been down-loaded to the MCU (Right mouse click on MCU module to show dialog box of these variables).
- Confirm that the user input values are set to either zero, or 'acceptable' values, which will not cause a current trip of the converter.
- Confirm that the converter 'switch' is set to OFF and the power supply is on (DC bus voltage present).
- Confirm that the motor is connected firmly and properly.

After completion of the Pre-Drive checklist, activate the program and confirm that the supply voltage source level shown in the digital display is 24 V. If not stop the program and restart. With the correct DC voltage level established, turn on the converter (using the `on` button, shown) and monitor the motor shaft and diagnostic scope. With the voltage and speed setting shown the motor should rotate with a shaft speed of 300 rpm. Note: if the motor shaft is not rotating, then initially reduce the shaft speed reference to zero and slowly increase the reference speed to the value shown. The reader should be aware of the fact that the motor is operating under open-loop voltage control, hence the rotating voltage vector produces a rotating magnetic field which must be synchronous with the field due to the permanent magnet. If the drive is turned on with a set reference speed then the rotating vector will accelerate (as defined by the user ramp value, in the 'operation' dialog box) to the set speed. Any motor has inertia, hence acceleration must match the torque level which can be applied. If acceleration is too high the motor will fall out of synchronous operation and subsequently stall. Note also that there is no current control, other than protection, hence care should be taken to keep the voltage reference vector to acceptable values (typically less than 2 V in this case).

A second diagnostic check which is important to undertake, is to allocate (in phase C) the test variables `Testch1`, `Testch2` to per unit variables $u^{\mathrm{n}}_{\alpha}, i^{\mathrm{n}}_{\alpha}$. Likewise, allocate test variables `Testch3`, `Testch4` to $u^{\mathrm{n}}_{\beta}, i^{\mathrm{n}}_{\beta}$. Subsequently, recompile (under phase C) to generate a new `.out` file and run (in phase C+) the motor with the same input conditions. The diagnostic scope results for both channel select button options are given in Fig. 3.18.

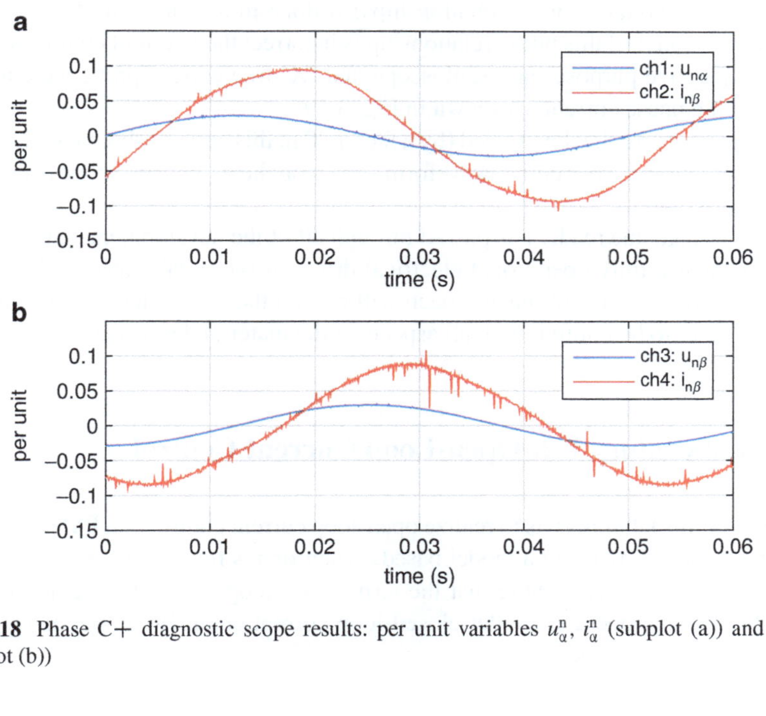

Fig. 3.18 Phase C+ diagnostic scope results: per unit variables u^{n}_{α}, i^{n}_{α} (subplot (a)) and u^{n}_{β}, i^{n}_{β} (subplot (b))

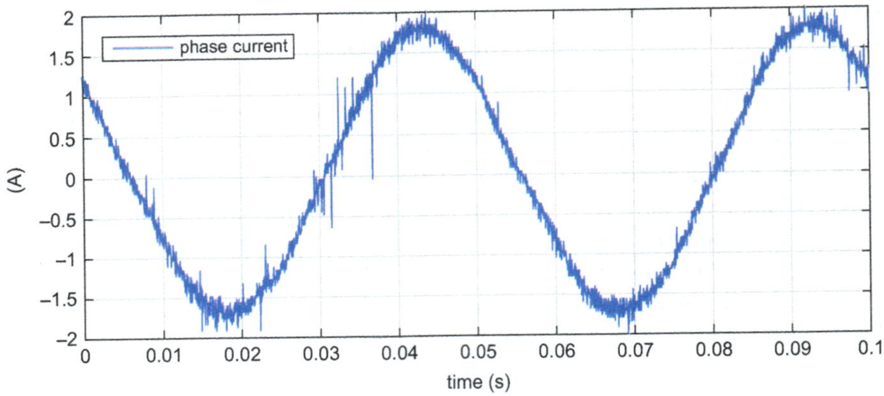

Fig. 3.19 Phase C+ experimental result: measured phase current waveform using a Tektronix DC-true current probe

Observation of Fig. 3.18, learns that the current waveform lags the voltage waveform. This result is what is verified here, given that the motor is inductive, hence a lagging current is expected. Note that a similar observation can also be made from the vector plot given in Fig. 3.10, where the current vector lags the voltage vector (both vectors turning anti-clockwise). This type of test (verification of the correct voltage/current relationships) is done in anticipation of moving to a current controller. If this phase relationship is incorrect then current control will fail. For comparative purposes an oscilloscope and DC-true current probe were used to measure the phase current, as shown in Fig. 3.19.

The phase current, when scaled (by a factor 20 in this case, as shown in Fig. 3.13), should represent the per unit i_{α}^{n} waveform shown on the diagnostic scope (Fig. 3.18), as indeed it does.

At this stage, the reader has passed through all of the development stages needed to implement a fully operational electrical drive. In future laboratories the control structure will change, but the approach will remain the same. Hence it is important for the reader to be familiar with all aspects of the material described in this section.

3.2 Laboratory 1:2: Open-Loop Current Control

Purpose of this laboratory is to realize open-loop current control of a PM motor. This may be achieved by using a model based, synchronous frame current controller, as discussed in Sect. 2.1.5. Note that the term 'open-loop' refers to operation where the PM machine operates (under closed-loop current control) without any form of rotor position feedback.

3.2.1 *Lab 1:2: Phase A*

This module considers the simulation of the PM motor drive in use, with an open loop current controller. In this example the switching process of the converter is specifically modeled, to allow the reader to comprehend the waveforms that typically occur in a drive. The following information is relevant for this laboratory component:

- Reference program [11]: `Lab1-2_LaunchXL- 069phaseAv2.plecs`.
- Description: Open loop current control of a PM motor.
- Equipment/Software: PLECS simulation program.
- Outcomes: Show the various waveforms that occur in the drive when using a converter with discrete power switches.

The basic structure of this laboratory is the same as discussed for the Voltage Controller: phase A, hence details of the converter, motor structure and ADC are not further discussed here. Instead, the reader is referred to the previous section, for details on these modules.

In this laboratory component, the objective is to create a controller that can generate a rotating current space vector as shown in Fig. 3.20. Inputs to the controller given in Fig. 3.21 are the reference direct/quadrature current components

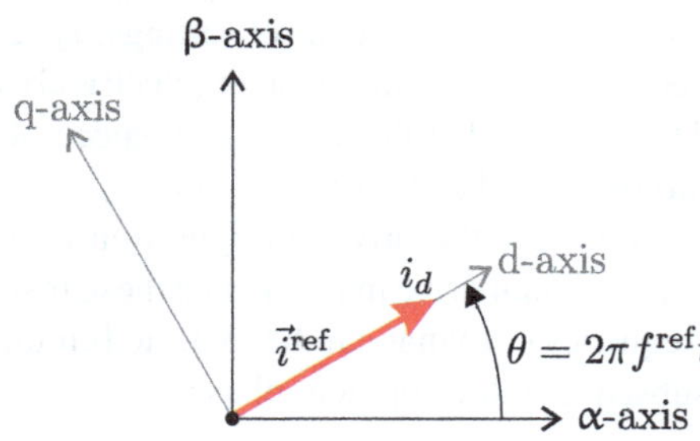

Fig. 3.20 Reference current vector

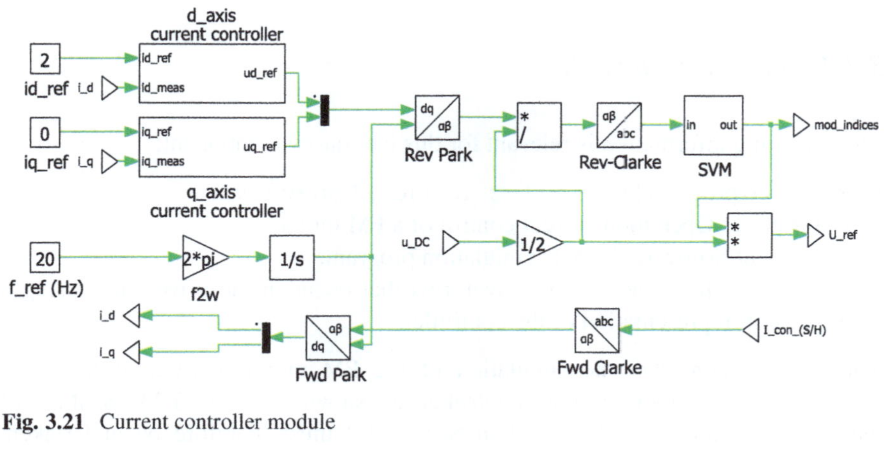

Fig. 3.21 Current controller module

i_dref, i_qref and the rotational speed f_{ref} of the current vector. Two model based PI controllers are used, as shown earlier in Fig. 2.18, which implies that current control is undertaken in a d, q synchronous reference frame. Hence the measured (Clarke transformed) phase currents are converted to variables i_d, i_q using a Forward Park transformation. The outputs of the PI controllers are the reference voltages u_dref,u_qref, where both are connected to a reverse Park module, which also makes used of the angle input $\theta = 2\pi f^{\rm ref} t$. The resultant reference voltages, in a stationary reference frame are then scaled by half the DC bus voltage and converted to three modulation index variables with the aid of a reverse Clarke transformation. A space vector modulator (SVM) is again used to maximize the available linear operating range of the modulator. The modulation index variables are then combined to a single column vector mod_indices. An additional output vector U_ref is also provided, which represents the three reference phase voltages, which are found by multiplying the modulation index vector by half the DC bus voltage as shown in Fig. 3.21.

The reader is again encouraged to undertake a detailed examination of this simulation using the PLECS Simulation platform and this specific file. Current control, as discussed in this section, is at the heart of so called 'field-oriented current control' as discussed at a later stage in this chapter. An example of drive operation is shown in Fig. 3.22. Shown in this figure (subplot (a)) is the measured phase current together with the corresponding reference phase voltage. Subplot (b) shows the actual phase voltage together with the reference voltage. This example corresponds to user control settings: $i_{\rm d}$=2.0 A, $i_{\rm q}$=0 A, $f^{\rm ref}$=20 Hz, hence a current vector of 2.0 A, rotating at an (electrical) speed of 1200 rpm is applied to the motor. This implies that the actual shaft speed of the motor will be 300 rpm (given that the motor in use has four pole pairs).

Note that the drive configuration used for this simulation closely matches the experimental drive in use, hence these results are representative (but at a lower PWM frequency) of what can be expected in the actual drive, as will become apparent in subsequent development phases.

3.2.2 Lab 1:2: Phase B

The following information is relevant for this laboratory component:

- Reference program [11]: lab1-2_LaunchXLphBv2.vsm.
- Description: Open loop current control of a PM motor.
- Equipment/Software: VisSim simulation program.
- Outcomes: Show the various waveforms that occur in the drive when using a fixed point representation of the controller.

This module considers the simulation of the PM motor drive in use, with a fixed point open loop current controller as shown in Fig. 3.23. Scaling of the currents/voltages as discussed in Sect. 1.4.2 must therefore be undertaken.

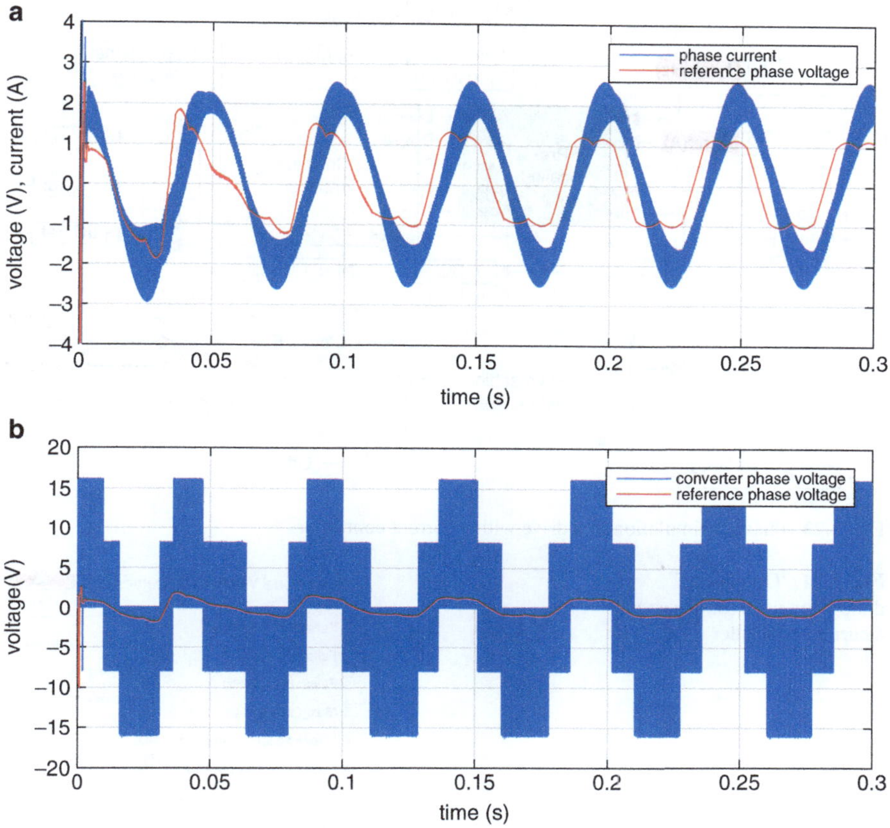

Fig. 3.22 Phase A: Typical waveforms obtained using a current controller

The outputs of the controller are the three reference modulation indices `mA_ref`, `mB_ref`,`mC_ref`. These are multiplied by the term $u_{DC}/2$ and connected to the 'converter', which in this case is simply represented by three limiters, that limit the voltage to the motor to $\pm u_{DC}/2$. A Forward Clarke transformation is again used to generate the voltage vector `U_s` for the machine model. A shaft speed dependent load module has been provided, which allows the user to apply a quadratic load characteristic to the machine. In this case a load of 100 mNm at 1000 rpm has been introduced, in order to provide some mechanical damping to the system.

As with the previous case a set of sliders is used to define the current vector, in terms of its amplitude and rotational speed. However in this case a 'pre-scaling' module (as discussed on Sect. 1.4.2) has been added to convert the floating point variables required for the controller to fixed point format. Furthermore the quadrature reference current value i_q^{ref} has arbitrarily been set to zero (see Fig. 3.20). An 'Operational Variables' dialog box (see Figure 3.24) is again used to assign full scale values for the voltage, current and frequency. However, in this case the current

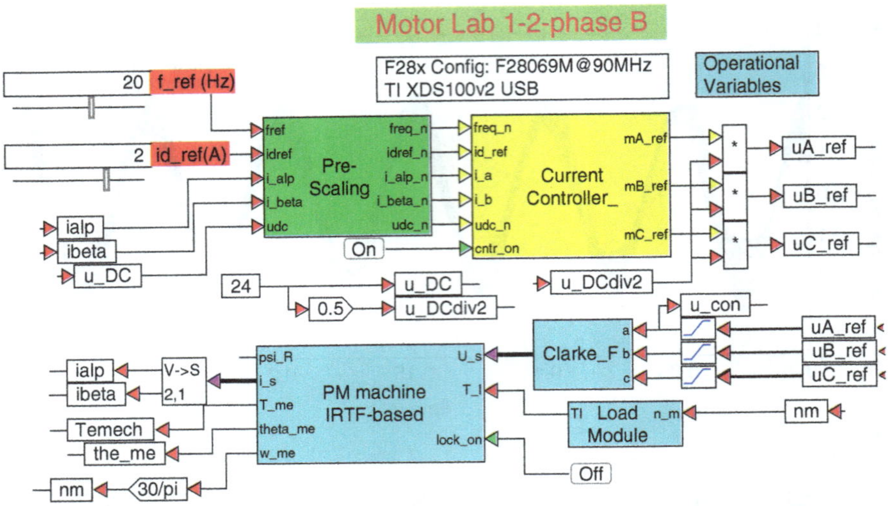

Fig. 3.23 Phase B simulation of a drive with a current controller

Fig. 3.24 'Operational' dialog box used for drive with a current controller

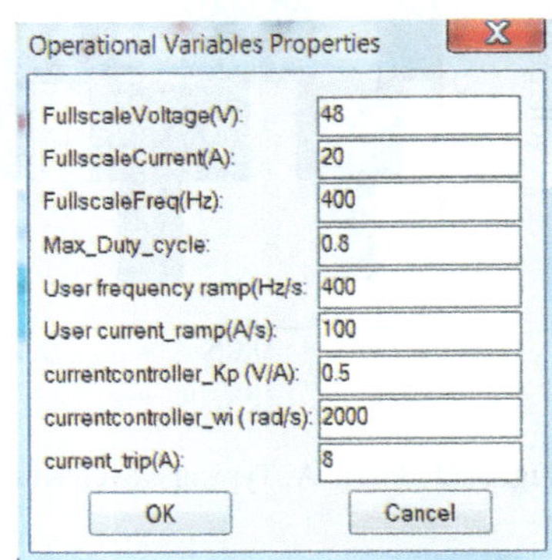

controller gain and bandwidth must also be assigned. These values are based on the motor in use, as was discussed in Sect. 2.1.5. A dialog box entry for the maximum allowable modulation index value m^{ref} must also be provided. This value should typically not exceed $2/\sqrt{3} \approx 1.15$, but a lower value may be needed if shunt resistors are used and no measures are taken to combat current sampling problems associated with a high modulation index value (see Sect. 2.1.6). Also assigned in this box is the rate of variable change for current and frequency, i.e. introduced as a safety measure to avoid erratic slider action from causing over-currents in the drive.

An example of drive operation is shown in Fig. 3.25, where the time step for the simulation was set to $T_s = 66.66\,\mu$s, which implies that a 15 kHz sampling frequency is used for the discrete controller model. Subplot (a) of Fig. 3.25, shows the reference voltage (per sample) waveform $u_A(t)$ (after the converter limiter, which limits the output to ± 12 V, in this case) together with the motor phase current i_A. In this case the simulation was run over a 0.3 s time interval. Subplot (b) of

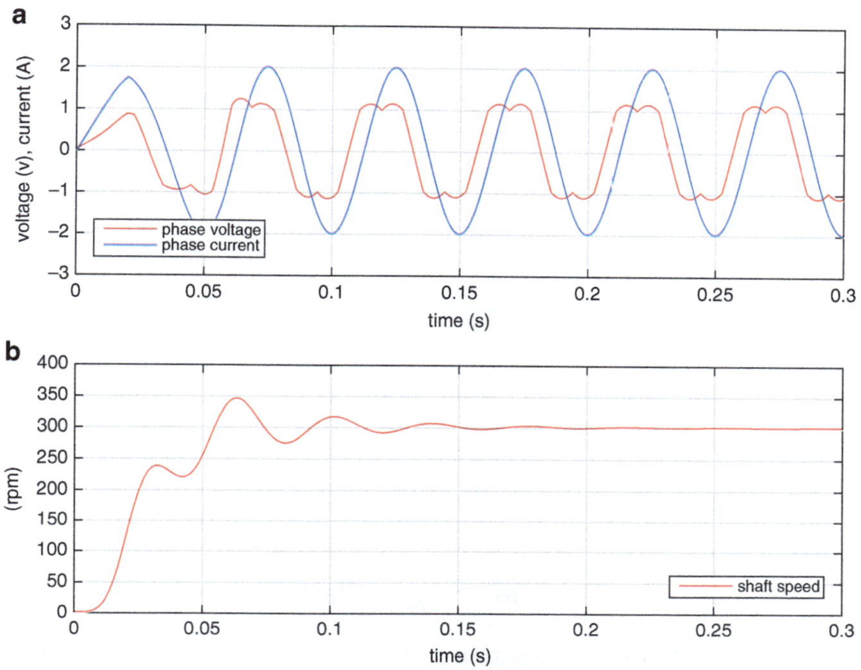

Fig. 3.25 Phase B: Typical waveforms obtained using a current controller

Fig. 3.25 shows the actual shaft speed during start-up, for a given frequency ramp setting and, in this case, operating frequency of 20 Hz, which corresponds to a steady state operating shaft speed of 300 rpm, as may be observed from Fig. 3.25. The corresponding vector plot for this drive is shown in Fig. 3.26. Shown in this figure are the steady-state vectors, which represent the reference current vector $\vec{i}_s$and voltage vector $\vec{u}_s$. In addition, a 'd-axis' (amplitude of '1') vector is introduced to show the orientation of the d,q reference frame, where the positive d-axis is displayed as a 'green' vector. In this example, the reference $i_q$value is set to zero, hence the resulting current vector is aligned with the d-axis vector, as is indeed the case. Note that the endpoint trajectory of the vectors is also shown in the vector plot.

The reader is again encouraged to undertake a detailed examination of this simulation using the VisSim Simulation platform and this specific file. Only by becoming familiar with each aspect of operation and the program used, can one hope to understand the modeling process of the drive as a whole. Note that further information on the controller implementation is given in the next, phase C, subsection.

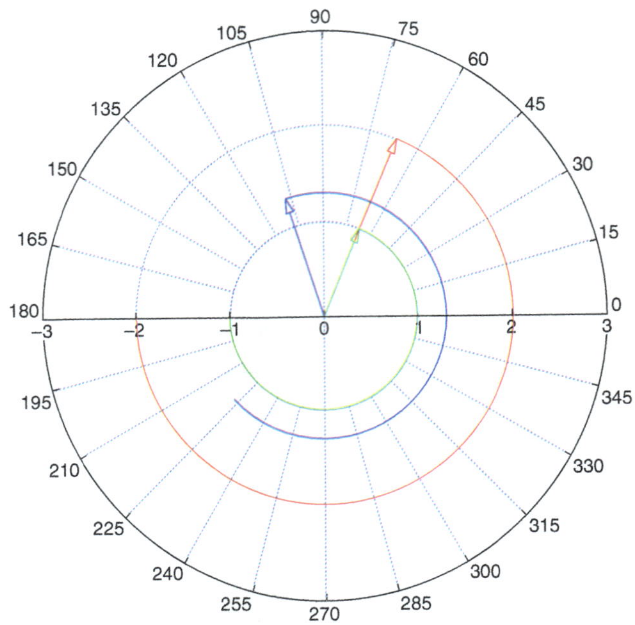

Fig. 3.26 Phase B: Vector plot obtained using a current controller with current vector $\vec{i}_s$ ('red'), voltage vector $\vec{u}_s$('blue') and unity d-axis vector ('green')

3.2.3 *Lab 1:2: Phase C*

In the previous phase, only knowledge of the control card to be used was required, in order to test the fixed point controller. For this laboratory the complete drive will be evaluated. The following information is relevant for this laboratory component:

- Reference program [11]: `lab1-2_LaunchXLphCv3.vsm`.
- Description: Open-loop current control of a PM motor.
- Equipment/Software: Texas Instruments LAUNCHXL-F28069M, with BOOSTXL-DRV8301 module ('aft' position) and VisSim simulation program.
- Outcomes: Develop a complete drive algorithm, which handles current control, voltage measurement, converter PWM and setup MCU timing, as required. Compile and download `.out` file for use in phase C+.

As mentioned above, a single boost pack located in the 'aft' position (furthest away from the USB connector) is used for single motor operation, in which case the following jumper and dip-switch positions on the LAUNCHXL-F28069M module are required:

- Jumpers `JP1` and `JP2` OPEN
- Jumpers `JP4` and `JP5` CLOSED
- Jumpers `JP3`,`JP6` and `JP7` CLOSED
- Dip-switches `SW1` to `SW3` ON

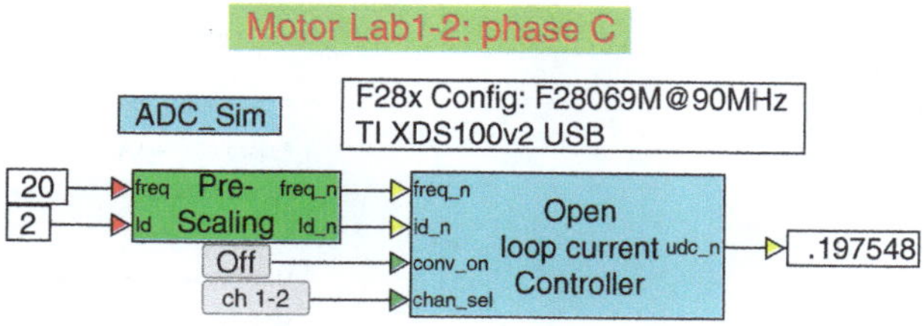

Fig. 3.27 Phase C simulation of a drive with a current controller

Fig. 3.28 One level down into the controller module

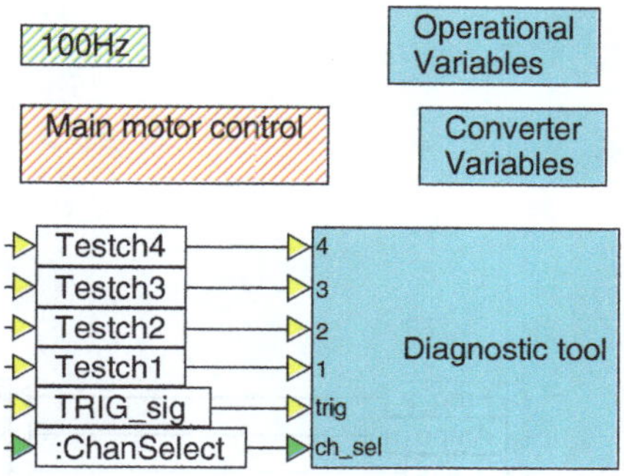

A phase C development stage, cannot be used to run a drive, but its primary task is to assemble all the modules needed for drive operation with a given controller configuration. The drive setup as given in Fig. 3.27, shows the 'open-loop Current Controller' module, which must be compiled to generate an `.out` file. Input to the controller module is the user reference current `i_d`, which is transformed to a 'per unit' value in the 'pre-scaling' module. The i_q reference value of the rotating current reference vector $\vec{i}^{\mathrm{ref}}$ has arbitrarily been set to zero. In addition, an electrical shaft speed reference input variable `freq` is used to set the rotational speed of the motor, and its value is converted to a 'per unit' value in the 'Pre-scaling' module. Both speed and current reference values are shown as 'constants' instead of sliders for convenience purposes only. Two additional logic inputs are provided, which allow the user to turn on the converter and to select which combination of diagnostic channels should be viewed. Output of the Controller module is the per unit bus voltage (see Fig. 3.27).

Moving one level down into the Controller module ('right' mouse click, after selecting module) leads to the set of modules/dialog boxes shown in Fig. 3.28. Two dialog boxes are present as given in Fig. 3.29, of which the 'Operational' dialog box contains the parameters as discussed in Fig. 3.24. A 'Converter' dialog box is again used to specify the voltage/current-ADC full-scale values for the board in use. A 'Diagnostic tool' module is used to buffer to variables, which can subsequently displayed in a graph when operating in phase C+. A two channel multiplexer is used to allow the user to either display variables `Testch1`, `Testch2` or `Testchch3`, `Testch4` by using a 'channel select' button discussed above. A '100 Hz' module generates a 'heart beat' flashing 'red' LED on the control card and undertakes MCU

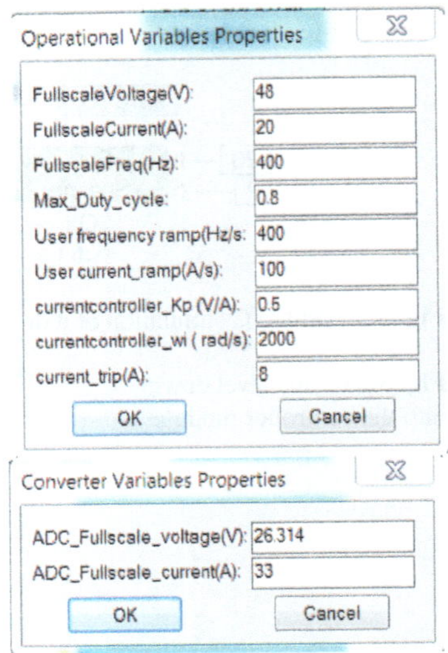

Fig. 3.29 Phase C dialog boxes used for drive with a current controller

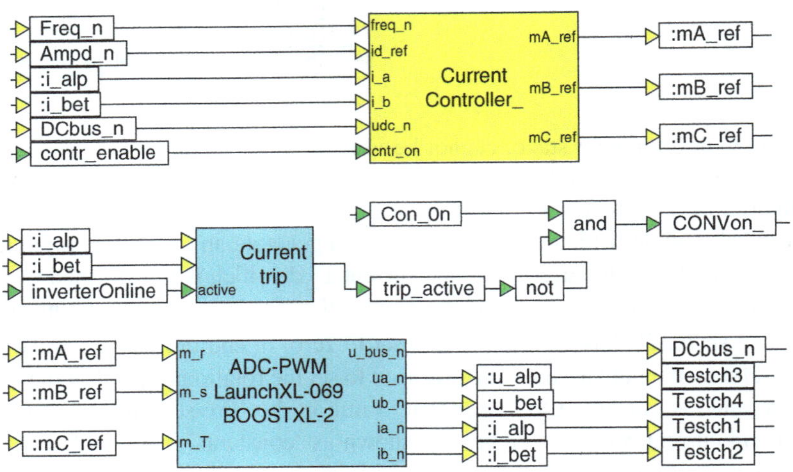

Fig. 3.30 One level down into the 'Main motor control' module

background tasks. A 'blue' LED is used to show a current trip condition, in which case it turns on when active. In addition, this module undertakes all the background tasks required for the drive. The 'green' hatching signifies a special function,namely that this module is configured to operate with a 'local sampling rate of (in this case) 100 Hz. A 'red' hatching is used for the 'Main motor Control' module, which signifies that the unit is executed upon receipt of a EOC pulse (as discussed in Sect. 2.1.6). Moving one level lower into the 'Main motor Control' module reveals a set of modules, as shown in Fig. 3.30. Of these, the 'Current controller' was introduced in phase B. The 'ADC-PWM' unit shown, is identical to the one used in the previous laboratory given that the same drive board is in use. Furthermore, a PWM frequency of 30 kHz is used, together with a ADC sampling time of 66.67 μs, that corresponds to a frequency of 15 kHz that is set in the VisSim 'System Properties' pull down menu (either as a frequency or time). Drive operation is controlled by a `CONVon_` signal, which is in turn generated by a logic AND output that has as inputs a user variable `Con_On` ('converter on') signal and a over-current trip signal `trip_active`. If the current trip is not active, the `CONVon_` flag will be set, when drive operation is required, in which case ADC offsets are determined, after which a `contr_enable` (control active) and `inverterOnline` flag are set. Outputs of the 'ADC-PWM' unit are the DC bus voltage `DCbus_n` and per unit phase currents i^{n}_{α}, i^{n}_{β}, represented by the variables `i_alp`, `i_bet` respectively. In addition, the per unit u^{n}_{α}, u^{n}_{β} (filtered) converter voltages are generated and represented by variables `u_alp`,`u_bet` respectively. The 'current trip' module monitors the per unit i^{n}_{α}, i^{n}_{β} variables and sets an over-current flag `trip_active` if the value set in the 'operations' dialog box is exceeded. Inputs to the 'ADC-PWM' module are the modulation indices generated by the current controller module. For diagnostic purposes the per unit current and voltage variables have been connected to variables `Testch1`, `Testch2` and `Testch3`, `Testch4` respectively, so that these can be displayed in phase C+.

Moving one level lower into the 'Current Controller' module reveals a set of modules as given in Fig. 3.31. In contrast to the floating point version given in Fig. 3.21, the PI controllers and limiter modules have now been combined in a single 'PI-kp-wi-lim'module. As in the previous laboratory, an integrator is used to generate a Cosine/Sine function, with variables `Cos_th`, `Sin_th` instead of the instantaneous angle of the current reference vector (see Fig. 3.20). These two variables are then consistently used for subsequent Park transformations, which have been adapted to allow the use of Cosine/Sine instead. A variable `TRIG_sig` is connected to the output of the integrator and is used by the diagnostic scope to

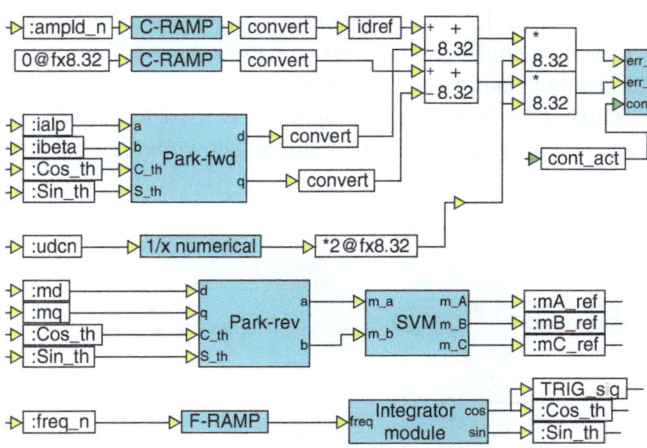

Fig. 3.31 Fixed point representation of the current controller

trigger the buffering process. A Reverse Park transformation is used to generate the modulation indices m_α, m_β, which are then processed by a 'Pulse Center (SVM)' module, which centers the phase modulation indices relative to the average value (zero in this case) of the PWM triangular waveform. A set of 'Rate Limiter' modules, entitled 'C-ramp' (sets the current ramp rate) and 'F-ramp' (sets the frequency ramp rate), have been introduced in Fig. 3.31 to avoid erratic changes to the user input variables (as actually used by the controller), which may damage the converter or motor. The rate of variable change is specified in the 'Operations' dialog box (see Fig. 3.29). A '1/x numerical' module is used to generate the inverse per unit bus voltage variable, which means that the 'd,q' outputs of the current controller are now in terms of the modulation index rather than a reference voltage.

3.2.4 *Lab 1:2: Phase C+*

Phase C+, is the drive operation component of the laboratory and is basically a run version of the `.out` file compiled and downloaded to the MCU in phase C (see previous subsection).
The following information is relevant for this laboratory component:

- Reference program [11]: `lab1-2_LaunchXLphCv3_d.vsm`.
- Description: Open loop current control of a PM motor.
- Equipment/Software: Texas Instruments LAUNCHXL-F28069M, with BOOSTXL-DRV8301 module ('aft' position), Texas Instruments LVSERVOMTR PM motor and VisSim simulation program.
- Outcomes: To achieve open-loop current control of a PM drive.

Note that the required jumper and dip-switch settings for the LAUNCHXL-F28069M module are given in phase C.

The run version as shown in Fig. 3.32 uses a VisSim run module which executes the `.out` file, shown in said module. Two sliders are used the set the current reference amplitude and instantaneous rotational speed (in this case represented in terms of the electrical frequency). A post-scaling module is used to convert the per-unit measured DC bus voltage to a reading in Volts, as shown with a numeric

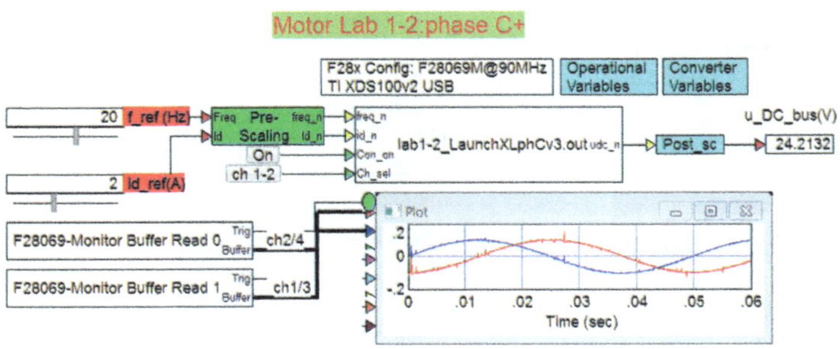

Fig. 3.32 Phase C+ operation of a drive with a current controller

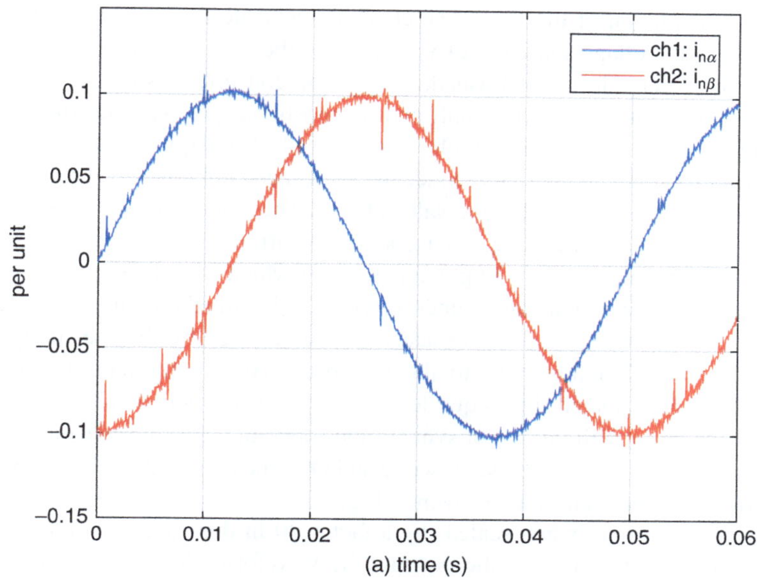

Fig. 3.33 Phase C+, scope results with channel select option: 'ch 1-2'

display. Two 'Monitor Buffer' modules are used to display four selected internal variables(using the channel select switch) connected to the `channel_sel` input). The VisSim scope module shows the per unit α, β currents and an enlarged view of these results is given in Fig. 3.33. Multiplication of these waveforms by the full scale current value of i_{fs}= 20 A gives actual current values.

Prior to activating the drive the following 'Pre-Drive' check list should be executed:

- Dialog boxes used in C+ model match those of C: the run version is compiled with the dialog boxes entry specified under phase C. The dialog boxes shown in C+ are used by the 'Pre/Post' scaling modules.
- Ensure that the sample time used is correct and latest (and correct) `.out` file has been downloaded to the MCU (Right mouse click on MCU module to show dialog box of these variables).
- Confirm that the user input values are set to either zero, or 'acceptable' values, which will not cause a current trip of the converter
- Confirm that the converter 'switch' is set to OFF and the power supply is on (DC bus voltage present).
- Confirm that the motor is connected firmly and properly.

After completion of the Pre-Drive checklist, activate the program and confirm that the supply voltage source is 24 V. If not stop the program and restart. With the correct DC voltage level established, turn on the converter (using the ON button, shown) and monitor the motor shaft and diagnostic scope. With the speed setting shown the motor should rotate with a shaft speed of 300 rpm. Note: if the motor shaft is not rotating then initially reduce the shaft speed reference to zero and slowly increase the reference speed to the value shown. The reader should be aware of the fact that the motor is operating under open-loop current control, hence the rotating current vector produces a rotating magnetic field which must be synchronous with the field due to the permanent magnet. If the drive is turned on with a set reference speed then the rotating vector will accelerate (as defined by the user ramp value, in the 'operation' dialog box) to the set speed. Any motor has inertia, hence acceleration must match the torque level which can be applied. If acceleration is too high the motor will fall out of synchronous operation and subsequently stall.
For comparative purposes an oscilloscope and DC-true current probe was connected to measure the phase current, as given in Fig. 3.34.

The phase current, when scaled (by a factor 20 in this case), should match (in terms of overall appearance) the per unit $i_{n\alpha}$ waveform shown on the diagnostic scope (see Fig. 3.33), as indeed it does.
The ability to achieve current control and manipulate a current vector, is the final step, needed to achieve 'field-oriented' control (FOC) of the PM motor. As will be discussed in the next laboratory.

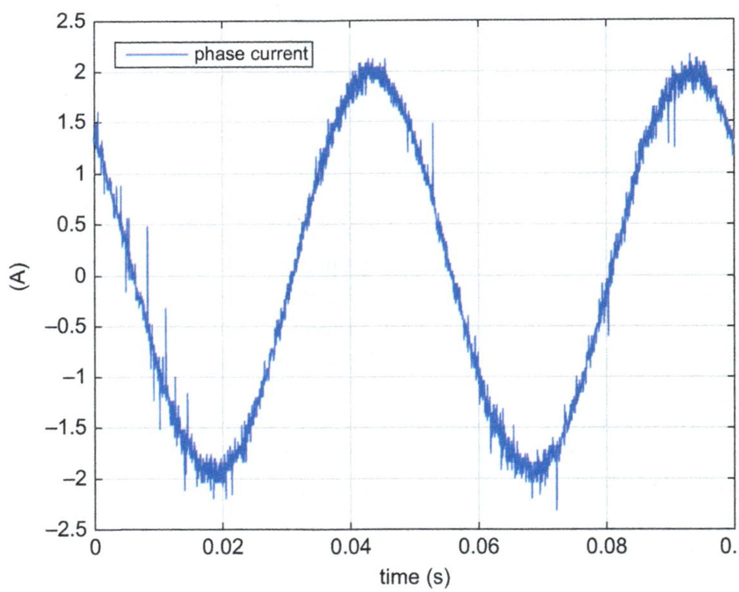

Fig. 3.34 Phase C+ experimental result: phase current waveform, measured with a Tektronix current probe

3.3 Laboratory 1:3: FOC Sensored Control of PM

Purpose of this laboratory is to realize field-oriented control of a PM motor. Use is made of the current control approach discussed in the previous laboratory. A shaft encoder will be used to provide the angle information needed to orientate the current vector relative to the magnetic field of the PM motor, as discussed in Sect. 2.1.3. The Texas Instruments LVSERVOMTR Teknic motor in use has an incremental shaft encoder, which generates 4000 counts/revolution. In addition, the shaft encoder produces an index pulse (one pulse/per revolution), which provides the ability to use the incremental encoder for absolute angle sensing purposes (this implies that this function becomes active once an index pulse has been identified). A speed controller has also been added to this laboratory, where the user can set the corresponding speed gains using sliders. For this laboratory, the phase A development phase is omitted given that the converter switching process does not need to be modeled.

3.3.1 Lab 1:3: Phase B

This module considers the simulation of the PM motor drive in use, with a fixed point current/speed controller as shown in Fig. 3.35. The following information is relevant for this laboratory component:

- Reference program [11]: `lab1-3_LauchXLphBv2.vsm`.
- Description: Field-oriented, sensored control of a PM motor.

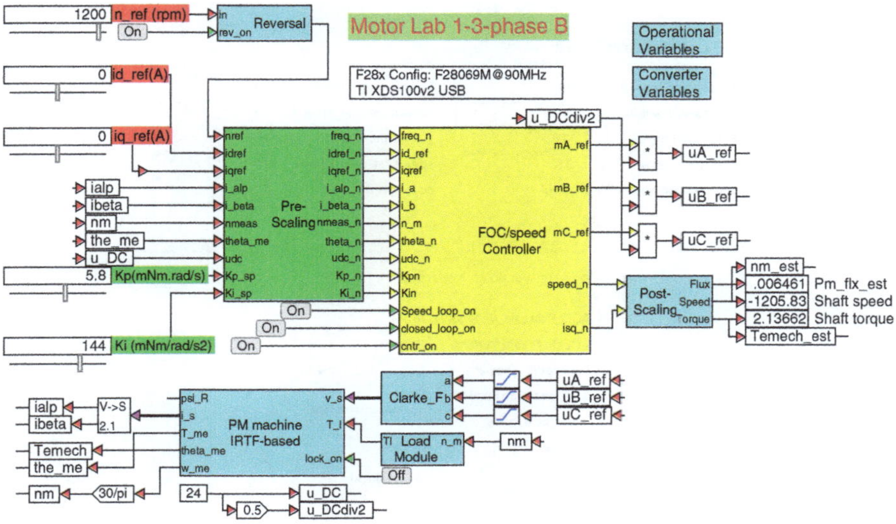

Fig. 3.35 Phase B simulation of a drive with a FOC/Speed controller

- Equipment/Software: VisSim simulation program.
- Outcomes: Show the various waveforms that occur in the drive when using a fixed point representation of the controller when operating under field oriented control using a shaft angle sensor.

Scaling of the currents/voltages as discussed in Sect. 1.4.2 is carried out using a 'Pre-scale' module. The outputs of the controller, which are the three reference modulation indices `mA_ref`, `mB_ref`, `mC_ref`, are multiplied by the term $u_{DC}/2$ and connected to the 'converter', which in this case is simply represented by three limiters, that limit the voltage to the motor to $\pm u_{DC}/2$. A Forward Clarke transformation is again used to generate the voltage vector for the machine model. Outputs of the machine model are the currents i_α, i_β, shaft torque T_m and shaft angle θ_m. The latter variable is used by the controller to realize proper alignment of the current vector relative to the rotor flux, otherwise referred to as field-oriented control, as discussed in Sect. 2.1.3.

As with the previous case a set of sliders is used to define the current reference vector, in terms of its direct and quadrature components i_d^{ref}, i_q^{ref} and rotational speed n_m^{ref}. A 'Pre-scaling' module (as discussed on Sect. 1.4.2) has been added to convert the floating point variables required for the controller to fixed point format. In addition, two sliders have been added which allow the user to set the proportional $K_{p_{sp}}$ and integral $K_{i_{sp}}$ speed gains of the controller. Note that these variables represented as `Kp`, `Ki` are shown in (milli) Nms/rad and (milli) Nm/rad respectively. A guideline for choosing these parameters is given in Sect. 2.1.2.

The two dialog boxes for this laboratory, as shown in Fig. 3.36, contain all the user settings required for this laboratory. The 'operational Variables' dialog box is again used to assign full scale values for the voltage, current and frequency. In addi-

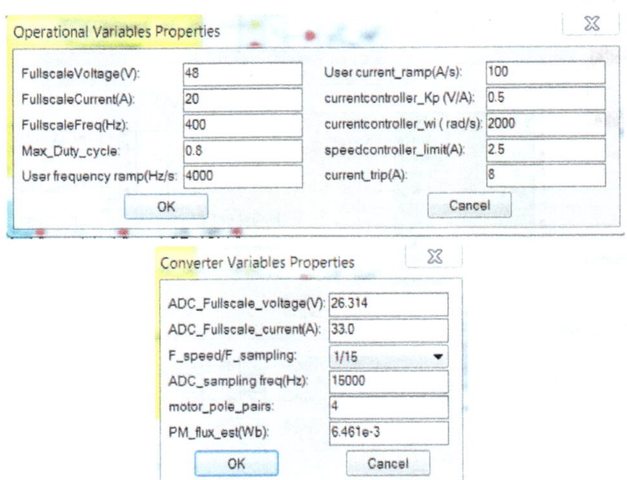

Fig. 3.36 Phase B: Dialog boxes used for drive with a FOC controller

tion, the current controller gain and bandwidth are also assigned here. These values are based on the motor in use, as was discussed in Sect. 2.1.5. Note the current gain value shown is in fact approximately 25 % of the value calculated using Eq. (2.23), which can be seen as an upper limit value in practical terms. A dialog box entry for the maximum allowable modulation index value m^{ref} must also be provided together with a speed controller limit value `speedcontroller_limit(A)`. This value is the maximum (absolute) quadrature current value which the speed controller is set to provide. Also assigned in this box is the rate of variable change for current and frequency, i.e. introduced as a safety measure to avoid erratic slider action from causing over-currents in the drive.

The 'Converter Variables' dialog box assigns the motor pole pair number and flux level of the motor in use. These are used to calculate the torque output and specify a user reference shaft speed. In addition, the sample frequency used for the speed controller module is set in terms of a ratio relative to the ADC sample frequency, also indicated in this dialog box for convenience. Note that the actual ADC sampling rate is set in the VisSim 'Systems property dialog box'. Finally the ADC full-scale voltage and current parameters are also set in this dialog box (but not used in this development phase), given that the same module is also used in subsequent development phases.

An example of drive operation is shown in Fig. 3.37. The sampling time T_s as defined in the program pull-down menu is set to 66.6 μs, i.e. a 15 kHz sampling frequency is used for the discrete controller model. Subplot (a) of Fig. 3.37, shows the motor phase current during start up and a speed reversal $n = 1200 \rightarrow -1200$ rpm, under sensored, field-oriented, speed control, which is initiated at $t = 0.2$ s. In this case the simulation was run over a 0.5 s time interval, with a motor inertia $J = 70 \cdot 10^{-6}\,\text{kgm}^2$ (taken to be the estimated combined inertia of the PM/IM motor unit). Operation under no-load is assumed. Subplot (b) of Fig. 3.37 shows the actual and reference shaft speed for a given frequency ramp setting (as set in the dialog box). The slider speed settings in use are based on the known inertia and an assumed speed bandwidth of $\omega_b = 50$ rad/s, as discussed in Sect. 2.1.2. Finally, subplot (c) shows the actual shaft torque of the machine over the 0.5 s simulation interval.

Some interesting observations can be made from the results shown in Fig. 3.37 during the speed reversal, namely:

- the maximum phase current is 2.5 A, which is expected as this is also the maximum $i_q^{\max}$ value set in the speed controller. Furthermore, i_d has been set to zero, hence the phase current amplitude will be equal to i_q.
- The phase current will be zero when the shaft speed is constant, because no mechanical load is attached.
- The reference set speed ramp is higher than achievable by the drive, given the inertia present.
- The maximum shaft torque during the reversal can be written as $1.5\, p\, \psi_{\text{PM}}\, i_q^{\max}$, which is equal to 97 mNm in this case.

Results shown above relate to operation under speed control, but this laboratory can also be operated in two alternative modes, which can be set by the reader namely:

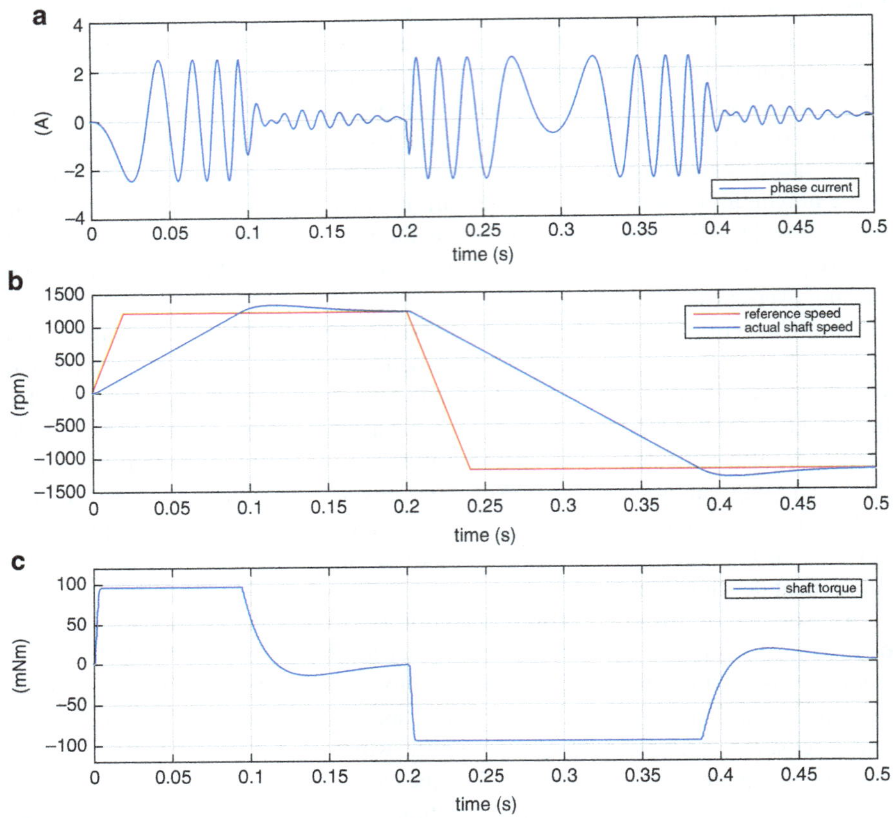

Fig. 3.37 Phase B: Typical waveforms obtained using a FOC controller

- `Speed_loop_on` button OFF: speed loop disabled, in which case torque control is activated, i.e. the i_q slider sets the torque. Note that a 'speed runway situation' (shaft speed increased to high value) can occur, as torque is (in theory) independent of the shaft speed.
- `Closed_loop_on` button OFF: operation without a shaft encoder, hence speed is dictated by the shaft reference, as undertaken in the previous laboratory.

In the sequel to this section a vector plot is shown of the drive in question operating under steady-state conditions, with a constant speed of $n^{\text{ref}}=400$ rpm and arbitrarily chosen mechanical load of 15.9 mNm. The corresponding vector plot given in Fig. 3.38, shows the reference current vector $\vec{i}_s$ and voltage vector $\vec{u}_s$. In addition, a unity 'd-axis' vector is shown to depict the orientation of the d,q reference frame, where the positive d-axis is displayed as a 'green' vector, which is aligned with the PM flux vector. In this example, a current i_q value is present given that the machine is producing torque to match the load torque applied. Consequently the current vector must be orthogonal to the d-axis vector (given that $i_d = 0$), as is

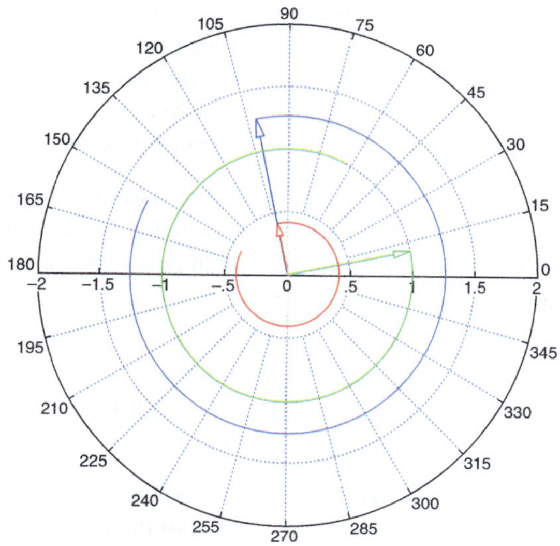

Fig. 3.38 Phase B: Vector plot obtained using a FOC/Speed control with current $\vec{i}_s$ ('red'), voltage vector $\vec{u}_s$ ('blue') and unity d-axis vector ('green')

indeed the case. The voltage vector is made up of a back EMF component that is in phase with the current vector and resistive voltage component, which is also aligned with said current vector. The third component of the voltage vector is due to the inductive component of the machine and its value will depend (among others) on the electrical frequency and will be oriented along the negative d axis (hence in the opposite direction of the shown d-axis vector). In this case its value is almost zero, given the low operating speed set for this example. However as speed is increased the voltage vector will phase advance relative to the current vector due to this third voltage term.

The reader is again encouraged to undertake a detailed examination of this simulation using the VisSim Simulation platform and this specific file. Only by becoming familiar with each aspect of operation and the program used, can one hope to understand the modeling process of the drive as a whole. Note that further information on the controller implementation is given in the next, phase C, subsection.

3.3.2 Lab 1:3: Phase C

In the previous phase, only knowledge of the control card to be used was required, in order to test the fixed point controller. The following information is relevant for this laboratory component:

- Reference program [11]: `lab1-3_LaunchXLphCv2.vsm`.
- Description: Field Oriented, sensored control of a PM motor.

- Equipment/Software: Texas Instruments LAUNCHXL-F28069M, with BOOSTXL-DRV8301 module ('aft' position) and VisSim simulation program.
- Outcomes: Develop a complete drive algorithm, which handles field-oriented current control, voltage measurement, converter PWM and setup MCU timing, as required. Compile and download `.out` file for use in phase C+.

As mentioned above, a single boost pack located in the 'aft' position (furthest away from the USB connector) is used for single motor operation, in which case the following jumper and dip-switch positions on the LAUNCHXL-F28069M module are required:

- Jumpers `JP1` and `JP2` OPEN
- Jumpers `JP4` and `JP5` CLOSED
- Jumpers `JP3`,`JP6` and `JP7` CLOSED
- Dip-switches `SW1` to `SW3` ON

Furthermore, the 'J4' encoder connector of the PM machine cable should be attached to encoder input `QEP_A` of the LAUNCHXL-F28069M module. Also note that the wiring sequence for the motor is now important because it ties the α, β reference frame to the encoder (wiring details shown in development phase C+, see Fig. 3.43). A phase C development stage, cannot be used to run a drive, but its primary task is to assemble all the modules needed for drive operation with a given controller configuration. The drive setup as given in Fig. 3.39, shows the 'Open/closed loop Current/Speed controller' module, which must be compiled to generate an `.out` file. Inputs to the controller module are (among others) the per unit user reference currents `id_n`,`iq_n`. In addition, an electrical shaft speed reference `Freq_n` variable is used to set the shaft speed of the drive. The remaining input variables, are the speed-loop gains, as discussed in the previous section. All non-logic inputs to the pre-scaling module are, for the sake of convenience shown as 'constants' whereas in the subsequent development phase they will be replaced by slider inputs. Four logic inputs are provided to the module, which allows the user to turn on the converter, select torque or speed control, open-loop or shaft encoder operation and choose which combination of diagnostic channels is to be viewed.

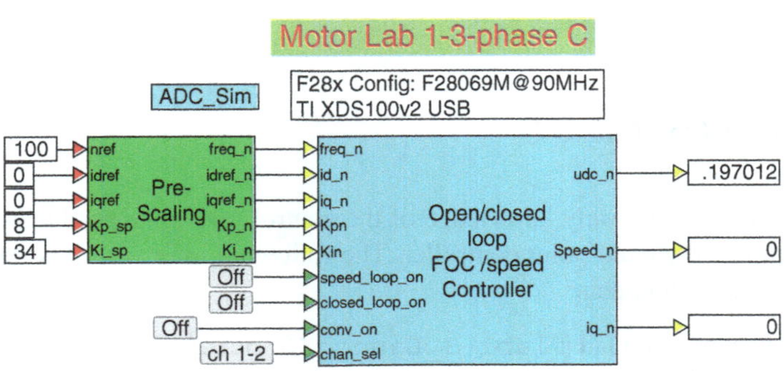

Fig. 3.39 Phase C simulation of a drive with a field-oriented current controller

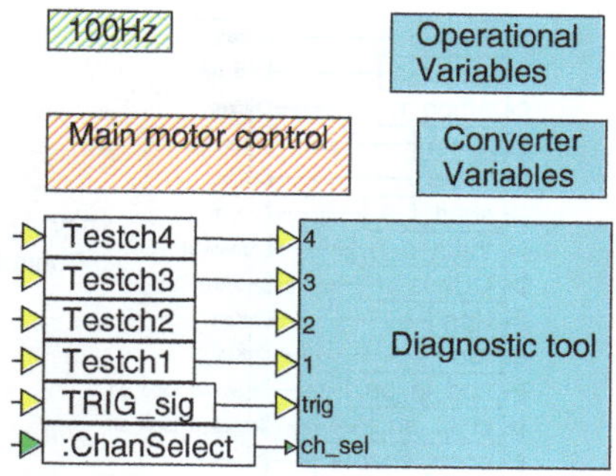

Fig. 3.40 One level down into the controller module

Outputs of the controller module are the per unit bus voltage together with the per unit shaft speed, provided by the encoder and measured quadrature current `iq_n`. The latter variable is used to calculate the torque, given that the PM flux of the machine is known in this case.

Moving one level down into the Controller module leads to the set of modules/dialog boxes shown in Fig. 3.40. The dialog boxes shown in this diagram have already been shown in Fig. 3.36, hence the reader is referred to this section for further details. A 'Diagnostic tool' module is again used to buffer to variables, which can subsequently displayed in a graph when operating in phase C+. A two channel multiplexer is used to allow the user to either display variables `Testch1`, `Testch2` or `Testch3`, `Testch4` by using the button connected to the `chan_sel` input of the controller.

A '100 Hz' module acts as a 'heart beat' and flashes a red LED on the LAUNCHXL-F28069M module. A blue LED on said module turns on if a current trip condition occurs. In addition, all the background tasks of the drive such as ADC offset control, initialization of the boost module are executed in this unit. Moving one level lower into the 'Main motor Control' module reveals a set of modules, as shown in Fig. 3.41. Of these, the 'FOC/Speed controller' was introduced in phase B. The 'ADC-PWM' unit shown, is identical to the one used in the previous laboratory given that the same drive board is in use. Furthermore, a PWM frequency of 30 kHz is used, together with a ADC sampling frequency of 15 kHz which corresponds to a sampling time of 66.67 μs, that is set in the VisSim 'System Properties' pull down menu. This module is controlled by a `CONVon_` signal, which is in turn generated by a logic AND output that has as inputs a user variable `Con_On` ('converter on') signal and a over-current trip signal `trip_active`. Relevant outputs (to this laboratory) of this unit are the DC bus voltage `DCbus_n`, and the per unit phase currents i_{α}^{n}, i_{β}^{n}, represented by the variables `i_alp`, `i_bet` respectively. An output `control_enable` (control active) activates the FOC/Speed controller, after the converter has been turned on and internal ADC/PWM initialization actions have been performed. An 'Encoder Speed/angle' module, generates the per unit electrical shaft speed `shaft_f_e` (which is equal to f_e = `shaft_f_e`f_{fs}, with $f_e = {}^{n_m p}/_{60}$ Hz and per unit shaft angle `theta_e_n` (which is equal to $\theta_e = 2\pi$ `theta_e_n` rad).

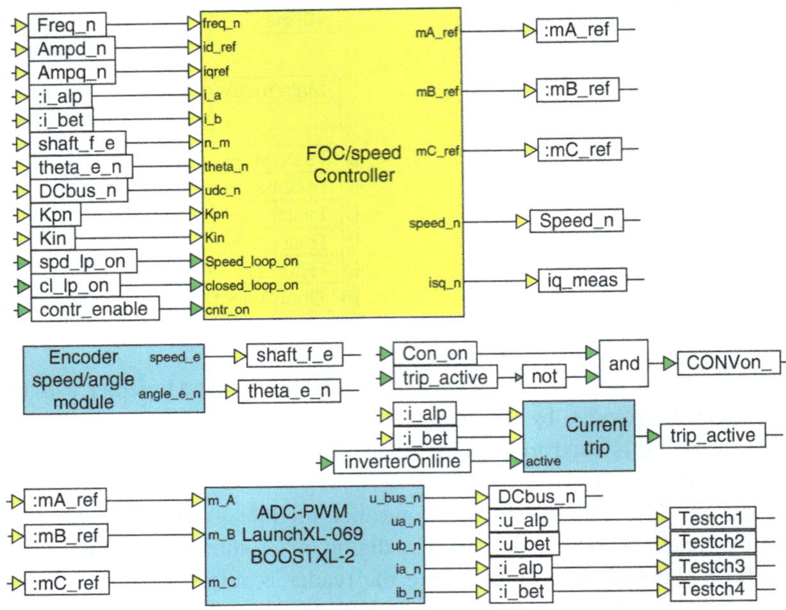

Fig. 3.41 One level down into the 'Main motor control' module

The inputs to the 'ADC-PWM' module are the modulation indices generated by the current/speed controller module. For diagnostic purposes the per unit voltage and current variables have been connected to variables `Testch1`, `Testch2` and `Testch3`, `Testch4` respectively, so that these can be displayed in phase C+.

Moving one level lower into the 'FOC/Speed Controller' module reveals a set of modules given in Fig. 3.42. Synchronous model based current control is undertaken by a 'PI-kp-wi-lim' module, which has as inputs the per unit quadrature and direct axis current error and the inverse DC bus voltage. The latter is generated by the '1/x numerical' module, which provides the per unit inverse DC bus voltage using a numerical routine. A Reverse Park transformation is used to generate the modulation indices m_α, m_β, from the outputs of the current controller, which are then processed by a 'Pulse Center(SVM)' module.

As in the previous laboratories a Cosine/Sine function, with variables `Cos_th`, `Sin_th` is used instead of the instantaneous angle of the current reference vector (see Fig. 3.20). These two variables are then consistently used for subsequent Park transformations, which have been adapted to allow the use of Cosine/Sine instead. A 'switch' module, activated by the `closloopON` signal (active if the shaft encoder is in use) is used to select the Cos/Sin variables generated by the encoder or integrator, which is in turn connected to the reference speed value n_m.

Inputs to the speed controller are the (rate limited) reference frequency `reffreq_n` and actual shaft frequency (from the shaft encoder) `shaft_freq_n`. Output of this module (when enabled) is the per unit quadrature reference current

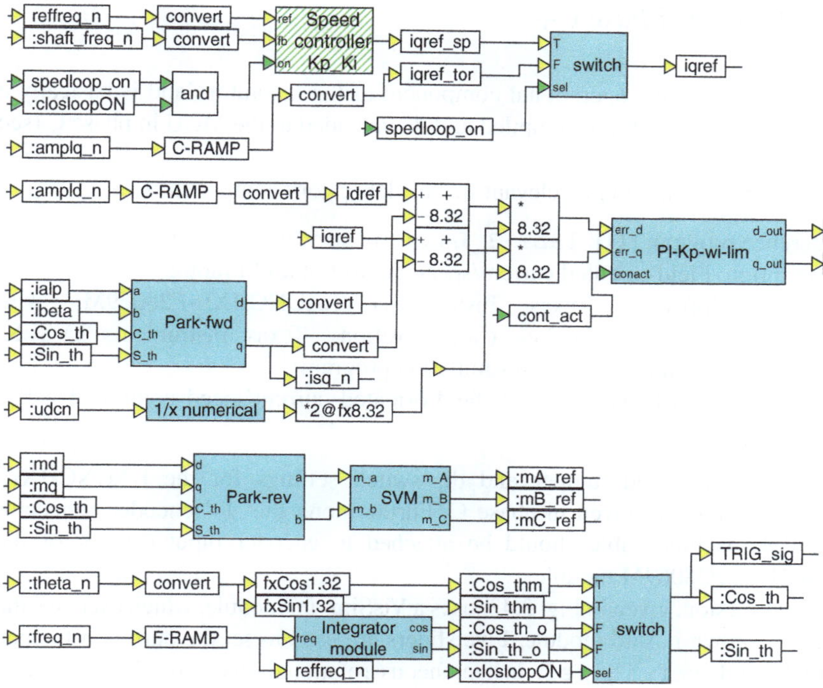

Fig. 3.42 Fixed point representation of the FOC/Speed controller

`iqref_sp`. Note that the speed controller module is shown 'hatched green', which implies that its sample rate is different with respect to the remaining part of the controller. In this case the sample frequency for the speed controller was set to 1 kHz. For less time demanding activities it is prudent to use a lower sampling frequency, which serves to avoid MCU overrun errors that can occur when insufficient time is available to execute all the required algorithms for a given ADC sample frequency. A 'switch' module with output `iqref` determines whether the quadrature reference current for the current controller is provided by the speed controller or the user i_q slider. The switch in question is controlled by the `speedloop_on` signal (active when closed speed-loop control is required, OFF implies operation in constant torque mode).

A variable `TRIG_sig` is connected to the Sine variable `Sin_th` and is used by the diagnostic scope to trigger the buffering process. A set of 'Rate Limiter' modules, entitled 'C-ramp' (sets the current ramp rate) and 'F-ramp' (sets the frequency ramp rate), have been introduced in Fig. 3.42 to avoid erratic changes to the user input variables (as actually used by the controller), which may damage the converter or motor. The rate of variable change is specified in the 'Operations' dialog box (see Fig. 3.36).

3.3.3 Lab 1:3: Phase C+

Phase C+, is the drive operational component of the laboratory and is basically a run version of the `.out` file compiled and downloaded to the MCU in phase C (see previous subsection).
The following information is relevant for this laboratory component:

- Reference program [11]: `lab1-3_LaunchXLphCv2_d.vsm`.
- Description: Field-oriented current/speed control of a PM motor.
- Equipment/Software: Texas Instruments LAUNCHXL-F28069M, with BOOSTXL-DRV8301 module ('aft' position), Texas Instruments LVSERVOMTRPM motor and VisSim simulation program.
- Outcomes: To achieve sensored, field-oriented current/speed control of a PM drive.

Note that the required jumper and dip-switch settings for the LAUNCHXL-F28069M module are given in phase C. Furthermore, the 'J4' encoder connector of the PM machine cable should be attached to encoder input `QEP_A` of the LAUNCHXL-F28069M module.

The run version, given in Fig. 3.43, uses a VisSim run module, which executes the `.out` file, shown in said module. Five sliders are used to set the current reference amplitudes `id_ref`, `iq_ref`, instantaneous rotational shaft speed `n_ref` and speed gains. Two post-scaling modules are used to process data from the run module and display the measured DC bus voltage, PM magnet flux, shaft speed and torque (both instantaneous and filtered using a 10 Hz low pass filter). The filtered torque signal is displayed on a torque meter in (milli)Nm. Two 'Monitor Buffer' modules are used to display two selected (using channel select button connected to the `channel_sel` input). The VisSim scope module shows the per unit α, β

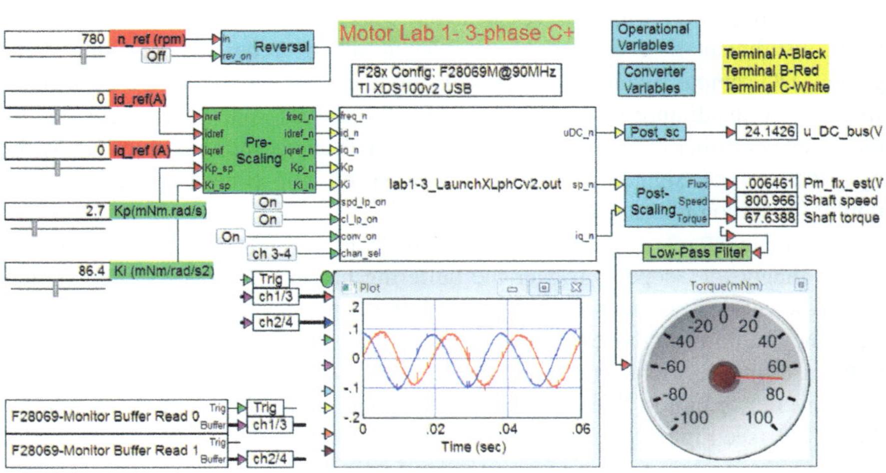

Fig. 3.43 Phase C+ operation of a drive with a field-oriented current/speed controller

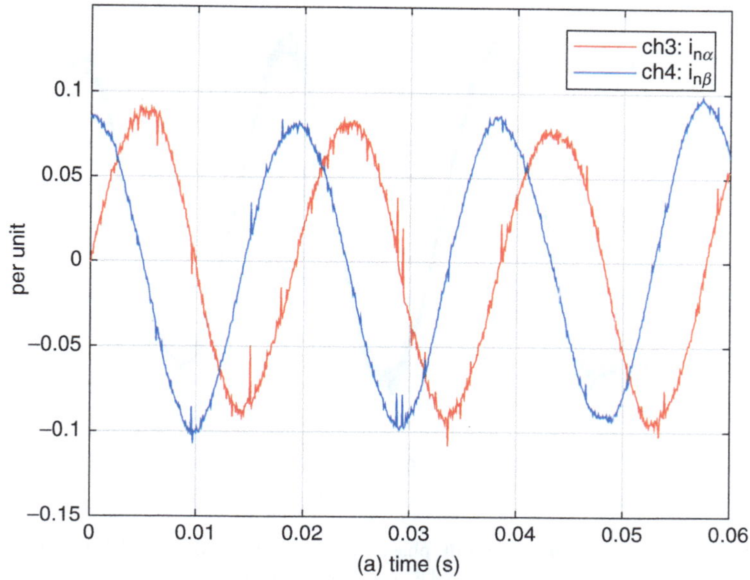

Fig. 3.44 Phase C+ operation of a drive: VisSim scope results with channel select option: 'ch 3-4'

currents and an enlarged view of these results is given in Fig. 3.44. Multiplication of these waveforms by the full scale current value of i_{fs}= 20 A gives actual current values.

Prior to activating the drive the following 'Pre-Drive' check list should be executed:

- Dialog boxes used in C+ model match those of C: the run version is compiled with the dialog box entries specified under phase C. The dialog boxes show in C+ are used by the 'Pre/Post' scaling modules.
- Ensure that the sample time used is correct and the latest (and correct) `.out` file has been downloaded to the MCU (Right mouse click on MCU module to show dialog box of these variables).
- Confirm that the user input values are set to either zero, or 'acceptable' values, which will not cause a current trip of the converter
- Confirm that the converter 'switch' is set to OFF and the power supply is on (DC bus voltage present).
- Confirm that the motor is connected firmly and properly.

After completion of the Pre-Drive checklist, activate the program and confirm that the supply voltage source is 24 V. If not stop the program and restart. With the correct DC voltage level established, turn on the converter (using the ON

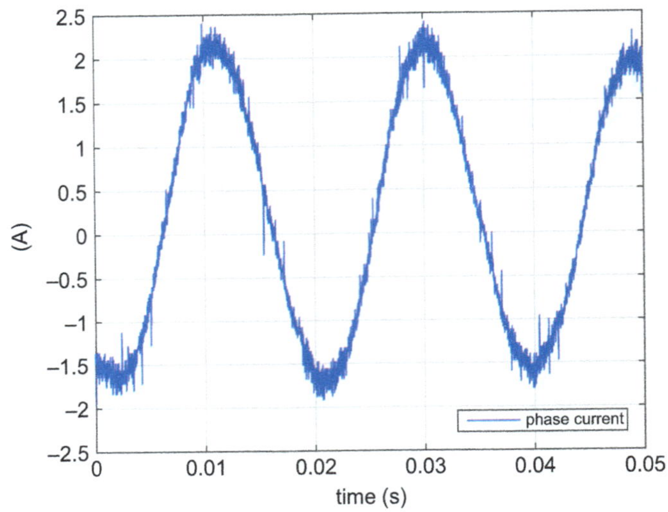

Fig. 3.45 Phase C+ experimental result: phase current waveform during FOC (sensored) operation under load, measured with Tektronix DC-true current probe

button, shown) and monitor the motor shaft and diagnostic scope, which in the example given (see Fig. 3.43) is set to show the per unit current components $i_{\alpha}^{n}, i_{\beta}^{n}$. Observation of Fig. 3.43 learns that the machine is currently operating under closed-loop speed control and with a load (hand induced friction), as is evident from the torque meter value and the (per unit) i_{α}^{n}, i_{β}^{n} currents displayed in the diagnostic scope. Operation with $i_{d} = 0$ is undertaken in this case, thus the maximum phase current value must also correspond to the reference i_{q} value (set by the speed controller).

Experimental verification of the results obtained via the VisSim scope was carried out with the aid of an oscilloscope and DC-true current probe. The latter was used to measure the phase current, as shown in Fig. 3.45.

3.4 Laboratory 1:4: Use of SpinTAC Motion Control Suite

In the previous laboratory a 'conventional' speed controller was used to generate the quadrature reference current for a field-oriented current controller. In this example the quadrature current will be provided via a set of modules known as the 'SpinTAC Motion Control Suite'[8]. These modules shown in Fig. 3.46 are located in the F28069M ROM code. As such they provide the reader with the ability to implement a sophisticated motion control capability for the drive as used, for example, in a washing machine. Also present in Fig. 3.46 is the 'current controller', which provides the reference modulation indices $m_{1}^{\mathrm{ref}}..m_{3}^{\mathrm{ref}}$, which are converted in the

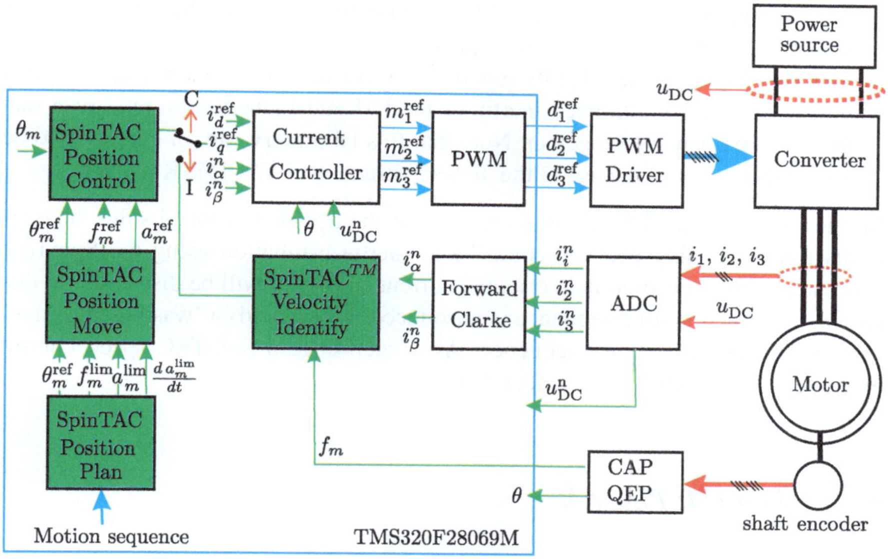

Fig. 3.46 Sensored drive setup using the SpinTAC motion control suite

'PWM' module to three duty-cycle reference values for use by the 'PWM driver' module that controls the power-electronic devices of the converter. The three-phase PM motor is connected to a shaft sensor which provides the shaft angle θ and shaft speed, shown here as the per unit shaft frequency f_{m}. An 'ADC' module is used to acquire the (per unit) currents i_{α}^{n}, i_{β}^{n}, and bus voltage $u_{\mathrm{DC}}^{\mathrm{n}}$ for use with the current controller. The 'SpinTAC Control Suite' [9] developed by LineStream Technologies [8] can be used for sensored or sensorless operation and a brief description of these modules (for position control) is given here:

- SpinTAC Position Control: generates the quadrature reference current value for the current controller when this module is active (switch in position 'C'). Inputs are the actual and reference angle values, shown here as variables θ_m, θ_m^{ref} respectively. Note that the variable θ_m is derived via a SpinTAC module from the encoder angle θ. In addition, the reference frequency f_m^{ref} and acceleration $a_{\mathrm{m}}^{\mathrm{ref}} = df_{\mathrm{m}}^{\mathrm{ref}}/dt$ are also required.
- SpinTAC Position Move: This module allows the user to execute a complex position versus time sequence, which is then translated to a set of outputs θ_m^{ref}, f_m^{ref} and $a_{\mathrm{m}}^{\mathrm{ref}} = df_{\mathrm{m}}^{\mathrm{ref}}/dt$ as discussed above.
- SpinTAC Position Plan: generates the reference shaft position θ_m^{ref} and limit variables f_m^{lim}, $a_{\mathrm{m}}^{\mathrm{lim}}$, $da_{\mathrm{m}}^{\mathrm{lim}}/dt$, which define the frequency, shaft acceleration and its derivative (known as 'jerk') respectively. Input to this module is the user defined 'motion sequence'. Note that the 'SpinTAC plan' module is not integrated in VisSim, hence an application example is given in this laboratory which provides

a set of inputs to the 'Position Move' module. However, 'SpinTAC plan' is integrated in the InstaSPIN motion package [9].

- SpinTAC Velocity Identify: Its function is to identify the system inertia, which in this case is the combined inertia of the PM and Induction motor, given that the two are connected together. Note that this laboratory can, of course, also be undertaken without the use of the Texas Instruments LVACIMTR machine.

In this laboratory a VisSim based example of the SpinTAC control suite modules is discussed. Notably position control and Inertia estimation using the SpinTAC approach, with a sensored field oriented current controller will be discussed. In this case a predefined motion sequence is introduced which yields a 'washing machine' type position/speed control sequence. At a later stage a SpinTAC speed control example will be discussed (see Sect. 6.3).

3.4.1 Lab 1:4: Phase B

This lab module considers the simulation of the PM motor drive in use, with a fixed point current/speed controller as shown in Fig. 3.47. The following information is relevant for this laboratory component:

- Reference program [11]: `lab1-4_LaunchXLphBv1.vsm`.
- Description: Field-oriented, sensored control of a PM motor using the SpinTAC control suite.

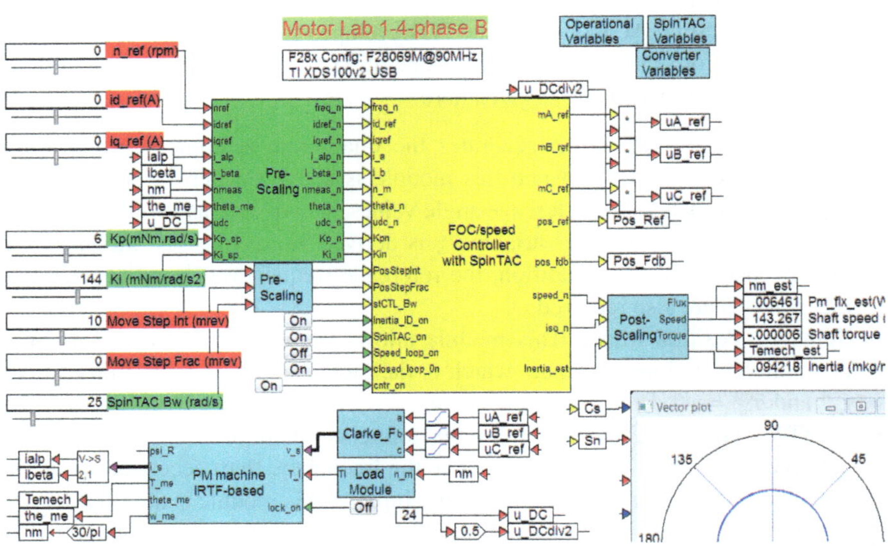

Fig. 3.47 Phase B simulation of a drive with a FOC/Speed controller with SpinTAC option

- Equipment/Software: VisSim simulation program and SpinTAC tool box.
- Outcomes: Show the various waveforms that occur in the drive when operating under field oriented control using a shaft angle sensor with SpinTAC motion control and inertia estimation.

The outputs of the controller, which are (among others) the three reference modulation indices `mA_ref`, `mB_ref`,`mC_ref` are multiplied by the term $u_{DC}/2$ and connected to the 'converter', which in this case is simply represented by three limiters that limit the voltage to the motor to $\pm u_{DC}/2$. A Forward Clarke transformation is again used to generate the voltage vector for the machine model. Furthermore, the controller provides an output which displays the estimated inertia of the drive and two outputs `Pos_Ref`, `Pos_Fdb` which represent the reference and actual position values respectively. These two variables should coincide during normal operation. The remaining outputs are the per unit shaft speed and quadrature current as discussed in the previous laboratory. Outputs of the machine model are the currents i_α, i_β, shaft torque T_m and shaft angle θ_m. Shown partly is a vector diagram, which is used to display the d-axis of machine. i.e. tied to the PM flux of the machine. Hence the rotational angle will be the electrical position which is the shaft angle times the number of pole pairs (four in this case).

As with the previous case, a set of sliders is used to define the current reference vector, in terms of its direct and quadrature components i_d^{ref}, i_q^{ref}, rotational speed n_m^{ref} and the proportional K_p^{sp} and integral K_i^{sp} speed gains of the conventional PI controller. Three additional sliders are present which are used to control the positional sequence in terms of the number of requested integer shaft rotations: `Move Step Int` and fractional rotations: `Move Step Frac`. The third slider is used to control the bandwidth: `SPinTAC BW` of the SpinTAC motion controller.

Five logic buttons are present which perform the following functions:

- `Inertia_ID_on`: activates the SpinTAC inertia estimation routine, which requires `closed_loop_on` control to be on and `Speed_loop_on` to be off.
- `SpinTAC_on`: position control using SpinTAC control when enabled. Conventional speed control using the speed gain sliders `K_p`, `K_i` becomes active when `SpinTAC_on` is disabled.
- `Speed_loop_on`: conventional PI or SpinTAC based Speed control when enabled. Torque control using the i_q slider when disabled.
- `closed_loop_on`: Field-Oriented sensored control when enabled. Open loop current control using the i_d, i_q sliders and speed reference slider when disabled.
- `cntr_on`: activates the current controller. This function remains ON in phase 'B'.

The three dialog boxes for this laboratory, as shown in Fig. 3.48, contain all the user settings required for this laboratory. The 'operational Variables' dialog box is (again) used to assign full scale values for the voltage, current and frequency. In addition, the current controller gain, bandwidth and the rate of variable change for current and frequency must also be assigned. A 'Converter Variables' dialog box assigns the ADC-fullscale voltage/current values for the board in use, as discussed

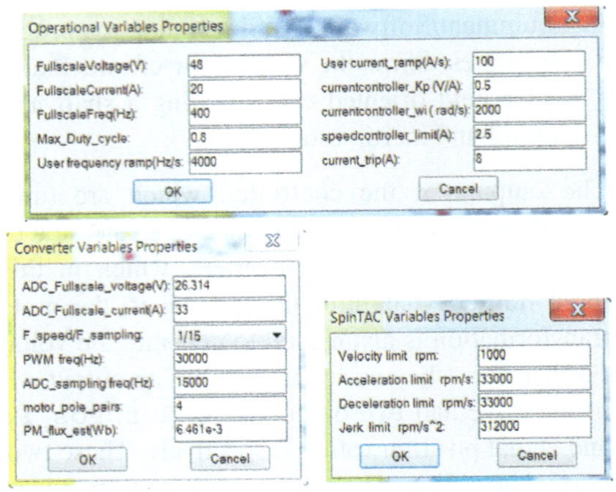

Fig. 3.48 Dialog boxes used for drive with a FOC/Speed controller and SpinTAC option

in previous laboratories. Also the sampling rate of the speed controller (relative to the chosen ADC sampling rate) is set in this dialog box. In this example the ratio is set to $1/15$, which implies a speed or position controller sampling frequency of 1000 Hz. A third dialog box is introduced in this laboratory which assigns critical SpinTAC parameters namely:

- `Velocity limit`: maximum allowable speed (rpm) of the pre-programmed motion sequence.
- `Acceleration limit rpm/s`: maximum rate of positive speed change (rpm/s) of the motion sequence.
- `Deceleration limit rpm/s`: maximum rate of negative speed change (rpm/s) of the motion sequence.
- `Jerk limit rpm/s^2`: maximum (absolute) rate of acceleration change (rpm/s^2).

An example of drive operation, during inertia estimation, is shown in Fig. 3.49. The sampling time T_s, as defined in the program pull-down menu is set to a 15 kHz. This is also the sampling frequency used for the discrete controller model. Subplot (a) of Fig. 3.49, shows the motor phase current during the estimation sequence. In this case the simulation was run over a 0.5 s time interval, with a motor inertia $J = 89 \cdot 10^{-6}\,\text{kgm}^2$ (taken to be the estimated inertia of the PM/IM motor combination). Operation under no-load is assumed with i_d=0. Therefore, the amplitude of the phase current is equal to the i_q value generated by the SpinTAC Identification module (see Fig. 3.46). Subplot (b) of Fig. 3.49 shows the actual shaft speed during Inertia estimation. In addition, subplot (c) shows the shaft torque (in mNm) of the machine. Note that the maximum shaft torque of the machine is in this case controlled by the maximum allowable output current of the speed controller which has been set to 2.5 A (see Fig. 3.48). Upon completion of the estimation routine the estimated value shown on the numeric display is 0.094 mNm/rad/s^2, which compares favorably with the $0.089 \cdot 10^{-3}$ value set in the machine model.

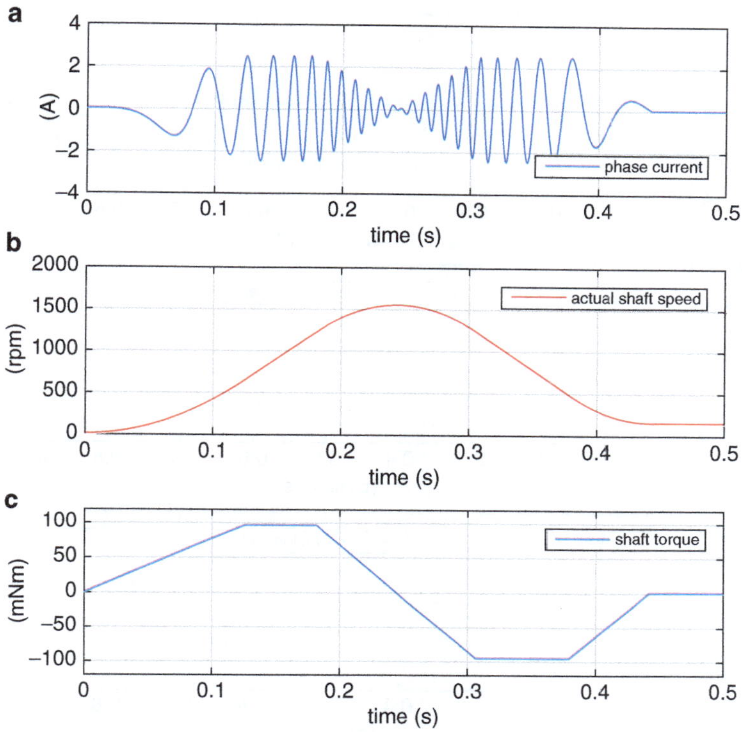

Fig. 3.49 Phase B: Typical waveforms of the drive during the 'Inertia Estimation phase'

A second example of drive operation using SpinTAC based position control is shown in Fig. 3.50. In this case a 'washing machine' cycle is chosen when the machine shaft rotates periodically, whereby the machine stops briefly after ten revolutions (a value set with the `Move Step Int` slider). Maximum speed and acceleration setting are defined in the SpinTAC dialog box, which implies that the maximum shaft speed is limited to 1000 rpm in this case. Subplot (a) given in Fig. 3.50, shows the motor phase current present during one 'ten revolution' cycle. Operation under no-load is assumed with $i_d = 0$. Subplot (b) shows the corresponding actual shaft speed. Note that the rate of reference speed change is set in the user 'operations' dialog box (see Fig. 3.48), while the actual rate of change is determined by the speed bandwidth set in the 'SpinTAC dialog box. The actual shaft torque of the machine over the 1.0 s simulation interval is given in subplot (c). Clearly observable is that the 'washing cycle' sequence is repetitive and in this case restarts after ≈ 0.75 s.

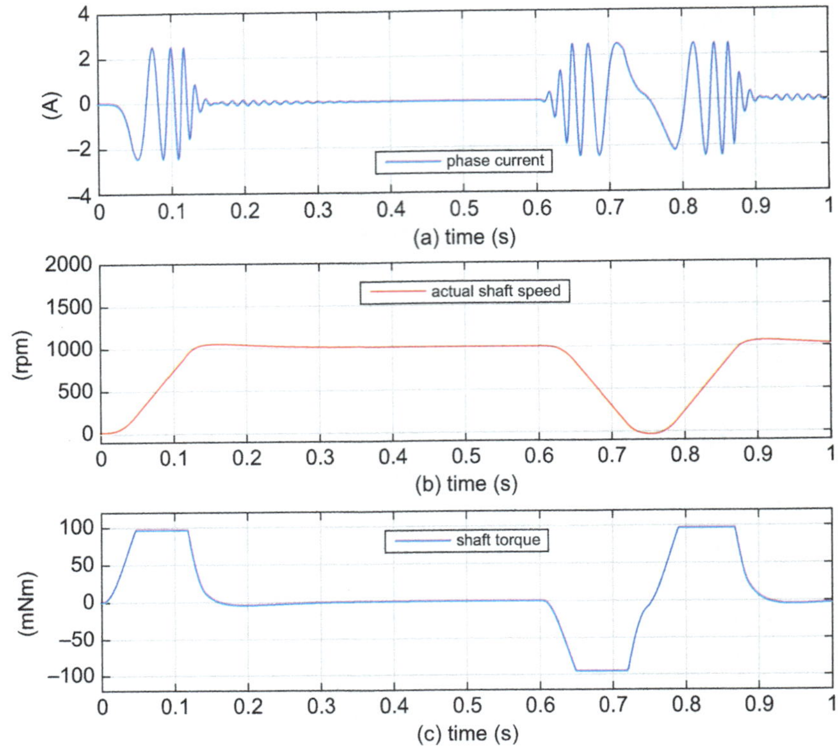

Fig. 3.50 Phase B: Waveforms of the drive over using the SpinTAC option with the selected motion sequence

3.4.2 Lab 1:4: Phase C

In the previous phase, only knowledge of the control card to be used was required, in order to test the fixed point controller. The following information is relevant for this laboratory component:

- Reference program [11]: `lab1-4_LaunchXLphCv2.vsm`.
- Description: Field-oriented, sensored control of a PM motor with SpinTAC control option.
- Equipment/Software: Texas Instruments LAUNCHXL-F28069M, with BOOSTXL-DRV8301 module ('aft' position) and VisSim simulation program.
- Outcomes: Develop a complete drive algorithm, which handles field-oriented current control with the SpinTAC Control Suite. Compile and download `.out` file for use in phase C+.

As mentioned above, a single boost pack located in the 'aft' position (furthest away from the USB connector) is used for single motor operation, in which case the following jumper and dip-switch positions on the LAUNCHXL-F28069M module are required:

- Jumpers `JP1` and `JP2` OPEN
- Jumpers `JP4` and `JP5` CLOSED
- Jumpers `JP3`, `JP6` and `JP7` CLOSED
- Dip-switches `SW1` to `SW3` ON

Furthermore, the 'J4' encoder connector of the PM machine cable should be attached to encoder input `QEP_A` of the LAUNCHXL-F28069M module. Also note that the wiring sequence for the motor is now important because it ties the α, β reference frame of the motor to the encoder (wiring details: 'Terminal A-black' wire etc., shown in development phase C+, see Fig. 3.55). A phase C development stage cannot be used to run a drive, but its primary task is to assemble all the modules needed for drive operation with a given controller configuration. The drive setup as given in Fig. 3.51, shows the 'FOC/Speed Controller with SpinTAC option' module, which must be compiled to generate an `.out` file. A right click on the controller module, invokes a step down to a lower level, as shown in Fig. 3.52.

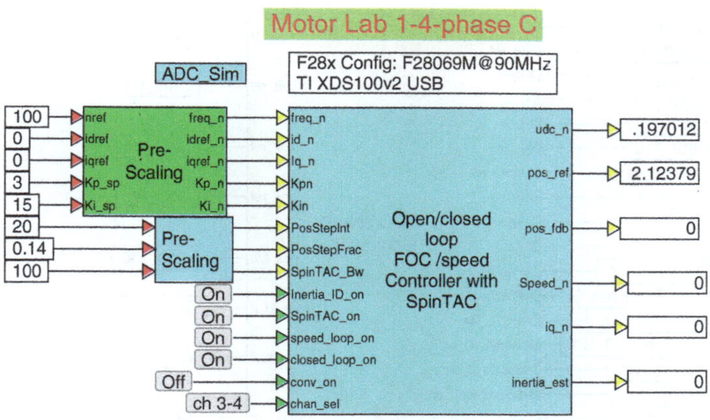

Fig. 3.51 Phase C simulation of a drive with a field-oriented current controller and SpinTAC option

Fig. 3.52 One level down into the controller module

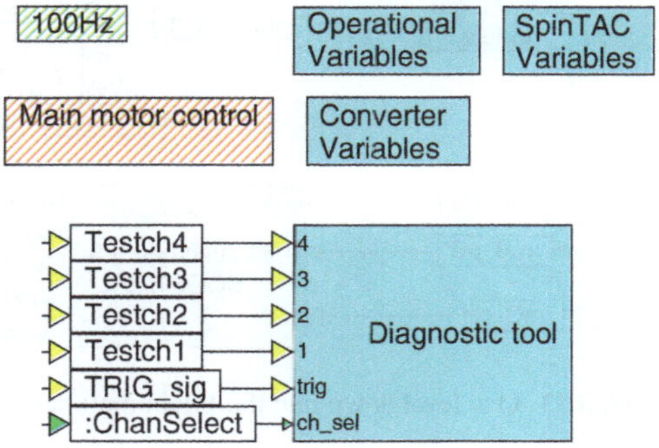

The dialog boxes shown in this diagram are as discussed in Fig. 3.48. A 'Diagnostic tool' module is again used to buffer/store four (user selected) variables, which can subsequently displayed in a graph when operating in phase C+. A two channel multiplexer is present to allow the user to either display variables `Testch1`, `Testch2` or `Testch3`, `Testch4` by using a button connected to the `chan_sel` input of the controller. A '100 Hz' module undertakes the house keeping tasks for the MCU and flashes a 'red' LED on the LAUNCHXL-F28069M module when active. A 'blue' LED remains ON on said module if a current trip condition occurs. Moving one level lower into the 'Main motor Control' module reveals a set of modules shown in Fig. 3.53. Of these, the 'FOC/Speed controller' was introduced in phase B. The 'ADC-PWM' unit shown, is identical to the one discussed in the previous laboratory given that the same drive board is in use. Furthermore, a PWM frequency of 30 kHz is used, together with a ADC sample frequency of 15 kHz, which corresponds to an ADC sampling time of 66 μs, set in the VisSim 'System Properties' pull down menu. This module is controlled by a `CONVon_` variable, which is in turn generated by a logic `and` output that has as inputs a user variable `Con_On` ('converter on') signal and an over-current trip signal `trip_active`.

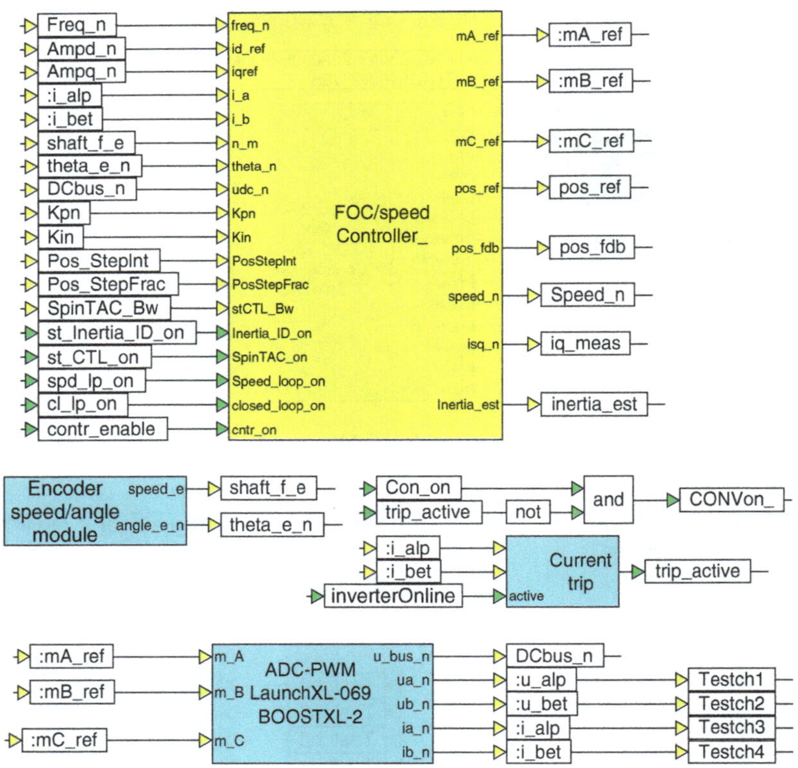

Fig. 3.53 One level down into the 'Main motor control' module

When the `CONVon_` variable is set, ADC offset compensation takes place after which the `contr_enable` flag is set, which enables controller operation. At the same time the variable `inverterOnline` is set which enables the 'current trip' module. Relevant (to this laboratory) outputs of the ADC-PWM unit are the DC bus voltage `DCbus_n`, and the per unit phase currents i_{α}^{n}, i_{β}^{n}, represented by the variables `i_alp`, `i_bet` respectively. The remaining inputs to this 'ADC-PWM' module are the modulation indices generated by the FOC/speed controller module. An 'Encoder Speed/angle' module, generates the per unit electrical shaft speed `shaft_fe` and per unit shaft angle `theta_e_n`. For diagnostic purposes the per unit current and voltage variables have been connected to variables `Testch3`, `Testch4` and `Testch1`, `Testch2` respectively, so that these can be displayed in phase C+.

Moving one level lower into the 'FOC/speed Controller with SpinTAC option' module reveals a set of modules given in Fig. 3.54. Synchronous model based current control is undertaken by the 'PI-kp-wi-lim' module, which has as inputs the per unit quadrature and direct axis current error and (indirectly) the inverse DC bus voltage. A Reverse Park transformation is used to generate the modulation indices m_{α}, m_{β}, from the outputs of the current controller, which are then processed by a 'Pulse Center (SVM)' module.

As in the previous laboratories a Cosine/Sine function, with variables `Cos_th`, `Sin_th` is used instead of the instantaneous angle of the current reference vector.

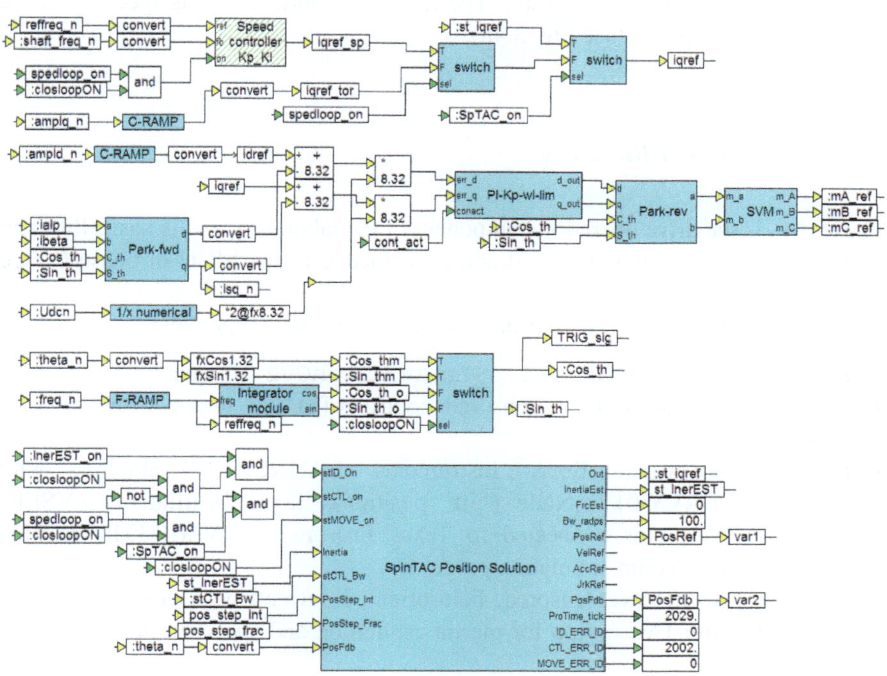

Fig. 3.54 Fixed point representation of the FOC/Speed controller with SpinTAC option

These two variables are then consistently used for subsequent Park transformations, which have been adapted to allow the use of Cosine/Sine variables. A 'switch' module, activated by the `closloopON` signal (active if the shaft encoder is in use) allows the variables `Cos_th`,`Sin_th` to be generated via the encoder or integrator. A 'switch' module with output `iqref` determines whether the quadrature reference current for the current controller is provided by either a speed controller (conventional or SpinTAC based) or the user i_q slider. The switch in question is controlled by the `speedloop_on` signal (active when closed speed-loop control is required, OFF implies operation in constant torque mode). A switch with logic input `SpTAC_on`, defines whether the quadrature reference is provided by the conventional 'Speed Controller' (with output `iqref_sp`) or 'SpinTAC Position Solution' module (with output `st_iqref`). The 'SpinTAC Position Solution' compound unit contains the SpinTAC modules shown in Fig. 3.46, together with a 'Plan' module which contains the 'washing machine cycle example' used in this laboratory. Inertia estimation is activated using the user logic signal `st_InerEST`, which activates the SpinTAC Identification module, within the 'SpinTAC' module. When active, the quadrature reference current for the current controller `iqref` is defined by variable `st_iqref`. A per unit inertia estimate is provided by variable `st_InerEST` of the 'SpinTAC Position Solution' module.

A set of 'Rate Limiter' modules, entitled 'C-ramp' (sets the current ramp rate) and 'F-ramp' (sets the frequency ramp rate), have been inserted in Fig. 3.54 to avoid erratic changes to the user input variables (as actually used by the controller), which may damage the converter or motor. The rate of variable change is specified in the 'Operations' dialog box (see Fig. 3.48).

3.4.3 Lab 1:4: Phase C+

Phase C+, is the drive operation component of the laboratory and is basically a run version of the `.out` file compiled and downloaded to the MCU in phase C (see previous subsection).

The following information is relevant for this laboratory component:

- Reference program [11]: `lab1-4_LaunchXLphCv2_d.vsm`.
- Description: Field-oriented, sensored control of a PM motor with SpinTAC control option.
- Equipment/Software: Texas Instruments LAUNCHXL-F28069M, with BOOSTXL-DRV8301 module ('aft' position), Texas Instruments LVSERVOMTR PM motor connected to Texas Instruments LVACIMTR machine (optional) and VisSim simulation program.
- Outcomes: To achieve sensored, field-oriented current/speed control of a PM drive, with SpinTAC option for inertia estimation and specified motion control sequence.

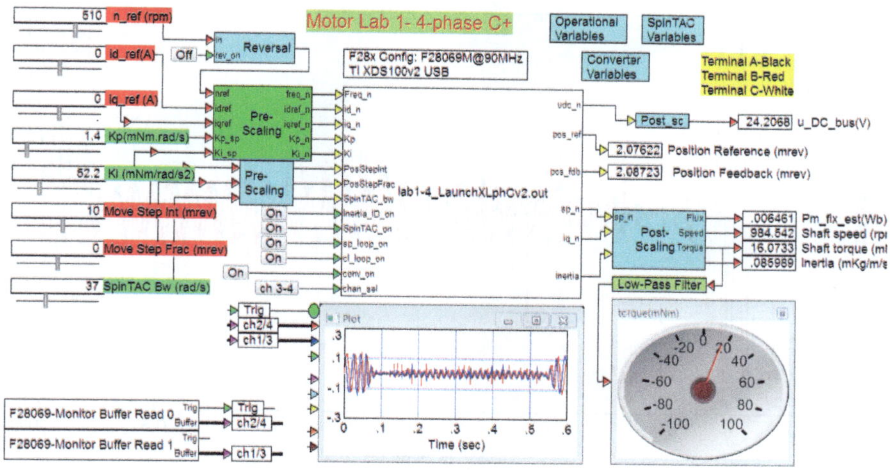

Fig. 3.55 Phase C+ operation of a drive with a field-oriented current/speed controller, with SpinTAC option

Note that the required jumper and dip-switch settings for the LAUNCHXL-F28069M module are given in phase C. Furthermore, the 'J4' encoder connector of the PM machine cable should be attached to encoder input `QEP_A` of the LAUNCHXL-F28069M module.

The run version as shown in Fig. 3.55 uses a VisSim run module, which executes the `.out` file, shown in said module. Five sliders are used to set the current reference amplitudes `id_ref`, `iq_ref`, shaft speed and speed control loop gains. Two post-scaling modules are used to display the measured DC bus voltage, PM magnet flux, shaft speed and torque (both instantaneous and filtered using a 10 Hz low pass filter). The filtered torque signal is displayed on a torque meter in (milli)Nm. In addition, the estimated inertia of the drive is shown. Finally two numerical displays are present which show the `Position Reference` and `Position Feedback` signals. These two must be equal for SpinTAC to work correctly. Two 'Monitor Buffer' modules are used to display two selected (using a button connected to the `chan_sel` input) internal variables. The VisSim scope module shows the per unit α, β currents and an enlarged view of these results is given in Fig. 3.56. Multiplication of these waveforms by the full scale current value of i_{fs}= 20 A gives actual current values. Note that the length of the monitor buffers used for the storage of the VisSim scope data has been increased (in the phase C file) from 901 → 9001. This implies that the VisSim scope now displays 9000 samples, which at a sample rate of 15 kHz corresponds to a time interval of 0.6 s.

Prior to activating the drive the following 'Pre-Drive' check list should be executed:

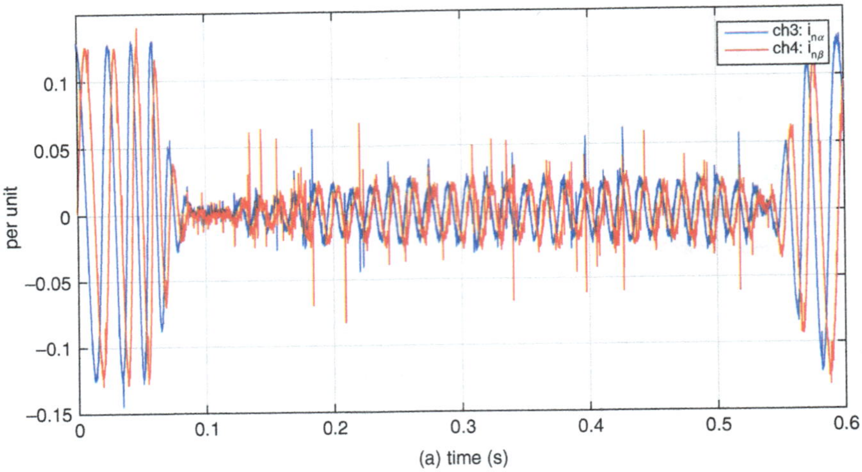

Fig. 3.56 Phase C+ operation of a drive: VisSim scope results with channel select option: 'ch 3-4'

- Dialog boxes used in phase C+ model match those of phase C: the run version is compiled with the dialog boxes entry specified under phase C. The dialog boxes show in C+ are used by the 'Pre/Post' scaling modules. Make sure to compile the phase C module first before executing the 'codegen' option in VisSim.
- Ensure that the sample time used is correct and latest (and correct) `.out` file has been downloaded to the MCU (Right mouse click on MCU module to show dialog box of these variables).
- Confirm that the user input values are set to either zero, or 'acceptable' values, which will not cause a current trip of the converter
- Confirm that the converter 'switch' is set to OFF and the power supply is on (DC bus voltage present).
- Confirm that the motor is connected firmly and properly.

After completion of the Pre-Drive checklist, activate the program and confirm that the supply voltage source is 24 V. If not stop the program and restart. With the correct DC voltage level established, turn on the converter (using the ON button, shown) and monitor the motor shaft and diagnostic scope, which in this example shown (Fig. 3.55) is set to show the per unit current components i_{α}^{n}, i_{β}^{n}. Note that the actual current values are found multiplying the per unit value with the full-scale current value (set to 20 A in this case).

Operation of this laboratory requires three basic steps namely:

- Step 1: `Inertia_ID_on` button OFF, `sp_loop_on` button ON, `SpinTAC_on` button OFF, `cl_loop_on` button ON: then activate the drive which should then operate at the set user reference speed. This test verifies that the incremental encoder is providing the correct signals to the FOC controller.
- Step 2: `sp_loop_on` button OFF, `Inertia_ID_on` button ON, `SpinTAC_on` button ON, '`cl_loop_on` button ON: this causes the controller to enter the inertia estimation mode, which it will continue to do sequentially whereby the new inertia estimate is displayed each time.
- Step 3: `sp_loop_on` button ON, `Inertia_ID_on` button ON, `SpinTAC_on` button ON, `cl_loop_on` button ON: which causes the drive to enter the pre-programmed SpinTAC position control sequence.

Observation of the results shown in Fig. 3.55 show a combined estimated inertia value of 85.9 μkgm^2 for the PM/IM machine combination which have been mechanically connected. The scope waveforms (see Fig. 3.56) show the (per unit) α, β currents during part of the motion sequence. Operation with $i_d = 0$ is undertaken in this case, hence the phase current value must also correspond to the reference i_q value (set by the SpinTAC controller).

An oscilloscope and DC-true current probe were also used to measure the phase current, when operating with the SpinTAC motion control sequence, as shown in Fig. 3.57. Behavior of the drive during the execution of the preprogrammed motion sequence is consistent with the phase B results given in Fig. 3.50. Note however,

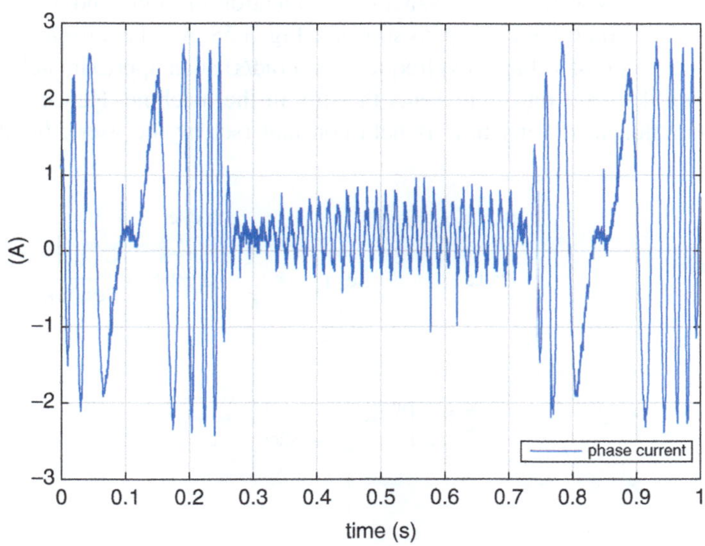

Fig. 3.57 Phase C+ experimental result: phase current waveform, with SpinTAC option active and pre-programmed motion cycle. Measured with Tektronix DC-true current probe

that the presence of the mechanically connected induction machine represents a friction load for the PM machine of approximately 15–20 mNm. Hence the current waveform during the interval (0.25 → 0.75 s in Fig. 3.57) has a non zero peak current amplitude due to the presence of a i_q current component required to produce the torque needed to overcome said friction. Note also that the average phase current shown is not zero, which is due to measurement error.

3.5 Laboratory 1:5: Voltage/Frequency and Speed Control of IM

Purpose of this laboratory is to realize voltage/frequency and speed control of an induction motor. Use is made of the voltage control concept introduced in laboratory 1:1 (see Sect. 3.1). This control approach is then modified to include a voltage/frequency controller as discussed in [4]. Speed control capability has also been added to this laboratory, where use is made of the shaft encoder connected to the PM machine, which is (mechanically) connected to the IM under consideration. For this laboratory, developmental phases: B, C and C+ will be discussed. The aim is to take the reader through the steps needed to implement a functional drive based on a given set of equations and corresponding PLECS based (or any other) controller diagram. This implies that the reader can examine all the steps needed to implement a typical drive control example from floating- to fixed point (which requires scaling) simulation, to experimental implementation. For this laboratory session use is made of the Texas Instruments LVACIMTR induction machine and a voltage over frequency (V/f) controller concept as shown in Fig. 3.58. A V/f controller essentially maintains the ratio of voltage and frequency (in rad/s) at an approximately constant value, given that said ratio represents the flux in the machine. Hence the (steady state) stator flux in the machine is held constant (set by the user). Inputs to the

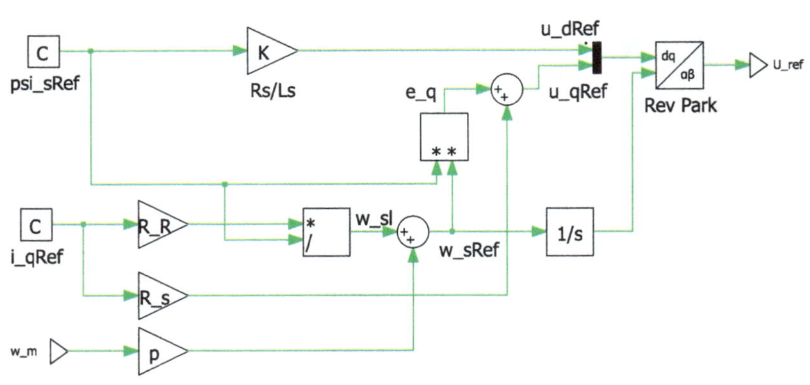

Fig. 3.58 V/f drive for induction machine, with reference stator flux ψ_s^{ref}, quadrature current i_q^{ref} inputs, which makes use of the measured shaft speed ω_m

controller are the reference stator flux `psi_sRef`$=\psi_s^{\rm ref}$ (Wb), measured electrical shaft speed `w_m`$=\omega_{\rm m}$ (rad/s) and reference quadrature current `i_qRef`$=i_{\rm q}^{\rm ref}$. Outputs are the reference voltages $u_\alpha^{\rm ref}$, $u_\beta^{\rm ref}$, as represented by the column vector `U_ref`. In the actual implementation this vector must be multiplied by a term $2/u_{\rm DC}$, where $u_{\rm DC}$ is the DC bus voltage, in order to generate the required modulation index vector for the modulator. The controller representation satisfies the following set of equations [4]

$$u_{\rm d}^{\rm ref} = \psi_{\rm s}^{\rm ref}\frac{R_{\rm s}}{L_s} \tag{3.4a}$$

$$u_{\rm q}^{\rm ref} = \omega_{\rm s}^{\rm ref}\psi_{\rm s}^{\rm ref} + i_{\rm q}^{\rm ref}R_{\rm s} \tag{3.4b}$$

$$\omega_{\rm s}^{\rm ref} \cong p\,\omega_{\rm m} + \frac{i_{\rm q}^{\rm ref}R_{\rm R}}{\psi_{\rm s}^{\rm ref}} \tag{3.4c}$$

where $L_{\rm s} = L_{\rm M} + L_\sigma$, represents the self inductance of the (four-parameter) machine (as discussed in Sect. 2.1.4) and $\omega_{\rm s}^{\rm ref}$ represents the stator frequency reference variable (rad/s).

The quadrature reference current $i_{\rm q}^{\rm ref}$ follows from torque expression

$$T_{\rm m} = \frac{3}{2}p\,i_{\rm q}\,\psi_s \tag{3.5}$$

where p represents the number of pole pairs of the machine and $T_{\rm m}$ the mechanical shaft torque. The speed controller, with gains $\mathrm{K_p}$, $\mathrm{K_i}$ can in the continuous time domain be represented as

$$T_{\rm m} = \left(\mathrm{K_p} + \frac{\mathrm{K_i}}{s}\right)\left(\omega_{\rm m}^{\rm ref} - \omega_{\rm m}\right) \tag{3.6}$$

where $\omega_{\rm m}$ is the actual shaft speed of the machine (rad/s). Use of said equation in combination with Eqs. (3.5), (3.6) yields an explicit expression for the reference quadrature current $i_{\rm q}^{\rm ref}$ namely

$$i_{\rm q}^{\rm ref} = \underbrace{\frac{2}{3p^2\psi_{\rm s}^{\rm ref}}}_{\mathrm{K_{pi}}}\left(\mathrm{K_p} + \frac{\mathrm{K_i}}{s}\right)\left(\omega_{\rm m}^{\rm ref} - \omega_{\rm m}\right) \tag{3.7}$$

where a gain parameter $\mathrm{K_{pi}}$ is introduced, which will be subsequently used for fixed point controller implementation. Expression (3.7) defines the quadrature current input for the V/f controller shown in Fig. 3.58 based on the actual (from an encoder in this case) and reference electrical shaft speed. This expression together with Eq. (3.4) must be scaled using the approach set out in Sect. 1.4.2, which after some manipulation gives

$$u_{\rm dn}^{\rm ref} = \left(\frac{R_{\rm s}\,\psi^{\rm fs}}{L_{\rm s}\,u_{\rm fs}}\right)\psi_{\rm sn}^{\rm ref} \tag{3.8a}$$

$$u_{\rm qn}^{\rm ref} = \left(\frac{2\pi\,\psi^{\rm fs}}{u_{\rm fs}}\right) f_{\rm sn}^{\rm ref}\,\psi_{\rm sn}^{\rm ref} + \left(\frac{R_{\rm s}\,i_{\rm fs}}{u_{\rm fs}}\right) i_{\rm qn}^{\rm ref} \tag{3.8b}$$

$$f_{\rm sn}^{\rm ref} \cong f_{\rm en} + \left(\frac{R_{\rm R}\,i_{\rm fs}}{2\pi\,\psi^{\rm fs}}\right)\frac{i_{\rm qn}^{\rm ref}}{\psi_{\rm sn}^{\rm ref}} \tag{3.8c}$$

$$i_{\rm qn}^{\rm ref}(t_k) \cong \left(\mathrm{K_{pi}}\frac{2\pi f_{\rm fs}}{i_{\rm fs}}\right)\left(\mathrm{K_p} + \mathrm{K_i}\,\mathrm{T_{sp}}\right)\Delta f_n + i_{\rm qnINT}(t_{k-1}) \tag{3.8d}$$

with $\Delta f_{\rm n} = f_{\rm en}^{\rm ref} - f_{\rm en}$. The discretization of Eq. (3.7) is implemented using the approach shown for Eq. (2.7) which leads to Eq. (3.8d). A variable $\mathrm{T_{sp}}$ is introduced in the latter expression, which represent the time step used for the discrete speed controller module.

Note that the controller requires access to the parameters of the machine in use. These are assumed to be known (measured using the InstaSPIN motor identification routine, to be discussed in the next chapter).

3.5.1 Lab 1:5: Phase B

This module considers the simulation of the IM motor drive in use, with a fixed point V/f and speed controller as shown in Fig. 3.59. The following information is relevant for this laboratory component:

- Reference program [11]: `lab1-5_LaunchXLphBv2.vsm`.
- Description: V/f and Speed control of a IM motor.
- Equipment/Software: VisSim simulation program.
- Outcomes: Show the various waveforms that occur in the drive when using a fixed point representation of the controller when operating under Voltage/frequency control.

A 'Pre-scale' module is used to convert and scale the floating point, user (slider) variables to fixed point format. Inputs to the V/f controller module are:

- `fref_n`: per unit reference frequency, which is linked to the reference shaft speed $n_{\rm m}^{\rm ref}$ by a gain factor $\left(\frac{p\,n_{\rm m}^{\rm ref}}{60}\right)$.

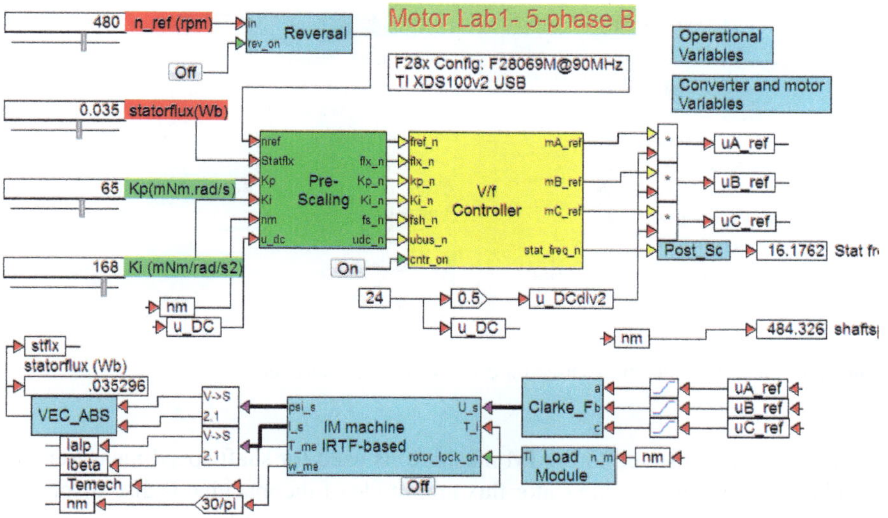

Fig. 3.59 Phase B simulation of a drive with a V/f and speed controller

- `flx_n`: per unit stator flux, which is scaled by the full scale flux ψ^{fs} set in the user dialog box (to be discussed).
- `kp_n`: scaled proportional controller speed gain, which is linked to the gain factor K_{pi} as shown in Eq. (3.8d). The resultant gain factor to be used is therefore equal to $\left(\frac{4\pi\, K_p f_{\text{fs}}}{3p^2\, \psi_s^*\, i_{\text{fs}}}\right)$.
- `ki_n`: scaled integral controller speed gain. The speed controller gains are in this lab example expressed in terms of the proportional and integral gain as shown by Eq. (2.3). This implies that a similar gain factor is also applicable here which includes the discrete time step T_{sp} used by the speed controller, which gives $\left(\frac{4\pi\, K_i\, T_{\text{sp}} f_{\text{fs}}}{3p^2\, \psi_s^*\, i_{\text{fs}}}\right)$.
- `fsh_n`: per unit electrical shaft frequency, which is linked to the shaft speed n_{m} by the gain factor $(p n_{\text{m}}/60)$.
- `ubus_n`: per unit bus voltage, scaled by the full scale voltage u_{fs} set in the user dialog box (to be discussed).
- `cntr_on`: control active, which enables the controller.

Outputs of the V/f module are the per unit stator frequency `stat_freq_n` together with the phase modulation indices: `mA_ref`, `mB_ref`, `mC_ref`, which are then multiplied by half the DC bus voltage and subsequently limited to $\pm 12\,\text{V}$ prior to their use by the induction motor model. Outputs of the machine model are the currents i_α, i_β, shaft torque T_{m} and shaft speed n_{m}. The latter variable is used by the controller to realize speed control. In reality, the encoder of the PM machine

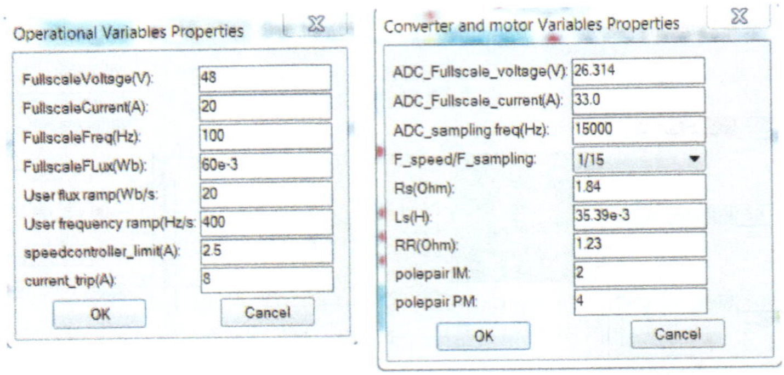

Fig. 3.60 Phase B: dialog boxes used for drive with a V/f controller

(mechanically connected to the IM machine) is used for shaft speed measurement. In addition to the above, the stator flux amplitude of the machine is also displayed, given that this variable is controlled by the controller in question.

The two dialog boxes shown in Fig. 3.60, contain all the user settings required for this laboratory. The 'Operational Variables' dialog box is used to assign full scale values for the voltage, current, frequency and (in this case) stator flux. Dialog box entries for the maximum allowable rate of change for flux and frequency are also provided, together with a speed controller limit value `speedcontroller_limit(A)`. This value is the maximum (absolute) quadrature current value, which the speed controller is set to provide.

The 'Converter and motor Variables' dialog box assigns the motor pole pair number and parameters of the LVACIMTR induction motor in use. In addition, the sampling frequency used for the V/f and speed controller modules (arbitrarily set to the same value) is set in this dialog box, relative to the ADC sampling frequency. Also added here, is the pole pair number of the attached PM motor, given that the electrical frequency of the encoder module used for speed control is set in terms of said machine. An example of drive operation is shown in Fig. 3.61. The sampling time T_s as defined in the program pull-down menu is set to 66.67 μs, i.e. a 15 kHz ADC sampling frequency is used for the discrete controller model. These results show the phase current (subplot(a)), reference/actual stator flux (subplot (b)) and reference/actual shaft speed (subplot(c)) for a time interval of 0.2 s when operating under no load conditions. The results shown in Fig. 3.61 confirm that both speed and stator flux control are achieved. Note that there is no current control, other than over current protection as is apparent by the transient current pulse upon start-up (see subplot (a), Fig. 3.61). The reader is again encouraged to undertake a detailed examination of this simulation using the VisSim Simulation platform and this specific file.

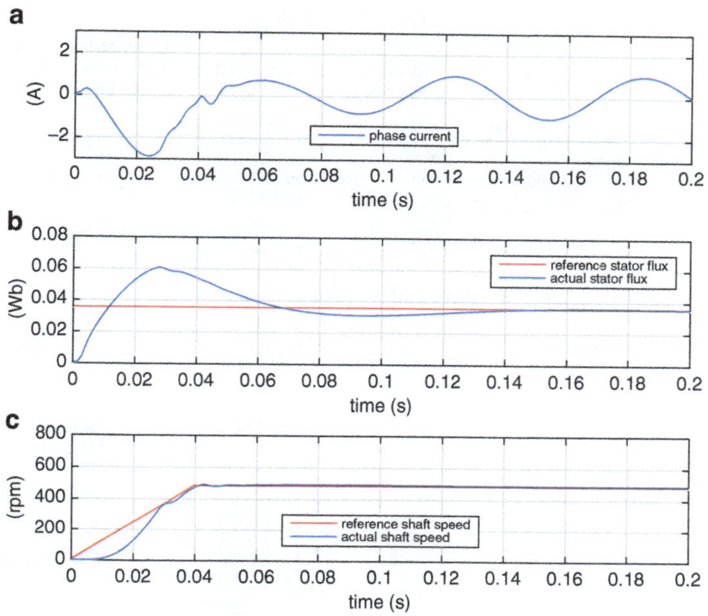

Fig. 3.61 Phase B: Typical waveforms obtained using a V/f and speed controller

3.5.2 Lab 1:5: Phase C

In the previous phase, only knowledge of the control card to be used was required, in order to test the fixed point controller. The following information is relevant for this laboratory component:

- Reference program [11]: `lab1-5_LaunchXLphCv2.vsm`.
- Description: Voltage/Frequency and speed control of a IM motor.
- Equipment/Software: LAUNCHXL-F28069M, with BOOSTXL-DRV8301 module ('aft' position)and VisSim simulation program.
- Outcomes: Develop a complete drive algorithm, which handles V/f and speed control. Compile and download an `.out` file for use in phase C+.

As mentioned above, a single boost pack located in the 'aft' position (furthest away from the USB connector) is used for single motor operation, in which case the following jumper and dip-switch positions on the LAUNCHXL-F28069M module are required:

- Jumpers `JP1` and `JP2` OPEN
- Jumpers `JP4` and `JP5` CLOSED
- Jumpers `JP3`, `JP6` and `JP7` CLOSED
- Dip-switches `SW1` to `SW3` ON

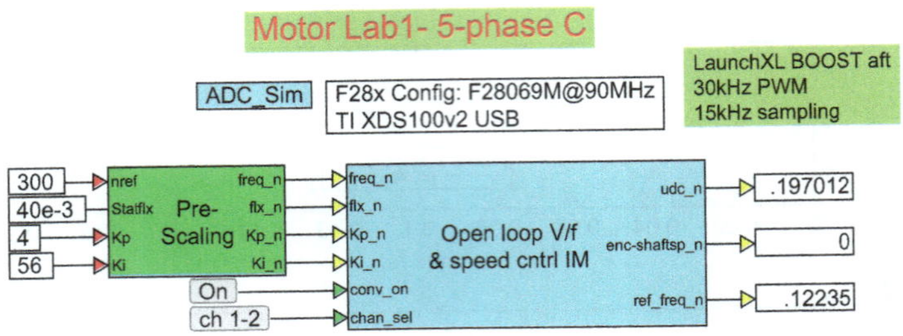

Fig. 3.62 Phase C simulation of a drive with a V/f and speed controller

Fig. 3.63 One level down into the controller module

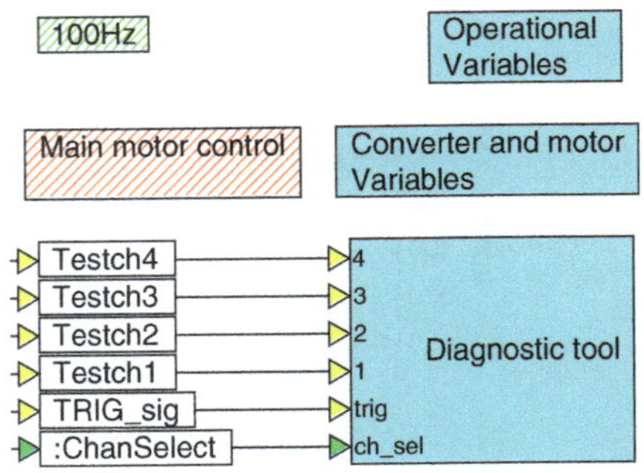

Furthermore, the 'J4' encoder connector of the PM machine cable should be attached to encoder input `QEP_A` of the LAUNCHXL-F28069M module. The encoder is in this case used to measure the speed of the induction machine. A phase C development stage, cannot be used to run a drive, but its primary task is to assemble all the modules needed for drive operation with a given controller configuration. The drive setup as given in Fig. 3.62, shows the 'Open loop V/f IM Controller' module, which must be compiled to generate an `.out` file. Inputs to the controller module via the pre-scaling unit are four variables (shown as 'constants' for convenience), which allow the user to set the shaft speed, stator flux amplitude and PI speed controller gains. When the logic input `conv_on` is set to ON a power on sequence occurs where the power converter is activated, ADC module offsets are determined and controller enabled. Outputs of the Controller module are the per unit DC bus-voltage, shaft speed (as measured with the shaft encoder connected to the PM motor) and stator frequency generated by the controller.

Moving one level down into the Controller module leads to the set of modules/dialog boxes shown in Fig. 3.63. The dialog boxes shown in this diagram have already been discussed in phase B (see Fig. 3.60). A 'Diagnostic tool' module is again used to buffer two variables, which are subsequently displayed in a graph when operating in phase C+. A two channel multiplexer is used to allow the user to either display variables `Testch1`, `Testch2` or `Testch3`, `Testch4` by using a button connected to `chan_sel` input of the module. A '100 Hz' module acts as a 'heart beat' and flashes a 'red' LED on the LAUNCHXL-F28069M module. A 'blue' LED remains ON if a current trip condition occurs. In addition, all the drive background tasks are undertaken by this unit.

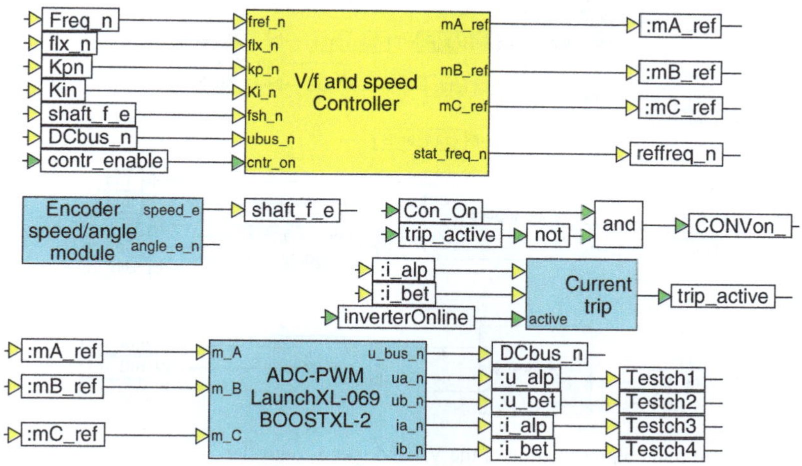

Fig. 3.64 One level down into the 'Main motor control' module

Moving one level lower into the 'Main motor Control' module reveals a set of modules, as shown in Fig. 3.64. Of these, the 'V/f and Speed Controller' was introduced in phase B. The 'ADC-PWM' unit shown, is identical to the one used in the previous laboratory given that the same drive setup is in use. Furthermore, a PWM frequency of 30 kHz is used, together with a 15 kHz ADC sampling frequency, that is set in the VisSim 'System Properties' pull down menu. This module is controlled by a `CONV_on_` signal, which is in turn generated by a logic and output that has as inputs a user variable `Con_On` ('converter on') signal and a over-current trip signal `trip_active`. Relevant (to this laboratory) outputs of this unit are the DC bus voltage `DCbus_n`, and the per unit phase currents i_{α}^{n}, i_{β}^{n}, represented by the variables `i_alp`, `i_bet` respectively. An output flag `contr_enable` activates the controller module, after the converter has been turned on and internal ADC/PWM initialization actions have been performed. An 'Encoder Speed/angle' module, generates the per unit electrical shaft speed `shaft_f_e` (which is equal to f_{e} = `shaft_f_e`f_{fs}), with $f_{\mathrm{e}} = {n_{\mathrm{m}}p}/{60}$ Hz. Note that the pole pair variable p in latter equation is tied to the PM machine. The inputs to this 'ADC-PWM' module are the modulation indices generated by the controller module.

Moving one level lower into the 'V/f and Speed Controller' module reveals a set of modules given in Fig. 3.65. Inputs to the speed controller are the reference electrical frequency `freq_n` and actual electrical shaft frequency (from the shaft encoder) `fshn`. Output of this module is the per unit quadrature reference current `iqn`. The reference shaft speed signal is connected to the `ref` input of the Speed controller module via a frequency rate limiter module 'F-RAMP' and switch (with input `con_act2`) which ensures that said variable is held at zero when the converter is disabled. As in previous laboratories a Cosine/Sine function, with variables `Cos_th`, `Sin_th` is used instead of the instantaneous angle of

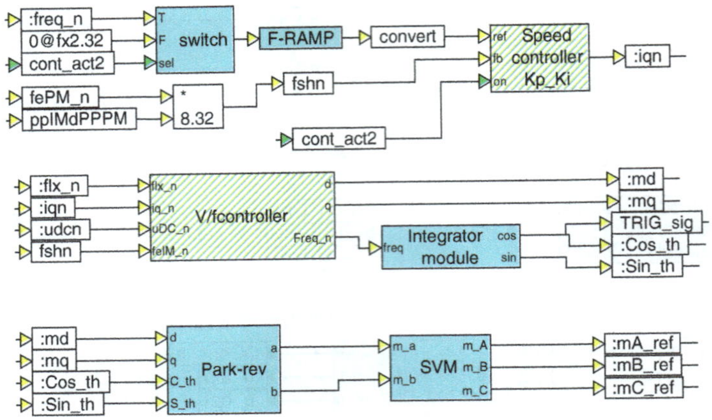

Fig. 3.65 Fixed point representation of the V/f and speed controller

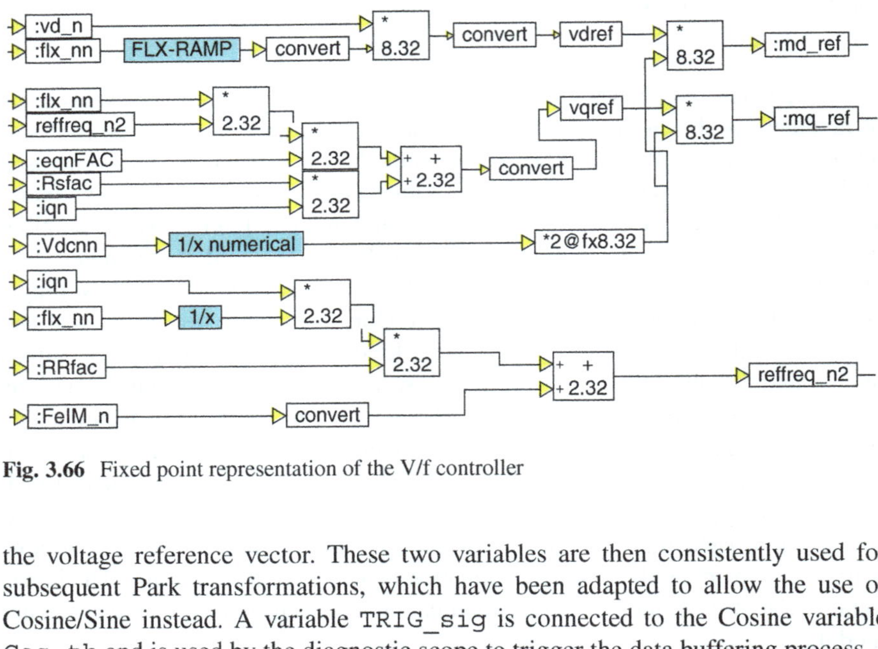

Fig. 3.66 Fixed point representation of the V/f controller

the voltage reference vector. These two variables are then consistently used for subsequent Park transformations, which have been adapted to allow the use of Cosine/Sine instead. A variable `TRIG_sig` is connected to the Cosine variable `Cos_th` and is used by the diagnostic scope to trigger the data buffering process.

The V/f controller module, is hatched 'green' because, like the speed module, it is executed at 1/15 of the ADC sampling frequency, i.e. at 1 kHz in this case. The basic structure of the fixed point controller as shown in Fig. 3.66 reflects the control structure of the floating point version (see Fig. 3.58). The difference being that the reverse Park module and integrator were also incorporated in the floating point module. Furthermore, the outputs of the fixed point module are given in terms of the modulation index. However, scaling as expressed by Eq. (3.8) is used to arrive at the specific structure shown in Fig. 3.66. A brief outline of the variables which appear in the discrete controller are given below:

- `vd_n`= $\frac{R_s}{L_s}\left(\frac{\psi^{fs}}{u_{fs}}\right)$, where ψ^{fs}, u_{fs} represent the full scale flux and voltage values respectively. Represented by bracketed term in Eq. (3.8a).
- `flxnn`= $\frac{\psi_s^{ref}}{\psi^{fs}}$, which is the rated scaled stator flux reference value.
- `reffreq_n2`= $\frac{f_s^*}{f_{fs}}$, which is the scaled reference stator frequency (in Hz). This variable is used as an input to the 'integrator' module of the voltage controller.
- `egnFAC`= $\frac{2\pi\psi^{fs}}{u_{fs}}$, which is a scaling factor used to scale the term $e_q = \omega_s \psi_s^{ref}$, as represented by the first RHS bracketed term in Eq. (3.8b).
- `Rsfac`= $\frac{R_s i_{fs}}{u_{fs}}$, which is the scaled stator resistance, as represented by the second RHS bracketed term in Eq. (3.8b).
- `iqn`= $\frac{i_q}{i_{fs}}$, which is the scaled quadrature current generated by the speed controller.
- `Vdcnn`= $\frac{u_{DC}}{u_{fs}}$, which is the per unit measured DC bus voltage generated by the ADC/PWM module.
- `RRFAC`= $\frac{R_R i_{fs}}{2\pi\psi^{fs}}$, which is a scaling factor used to scale the term $\omega_r = \frac{i_q R_R}{\psi_s^*}$, as represented by the RHS bracketed term in Eq. (3.8c).
- `FeIM_n`= $\frac{p n_m}{60 f_{fs}}$ per unit electrical shaft speed of the induction machine, where p is the number of pole pairs of said machine.

A set of 'Rate Limiter' modules, entitled 'F-ramp' (sets the frequency ramp rate) and 'FLX-ramp' (sets the flux ramp rate), with input the per unit user flux value `flx_n`, have been introduced in Figs. 3.65, 3.66 to avoid erratic changes to the user input variables (as actually used by the controller), which may damage the converter or motor. The rate of variable change is specified in the 'Operations' dialog box (see Fig. 3.60). Outputs of the V/f module are the modulation index variables `md_ref`, `mq_ref`, which are found by scaling the variables u_d^{ref}, u_q^{ref} by 2 times the inverse DC bus voltage. The third output is the per unit stator reference frequency `reffreq_n2`, which is connected to an integrator that generates the Cos/Sin variables for the reference voltage vector (see Fig. 3.58).

3.5.3 Lab 1:5: Phase C+

Phase C+, is the drive operation component of the laboratory and is basically a run version that makes use of the `.out` file compiled and downloaded to the MCU in phase C (see previous subsection).

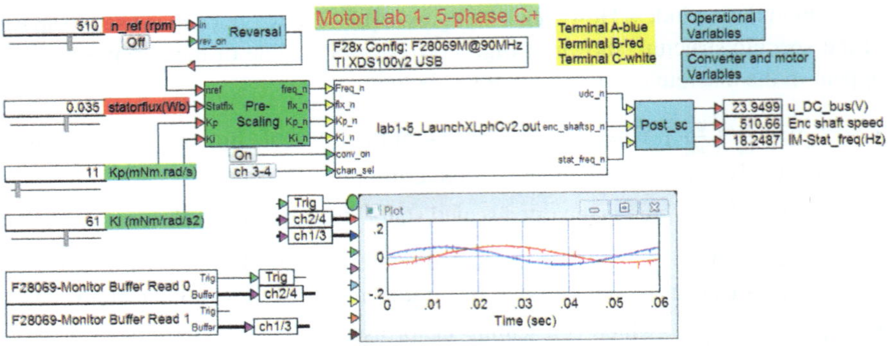

Fig. 3.67 Phase C+ operation of a drive with a V/f converter and speed controller

The following information is relevant for this laboratory component:

- Reference program [11]: `lab1-5_LaunchXLphCv2_d.vsm`.
- Description: Open loop voltage/frequency control of a IM motor.
- Equipment/Software: LAUNCHXL-F28069M, with BOOSTXL-DRV8301 module ('aft' position), Texas Instruments LVACIMTR IM machine connected to LVSERVOMTR PM machine motor and VisSim simulation program.
- Outcomes: To achieve closed loop speed control with a V/f controlled IM drive.

Note that the required jumper and dip-switch settings for the LAUNCHXL-F28069M module are given in phase C. Furthermore, the 'J4' encoder connector of the PM machine cable should be attached to encoder input `QEP_A` of the LAUNCHXL-F28069M module.

The C+ version shown in Fig. 3.67 uses a VisSim run module, which executes the `.out` file, shown in said module. Four sliders are used to set the reference flux ψ_s^{ref}, shaft speed and speed gains. Two 'Monitor Buffer' modules are used to display two selected (using a button connected to the `channel_sel` input) 'test' variables.

Note that the rotational direction of the induction machine, should be the same as the encoder, i.e. a positive reference speed of the controller, should show a positive encoder generated shaft speed. This is ensured by using the motor wiring sequence shown in Fig. 3.67.

The VisSim scope module shows the per unit α, β currents and an enlarged view of these results is given in Fig. 3.68. Multiplication of these waveforms by the full scale current value of $i_{fs}=20$ A gives actual current values.

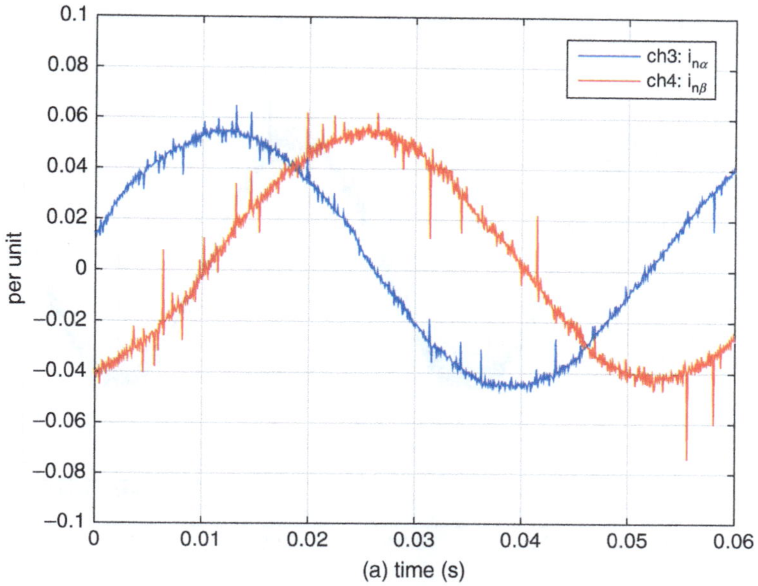

Fig. 3.68 Phase C+ operation of a drive: VisSim scope results with channel select option: 'ch 3-4'

Prior to activating the drive the following 'Pre-Drive' check list should be executed:

- Dialog boxes used in C+ model match those of C: the run version is compiled with the dialog box entries specified under phase C. The dialog boxes shown in C+ are used by the 'Pre/Post' scaling modules.
- Ensure that the sample time used is correct and latest (and correct) `.out` file has been downloaded to the MCU (Right mouse click on MCU module to show dialog box of these variables).
- Confirm that the user input values are set to either zero, or 'acceptable' values, which will not cause a current trip of the converter
- Confirm that the converter 'switch' is set to OFF and the power supply is on (DC bus voltage present).
- Confirm that the motor is connected firmly and properly.

After completion of the Pre-Drive checklist, activate the program and confirm that the supply voltage source is 24 V. If not stop the program and restart. With the correct DC voltage level established, turn on the converter (using the ON button, shown) and monitor the motor shaft and diagnostic scope, which in this example shown (Fig. 3.67) is set to show the per unit current components $i_{\alpha}^{n}, i_{\beta}^{n}$. Observation

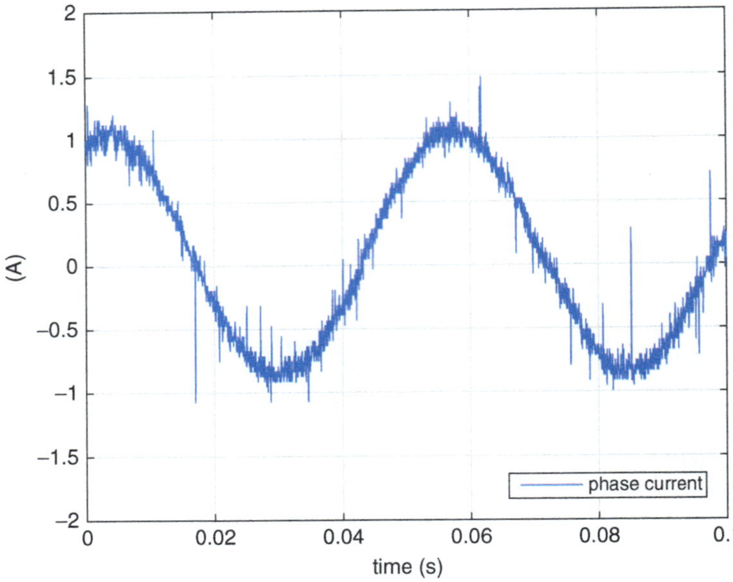

Fig. 3.69 Phase C+ experimental result: phase current waveform during no-load (friction only) operation. Measured with Tektronix DC-true current probe

of Fig. 3.67 learns that the machine is currently operating under no (external) load (friction only of both machines), as can be deduced from the stator frequency and shaft speed values. These show that the slip, defined as $s = {(n_{\mathrm{m}}^{\mathrm{syn}} - n_{\mathrm{m}})}/{n_{\mathrm{m}}^{\mathrm{syn}}}$, is equal to 0.067 per unit. Furthermore, the synchronous shaft speed of the machine $n_{\mathrm{m}}^{\mathrm{syn}} = {60 f_{\mathrm{s}}}/{p}$ is equal to 547.4 rpm.

Additional experimental evidence of drive operation is provided by the oscilloscope plot given in Fig. 3.69, which shows the phase current as measured with a DC-true current probe. Operation conditions are as indicated in Fig. 3.67. The results shown confirm that the frequency and amplitude of the measured current are in agreement with the scope results given in Fig. 3.67. Note that in the latter mentioned plots the current is scaled by a factor $1/20$ (20 A is the full-scale current value in use), given that the result is shown in per unit values.

3.6 Laboratory 1:6: FOC Sensored Control of IM

Purpose of this laboratory is to implement sensored rotor flux based field oriented control of an induction motor, with torque or speed control. The LVACIMTR induction machine used for this laboratory does not have a shaft sensor hence use must be made of the encoder attached to the LVSERVOMTR PM motor. Both machines must therefore be mechanically connected using the coupling assembly

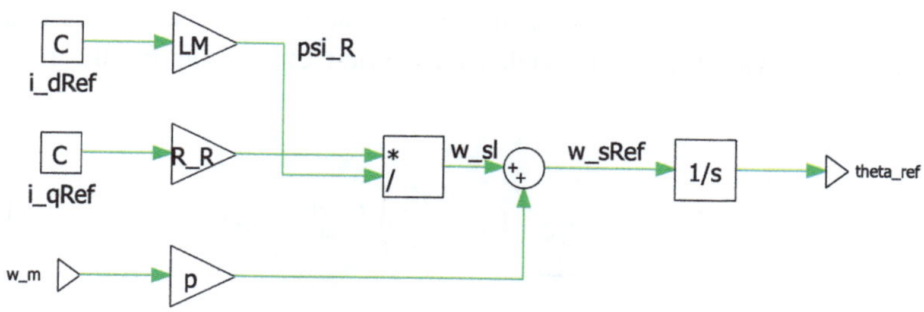

Fig. 3.70 Angle estimator for FOC induction machine control, with reference direct axis current $i_{\rm d}^{\rm ref}$, quadrature current $i_{\rm q}^{\rm ref}$inputs, which makes use of the measured electrical shaft speed $\omega_{\rm m}$

provided for this purpose. Use is made of the current control approach discussed in laboratory 1:2 (see Sect. 3.1). For rotor flux based field oriented control the rotor flux angle needs to be calculated, which can be realized by making use of (for example) the measured shaft speed [4]. The angle estimator shown (by way of a PLECS model) in Fig. 3.70 makes use of the reference direct axis current `i_dRef`$=i_{\rm d}^{\rm ref}$, quadrature reference current `i_qRef`$=i_{\rm q}^{\rm ref}$ and measured shaft speed w_m$=\omega_{\rm m}$. The angle `theta_ref` generated by the estimator is used by the synchronous model based current controller deployed previously for the FOC PM controller (see laboratory 1:3). In this case the direct axis reference current controls the flux in the machine, whilst the quadrature component controls the shaft torque. The angle estimator satisfies the following set of equations [4]

$$\psi_{\rm R}^{\rm ref} = i_{\rm d}^{\rm ref} L_M \tag{3.9a}$$

$$\omega_{\rm s}^{\rm ref} \cong p\,\omega_{\rm m} + \frac{i_{\rm q}^{\rm ref} R_{\rm R}}{\psi_{\rm R}^{\rm ref}} \tag{3.9b}$$

where L_M, represents the magnetizing inductance of the four-parameter model, as discussed in Sect. 2.1.4 and $\omega_{\rm s}^{\rm ref}$ represents the stator frequency reference variable (rad/s). The quadrature reference current $i_{\rm q}^{\rm ref}$ follows from the torque expression

$$T_{\rm m} = \frac{3}{2} p\, i_{\rm q}^{\rm ref}\, \psi_{\rm R}^{\rm ref} \tag{3.10}$$

where p represents the number of pole pairs of the machine and $T_{\rm m}$ the mechanical shaft torque. The speed controller, with gains $\rm K_p$, $\rm K_i$ can in the continuous time domain be represented as

$$T_{\rm m} = \left(\rm K_p + \frac{K_i}{\mathit{s}}\right)\left(\omega_{\rm m}^{\rm ref} - \omega_{\rm m}\right) \tag{3.11}$$

where ω_m is the actual shaft speed of the machine (rad/s). Use of said equation in combination with Eq. (3.10) yields an explicit expression for the reference quadrature current i_q^{ref} namely

$$i_q^{ref} = \underbrace{\frac{2}{3p^2\psi_R^{ref}}}_{K_{pi}} \left(K_p + \frac{K_i}{s}\right)\left(\omega_m^{ref} - \omega_m\right) \tag{3.12}$$

where a gain parameter K_{pi} is introduced, which will be subsequently used for fixed point controller implementation. Expression (3.12) defines the quadrature current input for the synchronous current controller based on the actual (from an encoder in this case) and reference electrical shaft speed when speed control is required. This expression together with Eq. (3.9) must be scaled using the approach set out in Sect. 1.4.2, which after some manipulation gives

$$f_{sn}^{ref} = f_{en} + \left(\frac{R_R}{2\pi L_M f_{fs}}\right)\frac{i_{qn}^{ref}}{i_{dn}^{ref}} \tag{3.13a}$$

$$i_{qn}^{ref}(t_k) \cong \left(K_{pi}\frac{2\pi f_{fs}}{i_{fs}}\right)\left(K_p + K_i T_{sp}\right)\Delta f_n + i_{qnINT}(t_{k-1}) \tag{3.13b}$$

with $\Delta f_n = f_{en}^{ref} - f_{en}$, where f_{en} is the per unit electrical shaft frequency. The discretization of Eq. (3.13a) makes use of expression (3.9), whilst Eq. (3.13b) is based on the use of expression (3.12). A variable T_{sp} is introduced in the latter expression, which represents the time step used for the discrete speed controller module.

Note that the controller requires access to the parameters of the machine in use. These are assumed to be known (measured using the InstaSPIN motor identification routine, to be discussed in the next chapter).

3.6.1 *Lab 1:6: Phase B*

This module considers the simulation of the IM motor drive in use, with a fixed point FOC and speed controller as shown in Fig. 3.71 The following information is relevant for this laboratory component:

- Reference program [11]: `lab1-6_LaunchXLphBv2.vsm`.
- Description: Field Oriented sensored control of an IM motor.
- Equipment/Software: VisSim simulation program.

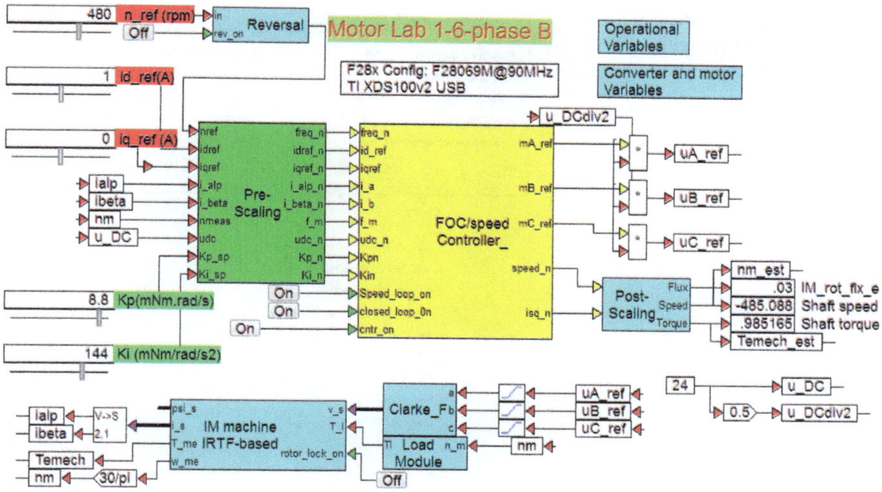

Fig. 3.71 Phase B simulation of a drive with a FOC/Speed controller

- Outcomes: Show the various waveforms that occur in the drive when using a fixed point representation of the controller when operating under field oriented control with a shaft speed sensor.

Scaling of the currents/voltages as discussed in Sect. 1.4.2 is carried out using a 'Pre-scale' module. The outputs of the controller are the three reference modulation indices `mA_ref`, `mB_ref`, `mC_ref`. These variables are then multiplied by the term $u_{DC}/2$ and connected to the 'converter', which in this case is simply represented by three limiters, that limit the voltage to the motor to $\pm u_{DC}/2$. A Forward Clarke transformation is again used to generate the voltage vector for the machine model. Outputs of the machine model are the currents i_α, i_β shaft torque T_m and shaft speed ω_m. The latter variable is used by the controller to calculate the appropriate slip frequency of the machine.

A set of sliders is used to define the current reference vector, in terms of its direct and quadrature components i_d^{ref}, i_q^{ref} and rotational speed n_m^{ref}. A 'pre-scaling' module (as discussed in Sect. 1.4.2) has been added to convert the floating point variables required for the controller to fixed point format. In addition, two sliders have been added which allow the user to set the proportional K_p^{sp} and integral K_i^{sp} speed gains of the controller. Note that said variables are shown in (milli)Nms/rad and (milli)Nm/rad respectively. A guideline for choosing these parameters is given in Sect. 2.1.2.

The two dialog boxes for this laboratory, as shown in Fig. 3.72, contain all the user settings required for this laboratory. The 'Operational Variables' dialog box is again used to assign full scale values for the voltage, current and frequency. In addition, the current controller gain and bandwidth are also assigned here. These values are based on the motor in use, as was discussed in Sect. 2.1.5. A dialog box entry for the maximum modulation index value and speed controller

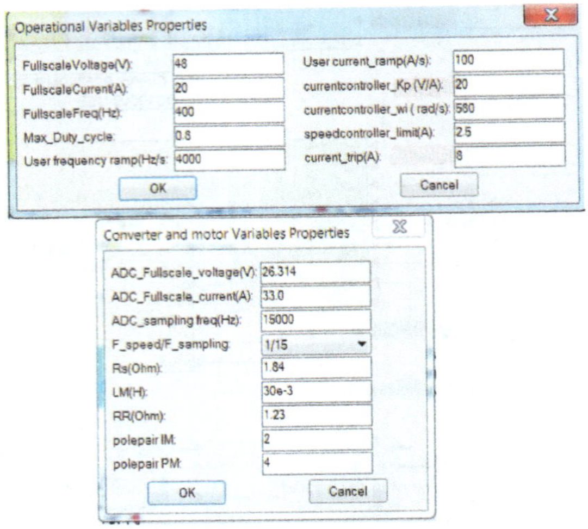

Fig. 3.72 Phase B: Dialog boxes used for drive with a FOC/speed IM controller

limit value `speedcontroller_limit(A)` are also shown. The latter value is the maximum (absolute) quadrature current value which the speed controller can provide. Also assigned in this box are the rate of variable change for current and frequency, i.e. introduced as a safety measure to avoid erratic slider action from causing over-currents in the drive.

The 'Converter Variables' dialog box assigns the motor pole pair number and required motor parameters (four parameter model representation) of the LVACIMTR motor in use. These are used by the angle estimator to calculate the rotor flux angle. In addition, the sample frequency used for the speed controller module (set in terms of a ratio relative to the ADC sample frequency) is also indicated in this dialog box for convenience. Note that the actual ADC sampling rate is set in the VisSim 'Systems property dialog box'. Finally the ADC full-scale voltage and current parameters are also set in this dialog box (but not used in this phase), given that the same parameter module is also used in subsequent development phases.

An example of drive operation is shown in Fig. 3.73. The sampling time T_s as defined in the program pull-down menu is set to 0.066 ms, i.e. a 15 kHz sampling frequency is used for the discrete controller model. Subplot (a) of Fig. 3.73, shows the motor phase current during start up and a speed reversal $n = 480 \rightarrow -480$ rpm, under sensored, field-oriented, speed control, which is initiated at $t \approx 0.25$ s. In this case the simulation was run over a 0.5 s time interval, with a motor inertia $J = 86.6 \cdot 10^{-6}\,\mathrm{kgm}^2$ (taken to be the estimated combined inertia of the PM/IM motor unit). Operation under no-load is assumed. Subplot (b) shows the actual and reference shaft speed for a given frequency ramp setting (as set in the dialog box). Finally, subplot (c) shows the actual shaft torque of the machine over the 0.5 s simulation interval.

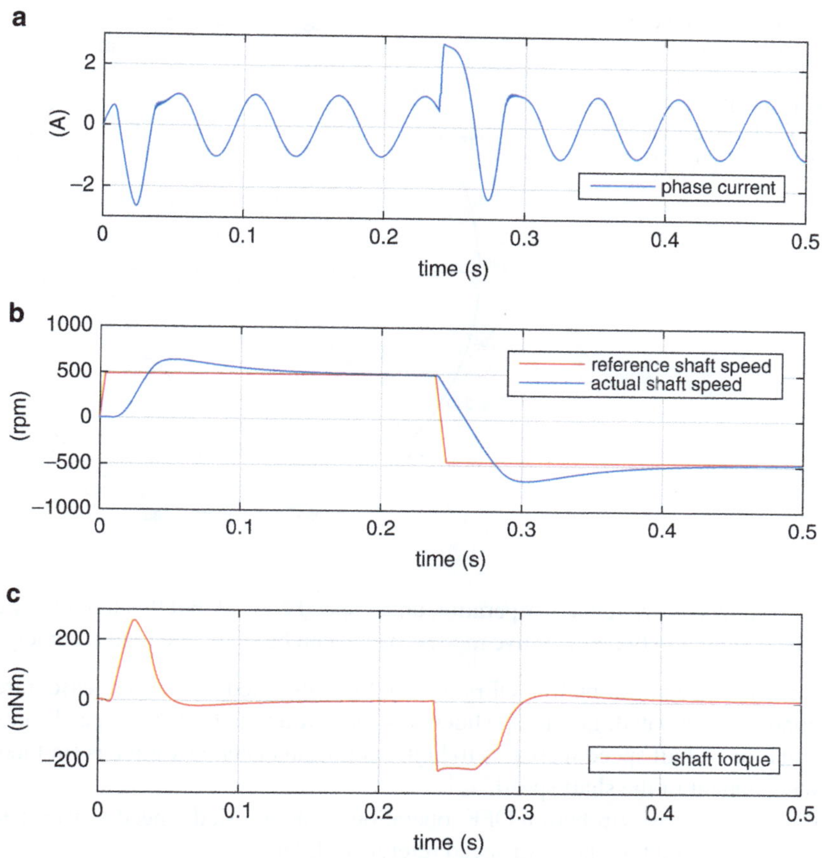

Fig. 3.73 Phase B: Typical waveforms obtained using a FOC/speed IM controller

Some interesting observations can be made from the results shown in Fig. 3.37 during the speed reversal, namely:

- the maximum phase current is determined by the maximum i_q^{max} value set in the speed controller and the chosen direct axis reference current value. Hence with $i_d^{ref} = 1.0$, $i_q^{max} = 2.5$ the maximum peak phase current will be equal to $i^{max} = \sqrt{1.0^2 + 2.5^2} = 2.69\,\text{A}$.
- The phase current will be equal to the selected direct axis reference current value when the shaft speed is constant, because no mechanical load is attached, i.e. 1.0 A.
- The reference speed ramp is higher than achievable by the drive, given the inertia present.
- The maximum shaft torque during the reversal can be written as $1.5\,p\,L_M\,i_d^{ref}\,i_q^{max}$, which is equal to 225 mNm in this case, given the chosen direct axis reference current setting of 1.0 A.

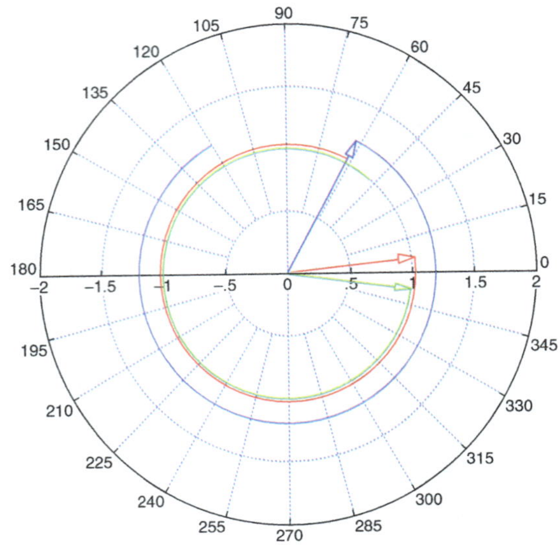

Fig. 3.74 Phase B: Vector plot obtained using a FOC/Speed controller with current $\vec{i}_s$('red'), voltage vector $\vec{u}_s/4$ ('blue') and unity d-axis vector ('green')

Results shown above relate to operation under speed control, but this laboratory can also be operated in two alternative modes, which can be set by the reader namely:

- `Speed_loop_on` button OFF: speed loop disabled, in which case torque control is activated, i.e. the i_q slider sets the torque. Note that a 'speed runway situation' (shaft speed increases to high value) can occur, as torque is (in theory) independent of the shaft speed.
- `Closed_loop_on` button OFF: operation without a shaft speed sensor, hence speed is dictated by the shaft speed reference slider.

In the sequel to this section a vector plot is shown of the drive in question operating under steady state conditions, with a constant speed of $n^{\text{ref}} = 480$ rpm and arbitrarily chosen mechanical load of 22.9 mNm. Said vector plot given in Fig. 3.74, shows the current vector $\vec{i}_s$ and voltage vector $\vec{u}_s$, whereby the latter is scaled by a factor 4. In addition, a unity 'd-axis' vector is introduced to show the orientation of the d,q reference frame, where the positive d-axis is displayed as a 'green' vector, which is aligned with the rotor flux vector. Its magnitude is set by the product of the direct axis current and magnetizing inductance L_M. In this example, a current i_q value is generated given that the machine is producing torque to match the load torque applied. Consequently the current vector orientation relative to the d-axis vector is defined by the direct axis current value i_d and the quadrature current i_q, as is indeed the case. The voltage vector is made up of a back EMF component that is orthogonal to the d-axis vector and a resistive voltage component, which is aligned with said current vector. The third component of the voltage vector is due to the inductive component of the machine which is orthogonal to the current vector and its value will depend (among others) on the electrical frequency.

The reader is again encouraged to undertake a detailed examination of this simulation using the VisSim Simulation platform and this specific file. Only by becoming familiar with each aspect of operation and the program used, can one hope to understand the modeling process of the drive as a whole. Note that further information on the controller implementation is given in the next subsection.

3.6.2 Lab 1:6: Phase C

In the previous phase, only knowledge of the control module to be used was required, in order to test the fixed point controller. The following information is relevant for this laboratory component:

- Reference program [11]: `lab1-6_LaunchXLphCv2.vsm`.
- Description: Field-oriented, sensored control of a IM motor.
- Equipment/Software: LAUNCHXL-F28069M, with BOOSTXL-DRV8301 module ('aft' position) and VisSim simulation program.
- Outcomes: Develop a complete drive algorithm, which handles field-oriented current control, voltage measurement, converter PWM and setup MCU timing, as required. Compile and download `.out` file for use in phase C+.

As mentioned above, a single boost pack located in the 'aft' position (furthest away from the USB connector) is used for single motor operation, in which case the following jumper and dip-switch positions on the LAUNCHXL-F28069M module are required:

- Jumpers `JP1` and `JP2` OPEN
- Jumpers `JP4` and `JP5` CLOSED
- Jumpers `JP3`, `JP6` and `JP7` CLOSED
- Dip-switches `SW1` to `SW3` ON

Furthermore, the 'J4' encoder connector of the PM machine cable should be attached to encoder input `QEP_A` of the LAUNCHXL-F28069M module. A phase C development stage, cannot be used to run a drive, but its primary task is to assemble all the modules needed for drive operation with a given controller configuration. The drive setup given in Fig. 3.75, shows the 'FOC & Speed Controller' module, which must be compiled to generate an `.out` file. Inputs to the controller module are (among others) the per unit user reference currents `id_n`, `iq_n`. In addition, an electrical shaft speed reference `freq_n` variable is used to set the shaft speed of the drive. The remaining input variables, are the speed-loop gains, as discussed in the previous section. All inputs to the pre-scaling module are, for the sake of convenience shown as 'constants' whereas in the subsequent phase they will be replaced by slider inputs. Four logic inputs are provided to the module, which allows the user to turn on the converter, select torque or speed control, open-loop or closed loop (with shaft speed sensor) operation and choose which combination of diagnostic channels is to be viewed.

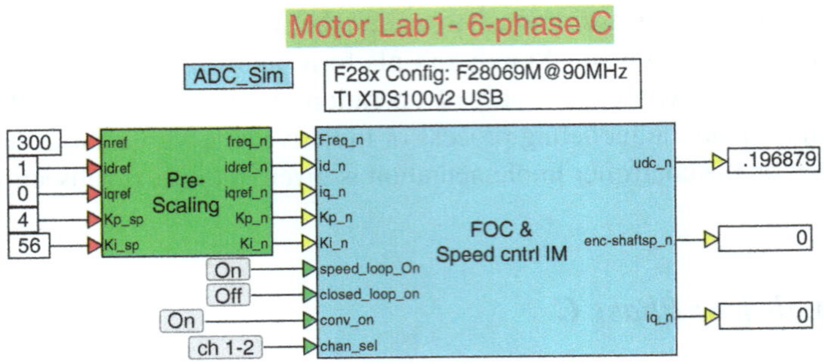

Fig. 3.75 Phase C simulation of a drive with a field-oriented current controller

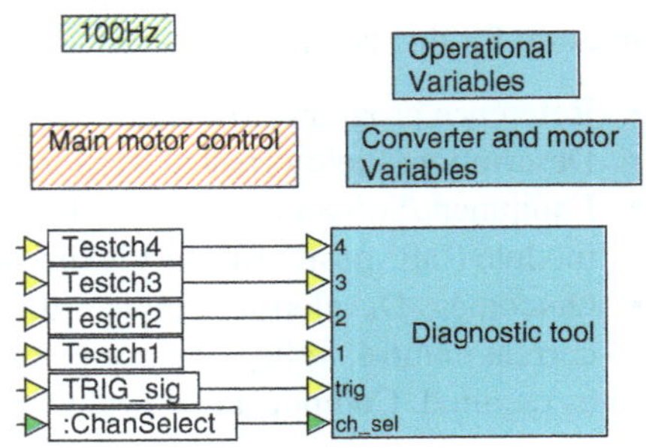

Fig. 3.76 One level down into the controller module

Outputs of the controller module are the per unit bus voltage together with the per unit shaft speed, provided by the encoder and measured quadrature current `iq_n`. The latter variable is used to calculate the torque, given that the rotor flux of the machine is calculated on the basis of the reference direct axis current and magnetizing inductance L_M.

Moving one level down into the controller module leads to the set of modules/dialog boxes shown in Fig. 3.76. The dialog boxes shown in this diagram have already been shown in Fig. 3.72, hence the reader is referred to this section for further details. A 'Diagnostic tool' module is again used to buffer to variables, which can subsequently be displayed in a graph when operating in phase C+. A two channel multiplexer is used to allow the user to either display variables `Testch1`, `Testch2` or `Testch3`, `Testch4` by a button connected to the `ch_sel` button input of the module.

A '100 Hz' module acts as a 'heart beat' and flashes a 'red' LED on the LAUNCHXL-F28069M board. A 'blue' LED on said board remains ON if a current trip condition occurs. In addition, all the background tasks of the drive such as ADC offset control is executed in this unit. Moving one level lower into the 'Main motor Control' module reveals a set of modules, shown in Fig. 3.77. Of these, the 'FOC/Speed controller' was introduced in phase B. The 'ADC-PWM' unit shown, is identical to the one used in the previous laboratory given that the same drive board is in use. Furthermore, a PWM frequency of 30 kHz is used, together with an ADC sampling frequency of 15 kHz that is set in the

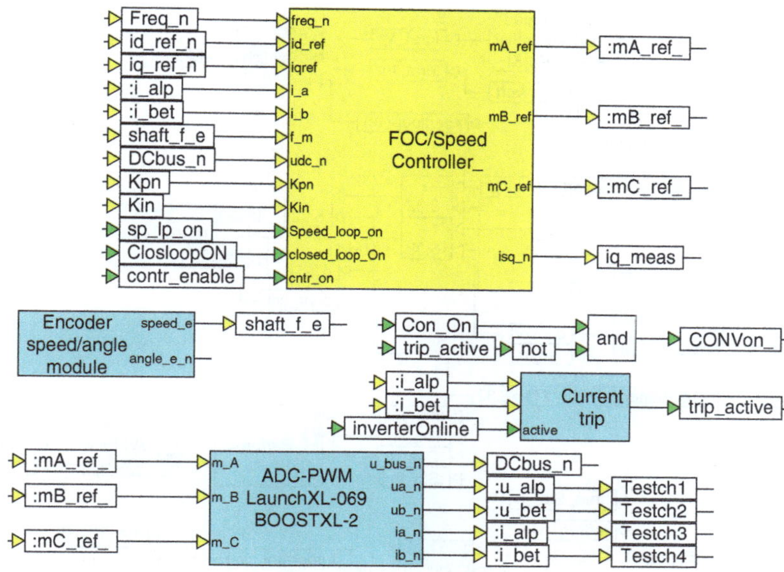

Fig. 3.77 One level down into the 'Main motor control' module

VisSim 'System Properties' pull down menu. This module is controlled by a `CONVon_` signal, which is in turn generated by a logic `and` output that has as inputs a user variable `Con_On` ('converter on') signal and a over-current trip signal `trip_active`. Relevant (to this laboratory) outputs of the 'ADC-PWM' unit are the DC bus voltage `DCbus_n`, and the per unit phase currents i_{α}^{n}, i_{β}^{n}, represented by the variables `i_alp`,`i_bet` respectively. An output flag `contr_enable` (control active) activates the FOC/Speed controller, after the converter has been turned on and internal ADC/PWM initialization actions have been performed. An 'Encoder Speed/angle' module, generates the per unit electrical shaft speed `shaft_f_e` (which is equal to f_e = `shaft_f_e` f_{sc}, with $f_e = n_m p/60$ Hz. Inputs to the 'ADC-PWM' module are the modulation indices generated by the current/speed controller module. For diagnostic purposes the per unit current and voltage variables have been connected to variables `Testch1`, `Testch2` and `Testch3`, `Testch4` respectively, so that these can be displayed in phase C+.

Moving one level lower into the 'FOC/Speed Controller' module reveals a set of modules given in Fig. 3.78. Synchronous model based current control is undertaken by a 'PI-kp-wi-lim' module, which has as inputs the per unit quadrature and direct axis current error and (indirectly) the inverse DC bus voltage. The latter is generated by the '1/x numerical' module, which generates the per unit inverse DC bus voltage using a numerical routine. A Reverse Park transformation is used to generate the modulation indices m_{α}, m_{β}, from the outputs of the current controller, which are then processed by a 'Pulse Center (SVM)' module. As in the previous laboratories a Cosine/Sine function, with variables `Cos_th`, `Sin_th` is used instead of the instantaneous angle of the current reference vector. These two variables are then

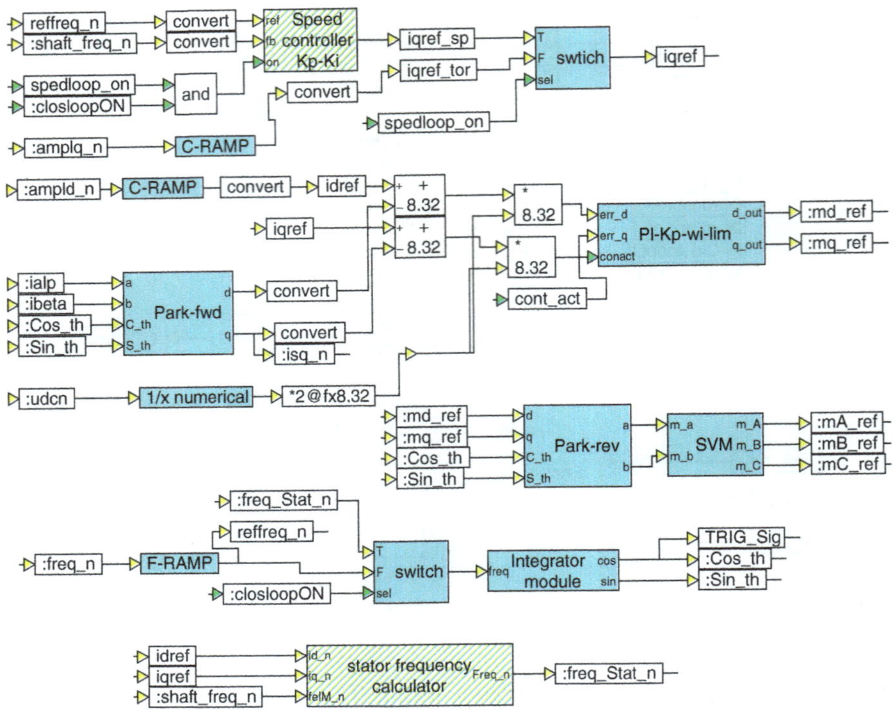

Fig. 3.78 Fixed point representation of the FOC/Speed controller

consistently used for subsequent Park transformations, which have been adapted to allow the use of Cosine/Sine instead. A 'switch' module, activated by the `closloopON` signal (active if the shaft encoder is in use) is used to select the integrator input variable. When `closloopON` is active the integrator input is provided by the 'stator frequency calculator' module. This module implements the per unit stator frequency output according to Eq. (3.13a). This module is shown 'hatched green', which implies that its sample rate is different to the remaining part of the controller. In this case the sample frequency is set to 1 kHz. Inputs to the speed controller are the (rate limited) reference frequency `reffreq_n` and actual shaft frequency (from the shaft encoder) `shaft_freq_n`. Output of this module (when enabled) is the per unit quadrature reference current `iqref_sp`. Note that the speed controller module is also shown 'hatched green', which implies that its sample rate is different to the remaining part of the controller. In this case the sample frequency for the speed controller was also set to 1 kHz. For less time demanding activities it is prudent to use a lower sampling frequency, which serves to avoid MCU overrun errors that can occur when insufficient time is available to execute all the required functions, given the available sampling time T_s. A 'switch' module with output `iqref` determines whether the quadrature reference current for the current controller is provided by the speed controller or the user i_q slider. The switch in question is controlled by the `speedloop_on` signal (active when closed speed-loop control is required, OFF implies operation in constant torque mode). A variable `TRIG_sig` is connected to the variable `Cos_th` and is used by the diagnostic scope to trigger the buffering process. A set of 'Rate Limiter' modules, entitled 'C-ramp' (sets the current ramp rate) and 'F-ramp' (sets the frequency ramp rate), have been introduced in Fig. 3.78 to avoid erratic changes to the user input variables (as actually used by the controller), which may damage the converter or motor. The rate of variable change is specified in the 'Operations' dialog box (see Fig. 3.72).

3.6.3 Lab 1:6: Phase C+

Phase C+, is the drive operational component of the laboratory and is basically a run version of the `.out` file compiled and downloaded to the MCU in phase C (see previous subsection).
The following information is relevant for this laboratory component:

- Reference program [11]: `lab1-6_LaunchXLphCv2_d.vsm`.
- Description: Field-oriented current/speed control of a IM motor.
- Equipment/Software: LAUNCHXL-F28069M, with BOOSTXL-DRV8301 module ('aft' position), Texas Instruments LVACIMTR IM machine connected to LVSERVOMTR PM machine and VisSim simulation program.
- Outcomes: To achieve sensored, field-oriented current/speed control of an IM drive.

Note that the required jumper and dip-switch settings for the LAUNCHXL-F28069M module are given in phase C. Furthermore, the 'J4' encoder connector of the PM machine cable should be attached to encoder input `QEP_A` of the LAUNCHXL-F28069M module.

The run version, shown in Fig. 3.79, uses a VisSim run module, which executes the `.out` file, shown in said module. Five sliders are used to set the current reference amplitudes ($i_{\mathrm{d}}^{\mathrm{ref}}$, $i_{\mathrm{q}}^{\mathrm{ref}}$), instantaneous rotational shaft speed and speed slider gains. Two post-scaling modules are used to display the measured DC bus voltage, rotor flux, shaft speed and torque (both instantaneous and filtered using a 10 Hz low pass filter). The filtered torque signal is displayed on a torque meter in (milli)Nm. Two 'Monitor Buffer' modules are used to display two selected (using a button connected to the `channel_sel` input). Note that wiring instructions ('Terminal A-blue' wire, etc.) are included in Fig. 3.79, given that use is made of a shaft encoder. This means that the chosen positive rotational direction of the shaft encoder

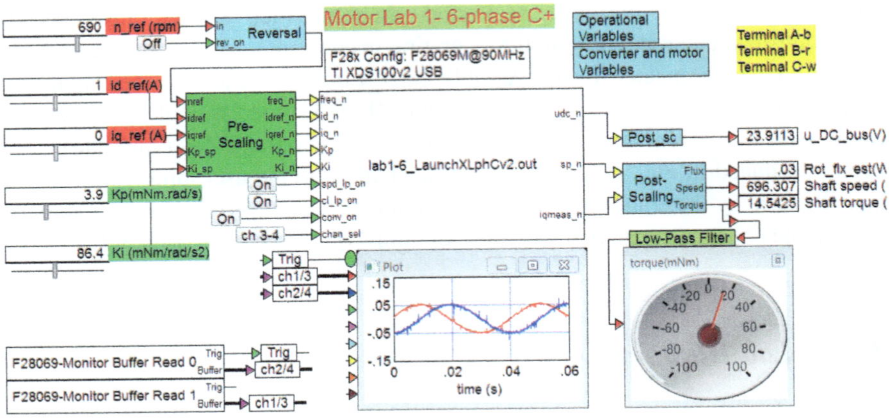

Fig. 3.79 Phase C+ operation of an IM drive with a field-oriented FOC/speed controller

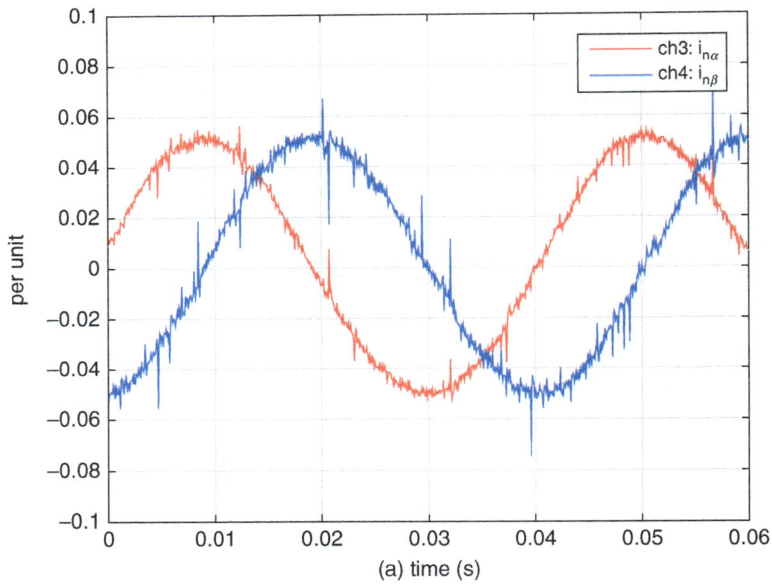

Fig. 3.80 Phase C+ operation of a drive: VisSim scope results with channel select option: 'ch 3-4'

must match that of the induction machine. The VisSim scope module shows the per unit α, β currents and an enlarged view of these results is given in Fig. 3.80. Multiplication of these waveforms by the full scale current value of $i_{fs} = 20$ A gives actual current values.

Prior to activating the drive the following 'Pre-Drive' check list should be executed:

- Dialog boxes used in C+ model match those of C: the run version is compiled with the dialog box entries specified under phase C. The dialog boxes show in C+ are used by the 'Pre/Post' scaling modules.
- Ensure that the sample time used is correct and latest (and correct) `.out` file has been downloaded to the MCU (Right mouse click on MCU module to show dialog box of these variables).
- Confirm that the user input values are set to either zero, or 'acceptable' values, which will not cause a current trip of the converter
- Confirm that the converter 'switch' is set to OFF and the power supply is on (DC bus voltage present).
- Confirm that the motor is connected firmly and properly.

After completion of the Pre-Drive checklist, activate the program and confirm that the supply voltage source is 24 V. If not stop the program and restart. With the correct DC voltage level established, turn on the converter (using the ON button, shown) and monitor the motor shaft and diagnostic scope, which in this example shown (Fig. 3.79) is set to show the per unit current components $i^{\mathrm{n}}_{\alpha}, i^{\mathrm{n}}_{\beta}$. Note that the actual current values are found multiplying the per unit value with the full-scale current value set to 20 A in this case. Observation of Fig. 3.79 learns that the machine is currently operating under closed-loop speed control and no external load (except friction), as is evident from the torque meter value and the (per unit) i^{n}_{α}, i^{n}_{β} currents displayed in the diagnostic scope. Operation with $i^{\mathrm{ref}}_{\mathrm{d}} = 1.0$ A is undertaken in this case. Additional experimental verification of this laboratory was carried out with the aid of an oscilloscope and DC-true current probe. The latter was used to measure the phase current, as shown in Fig. 3.81 with the drive undertaking a speed reversal. The conditions used were similar to those used in phase B (see Fig. 3.73). The result shown in Fig. 3.81 is consistent with those observed earlier during the phase B simulation. Note that the measured results given in Fig. 3.81 display a DC offset, which is due to the measuring equipment.

3.7 Laboratory 1:7: Dual Control of a IM and PM Machine

The final laboratory of this chapter considers dual operation of an induction and permanent magnet machine. For the PM machine a sensored field-oriented controller is used, as discussed in lab 1:3 (see Sect. 3.3), which will be configured to operate under torque control only (given its use as a dynamometer). The PM machine is mechanically connected to the induction machine, which is connected to a converter operating with a sensored field-oriented controller and speed controller (using shaft speed data from the encoder connected to the PM machine) as discussed in lab 1:6 (see Sect. 3.6).

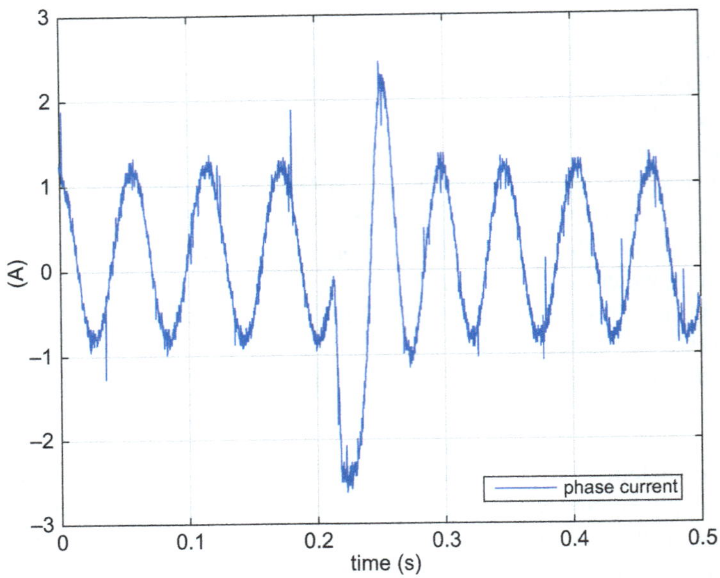

Fig. 3.81 Phase C+ experimental result: phase current waveform during IM FOC (sensored) operation under no external load (friction only). Measured with Tektronix DC-true current probe

The purpose of this laboratory is to explore dual machine operation, where, for example, one machine is acting as a load and the other as a motor. Care should be taken in this case to ensure that both machines are not operating as motors simultaneously, given that a speed runaway situation could then occur. In this laboratory use is made of a LAUNCHXL-F28069M board, with two BOOSTXL-DRV8301 modules (see Fig. 3.82), which are connected to a Texas Instruments LVACIMTR induction machine and LVSERVOMTR PM machine respectively. The 'forward' (closest to the USB connector) BOOSTXL-DRV8301 module is typically used to implement the 'dynamometer' function, i.e. where use is made of a FOC sensored PM machine. Also shown in this figure are a DC power supply (which can be replaced by a 24 V mains adapter), DC-true current probe and oscilloscope which are used to verify experimental results. Development phases B, C and C+ are shown in this laboratory, given that the switching process does not fundamentally affect the control behavior of the drives. The encoder connected to the PM machine provides shaft speed and shaft angle information for both FOC drives.

3.7.1 Lab 1:7: Phase B

This module considers the simulation of the dual machine drive in use, with two fixed point FOC controllers, as shown in Fig. 3.83. Only a brief summary of this

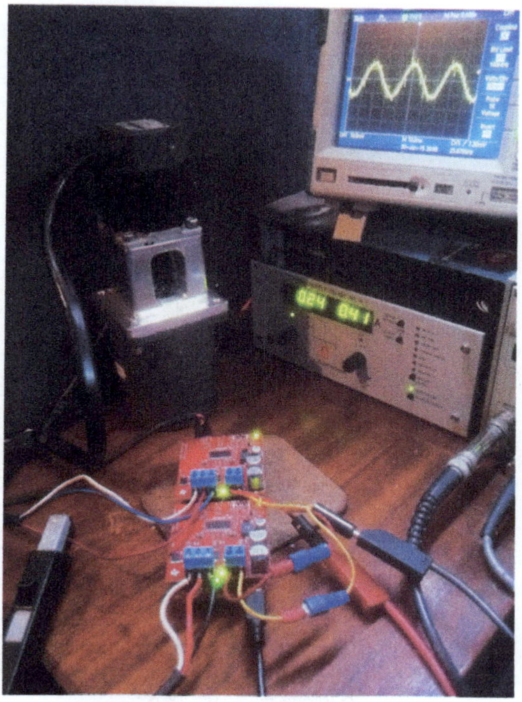

Fig. 3.82 Typical experimental setup for dual machine operation

model is provided here, given that extensive treatment of both controllers has been provided earlier (see lab 1:3 and 1:6 in Sects. 3.3, 3.6 respectively). The following information is relevant for this laboratory component:

- Reference program [11]: `lab1-7_LaunchXLphBv3.vsm`.
- Description: Dual FOC/speed control of a IM and PM machine.
- Equipment/Software: VisSim simulation program.
- Outcomes: Show the various waveforms that occur in the drive when using a fixed point representation of a dual PM/IM controller.

Central to Fig. 3.83 are the two controller modules 'PM FOC controller' and 'IM FOC/speed Controller', which are connected to the 'PM machine' and 'IM machine' models respectively. A non-flexible coupling is assumed between the two machines, in which case the relation between torque and shaft speed/angle can be written as

$$T_{\rm m1} + T_{\rm m2} = J\frac{d\omega_{\rm m}}{dt} \tag{3.14}$$

where $T_{\rm m1}$, $T_{\rm m2}$ represent the shaft torque of the induction and PM machine respectively. Furthermore, J, $\omega_{\rm m}$ represents the combined machine inertia and shaft speed (rad/s) of the drive respectively. The 'Machine Coupling' module

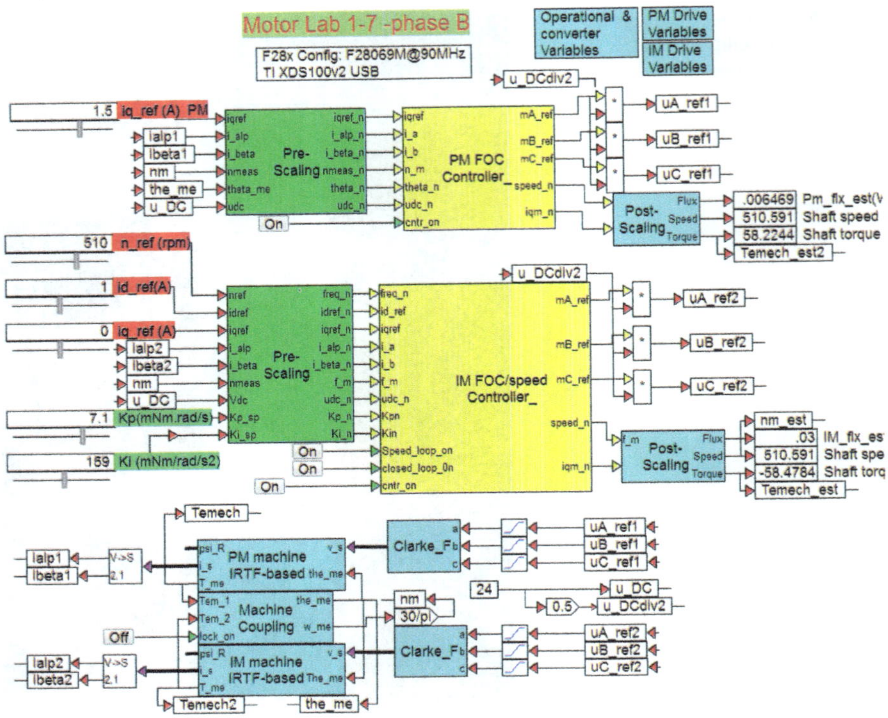

Fig. 3.83 Phase B simulation of a dual FOC based IM/PM machine drive

(see Fig. 3.83) adheres to Eq. (3.14) and also generates the shaft angle $\theta_m = {d\omega_m}/{dt}$. A `lock_on` function is available to simulate a 'locked rotor' condition. Note that the PM/IM machine models used in this lab, differ from those used in previous laboratories, given that a combined inertia value is now assigned in the 'Machine Coupling' module. Two 'Pre-scaling' modules are used to convert the floating point user variables to fixed point format for each controller. One slider is allocated to the PM controller, which sets the reference quadrature current value. This implies this slider can be used to realize motor or generator (in which case it acts as a load for the IM drive) operation of the PM machine. The corresponding PM direct axis reference current is set to zero. In addition, one logic input is used to enable the PM controller (`cntr_on` input). Five sliders and three control buttons are allocated to the IM FOC controller, which are used to set the direct/quadrature reference currents, speed reference and proportional/integral speed gains. Note that the IM drive requires a direct axis current reference given that the latter controls the rotor flux magnitude. A single 24 V DC supply is used for both converters, whilst two 'Post-scaling' modules are used to convert the per unit converter outputs to SI values. Both controllers provide numerical displays, which show shaft speed (common to both machines) and shaft torque (milli-Nm). In addition, the PM flux and IM rotor flux values are also shown. The three dialog boxes given in Fig. 3.84, contain all the user

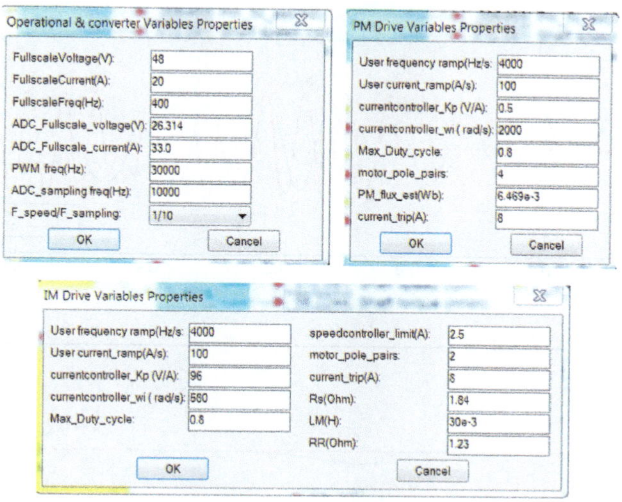

Fig. 3.84 Phase B; Dialog boxes used for the dual machine drive

settings required for this laboratory. The 'Operational Variables' dialog box is used to assign common (to both controllers) full scale values for the voltage, current, and frequency. In addition, the ADC full-scale current/voltage values (not used in phase B) and sampling frequency (as a ratio relative to the ADC sample frequency) for both speed loops are also assigned here. The dialog box 'PM Drive Variables', assigns the parameters needed for the PM controller. Hence, current controller gains, maximum modulation index and PM flux amplitude must be provided by the user. Likewise, a dialog box 'IM Drive variables' (also shown in Fig. 3.84) is used to assign the relevant parameters for the FOC controller of the IM drive. For both controllers the rate of change of control variables and pole pair number (which is different for both motors by a factor 2) is set in the corresponding PM/IM dialog boxes. Also the current controller gain and bandwidth for both controller is different, given that these are governed by the inductance and resistance of the machines in question, as was discussed in Sect. 2.1.5.

An example of drive operation as given in Fig. 3.85, corresponds to the following load scenario:

- time sequence $0 \rightarrow 0.2$ s: reference i_q for PM at zero.
- time sequence $0.2 \rightarrow 0.4$ s: reference $i_q = -1.5$ A for PM.
- time sequence $0.4 \rightarrow 0.6$ s: reference $i_q = 1.5$ A for PM.

The direct axis reference current for the induction machine is set to 1.0 A. Furthermore, a speed reference setting of 510 rpm is assumed. The corresponding phase current, actual/reference shaft speed and IM/PM shaft torque waveforms which appear for this drive scenario are shown in Fig. 3.85. The sampling time T_s as defined in the program pull-down menu is set to 0.1 ms, i.e. a 10 kHz sampling frequency is used for both discrete controller models whilst the speed module and

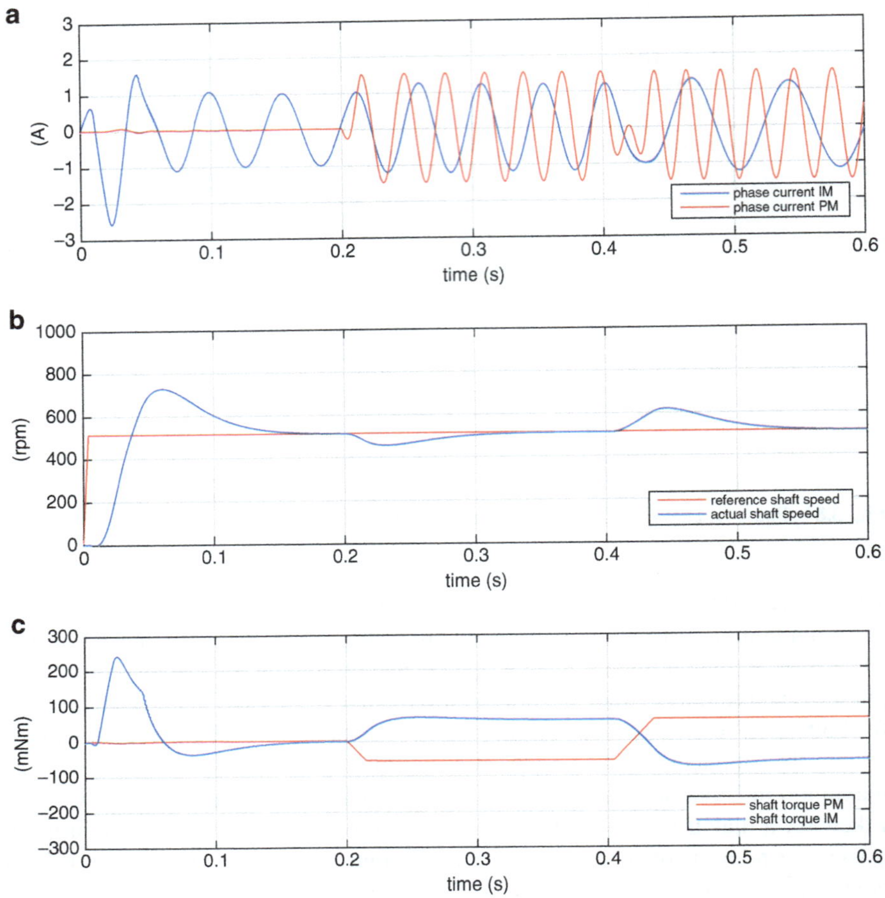

Fig. 3.85 Phase B: results for dual IM and PM drive

IM stator frequency calculator module are set to operate at 1 kHz. Some pertinent observations can be made from the results shown in simulation model (Fig. 3.83) and results (Fig. 3.85) namely:

- The operating speed of the drive closely matches the reference speed of $n_{\mathrm{m}}^{\mathrm{ref}}$= 510 rpm, set by the IM speed controller. Clearly noticeable are the effects of load changes, in which case the quadrature current of the induction machine is adjusted accordingly.
- During time interval 0.2 → 0.4 s the PM machine acts as a load, hence the slip of the IM machine can be calculated using the stator frequency f_{s}= 21.243 Hz (from observation of the current waveform), shaft speed n_{m} =510.0 rpm and IM pole pair number $p = 2$, which corresponds to a slip value of s= 0.1998 .

- During quasi-steady state operation the period time of the PM current waveform can be calculated using $T_{PM} = {}^{60}/_{(p\, n_m)} = 26.4$ ms, (p is the PM pole pair number), which agrees with the results shown in Fig. 3.85. Furthermore, the amplitude of the PM current waveform is 1.5 A, which agrees with the i_q^{ref} user slider setting, given that $i_d^{ref} = 0$.
- The PM shaft torque amplitude should be in accordance with torque expression $T_{PM} = 1.5\, p\, \psi_{PM}\, i_q^{ref} = \pm 58.2$ mNm, which is indeed the case.
- The torque produced by the IM will be the same when quasi-steady state operation is achieved (given that the friction torque of the machines is zero in this case), hence the quadrature current value generated by the speed controller can be calculated using $T_{IM} = 1.5\, p\, L_M\, i_d^{ref}\, i_q^{ref}$. The direct axis current is set to $i_d^{ref} = 1.0$ A which leads to a quadrature current of $i_q^{ref} = \pm 0.64$ A. Note that this value could also have been found by observation of the peak IM current $\hat{i}_{IM}$ in Fig. 3.85 and using $\hat{i}_{IM} = \sqrt{\left(i_d^{ref}\right)^2 + \left(i_q^{ref}\right)^2}$.

In the sequel to this section we also consider the quasi-steady state vector plots of both machines at time mark $t = 0.35$ s, which according to Fig. 3.83 implies that the induction and PM machine operate as a motor and generator respectively. Some interesting observations can be made with respect to the results shown in Fig. 3.86 namely:

- The (anti-clockwise) rotational speeds of the IM/PM vector plots are determined by the electrical frequencies in use for both machines and are in this case equal to 1273 rpm and 2040 rpm respectively.
- The induction machine current vector leads the unity d-axis vector (which is aligned with the rotor flux) hence its operates as a motor.

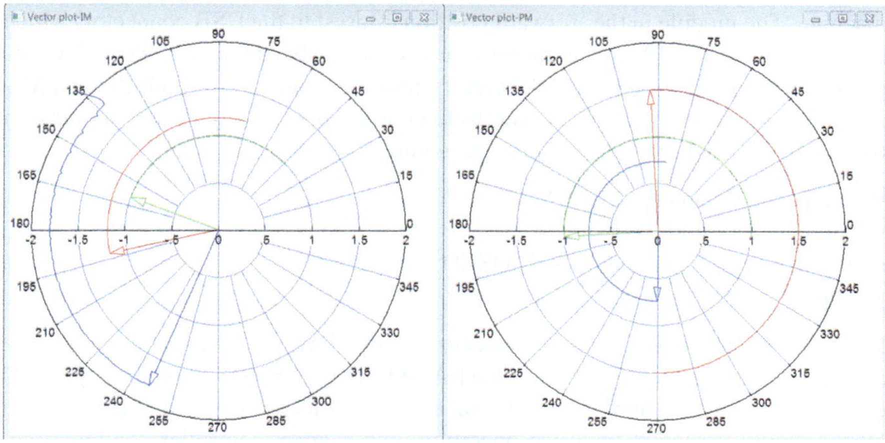

Fig. 3.86 Phase B: Vector plots of the IM (left) and PM machine, voltage vector ('blue'), current vector ('red'), d-axis vector (aligned with the flux) shown 'green'

- The PM machine current vector lags the unity d-axis vector (which is aligned with the PM flux) hence its operates as a generator, i.e machine is operating with a negative quadrature current.
- the voltage vector for the IM machine is attenuated by a factor 4. Both PM and IM voltage vectors lead the unity d-axis vector, given that they are governed by the EMF vector with amplitude $\omega_e \psi$, where ψ is the flux and ω_e the electrical frequency (rad/s).

The reader is again encouraged to undertake a detailed examination of this simulation using the VisSim Simulation platform and this specific file.

3.7.2 Lab 1:7: Phase C

In the previous phase, only knowledge of the control card to be used was required, in order to test the fixed point controller. In this lab a LAUNCHXL-F28069M board is used, together with a PM and IM machine. The following information is relevant for this laboratory component:

- Reference program [11]: `lab1-7_LaunchXLphCv4.vsm`.
- Description: Dual control of a IM and PM machine.
- Equipment/Software: LAUNCHXL-F28069M, BOOSTXL-DRV8301 module ('aft' position), BOOSTXL-DRV8301 module ('forward' position), DC power supply and VisSim simulation program.
- Outcomes: Develop a complete drive algorithm, which handles dual machine control. Compile and download and `.out` file for use in phase C+.

As mentioned above, two boost packs are used in this case, where the module in the 'aft' position (furthest away from the USB connector) is attached to the LVACIMTR machine. The module in the 'forward position (closest to the USB connector) which is connected to the PM machine will be used as a dynamometer. The 24 V DC power supply to be used for this laboratory must be connected to both BOOSTXL-DRV8301 modules in this case. The following jumper and dip-switch settings on the LAUNCHXL-F28069M module are required:

- Jumpers `JP1` and `JP2` OPEN
- Jumpers `JP4` and `JP5` OPEN
- Jumpers `JP3`, `JP6` and `JP7` CLOSED
- Dip-switches `SW1` to `SW3` ON

Furthermore, the 'J4' encoder connector of the PM machine cable should be attached to encoder input `QEP_A` of the LAUNCHXL-F28069M module. A phase C development stage, cannot be used to run a drive, but its primary task is to assemble all the modules needed for drive operation with a given controller configuration. The drive setup as given in Fig. 3.87, shows the 'drive 2: FOC IM, drive 1: FOC PM' module, which must be compiled to generate an `.out` file. Inputs to the

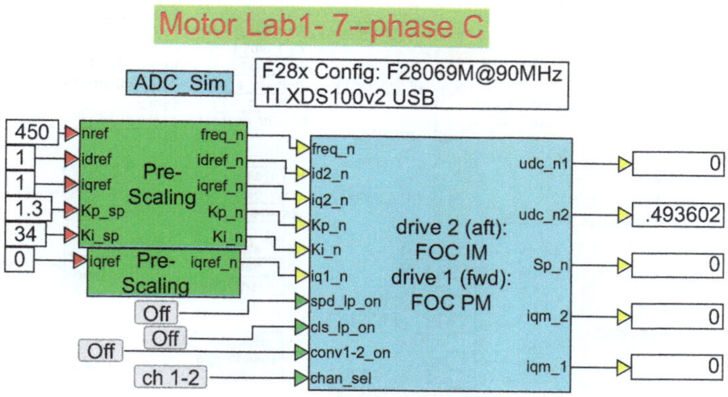

Fig. 3.87 Phase C simulation of a dual sensored IM/PM machine drive

Fig. 3.88 One level down into the controller module

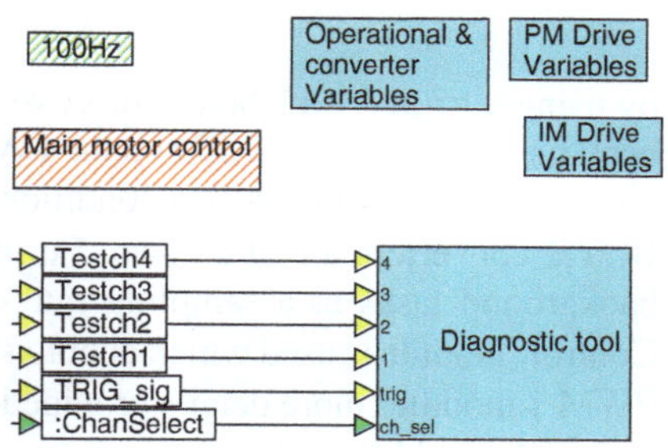

controller module (via the pre-scale modules) are six variables (shown as 'constants' for simplicity), which allow the user to set the direct axis/quadrature axis current references, speed reference and speed reference/control loop gains (IM drive only) for the IM drive and the quadrature axis reference current for the PM drive as discussed in the previous subsection. Four logic inputs are used of which one is assigned to enabling speed/torque control (via input `spd_lp_on`) and use of open or closed loop control (via input `cls_lp_on`) for the IM drive. The input `conv1-2_on` activates both boost packs and initializes the ADC module, after which both controllers are sequentially enabled. A button connected to `chan_sel` allows the user to toggle the diagnostic scope variables to be displayed on the diagnostic scope (phase C+). Five outputs of the Controller module show the per unit DC bus-voltages of both boost packs (`udc_n1` fwd boost, `udc_n2` aft boost), shaft speed, PM torque and IM torque (torque is calculated by making use of the per unit quadrature current values `iqm_1`, `iqm_2`).

Moving one level down into the Controller module leads to the set of modules/dialog boxes shown in Fig. 3.88. The dialog boxes shown in this diagram, have been discussed in the previous section (see Fig. 3.84). A 'Diagnostic tool' module is again used to buffer to variables, which can subsequently displayed in a graph when operating in phase C+. A two channel multiplexer is used to allow the user to either display variables `Testch1`, `Testch2` or `Testch3`, `Testch4`

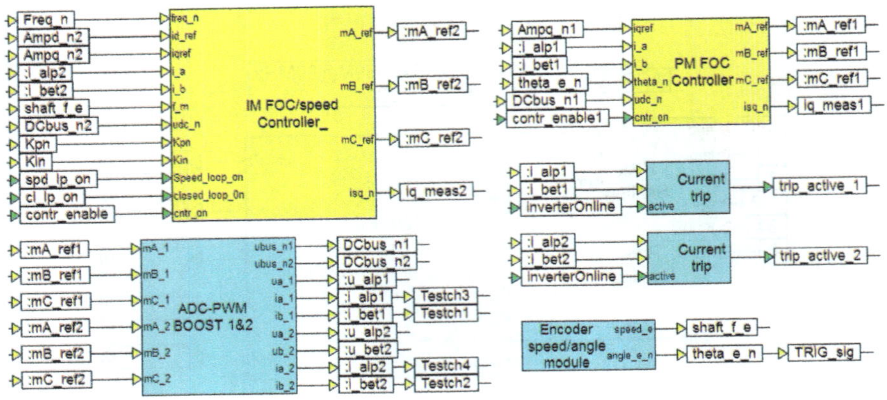

Fig. 3.89 One level down into the 'Main motor control' module

by using a `chan_sel` button discussed above. A '100 Hz' module acts as a 'heart beat' and flashes a 'red' LED on the LAUNCHXL-F28069M module. A 'blue' LED remains ON if a current trip condition occurs on either boost pack in which case BOTH converters are shutdown. Furthermore, this module executes all the drive background tasks at a sampling rate of 100 Hz. Note that a single 'Main motor Control' module is used which contains both drive controllers and is triggered by the PWM 1 module (more details provided in Sect. 4.5). An alternative implementation approach could have been to use two motor control modules (two 'red hatched' modules instead of one), each triggered by a PWM module linked to a boost pack. This approach is not chosen here as it doubles the interrupt overhead time for the MCU.

Moving one level lower into the 'Main motor Control' module reveals a set of modules, shown in Fig. 3.89. Of these, the 'IM FOC/speed controller' and 'PM FOC controller' are used to implement sensored FOC speed/torque control and torque control for the IM and PM machine respectively, as discussed in laboratories 1:3 and 1:6 respectively. The 'ADC-PWM' unit shown, generates the α, β current variables and per unit DC bus voltage for both converters. The per unit filtered α, β voltage variables `u_alp2`, `u_beta2` of the 'aft' boost (connected to the IM machine in this case) are also provided given that this module is also used for sensorless operation. For diagnostic purposes a single per unit low pass filtered converter voltage `u_alp1` of the 'fwd' boost is also generated. Furthermore, a PWM frequency of 30 kHz is used, together with a sampling frequency of 10 kHz that corresponds to an ADC sampling time of 0.1 ms, set in the VisSim 'System Properties' pull down menu (either frequency or sampling time can be set). Both converters are activated by the button connected to the `con1-2_on` input, after which ADC offset correction takes place. After the boost packs are activated the `inverterOnline` flag is set which arms the 'current trip' modules for both boost packs and sequentially enables both controllers. A current trip on either drive will shut down BOTH boost packs and sets the 'blue' trip LED on the LAUNCHXL-F28069M module (underneath the 'fwd' boost pack). Reset is possible via the `conv1-2` button. Relevant (to this laboratory) outputs of the 'ADC-PWM' module are the two DC bus voltages `DCbus_n1`, `DCbus_n2` which are used by the respective controllers. An 'Encoder Speed/angle' module, generates the per unit electrical shaft speed/angle variables `shaft_f_e`, `theta_e_n`, which are used for speed/FOC control of the IM and sensored FOC control of the PM machine respectively. The inputs to the 'ADC-PWM' module are the modulation indices generated by both controller modules. Note that in this case the ADC module samples six phase currents and 6 converter voltages, where use is made of so called 'dual sampling', which implies that two channels are sampled simultaneously (more information provided in the 'commissioning' laboratory, see Sect. 4.5). In addition, the DC busvoltages of both boost modules are also sampled by this ADC module. For diagnostic purposes the variables `TESTch3`, `TESTch4` have been allocated to per unit α current components of the IM and PM machine respectively. Likewise, the variables `TESTch1`, `TESTch2` have been allocated to per unit β current components of the PM and IM machine.

3.7.3 *Lab 1:7: Phase C+*

Phase C+, is the drive operational component of the laboratory and is basically a run version of the `.out` file compiled and downloaded to the MCU in phase C (see previous subsection).

The following information is relevant for this laboratory component:

- Reference program [11]: `lab1-7_LaunchXLphCv4_d.vsm`.
- Description: Dual FOC operation of a IM and PM drive.
- Equipment/Software: LAUNCHXL-F28069M, BOOSTXL-DRV8301 module ('aft' position) connected to Texas Instruments LVACIMTR machine, BOOSTXL-DRV8301 module ('forward' position) connected to LVSERVOMTR PM machine, DC power supply and VisSim simulation program.
- Outcomes: To investigate dual machine drive operation, using a sensored FOC PM and IM controller.

Note that the required jumper and dip-switch settings for the LAUNCHXL-F28069M module are given in phase C. Furthermore, the 'J4' encoder connector of the PM machine cable should be attached to encoder input `QEP_A` of the LAUNCHXL-F28069M module.

The run version shown in Fig. 3.90 uses a VisSim run module, which executes the `.out` file, shown in said module. The VisSim scope module shows the per unit α currents of both machines and an enlarged view of these results is given in Fig. 3.91. Multiplication of these waveforms by the full scale current value of i_{fs}= 20 A gives actual current values.

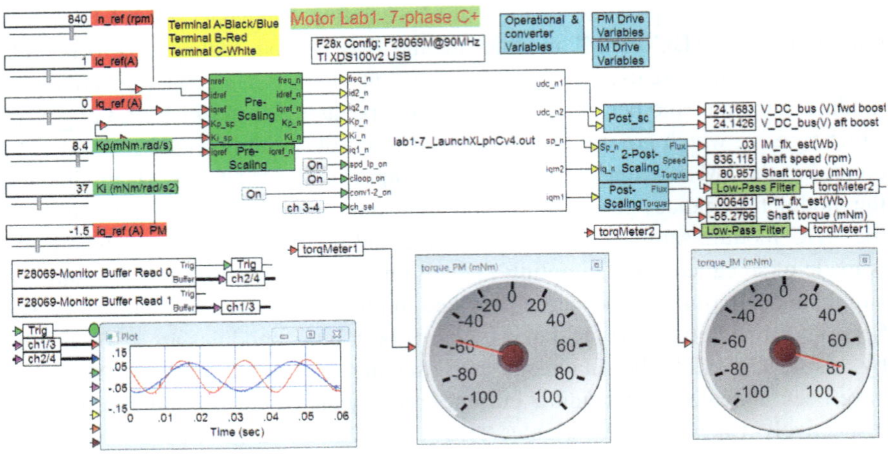

Fig. 3.90 Phase C+ dual sensored operation of a IM/PM drive

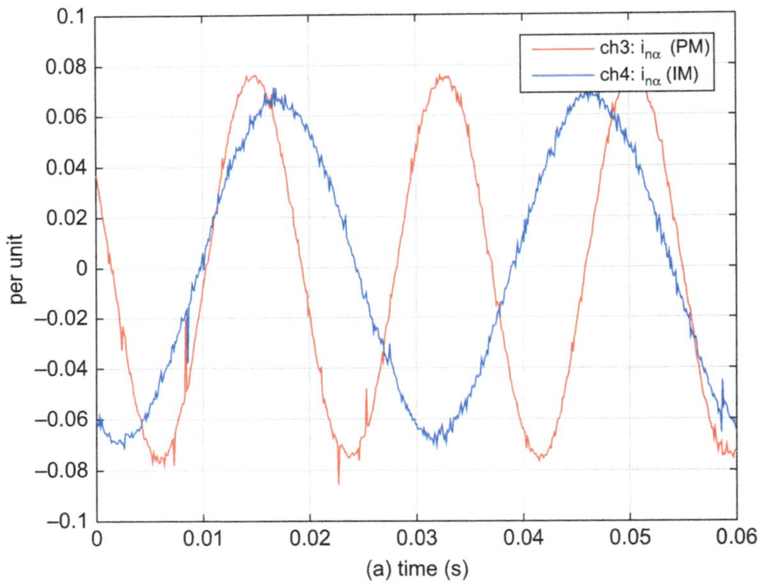

Fig. 3.91 Phase C+ operation of a drive: scope results with channel select option: 'ch 3-4'

Prior to activating the drive the following 'Pre-Drive' check list should be executed:

- Dialog boxes used in C+ model match those of C: the run version is compiled with the dialog box entries specified under phase C. The dialog boxes shown in C+ are used by the 'Pre/Post' scaling modules.
- Ensure that the sample time used is correct and latest (and correct) `.out` file has been downloaded to the MCU (Right mouse click on MCU module to show dialog box of these variables).
- Confirm that the user input values are set to either zero, or 'acceptable' values, which will not cause a current trip of the converter. Specifically check that the current reference slider for the PM machine is set to zero This ensures that the IM machine is not activated with an external load upon start-up.
- Confirm that the dual converter switch `conv1-2_on` is set to OFF and the power supply is on (DC bus voltage present on both boost packs).
- Confirm that the motors are connected (to the appropriate converter) firmly and properly.

After completion of the Pre-Drive checklist, activate the program and confirm that the supply voltage source is 24 V on both boost packs. If not stop the program and restart. With the correct DC voltage level established, turn on both converters (using the `conv1-2_on` button). Then monitor the motor shaft torque meters and diagnostic scope, which in this example (Fig. 3.90) is set to show the per unit α current components of the PM and IM machine. Note that the actual current values are found by multiplying the per unit value with the full-scale current value set to 20 A in this case. These results are consistent with those found in development phase B (see Fig. 3.85).

The screen shot given in Fig. 3.90 shows that the IM machine is currently operating under FOC based speed control, while the PM machine, operating under FOC torque control, is acting as a load, i.e. operating in generator mode with a torque of $\approx -60\,\text{mNm}$. Furthermore, the induction machine is acting as a motor with a torque output of $\approx 80\,\text{mNm}$. The difference between the filtered torque reading is the friction torque of both machines, which is $\approx 20\,\text{mNm}$ (note that the readings fluctuate). In the event of zero friction (as is the case for the Phase B model), the two torque readings would be equal in magnitude but opposite in terms of polarity. Hence in this case the machine which acts as motor must provide the friction torque of the drive as well as the load torque of the machine acting as a generator. Note the importance of connecting the DC bus of both boost packs together given that in this example energy from the induction machine (acting as a motor) is transferred mechanically (via the coupling) to the PM machine (acting as a generator). Part of this energy (minus the losses in the machines and converters) is returned via the electrical DC bus link from ‘fwd’ to ‘aft’ boost pack. Consequently the DC power supply only provides the losses of both drives in this case.

Note that the dual boost pack approach as discussed in this section, effectively provides the reader with a low cost laboratory setup that is equivalent to a dynamometer, where the sensored PM FOC drive is used to control and measure the shaft torque and speed of an attached machine.

Chapter 4
Laboratory Sessions: Module 2

A series of laboratory modules will be discussed in this chapter, which have been designed to lead the reader through the development phases needed to understand and work with a sensorless field-oriented PM drive. Sensorless control implementation is based on the use of the InstaSPIN algorithm, which will be used in different modes of operation. This implies use of the algorithm as a 'software encoder' as well as full motor identification, online resistance estimation and field-weakening. Implementation phases B, C and C+ are discussed in this chapter. For phase B development a so called 'processor in the loop' (PIL) approach is introduced in which case the processor is part of the simulation process. This approach is required because the sensorless algorithm resides in hardware only.

Dual drive operation, where the PM sensored drive acts as a load/torque transducer for a sensorless controlled PM drive will provide the reader with the opportunity to explore motor/generator operation.

Simultaneously, the reader will be guided through the hardware and software steps needed to achieve sensorless control using the InstaSPIN-FOC approach.

4.1 Importance of Sensorless Control and Introduction to InstaSPIN-FOC

Sensorless or more accurately, encoderless (without use of a physical shaft encoder) control of drives remains desirable for the following reasons:

Electronic supplementary material: The online version of this chapter (doi: 10.1007/978-3-319-20043-9_4) contains supplementary material, which is available to authorized users.

D.W.J. Pulle et al., *Applied Control of Electrical Drives*,
Power Systems, DOI 10.1007/978-3-319-20043-9_4

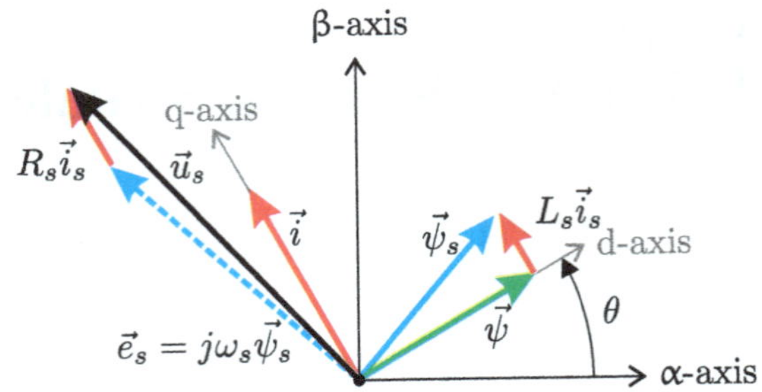

Fig. 4.1 Space vector diagram for PM motor

- Cost reduction: the ability to remove the encoder can be highly beneficial in terms of reducing the overall cost of a drive application.
- Smaller drive footprint: removing the encoder from the motor means that less space is required for the application and space saving up to 30 % is possible.
- Reliability: the use of encoder inevitably requires the use of a thin sensor cable (in addition to the, often bulky, three-phase motor leads) to connect the encoder to the drive. Such a sensor cable is easily damaged, which can lead to lengthy (costly) interruptions in drive operation. Elimination of the encoder (plus cable) improves reliability and simplifies motor installation.
- Redundancy: in applications where an encoder is required, a sensorless 'back-up' solution may still be desirable, to ensure that a drive remains functional when an encoder related fault occurs.
- Dual shaft operation: for applications where both ends of the motor shaft are connected to a load, the use of a standard shaft encoder is not possible, hence encoderless operation is required.

Given the above, there has been extensive research undertaken to find a sensorless algorithm, that makes use of the electrical voltage and/or phase currents to generate an estimate for the instantaneous shaft angle of the PM motor. Some insight, into the problems associated with such algorithms can be given by considering so called 'Direct Field Oriented Control (DFO) with voltage and current transducers' [4]. This DFO algorithm is briefly outlined below, where use is made of Fig. 4.1. For rotor flux oriented control (as discussed in Sect. 2.1.3) the PM flux ψ is aligned with the d axis of a synchronous reference frame. If field-weakening is not in use, the current vector $\vec{i}$ must be set orthogonally (at right angles) to the flux vector. Hence at any time instance the rotor angle θ needs to be known. The DFO approach estimates the voltage EMF vector $\vec{e}_s$, which is defined as

$$\vec{e}_s = \vec{u}_s - R_s\vec{i} \tag{4.1}$$

This requires measurement of the motor current $\vec{i}$ and voltage $\vec{u}_s$ vectors. In addition, the stator resistance R_s must be known and subsequently tracked, given that its value changes as the motor heats up. In a converter the voltage vector $\vec{u}_s$ is typically made up of two active vectors and a 'zero' vector during each sampling interval T_s, as was discussed in Sect. 2.1.6. Once the EMF vector is identified, the corresponding stator flux vector $\vec{\psi}_s$ can be found using

$$\vec{\psi}_s = \frac{\vec{e}_s}{j\omega_s} \tag{4.2}$$

where ω_s is the rotational speed on the EMF vector, which is directly tied to the shaft speed. Hence the immediate problem, with this approach is that estimation of this flux vector using Eq. (4.2) becomes difficult to evaluate, if not impossible, at low shaft speed. Finally, the magnitude and orientation of the PM rotor flux vector $\vec{\psi} = \psi$ can be found using

$$\vec{\psi} = \vec{\psi}_s - L_s \vec{i} \tag{4.3}$$

where L_s represents the stator inductance of the machine, hence its value must be known for correct identification of the PM flux vector $\vec{\psi}$. In general a 'good' sensorless algorithm must be able to meet the following guidelines:

- Tracking of the rotor angle θ: ability to calculate this angle over a wide speed range, including operation at ultra-low stator frequency.
- Fast response: ability to track near-instantaneous changes in $d\theta/dt$.
- Robust: ability to provide an accurate rotor angle estimate in the face of motor parameter variations, notably the stator resistance R_s.
- Flux tracking: ability to track the magnitude of the rotor flux over a wide speed range in order to estimate the shaft torque of the machine
- Motor parameters: ability to accurately identify the parameters needed for accurate rotor angle estimation.

The sensorless algorithm InstaSPIN-FOC developed by Texas Instruments [9] meets the guidelines set out above. Furthermore, this is the only algorithm available that can be used in conjunction with a graphical fixed point software environment such as VisSim. Hence, its use in this book for both sensorless PM and induction machine (IM) drive operation.

Central to the use of this algorithm is the Texas Instruments TMS320F2806xF device given in Fig. 4.2. This micro controller unit (MCU) houses the so called FAST software module, which is located in the ROM. Inputs to this module are the (per unit) currents (i_α, i_β), voltages (u_α, u_β) and bus voltage u_{DC}. Outputs of the FAST modules are estimated values for: rotor flux (due to the magnets in this case) ψ, rotor angle (orientation of the rotor magnets relative to the stator) θ, instantaneous shaft speed, expressed as the frequency f_m, shaft Torque T_m and inverse DC bus voltage $1/u_{DC}^n$. Furthermore, the FAST module requires access to the estimated (or measured) stator resistance R_s and stator inductance L_s. Also shown in this figure are the speed/current controllers, which provide speed control and generate the reference modulation indices $m_1^{ref}..m_3^{ref}$, based on the 'model based current control' approach, as discussed in the previous chapter. The 'PWM' module contains the 'pulse centering module' (as discussed in Sect. 2.1.6) and generates the duty cycle values required for the converter. A 'PWM Driver' module is used to interface the 'PWM' module to the power electronic devices of the converter. Voltage and current sensors are used to measure the phase currents/voltages and bus voltage (as discussed previously, see Fig. 3.5).

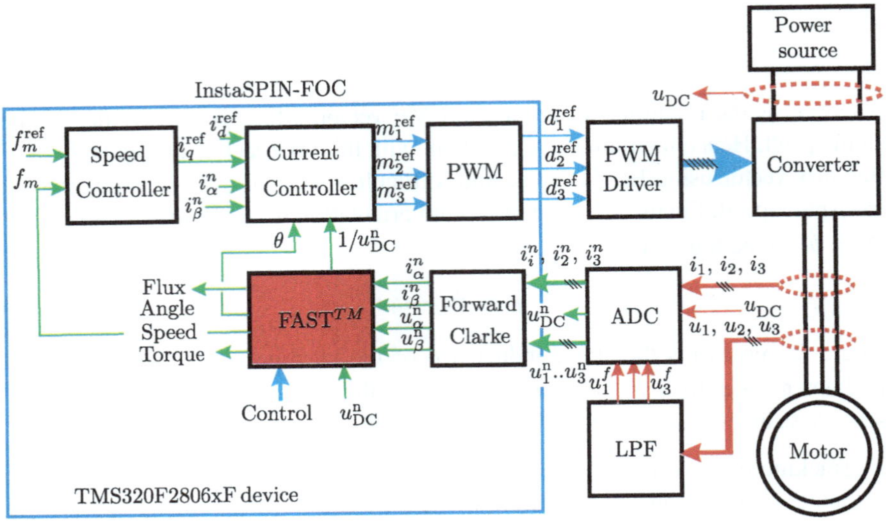

Fig. 4.2 Sensorless drive setup using an InstaSPIN-FOC ROM module

A number of 'Control' inputs (shown in Fig. 4.2 as a single arrow) allow the user to configure the functionality of this module. Three basic functions are available namely:

- FAST encoder: use of the FAST module as a software encoder, which implies that the user provides the algorithms for the speed/current controllers and generates the modulation indices for the PWM unit. These algorithms can then be located in either the FLASH or RAM section of the MCU.
- Motor identification: Use of the FAST module to estimate the required motor parameters needed for the sensorless algorithm. In addition, the current controller parameters are identified in this phase. For this function the speed/current controller located in the ROM module are needed.
- On-line stator resistance estimation: This function tracks the stator resistance, for use with the sensorless algorithm.

In the following laboratory sessions the use of the TMS320F2806xF device (to realize sensorless control), which is located on the LaunchXL-F28069M card, will be discussed. In addition, the other modules shown in Fig. 4.2 (with exception of the motor and power source) that are located on the BOOSTXL-DRV8301 module will be considered in this chapter. Central to these laboratories is the aim to familiarize the reader with the parameters and use of the InstaSPIN module. For this purpose use is made of two special VisSim InstaSPIN modules that handle (internally) all the communication with the TMS320F2806xF MCU.

4.2 Laboratory 2:1: FOC Sensorless Control of a PM Machine

A field-oriented controlled drive as discussed in Sect. 3.3 is adapted to sensorless operation by making use the FAST module shown in Fig. 4.2. The latter produces a shaft angle estimate, which will replace the shaft angle input from the encoder. However, the encoder is still used in this drive example (for phases C and C+ only) to allow the user to operate under either sensorless or sensored control. Primary objective of the laboratory is to familiarize the reader with the parameter required to achieve sensorless operation with KNOWN motor parameters, i.e. it is assumed that these have been measured (as indeed they have) prior to this laboratory.

4.2.1 Lab 2:1: Phase B

The purpose of this introductory laboratory is to demonstrate operation of the FAST based sensorless drive with the aid of a set of vector plots which show estimated and actual (machine model) rotor flux vectors. In addition, the flux current and voltage vectors can be observed during operation of the drive under speed or torque control. The basic FOC drive is identical to that discussed for FOC sensored control (see lab 1:3 in Sect. 3.3). The key difference is that rotor angle, rotor flux, shaft speed and torque data are now provided indirectly by a FAST module located within the 'Drive Controller' shown in Fig. 4.3. The term 'indirectly' refers to the fact that said FAST module contains a set of function calls which communicate with the TMS320F28069M unit located on the LAUNCHXL board. This so called 'Processor In the Loop' (PIL) approach makes it possible to evaluate drive performance under simulated conditions using the actual FAST algorithm located in the hardware.

The following information is relevant for this laboratory component:

- Reference programs [11]: `lab2-1_LaunchXLphBv4.vsm` and `FASTv4.vsm`.
- Description: Sensorless control of a PM motor using FAST module as an observer.
- Equipment/Software: Texas Instruments LAUNCHXL-F28069M and VisSim simulation program.
- Outcomes: Show the various vector plots that are present in the drive when using a fixed point representation of the controller when operating under field oriented control using the FAST sensorless algorithm.

No boost packs are used in the laboratory, nor is an external 24 V power supply required, given that power to the MCU is directly provided via the USB cable connected to the laptop. In this case the following jumper and dip-switch positions on the LAUNCHXL-F28069M module are required:

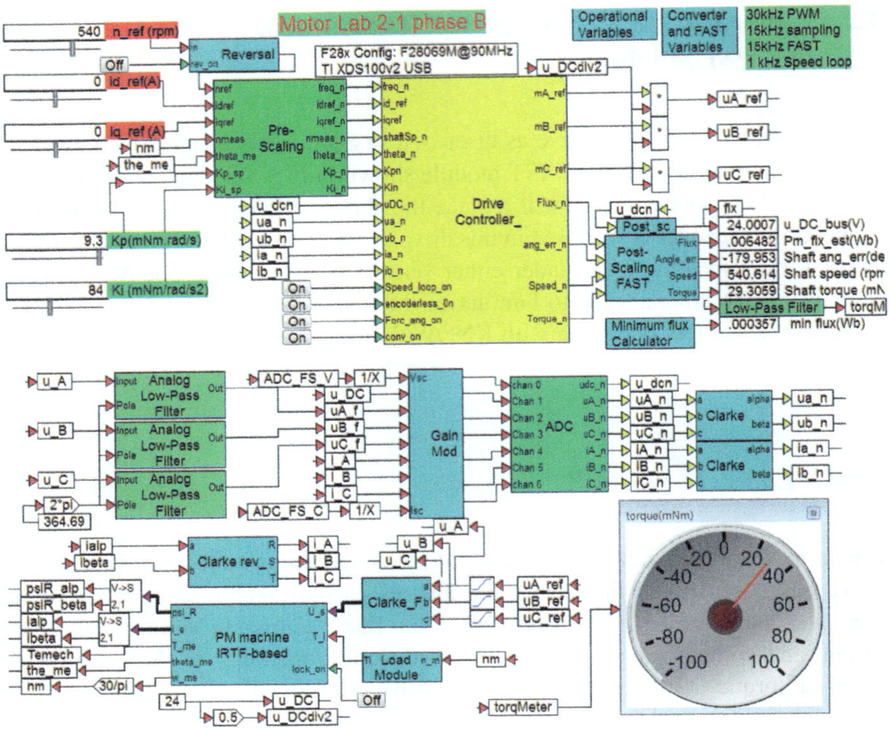

Fig. 4.3 Phase B simulation of a FAST based encoderless field-oriented controlled (FOC) drive

- Jumpers `JP1` and `JP2` CLOSED
- Jumpers `JP4` and `JP5` OPEN
- Jumpers `JP3`, `JP6` and `JP7` CLOSED
- Dip-switches `SW1` to `SW3` ON

Note that there are two reference programs required for this lab: the first is the main simulation model that contains a PIL 'interface module' which in turn makes use of a second program `FASTv4.vsm`. The outputs of the 'Drive Controller' shown in Fig. 4.3 are the three reference modulation indices `mA_ref`, `mB_ref`, `mC_ref`, that must be multiplied by the term $u_{DC}/2$, which yield the reference voltages for the drive. These three outputs `uA_ref`, `uB_ref`, `uC_ref` are connected to a 'converter', which in this case is simply represented by three limiter modules, that limit the motor voltages to $\pm u_{DC}/2$. The phase voltages `u_A`, `u_B`, `u_C` of the machine are used by the machine model and by the FAST algorithm. For the latter case the converter outputs are filtered using 'analog low-pass filter' modules that replace the electrical circuit models which are present in the actual hardware. The filtered voltages are then scaled with the aid of a 'Gain module', by a factor 3.3/`ADC_FS_V`, where `ADC_FS_V` is the full scale ADC voltage set to 26.314 V for the boost pack. A similar scaling is also used for the machine phase currents `i_A`, `i_B`, `i_C`, in which case the scaling factor is equal to 3.3/`ADC_FS_C`, where `ADC_FS_C` is the full scale ADC current value set to 33.0 A. A detailed model of the 12 bit AD converter is used to convert the scaled voltage/current inputs value to per unit outputs for use with the current controller and FAST module.

A set of sliders is used to define the current reference vector, in terms of its direct and quadrature components $i_{\mathrm{d}}^{\mathrm{ref}}$, $i_{\mathrm{q}}^{\mathrm{ref}}$ and rotational speed $n_{\mathrm{m}}^{\mathrm{ref}}$. A 'pre-scaling' module has been added to convert the floating point variables required for the controller to fixed point format. In addition, two sliders have been added which allow the user to set the proportional $\mathrm{K_p}$ and integral $\mathrm{K_i}$ speed gains of the controller. Note that the variables $\mathrm{K_p}$, $\mathrm{K_i}$ are shown in (milli)Nms/rad and (milli)Nm/rad respectively.

The two dialog boxes for this laboratory, given in Fig. 4.3, contain all the user settings required for this laboratory. For example, the 'operational Variables' dialog box is again used to assign full scale values for the voltage, current and frequency. In addition, the current controller gain and bandwidth are also assigned here. These values are based on the motor in use, as was discussed in Sect. 2.1.5. Also assigned in this box is the rate of variable change for current and frequency, i.e. introduced as a safety measure to avoid erratic slider action from causing over-currents in the drive. The 'Converter and FAST variables' dialog box contains all the parameters which must be set for the FAST algorithm, which includes among others the full-scale ADC values and low-pass filter corner frequency. A more detailed discussion on these parameters will be given in the 'phase C' development stage.

Central to this laboratory is the ability to use the VisSim PIL software tool, which provides access to the FAST algorithm in the hardware. Located within the controller module (two levels down) is a 'Target Interface module' given in Fig. 4.4, which is used to communicate with the FAST algorithm. Inputs to this module are the per unit α, β voltage and currents generated by the ADC unit and a set of control inputs. Key outputs are the per unit inverse DC bus voltage `invB` (used by the current controller), rotor flux estimate `Flux`, rotor angle estimate `Ang` and Speed/Torque estimates. A set of `Debug` variables are used to monitor the current/voltage waveforms. The `LED_in`/`LED_out` variables are used to activate a 'blue' LED on the LAUNCHXL board which flashes when PIL operation is active. Of central importance in the dialog box shown in Fig. 4.4 is the `Target Execution File`, which in this case is set to `FASTv4.out`. When the main simulation program is activated (via the 'green' arrow button), this `.out` file will be downloaded into the MCU prior to the start of the actual simulation. Once downloaded the rate of data exchange between MCU and VisSim simulation will be dictated by the `Sample Rate (Hz)` entry (set to 200 Hz in this case) provided that the `Sync Target to This Block` option has been activated. Note that the simulation frequency (set in the system properties) is set to 15 kHz given that this is the sampling frequency specified. However, actual run time during PIL operation is dictated by the `Sample Rate (Hz)` entry mentioned above.

The `FASTv4.out` file is generated via separate compilation of the module 'FAST EST', with file name `FASTv4.vsm` (see Fig. 4.5). This module contains all the software code needed to access the FAST ROM module. More details on the content of the 'FAST EST' module will be given in development phase C.

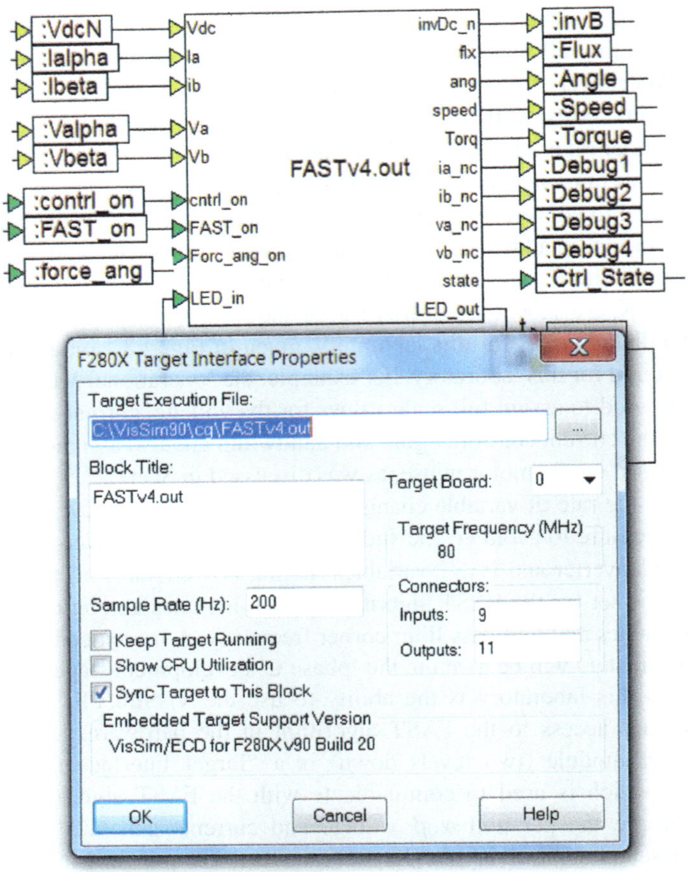

Fig. 4.4 Target interface module with corresponding dialog box entry

Starting this simulation is done via the usual 'green' button in which case the `.out` file for the PIL process is downloaded first. Then the ADC offsets are determined within the ADC unit, which takes $\approx 1.4\,\mathrm{s}$, (simulation time) AFTER that period the PIL should become active (flashing 'blue' LED on LaunchXL board) and drive operation should become apparent from observation of the vector plots. A vector plot example of the drive in question operating under steady state conditions, with a constant speed of $n_{\mathrm{m}}^{\mathrm{ref}} = 540\,\mathrm{rpm}$ and arbitrarily chosen mechanical load of 29 mNm is shown in Fig. 4.6. The sampling time $\mathrm{T_s}$ as defined in the program pull-down menu, is set to 66.6 µs, i.e. a 15 kHz sampling frequency is used for the discrete controller model. The vector plot given in Fig. 4.6, shows the machine current vector $\vec{i}_{\mathrm{s}}$ and voltage vector $\vec{u}_{\mathrm{s}}$. In addition, a unity 'd-axis' vector is introduced to show the orientation of the (d,q) reference frame. The ('green') 'd-axis' vector is aligned with the estimated PM flux vector. In this example, a current i_{q} value is generated given that the machine is producing torque to match the load

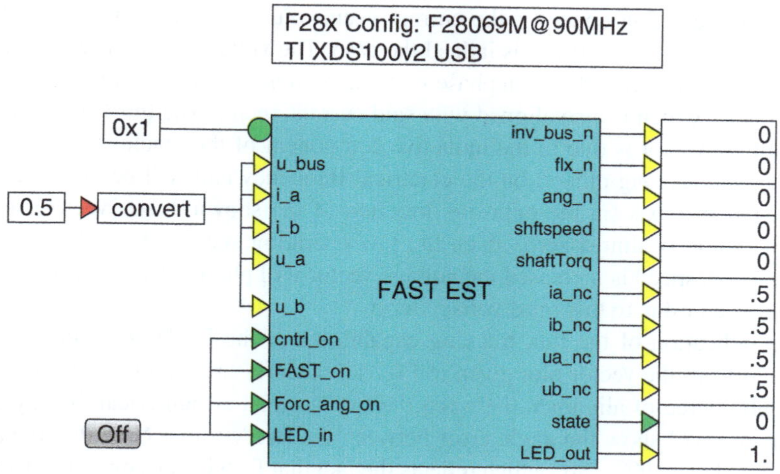

Fig. 4.5 FAST estimation module

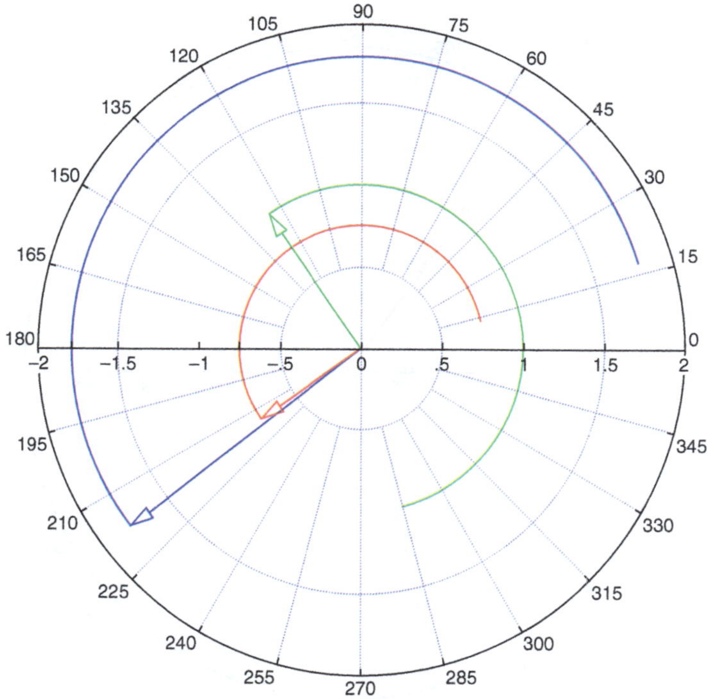

Fig. 4.6 Phase B: Vector plot obtained using a FOC/Speed sensorless control with current $\vec{i}_s$ ('red'), voltage vector $\vec{u}_s$ ('blue') and unity d-axis vector ('green')

torque applied. Consequently, the current vector must be orthogonal to the d-axis vector (given that $i_d = 0$), as is indeed the case. The voltage vector is made up of a back EMF component that is in phase with the current vector and a resistive voltage component, which is also aligned with said current vector. The third component of the voltage vector is due to the inductive component of the machine and its value will depend (among others) on the electrical frequency and will be oriented along the negative d-axis (in the opposite direction of the shown d-axis vector). In this case its value is almost zero, given the low operating speed set for this example. However, as speed is increased the voltage vector will phase advance relative to the current vector due to this third voltage term.

An indication of the flux tracking capabilities of the FAST algorithm may be deduced from the vector plot given in Fig. 4.7. Clearly observable from Fig. 4.7 is the high degree of alignment between the two vectors. A numerical display given in Fig. 4.3 will show the angle error between the two vectors. Note that the angle error may be 0°, 90°, 180° which reflects the fact that FAST locks onto the electrical angle. Furthermore, a four pole pair machine is used, hence the sensorless algorithm will lock on to the nearest flux vector.

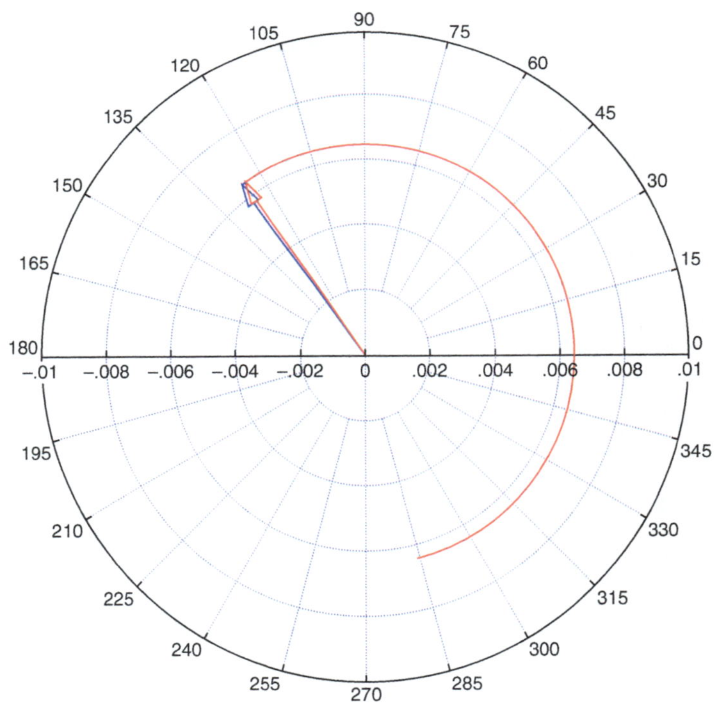

Fig. 4.7 Phase B: Flux vectors generated by the machine model $\vec{\psi}_R$ ('red'), and FAST estimator $\vec{\psi}_R^{\mathrm{est}}$ ('blue')

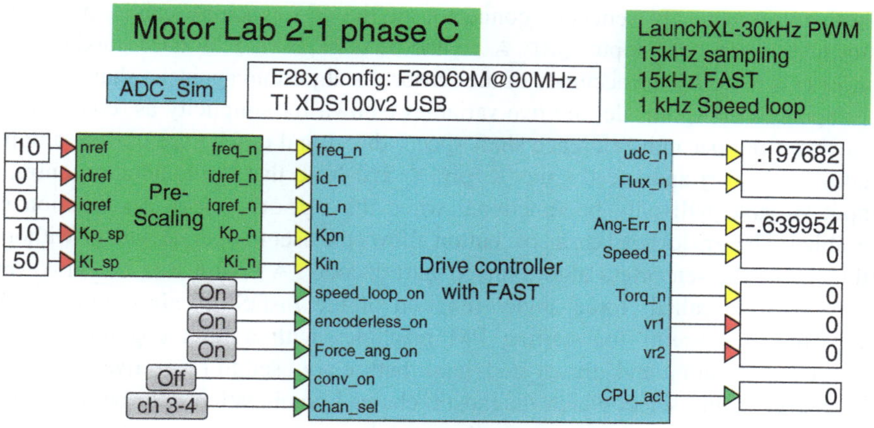

Fig. 4.8 Phase C simulation of a FAST based encoderless field-oriented controlled (FOC) drive

The reader is encouraged to explore all modes of operation as to obtain a better understanding of the sensorless control approach presented here, before moving on to a more in depth discussion on the actual experimental implementation.

4.2.2 Lab 2:1: Phase C

Prior to being able to work with an operational drive, an `.out` file must be generated of a 'Drive Controller with FAST' module, shown in Fig. 4.8, which represents the drive under consideration. Hence details of the drive structure and corresponding dialog box parameter assignments will be discussed in this subsection.

The following information is relevant for this laboratory component:

- Reference program [11]: `lab2-1_LaunchXLphCv4.vsm`.
- Description: Sensorless control of a PM motor using FAST module as an observer.
- Equipment/Software: Texas Instruments LAUNCHXL-F28069M, with BOOSTXL-DRV8301 module ('aft' position) and VisSim simulation program.
- Outcomes: Generate the `.out` file needed to represent the drive structure under investigation.

As mentioned above, a single boost pack located in the 'aft' position (furthest away from the USB connector) is used for single motor operation, in which case the following jumper and dip-switch positions on the LAUNCHXL-F28069M module are required:

- Jumpers `JP1` and `JP2` OPEN
- Jumpers `JP4` and `JP5` CLOSED
- Jumpers `JP3`, `JP6` and `JP7` CLOSED
- Dip-switches `SW1` to `SW3` ON

Furthermore, the 'J4' encoder connector of the PM machine cable should be attached to encoder input QEP_A of the LAUNCHXL-F28069M module. Its purpose is to verify the accuracy of the sensorless algorithm against a shaft sensor.

Inputs to the Controller are five variables (shown for simplicity as 'constants'), which set the per unit reference shaft speed, direct and quadrature current amplitudes i_d, i_q and speed gains K_pn, K_in. Note that the quadrature current input variable will only be functional when the speed-loop_on function is disabled. The encoderless_on button allows the user to operate with or without (thus using the sensorless algorithm) a shaft encoder. A button Force_ang_on activates a so called 'force angle' function of the InstaSPIN algorithm, which is particularly useful for starting PM machines with a high cogging torque. A conv_on button and chan_sel input button are used to respectively activate the converter and select which diagnostic channel combination is to be displayed in phase C+. Outputs of the controller module are (per unit), measured DC bus voltage: u_dc_n, rotor flux: Flux_n error between the estimated and actual (as measured with the shaft encoder) rotor angle. Note that the angle error can be $0°, 90°, 180°, 270°$, because FAST estimates the electrical flux angle (four pole pair machine in use). Additional outputs are, shaft speed: Speed_n, which is the estimated value when sensorless operation is in use, or the value derived from the shaft encoder and shaft torque: Torq_n, which is proportional to the quadrature current and magnetic flux (see Eq. (2.8)). Three additional outputs vr1, vr2 and CPU_act are used for diagnostic purposes, hence variables of interest can be allocated before compilation, so that they can be subsequently (in Phase C+) shown on a numerical display. The variable CPU_act provides information on MCU usage, which hitherto has not been shown given that CPU utilization was relatively low. With increased laboratory complexity, this variable becomes more important, hence its inclusion in key laboratories. Moving one level lower into the 'Drive Controller with FAST' module reveals the modules shown in Fig. 4.9. Of these shown in the figure, the 'Diagnostic tool' module has been discussed in previous laboratories. A '100 Hz' module is used to execute tasks which are non time critical. Of interest are the 'Operational' and 'Converter and FAST variable' dialog boxes, shown in Fig. 4.9. These contain all the data entries that must be provided by the user in order to achieve successful sensorless drive operation (see Fig. 4.10).

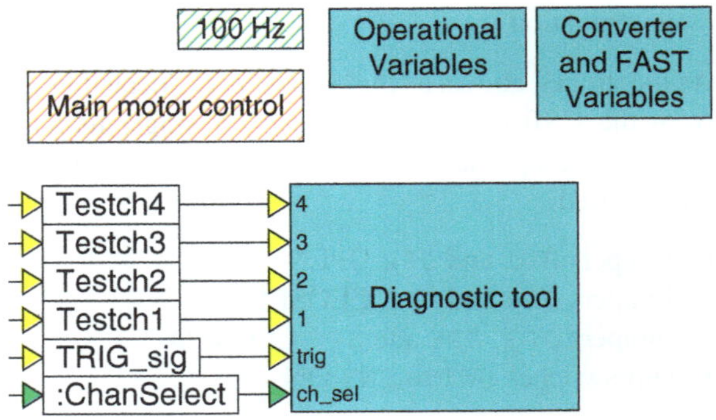

Fig. 4.9 Phase C simulation of a FAST based encoderless field-oriented controlled (FOC) drive: one level into the drive controller module

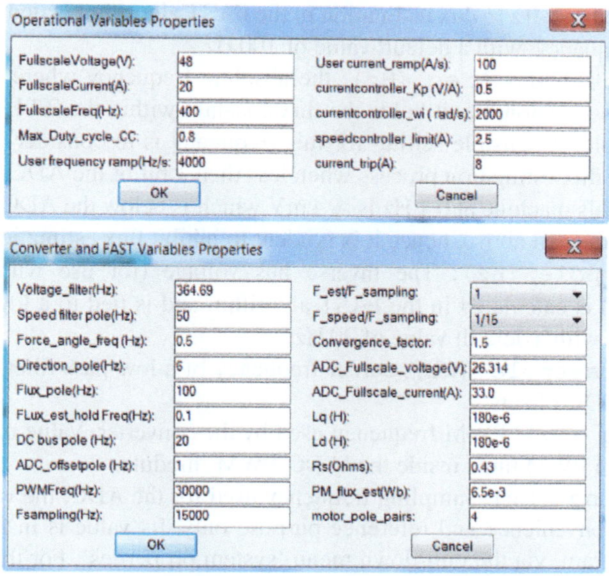

Fig. 4.10 Phase C simulation of a FAST based encoderless field-oriented controlled (FOC) drive: Dialog box entries

The 'Operational Variables' dialog box, contains all the parameters needed for scaling, rate limit values for user inputs and gain/limit settings for current/speed control. These are identical to those discussed for laboratory 1:3 (see Sect. 3.3). The meaning of the variables and values to be assigned on the 'left' column of the 'Converter and FAST Variables' dialog box are itemized as follows:

- `Voltage_filter (Hz)`: corner frequency value (in Hz) of the low-pass filter used to measure the phase voltages, which can be found with the aid of Eq. (3.1b).
- `Speed_filter pole (Hz)`: corner frequency (in Hz) of a low-pass filter that is used to filter the estimated shaft speed value. Typically its value is a factor 5 higher then the bandwidth of the speed-loop controller.
- `Force_angle_freq (Hz)`: force angle function is used to overcome the cogging torque of the machine during start up. The force angle frequency value shown specifies the absolute frequency range within, which this function is active. Default value is 1 Hz and its value is typically chosen to ensure that FAST makes use of at least 1-2 bits of the ADC before this function is switched off. In this case the EMF amplitude at 0.5 Hz will be $\approx 20\,\mathrm{mV}$ and the ADC resolution is $^{26.314}/_{4096} = 6.4\,\mathrm{mV}$ (ADC fullscale voltage is 26.314 V), hence sufficient ADC resolution is present when force-angle is turned off.
- `Direction_pole (Hz)`: motor direction is identified by the FAST algorithm, and this function is tied to a low-pass corner frequency, with a default value of 1 Hz.

- `Flux_pole (Hz)`: flux estimation in the FAST algorithm is tied to a low-pass corner frequency, with a default value of 100 Hz.
- `Flux_est_hold Freq (Hz)`: the absolute frequency where the estimator 'freezes' estimation, i.e. it holds the flux constant within the 0.1 Hz band set by this variable. A sensible setting for this frequency is to consider the ADC and freeze the flux estimation process when less then 1 bit of the ADC is active. The EMF for this machine at 0.1 Hz is ≈ 4 mV which is below the ADC resolution of 6.4 mV indicated above, hence it is prudent to inhibit flux estimation.
- `DC bus_pole (Hz)`: The inverse bus voltage (for use with the current controller) is calculated in the FAST algorithm and is tied to a low-pass corner frequency with a default value of 20 Hz.
- `ADC_offsetpole (Hz)`: corner frequency of a low-pass filter used to measure the ADC offsets.
- `PWMfreq (Hz)`: PWM frequency used by the converter. Value must be set via the VisSim PWM units inside the 'ADC-PWM' module.
- `Fsampling (Hz)`: sampling frequency used for the ADC, the value is shown here for convenience and reference purpose only. Its value is in fact set in the main program, via the pull-down menu 'system properties'. For this application it is set to 15 kHz.

The variables and values to be assigned on the 'right' column of the 'Converter and FAST Variables' dialog box are more self explanatory namely:

- `F_est/F_sampling`: ratio between sampling frequency used by the FAST estimator algorithm and ADC sampling frequency. The ratio is an integer value, hence the user can select the ratio from a dialog box pull-down menu. For example, a choice of `1/1` implies an estimator sampling frequency of 15000 Hz, given that the ADC sampling frequency is set to 15000 Hz.
- `F_speed/F_sampling`: ratio between sampling frequency used by the speed controller and ADC sampling frequency. The ratio is an integer value, hence the user can select the ratio from a dialog box pull-down menu. For example, a choice of `1/15` implies an speed controller sampling frequency of 1000 Hz, given that the ADC sampling frequency is set to 15000 Hz.
- `Convergence_factor`: this variable governs the transient ability of the algorithm to 'home in' on the actual rotor angle, its default value is 1.5.
- `ADC_Fullscale_voltage (V)`: a value defined by the voltage signal conditioning circuit and can be calculated using Eq. (3.3a).
- `ADC_Fullscale_current (A)`: a value defined by the current signal conditioning circuit and can be calculated using Eq. (3.3b).
- `L_q (H)`: the 'quadrature' axis inductance (H) of the machine. If a so called 'salient' PM machine is used the quadrature value will differ from the direct axis value mentioned in the next entry. For this lab, the value must be known, via a data sheet, or measured (as can be done in a subsequent laboratory), or with the aid of an L-C-R meter.
- `L_d (H)`: the 'direct' axis inductance (in H) of the machine, which is the stator inductance in the direction of the magnetic field, i.e. the d-axis, of the d, q

synchronous coordinate system tied to the rotor. Its value must be known for this laboratory, via a data sheet, or measured (as can be done in a subsequent laboratory), or with the aid of an L-C-R meter.

- `Rs(Ohms)`: the stator resistance (in Ω) of the machine, which must be known for this laboratory, via a data sheet, or measured (as can be done in a subsequent laboratory), or with the aid of an L-C-R meter.
- `PM_flux (Wb)`: an estimate for the rotor flux (in Wb). In this example its value is set to 6.5 mWb, which is the known value (as derived in the next laboratory) for the LVSERVOMTR PM machine in use.
- `motor_pole_pairs`: the number of motor poles in use. Its value must be derived from the data sheet or measured. For example, by doing an open loop voltage control (see previous chapter) experiment and adjusting the input electrical frequency in such a manner that the shaft rotates at 1 rotation per revolution (which can be observed visually). The pole pair number is then found from the ratio of user frequency and observed shaft speed. The machine in use has eighth poles (four pole pairs $p = 4$).

Note that the LVSERVMTR PM machine in use is non-salient, hence the direct and quadrature stator inductance values are equal and found to be $L_s = L_{sd} = L_{sq} = 0.18\,\text{mH}$. Moving one level lower into the 'Main motor control' module, shows the set of modules given in Fig. 4.11. Of these, all with exception of the 'Drive controller' have, in fact, been identified in earlier laboratories. Inputs to the controller module, are those identified and discussed in laboratory 1:3. However, for the sensorless controller considered here, both (per unit) voltage and current α, β variables are now required. Moving yet one level lower into the 'Drive Controller' module, reveals a set of modules given in Fig. 4.12. Hopefully familiar, are the modules linked to current and speed control. Central to the figure is the InstaSPIN module, which in this case is configured to operate as a software encoder. Input variables to this module are the (per unit) α, β voltages/currents and DC bus voltage. Outputs of the current controller are the three phase modulation indices. The InstaSPIN module provides the following instantaneous (per unit) output variables:

- `invVdc`: inverse DC bus voltage.
- `ROTORFLX`: rotor flux estimate.
- `ang`: estimate for electrical rotor angle.
- `rotfreq`: filtered estimate for shaft frequency.
- `elecTT`: estimate for shaft torque.

Observation of Fig. 4.12 learns that use is made of the `encoderless_on` (user activated) function to activate a 'switch' module that ensures that the speed feedback signal for the 'Speed controller' module is provided by the FAST output variable `rotfreq`. At the same time the variables `Cos_th`, `Sin_th` (used for Park transformations) are switched to the variables `FCos_th`, `FSin_th`, which are the Sin/Cos functions that make use of the FAST generated variable `ang`. The latter variable is used in combination with the encoder generated angle `theta_n`

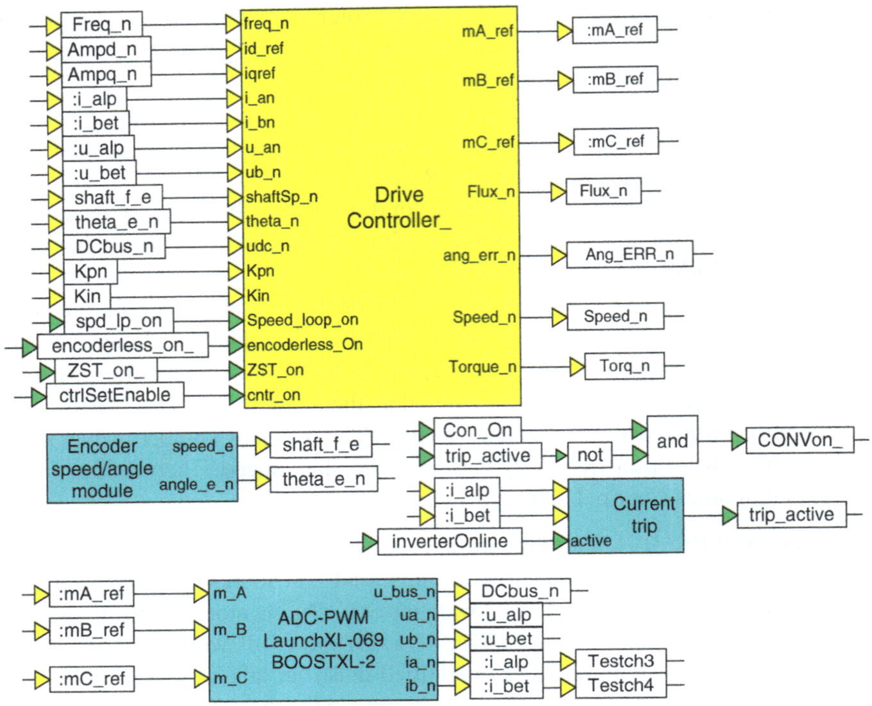

Fig. 4.11 Phase C simulation of a FAST based encoderless field-oriented controlled (FOC) drive: one level into the 'Main motor control' module

to generate a variable `ang_err_n`, which represents the difference between the two angles. Note that in Fig. 4.12 two test signals `testch1`, `testch2` have been attached to the cosine of the angle generated by the encoder and InstaSPIN respectively. Hence these can be observed in development phase C+ via the scope module. Using this tool the encoder offset (inside the 'encoder speed/angle' module) was adjusted to ensure complete phase alignment between the two cosine functions.

If the logic variable `speed_loop_on` is zero (speed loop not active) the quadrature reference variable `iqref` will be connected to the variable `amplq_n`, which is set by the `i_q` slider. Under them conditions, the drive is operating under 'torque control' and care should be taken to MAINTAIN a load on the machine, as a speed runaway situation can occur (the machine will try and maintain a constant torque value, defined by the user set `i_q` value, independent of speed).

Moving one level into the FAST module, as shown in Fig. 4.13, reveals the basic VisSim 'InstaSPIN Rotor Angle Estimator' and 'Estimator Calculator' modules. The 'InstaSpin Rotor Angle Estimator' module communicates directly with the InstaSPIN ROM (inside the MCU) and returns the rotor angle estimate at the sampling rate set in the user dialog box (see Fig. 4.10). Inputs to this module are the per unit α, β currents/filtered voltages and DC bus voltage, which are converted to `IQ24` format. This module has an extensive dialog box, which holds

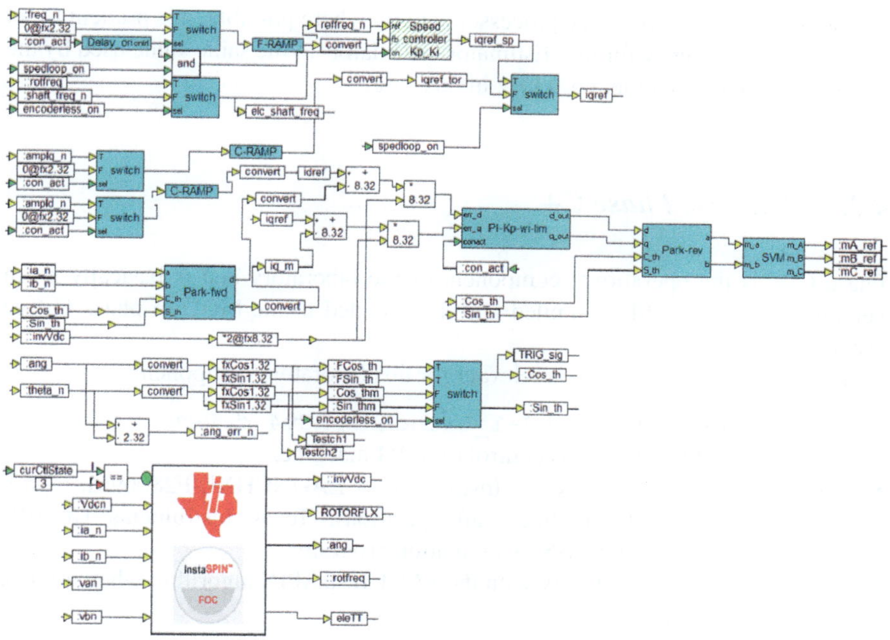

Fig. 4.12 Phase C simulation of a FAST based encoderless field-oriented controlled (FOC) drive: one level into the 'Drive Controller' module

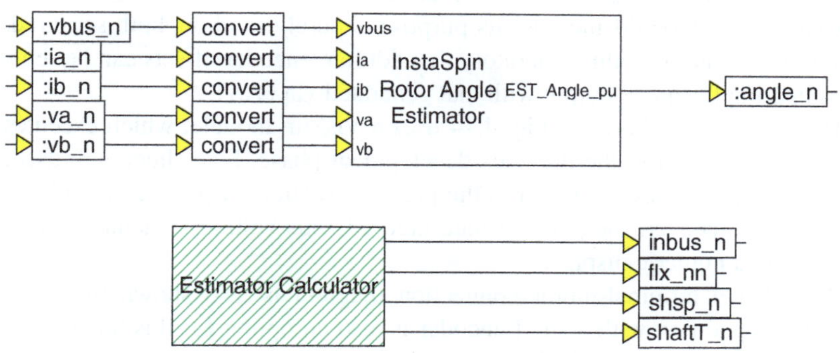

Fig. 4.13 Phase C simulation of a FAST based encoderless field-oriented controlled (FOC) drive: one level into the 'FAST' module

(among others) the dialog box parameters introduced in Fig. 4.10. However, more detailed parameter settings are also available. Note that a number of less time critical estimator tasks have been allocated to the '100 Hz' module shown in Fig. 4.9. An 'Estimator calculator' compound module, shown in Fig. 4.13, is used to derive the per unit inverse-bus voltage, rotor-flux, shaft-speed and shaft torque values from the ROM. The use of a hatched surface, indicates that the sampling rate has been set independently of the main simulation frequency, which is typically the ADC sampling frequency. To reduce processor activity where possible, it is prudent to set the sampling frequency for the 'Estimator Calculator' to the same value used by the 'Speed controller' module, i.e. 1 kHz in this case.

4.2.3 *Lab 2:1: Phase C+*

Phase C+, is the operational component of the laboratory and is basically a run version of the `.out` file compiled and downloaded to the MCU in phase C (see previous subsection).

The following information is relevant for this laboratory component:

- Reference program [11]: `lab2-1_LaunchXLphCv4_d.vsm`.
- Description: FOC sensorless control of a PM machine.
- Equipment/Software: Texas Instruments LAUNCHXL-F28069M, with BOOSTXL-DRV8301 module ('aft' position), Texas Instruments LVSERVOMTR PM motor and VisSim simulation program.
- Outcomes: to gain familiarity with the FAST InstaSPIN algorithm when used as a 'software encoder'.

Note that the required jumper and dip-switch settings for the LAUNCHXL-F28069M module are given in phase C. Furthermore, the 'J4' encoder connector of the PM machine cable should be attached to encoder input `QEP_A` of the LAUNCHXL-F28069M module. Its purpose is purely to allow both sensored and sensorless operation in this laboratory. In addition, encoder offsets can be removed by comparing the angle output with that generated via FAST.

The run version shown in Fig. 4.14 uses a VisSim module, which executes the `.out` file compiled in the previous development phase. Five sliders are used, the purpose of which was discussed in the previous section. A post-scaling module is again used to convert the per unit measured DC bus voltage to actual voltage, as shown with a numeric display.

Note that a terminal-motor connection diagram is also shown in Fig. 4.14, given that operation with a shaft encoder is also an option for this laboratory. For sensorless operation a terminal-motor diagram is not required, except to ensure that the machine rotation direction, matches the envisaged (user application defined) reference direction. An additional 'Post scaling FAST' module is used, where a number of scaling factors are used to generate the required outputs from the inputs provided by the `lab2-1_LaunchXLphCv4.out` target module. The computation of the scaling factors for the output variables shown is as follows:

- `PM_flx_est(Wb)`: the estimated magnetic flux (Wb) of the machine in use. Calculated using the flux output `flux_n` from the target module, which is not scaled.

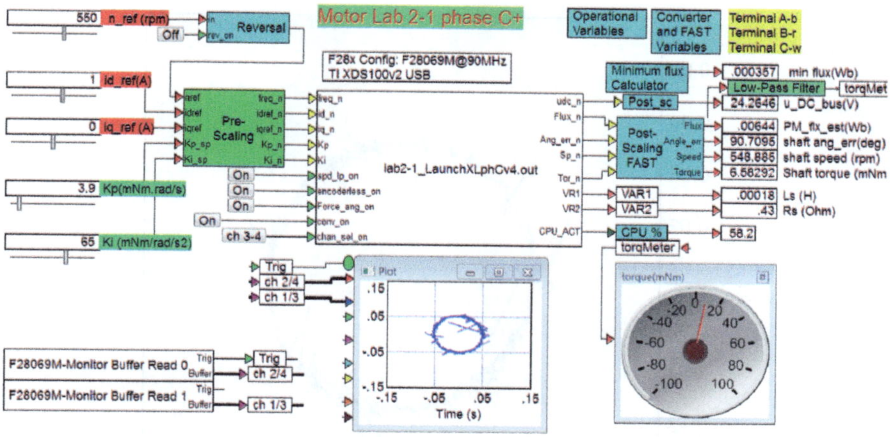

Fig. 4.14 Phase C+ simulation of a FAST based encoderless field-oriented controlled (FOC) drive

- `shaft ang_err(deg)`: the error between the FAST estimated and shaft-encoder generated rotor angles (in degrees). Scaled by a factor $^{360}/_{p}$, where p is the pole pair number.
- `shaft speed (rpm)`: shaft speed of the motor as calculated by the FAST algorithm or shaft encoder. The value displayed is set by the logic signal `encoderless_on` (sensorless operation active). Hence, when sensorless control is active the value shown is derived from FAST. Scaled by a factor `F_fs*60/p`, where `F_fs` is the full scale frequency.
- `shaft torque (mNm)`: shaft torque of the machine in use, where a scaling factor of 1000 is used to generate a result in terms of (milli-Nm). A low-pass filter with a 5 Hz corner frequency, is used to filter this signal for display in the torque meter (see Fig. 4.14).

Two 'Monitor Buffer' modules are used to display two selected (using an on/off button connected to the `channel_sel_on` input) diagnostic signals.

The VisSim scope module (set to 'xy mode') shows the per unit α, β currents of the machine and an enlarged view of these results is given in Fig. 4.15. Multiplication of these waveforms by the full scale current value of i_{fs}= 20 A gives actual current values. In this example the reference i_d has been arbitrarily set to 1.0 A.

Prior to activating this type of lab component, the reader needs to be aware of the fact that he or she is about to activate a complex electrical system, with live voltages. Hence it is prudent, **always**, to execute the following 'Pre-Drive' check list:

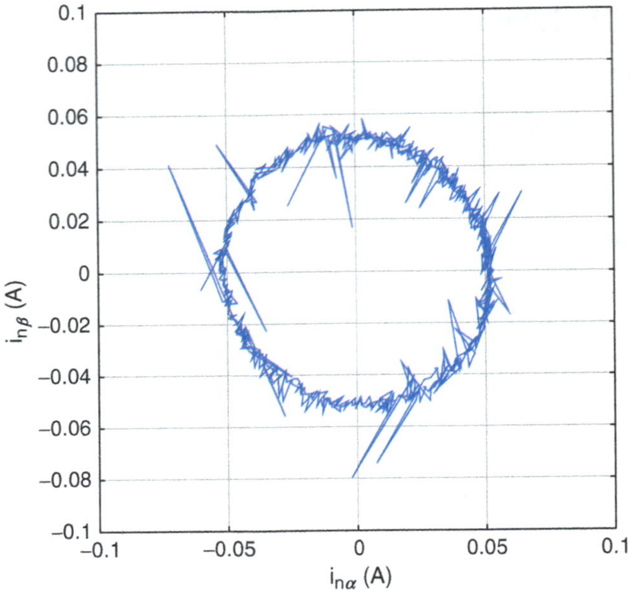

Fig. 4.15 Phase C+ operation of a drive under no-load (friction only): scope results with channel select option: 'ch 3-4'

- Dialog boxes used in C+ mode, match those of phase C: the run version is compiled with the dialog box entries specified under phase C. The dialog boxes shown in phase C+ are used by the 'Pre/Post' scaling modules.
- Ensure that the sample time used is correct and latest (and correct) `.out` file has been downloaded to the MCU (Right mouse click on MCU module to show dialog box of these variables).
- Confirm that the user input values are set to either zero, or 'acceptable' values, which will not cause a current trip of the converter.
- Confirm that the converter 'switch' is set to OFF and the power supply is on (DC bus voltage present).
- Confirm that the motor is connected firmly and properly.

After completion of the Pre-Drive checklist, activate the program and confirm that the supply voltage source reading shown on the digital display is 24 V. If not, stop the program and restart. With the correct DC voltage level established, turn on the converter (using the ON button) and monitor the motor shaft and diagnostic scope. Observation of Fig. 4.14 learns that the machine is currently operating under closed-loop speed control and without a load, as is evident from the torque meter value and the (per unit) i_{α}^{n}, i_{β}^{n} currents displayed in the diagnostic scope. Operation with i_d= 1.0 A is undertaken in this case, hence the maximum phase current value must at least correspond to the reference i_d value. In reality the current vector amplitude is slightly higher due to the presence of friction load which causes an additional i_q^{ref} component that is generated by the speed controller. Sensorless operation is in use, as is evident by the button: `encoderless_on` which is set to ON. Note that the results displayed will be identical, with either sensored or sensorless operation. The variables `VR1`, `VR2` were in this example configured to show the stator inductance and resistance values used by the estimator (user supplied in this lab via the appropriate dialog box entries). Confirmation of phase alignment between the angle generated by InstaSPIN and the encoder may be undertaken by setting the channel select button to `ch1-2`, in which case the Cosine function of the encoder and InstaSPIN generated angles will be shown. These must be fully aligned, as is indeed the case as may be observed from Fig. 4.16. Note that in practice the cos function generated by InstaSPIN serves as a good reference for removing any encoder offsets, i.e. one can change the encoder offset in the 'encoder speed/angle module' (see Fig. 4.11) to ensure that alignment as shown in Fig. 4.16 is achieved.

In conclusion to this laboratory a speed reversal from n_m=1450→ −1450 rpm is undertaken under no-load conditions (sensorless FOC). The results in the form of the phase current measured with a Tektronix DC-true current probe and scope is shown in Fig. 4.17. Observation of Fig. 4.17 learns that the phase current is limited to ±2.5 A by the speed controller (a user defined setting in the dialog box, see Fig. 4.10). Note also that before the speed reversal the phase current amplitude is not zero but approximately 0.4 A, which is the i_q current needed to generate a torque

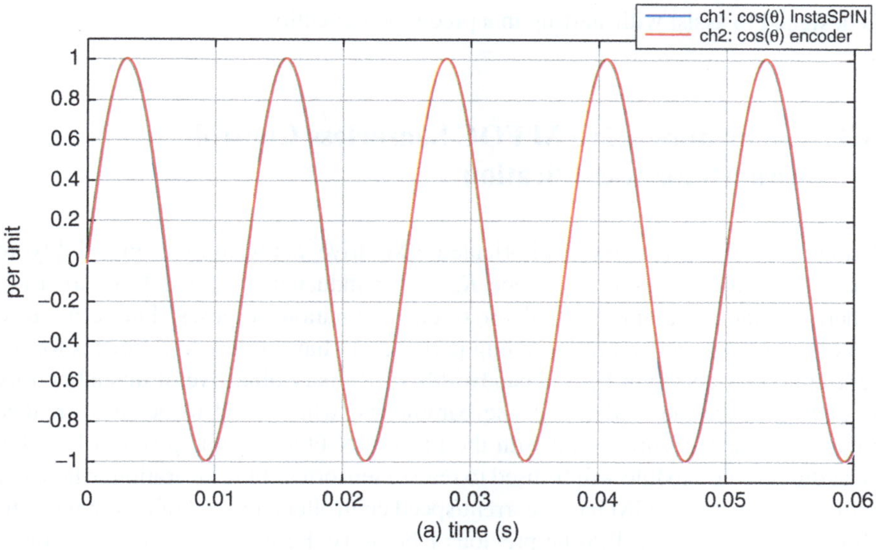

Fig. 4.16 Phase C+ simulation output, with 'ch1-2' button active: comparison of cosine functions generated by the encoder and FAST

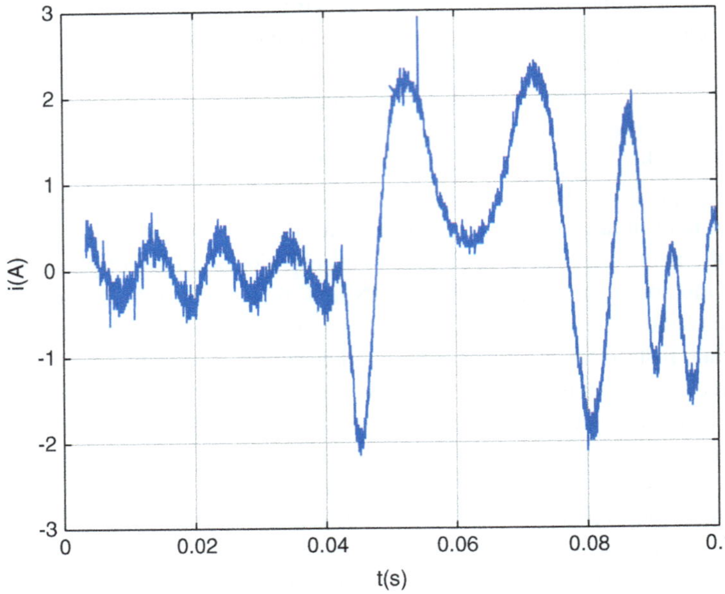

Fig. 4.17 Phase C+ output: phase current during speed reversal measured with Tektronix DC-true current probe and oscilloscope

value that matches the friction load torque present in the drive. Furthermore, the 'force angle' button should be OFF when undertaking speed reversals. This function is designed to assist with start-up in a predefined direction.

4.3 Laboratory 2:2: PM FOC Sensorless Control with Motor Identification

Critical to any sensorless field-oriented PM drive application is the ability to correctly estimate: stator resistance R_s, stator inductance L_s and PM flux ψ_{PM}, where the latter variable is used for torque estimation purposes. Furthermore, it is helpful for current controller tuning purposes, to have access to estimates for the (current controller) gain K_p and bandwidth ω_i. Primary objective of this laboratory is to familiarize the reader with the parameters, which need to be set to realize motor parameter identification with the InstaSPIN-FOC module given in Fig. 4.2. In addition, said module will be used to realize sensorless FOC operation, where use will be made of the ROM based current/speed controllers, instead of external (to the ROM) modules, as used in the previous laboratory. Furthermore, stator resistance features such as recalibration and online (during normal drive operation) estimation will also be discussed. Finally, so called 'goodness' plots will be introduced at the end of this section, which provide guidance, in terms of the parameters used for motor identification.

4.3.1 Lab 2:2: Phase B

The purpose of this laboratory is to demonstrate operation of the FAST based sensorless drive with motor identification. In particular the various stages of identification will be discussed. For this purpose the VisSim processor in the loop (PIL) approach will again be used, which communicates with the TMS320F28069M unit located on the LAUNCHXL board.

The following information is relevant for this laboratory component:

- Reference programs [11]: `lab2-2_LaunchXLphBv4.vsm` and `InstaSPINv4.vsm`.
- Description: Sensorless control of a PM motor with motor parameter identification.
- Equipment/Software: Texas Instruments LAUNCHXL-F28069M and VisSim simulation program.
- Outcomes: To familiarize the reader with the InstaSPIN motor identification algorithm for PM drives.

No boost packs are used in the laboratory, nor is an external 24 V power supply required, given that power to the MCU is directly provided via the USB cable connected to the laptop. In this case the following jumper and dip-switch positions on the LAUNCHXL-F28069M module are required:

- Jumpers `JP1` and `JP2` CLOSED
- Jumpers `JP4` and `JP5` OPEN
- Jumpers `JP3`, `JP6` and `JP7` CLOSED
- Dip-switches `SW1` to `SW3` ON

Note that there are two reference programs required for this lab: the first is the main simulation model which contains a PIL 'interface module' which in turn makes use of a second program `InstaSPINv4.vsm`. The outputs of the 'Drive Controller' shown in Fig. 4.18 are the three reference modulation indices `mA_ref`, `mB_ref`, `mC_ref`, which must be multiplied by the term $u_{DC}/2$. The resultant reference voltages `uA_ref`, `uB_ref`, `uC_ref` are connected to the 'converter', which in this case is simply represented by three limiters, that limit the voltage to the motor to $\pm u_{DC}/2$. The phase voltages `u_A`, `u_B`, `u_C` of the machine are used by the machine model and by the InstaSPIN algorithm. For the latter case the converter outputs are filtered using 'analog low-pass filter' modules (see Fig. 4.18) that replace the electrical circuit models which are present in the actual hardware. The filtered voltages are then scaled with the aid of a 'Gain module', by a factor

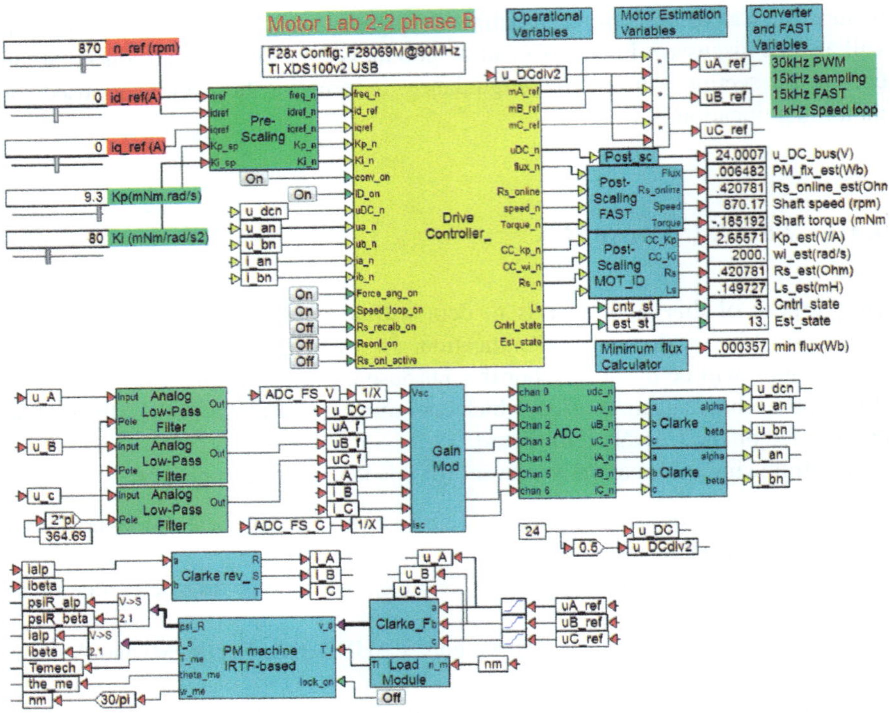

Fig. 4.18 Phase B simulation of a FOC PM drive with motor identification capability: screen shot taken after completion of parameter estimation

3.3/ADC_FS_V, where ADC_FS_V is the full scale ADC voltage set to 26.314 V for the boost pack. A similar scaling is also used for the machine phase currents i_A, i_B, i_C, in which case the scaling factor is equal to 3.3/ADC_FS_C, where ADC_FS_C is the full scale ADC current value set to 33.0 A. A detailed model of the 12 bit AD converter is again used to convert the scaled voltage/current inputs value to per unit outputs for use with the current controller and InstaSPIN module.

A set of sliders is used to define the current reference vector, in terms of its direct and quadrature components $i_{\mathrm{d}}^{\mathrm{ref}}$, $i_{\mathrm{q}}^{\mathrm{ref}}$ and rotational speed $n_{\mathrm{m}}^{\mathrm{ref}}$. A 'pre-scaling' module has been added to convert the floating point variables required for the controller to fixed point format. In addition, two sliders have been added which allow the user to set the proportional $\mathrm{K_p}$ and integral $\mathrm{K_i}$ speed gains of the controller.

The three dialog boxes for this laboratory, shown in Fig. 4.18, contain all the user settings required for this laboratory. For example, the 'Operational Variables' dialog box is again used to assign full scale values for the voltage, current and frequency. In addition, the current controller gain and bandwidth are also assigned here. Also assigned in this box is the rate of variable change for current and frequency, i.e. introduced as a safety measure to avoid erratic slider action from causing over-currents in the drive. The 'Converter and FAST variables' dialog box contains all the parameters which must be set for the InstaSPIN algorithm, which includes among others the full-scale ADC value and low-pass filter corner frequency. A third dialog box 'Motor Estimation Variables' is also shown, which contains the parameters that define the motor identification sequence. A more detailed discussion on these dialog boxes will be given in the 'phase C' development stage.

Central to this laboratory is the VisSim 'processor in the loop' (PIL) software tool, which provides access to the InstaSPIN ROM in the hardware. Located within the controller module (two levels down) is a 'Target Interface module' given in Fig. 4.19 which is used to communicate with the InstaSPIN ROM.

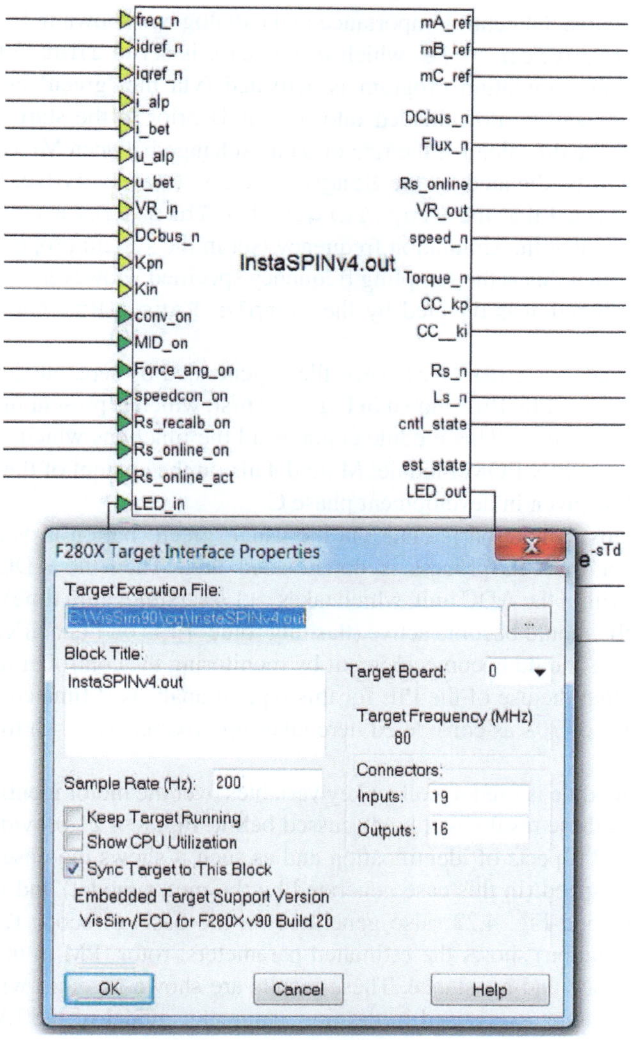

Fig. 4.19 Target interface module with corresponding dialog box entry

Inputs to this module are the per unit α, β voltage and currents generated by the ADC unit and a set of control inputs. Of these the logic input `MID_on` is used to activate the parameter identification process. Further details on all inputs will be given in the next lab section. Key outputs are the modulation indices `mA_ref` to `mC_ref`, DC bus voltage `DCbus_n` , rotor flux estimate `Flux_n`, and (among others) the estimated stator inductance `Ls_n` and stator resistance `Rs_n`. Furthermore, estimates are provided for the current controller gain `CC_kp` and bandwidth `CC_wi`. To monitor operation during the identification sequence the controller state `cntrl_state` and parameter estimation state `est_state` are also generated by this module. The `LED_in`/`LED_out` variables are used to activate a 'blue' LED on the LAUNCHXL board which flashes when PIL operation is active. Of central importance in the dialog box shown in Fig. 4.19 is the `Target Execution File`, which in this case is set to `InstaSPIN4.out`. When the main simulation program is activated (via the 'green' arrow button), this `.out` file will be downloaded into the MCU prior to the start of the actual simulation. Once downloaded the rate of data exchange between MCU and VisSim simulation will be dictated by the `Sample Rate (Hz)` entry (set to 200 Hz in this case) provided that the `Sync Target to This Block` option has been activated. Note that the simulation frequency (set in the system properties) is set to 15 kHz given that this is the sampling frequency specified. However, actual run time during PIL operation is dictated by the `Sample Rate (Hz)` entry mentioned above.

The required `InstaSPINv4.out` file is generated by separate compilation of the module 'InstaSPIN PIL' shown in Fig. 4.20 first, which is present in the program `InstaSPINv4.vsm`. This module contains all the functions which are needed to access the InstaSPIN ROM module. More details on the content of the 'InstaSPIN' module will be given in development phase C.

Starting this simulation is done via the usual 'green' button in which case the `.out` file for the PIL process is downloaded first. Then the ADC offsets are determined within the ADC unit, which takes ≈ 1.4 s, (simulation time) AFTER that period the PIL should become active (flashing 'blue' LED on LaunchXL board) and drive operation should become apparent by monitoring the controller and estimator states. Note that the use of the PIL for this type of analysis is time consuming as a time sequence of 70 s as considered here takes approximately ≈ 4 h to complete in real time.

A scope module is used to collect key variables over the motor identification time sequence and these results will be discussed below. Figure 4.21 provides details on the operational aspects of identification and as such it shows the observable phase current, shaft speed (in this case generated by the motor model) and the estimator state. The second Fig. 4.22 (also generated on the basis of scope results for the phase B simulation) shows the estimated parameters: rotor (PM in this case) flux, stator inductance and resistance. These results are shown together with the actual (reference) parameters as used in the four parameter model of the LVSERVMTR PM machine.

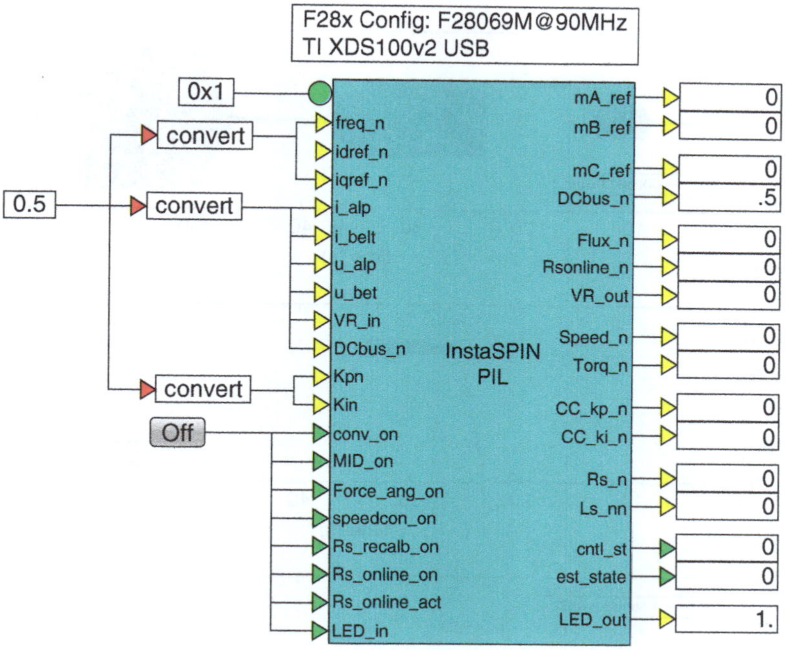

Fig. 4.20 InstaSPIN PIL module

On the basis of these two figures an overview will be given of the motor identification cycle as identified by the estimator state. Events shown, assume activation of the converter at $t = 0$ s with motor identification active (button `MID_on` ON). Other buttons such as `Rs_recalb_on`, `Rs_online_on` and `Rs_online_act` (used for resistance estimation purposes outside the 'normal' identification sequence) are set to OFF. Furthermore, the identification cycle makes use of the dialog box settings specific to this example. No mechanical loading of the machine is assumed in this example. Note also that not all estimator state numbers appear below, as some refer to induction machine identification hence not relevant to this discussion.

The following estimator states are present during PM machine identification given below:

- State 1: referred to as 'idle' because the estimator is not in use. After converter start up (button `conv_on` ON) an ADC offset measurement sequence takes place to remove any DC offsets in the ADC converter from the measurement process. Motor should **not** rotate during this phase.
- State 2: referred to as 'RoverL'. The current vector amplitude and frequency are set to 0.5 A and 300 Hz respectively. Note that current amplitude used is half the value used in the next phase. During this measurement sequence the so called high frequency (HF) stator resistance R_{HF} and stator inductance L_{HF} estimated values are found, which determine the current controller gain/bandwidth. Motor

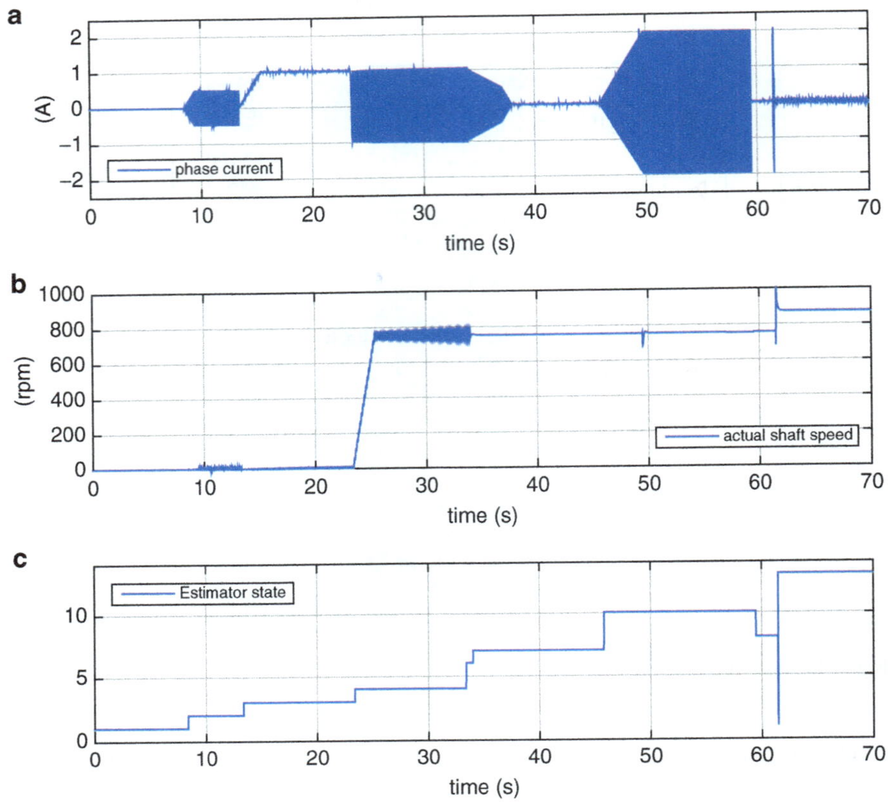

Fig. 4.21 Phase B simulation of a FOC PM drive with motor identification capability: operation cycle

should **not** rotate during this phase. Values generated should provide a realistic approximation of the inductance and resistance of the machine and these are then used to dimension the current controller, as discussed in Sect. 2.1.5.

- State 3: referred to as 'Rs'. DC current level is set tot 1.0 A. Note that the measured phase current may be unequal to the set (vector) amplitude value, as it depends on which phase is measured by the current probe, i.e. one phase can (for example) show 1.0 A, while the other two will then show 0.5 A each. Measurement of the stator resistance takes place as may be observed from Fig. 4.22. Motor should **not** rotate during this phase.
- State 4: referred to as 'Ramp up', which undertakes open-loop speed control, with a current vector amplitude set to 1.0 A. Motor should increase speed which matches the rotational frequency 50 Hz of the current vector as may be observed from Fig. 4.21. Hence, the rotational speed will be $^{50*60}/_{p} = 750$ rpm, where $p = 4$ is the pole pair number of the machine in use. Note that during open-loop speed control, stable speed operation is often impaired by the lack of

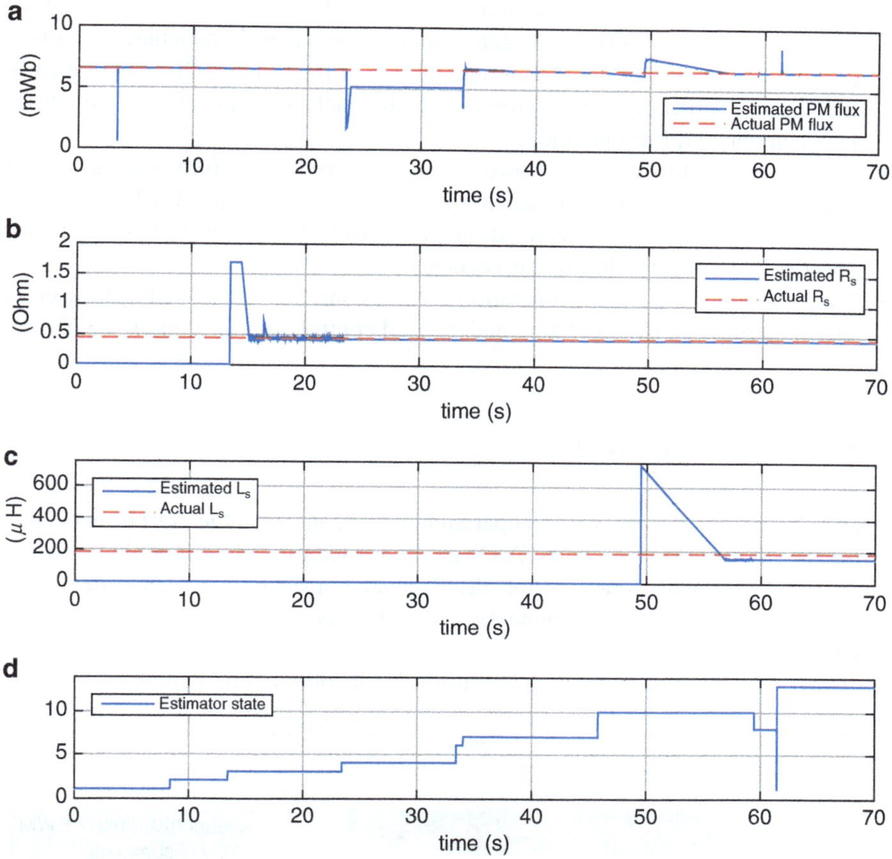

Fig. 4.22 Phase B simulation of a FOC PM drive with motor identification capability: parameter measurement

electrical damping. In practice mechanical damping caused by friction is usually sufficient to ensure stable operation during this sequence.

- State 6: referred to as 'open-loop flux'. During this time interval the flux during open-loop speed control is estimated.
- State 7: referred to as 'Rated flux', which undertakes closed-loop speed control. Current amplitude is determined by the load present, which should be friction only (no external load). In this case the phase current is zero because no friction is present here, as may be observed from Fig. 4.21. Shaft speed is set by the flux estimation frequency of (in this case) 50 Hz, hence the operational speed will be 750 rpm for the LVSERVOMTR PM motor. The rated rotor flux (in this case the flux due to the permanent magnets) is measured.
- State 10: referred to as 'Inductance'. A direct axis current of 2.0 A is applied whilst the drive is operating under closed-loop speed control. Shaft speed is determined by selected frequency 50 Hz and motor pole pair, hence 750 rpm.

The stator inductance is measured as may be observed from Fig. 4.22. Note that there is a 18 % error between estimated and actual inductance which is attributed to the use of an ADC model instead of the actual ADC. Consequently the sampling process and the compensation for such errors in the algorithm differ between real and PIL operation.

- State 8: referred to as 'ramp down', current ramp down before starting normal operation. In this example no friction is present hence the speed will not change when the current is set to zero. In practice, friction will ensure that the motor speed reduces to zero during this sequence.
- State 13: referred to as 'user operation', where control is transferred to the user. In this example the drive operating under speed control with a speed of ≈ 870 rpm.

4.3.2 Lab 2:2: Phase C

Prior to being able to work with an operational drive, an `.out` file must be generated using the 'Drive Controller with FAST and Motor Identification' module, as shown in Fig. 4.23, which represents the drive under consideration. Hence details of the drive structure and corresponding dialog box parameter assignments will be discussed in this subsection.

The following information is relevant for this laboratory component:

- Reference program [11]: `lab2-2_LaunchXLphCv11.vsm`.

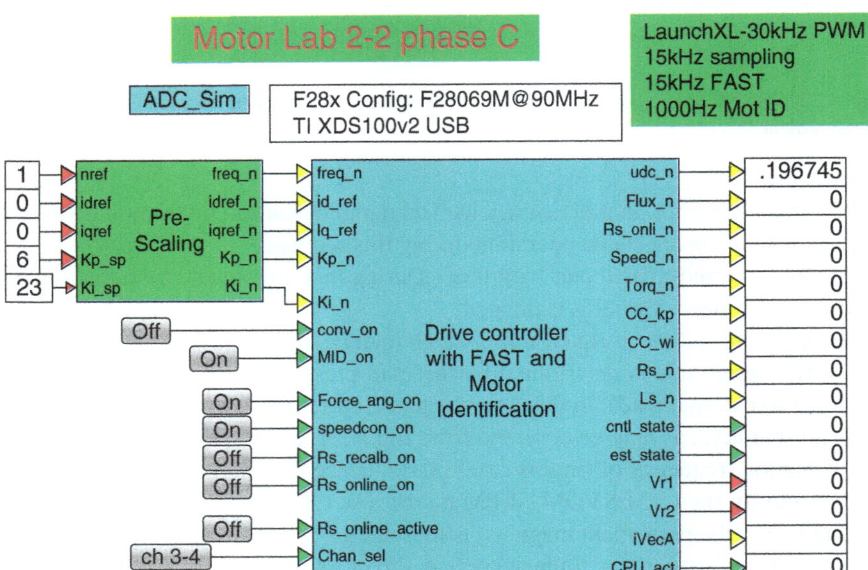

Fig. 4.23 Phase C simulation of a InstaSPIN FAST based encoderless field-oriented controlled (FOC) PM drive, with motor identification capability

- Description: Motor Parameter Identification and sensorless control of a PM motor using the InstaSPIN module.
- Equipment/Software: Texas Instruments LAUNCHXL-F28069M, with BOOSTXL-DRV8301 module ('aft' position) and VisSim simulation program.
- Outcomes: Generate the `.out` file needed to represent the drive structure under investigation.

As mentioned above, a single boost pack located in the 'aft' position (furthest away from the USB connector) is used for single motor operation (induction machine is not used electrically), in which case the following jumper and dip-switch positions on the LAUNCHXL-F28069M module are required:

- Jumpers `JP1` and `JP2` OPEN
- Jumpers `JP4` and `JP5` CLOSED
- Jumpers `JP3`, `JP6` and `JP7` CLOSED
- Dip-switches `SW1` to `SW3` ON

Inputs to the Controller are five variables, which set the reference shaft speed, direct/quadrature reference currents, and speed controller gains. Four logic buttons are used to select speed/torque control, diagnostic channels used, force angle on/off and on/off control of the converter, as discussed in the previous laboratory. In this laboratory, four additional buttons are present, which are associated with motor parameter identification and stator resistance measurement namely:

- `MID_on`: when activated, executes the identification sequence, after the converter has been activated (using the `conv_on` button) that identifies:
  - `CC_kp/wi`: the proportional gain K_p (V/A) and bandwidth ω_i (rad/s) used for the current controller. During this identification mode the machine must be at standstill.
  - `Rs`: the stator resistance R_s of the machine.
  - `FLux_n`:estimate of the PM flux for the machine.
  - `Ls_n`: estimate of the stator inductance L_s of the machine.
- `Rs_recalb_on`: when activated, measures the stator resistance using the 'offline' identification, i.e. after converter start-up. Upon completion, normal user operation resumes.
- `Rs_online_on`: when activated, measures the stator resistance when 'online', i.e. when the drive is operating.
- `Rs_online_active`: when activated, forces the FAST sensorless algorithm to use the 'online' estimate. Note that this function should be enabled provided that the online output value is in reasonable agreement with the 'offline value'.

Note that motor identification and online stator resistance measurement requires access to the current/speed controllers inside the MCU module shown in Fig. 4.2. Hence the approach used in this laboratory is preferred when motor identification together with sensorless FOC and online stator resistance estimation is required for a given application.

Outputs of this module are (among others) the estimated parameters: flux `Flux_n`, stator resistance `Rs_n` and stator inductance `Ls_n`. Also provided are the measured DC bus voltage `udc_n`, online stator resistance `Rs_onli_n`, shaft speed `Speed_n` and the controller/estimator states `contl_state/est_state`. Also added are two diagnostic variables `Vr1`, `Vr2` which have been assigned to show the HF estimated resistance and inductance (obtained during estimator stage 2). To assist with the observation of the identification process a variable `iVecA` is provided which represents the per unit amplitude of the measured current vector $\vec{i} = \sqrt{i_{\alpha}^2 + i_{\beta}^2}$. The variable `CPU_act` provides an indication of percentage CPU activity.

A PWM frequency of 30 kHz and a ADC sampling frequency of $f_{\mathrm{ADC}} = 15\,\mathrm{kHz}$ has been selected for this laboratory. The current/speed controllers have also been set to operate at the ADC sampling frequency.

> The FAST and Motor Identification algorithms must be set to operate at sampling frequencies of $15000/i$ kHz and $15000/j$ kHz respectively, where i, j are integer values only.

For this lab the values used are $i = 1$ and $j = 15$ which implies that the sampling frequency for the FAST and Motor ID algorithm have been set to 15000 Hz and 1000 Hz respectively. These frequencies are set in the appropriate dialog boxes as will be discussed below. Allocating different computational step times to less time critical algorithms (as discussed in Sect. 2.1.6) reduces the risk of processor interrupt overrun (allocated tasks cannot be carried out within one sample period). Moving one level lower into the 'Drive Controller' module reveals the modules shown in Fig. 4.24. Of these shown in the figure, the '100 Hz' and 'Diagnostic tool' modules have been discussed in previous laboratories. The dialog box parameters for this laboratory are given in Fig. 4.25. The parameters associated with the 'Operation variables' and 'Converter and FAST Variables' dialog boxes, have been discussed in the previous laboratory 2:1. Of interest here is therefore, dialog box 'Motor Estimation Variables', where the following motor identification parameters must be assigned:

- `F_motorID/F_sampling`: ratio between sampling frequency used by the Identification algorithm and ADC sampling frequency. The ratio is an integer value, hence the user can select the ratio from a dialog box pull-down menu. For example, a choice of `1/15` implies that motor identification algorithm is executed with a frequency of 1000 Hz, given that the ADC sampling frequency is set to 15000 Hz.
- `R/L sense freq (Hz)`: frequency used to estimate the current controller gain and bandwidth during estimator state 2.
- `Flux est_freq (Hz)`: frequency used to estimate the PM flux and stator inductance during estimator states: 4, 6, 7 and 10.

- `Freq_est-ramp (Hz/s)`: rate of frequency change used for motor identification purposes.
- `Current_est_ramp (A/s)`: rate of current change used for motor identification purposes.
- `2xR/L_est,Rs_es_current (A)`: current amplitude used for current controller gain/bandwidth and stator resistance estimation during estimator states 2 and 3.
- `FLx-Inductance_est_current(A)`: current amplitude used for PM flux and stator inductance estimation during estimator states 4 and 10. Value used during estimator state 4 is chosen to be half the value used during state 10.
- `OnlineRs_est_current (A)`: current amplitude used to modulate the direct axis current, as required to achieve online stator resistance estimation.

Moving one level lower into the 'main motor control' module, reveals a set of modules given in Fig. 4.26. Of these shown, all with exception of the 'InstaSPIN-FOC module' have, in fact, been identified in earlier laboratories. Key inputs to the controller module are the per unit α, β current/voltage variables: `i_alp`, `i_bet`, `u_alp`, `u_bet` and DC bus voltage variable `DCbus_n`, generated by the ADC-PWM module. User reference variables: `Freq_n`, `Ampd`, `Ampq_n` and `Kpn`, `Kin` represent the scaled users sliders for shaft-speed, direct/quadrature currents and speed controller gains. Logic input `doMotorID` is used to activate the motor identification sequence. Outputs of the 'InstaSPIN-FOC module' are (among others) the modulation indices: `mA_ref`, `mB_ref`, `mC_ref` generated by the internal current controller. The remaining outputs are those which have been named above. Moving one level into the InstaSPIN-FOC module, as shown in Fig. 4.27, reveals the basic VisSim 'InstaSPIN Motor Control' and 'estimator calculations' modules. The 'InstaSPIN Motor Control' module communicates directly with the InstaSPIN ROM module, which returns the modulation indices for use with the ADC/PWM

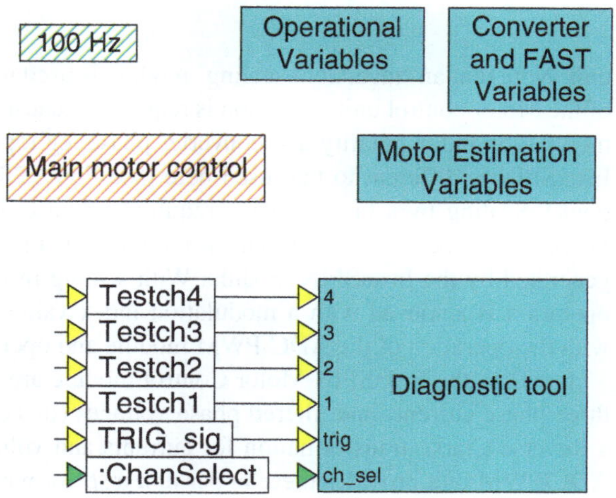

Fig. 4.24 Phase C simulation of a FAST based encoderless FOC drive, with motor identification: one level into the Drive Controller module

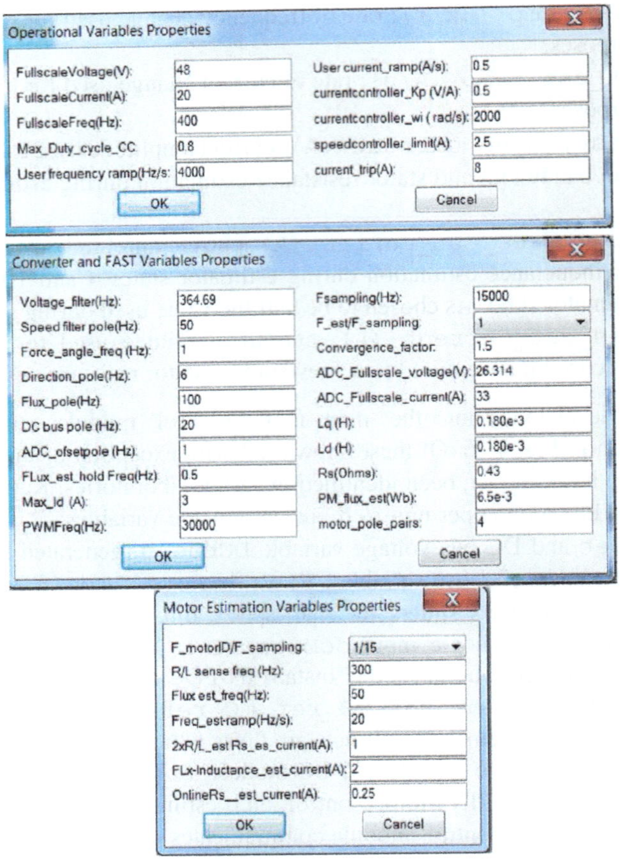

Fig. 4.25 Phase C simulation of a FAST based encoderless FOC drive, with motor identification: Dialog box entries

unit. Note that an 'inversion+scaling' module is used to invert and scale the outputs of the Motor control unit. Inversion is required because the relationship between the modulation index polarity and converter phase leg top/bottom switches as used by InstaSPIN is different to that used here (see Fig. 2.23 for convention used in this book). Scaling by a factor $\sqrt{3}/2$ is required because the FAST module has a gain factor of $2/\sqrt{3}$ between the outputs of the internal current controller and the outputs generated by the InstaSPIN module. With scaling in place, the boundary of linear operation is achieved with a modulation index value of $2/\sqrt{3}$, which is consistent with the operation of the ADC-PWM module and operation as shown in Fig. 2.21.

Inputs to the InstaSPIN Motor Control module are the per unit DC bus voltage, three phase currents and filtered phase voltages (in `IQ24` format). Use is made of a Reverse Clarke transformation for currents and voltage variables, given that the ADC/PWM unit provides these in per unit α, β format. The Motor Control module has an extensive dialog box, which holds (among others) the dialog box parameters set in Fig. 4.25. An 'Estimator calculations' compound module, also shown in Fig. 4.27, is used to derive the rotor-flux, shaft-speed and shaft torque values from the ROM. In addition, the per unit reference frequency `freq_n` and reference direct axis current `idref_n`/quadrature axis current `iqref_n` input variables are passed to the ROM via the 'Estimator calculations' module. Computation of the current vector amplitude `IVECA` as used to observe the identification process is also calculated in this module. Note the content of said module is almost identical to the content of the InstaSPIN PIL module shown in phase B (see Fig. 4.20). The use of a hatched surface, indicates that the sampling rate has been set independently of the main simulation frequency, which is typically the ADC sampling frequency. Practical, is to set the sampling frequency for the 'Estimator Calculator' to the same value used by the speed controller of the module, i.e. 1000 Hz in our case. Note that not all the input/output variables used by the InstaSPIN module shown in Fig. 4.26 appear in Fig. 4.27. The reason for this is that variables such as, for example, `Kpn`, `Kin`, `Rs_n` etc. do not need to be determined at the estimator frequency sampling rate, instead they are executed in the '100 Hz' module, i.e. they are sampled at 100 Hz instead of 15 kHz to reduce CPU usage. For diagnostic purposes the test variables: `TESTch1`, `TESTch2` have been connected to per unit variables: i_α^n and i_β^n respectively. Furthermore, test variables: `TESTch3`, `TESTch4` have been connected to per unit variables: $u_{n\alpha}$ and $u_{n\beta}$ respectively.

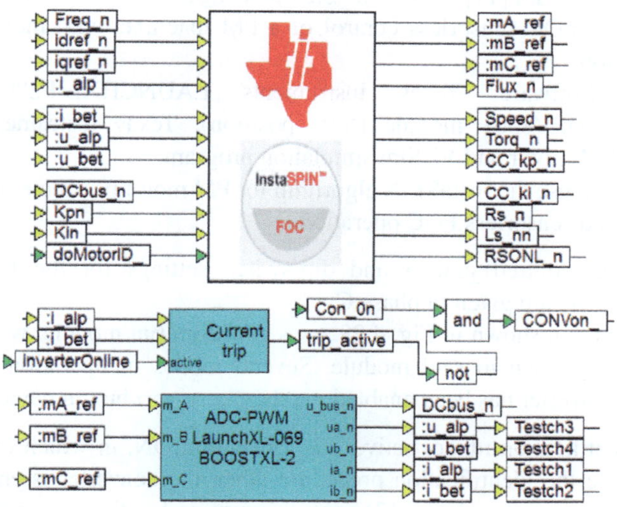

Fig. 4.26 Phase C simulation of a FAST based encoderless field-oriented controlled (FOC) drive, with motor identification: one level into the 'Main Motor Control' module

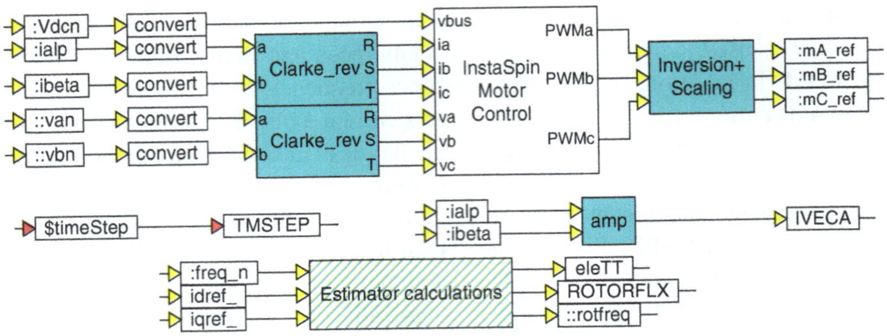

Fig. 4.27 Phase C simulation of a FAST based encoderless field-oriented controlled (FOC) drive, with motor identification: one level into the InstaSPIN-FOC module

4.3.3 Lab 2:2: Phase C+

Phase C+, is the operational component of the laboratory and is basically a run version of the `.out` file compiled and downloaded to the MCU in phase C (see previous subsection).

The following information is relevant for this laboratory component:

- Reference program [11]: `lab2-2_LaunchXLphCv11_d.vsm`.
- Description: FOC sensorless control of a PM machine, with motor parameter identification.
- Equipment/Software: Texas Instruments LAUNCHXL-F28069M, with BOOSTXL-DRV8301 module ('aft' position), Texas Instruments LVSERVOMTR PM motor and VisSim simulation program.
- Outcomes: to use the InstaSPIN algorithm for PM motor parameter identification purposes and sensorless FOC operation.

Note that the required jumper and dip-switch settings for the LAUNCHXL-F28069M module are given in phase C.

The run version shown in Fig. 4.28, uses a VisSim run module, which executes the `.out` file, shown in said module. Several modes of operation are possible AFTER the converter has been enabled via the `conv_on` button namely:

- Motor Identification mode: active, `ID_on` button ON, in which case the drive will execute the identification procedure when the converter is enabled. Upon completion (estimator status: 12) control is returned to the user and the parameters are shown on the numerical displays (see Fig. 4.28).
- Motor Identification mode: not active, `ID_on` button OFF, in which case drive operation will depend on the status of the `Rs_recalb_on` button, hence there are two options:

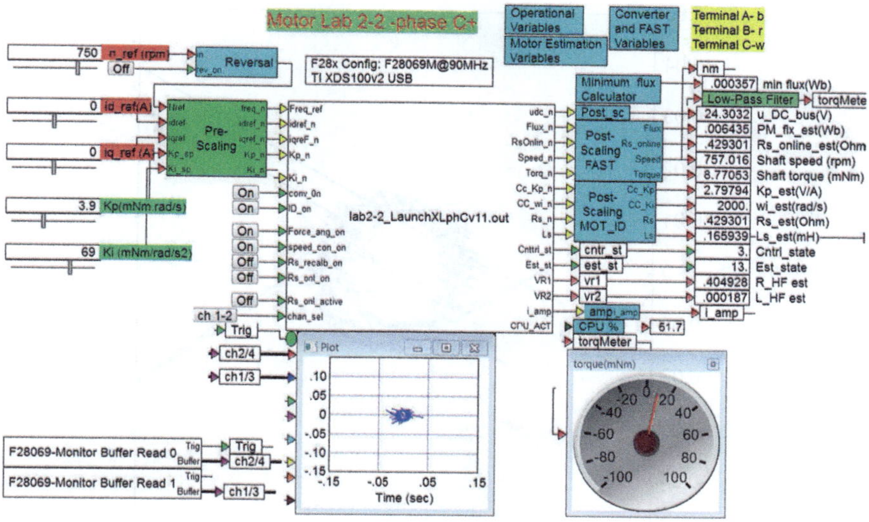

Fig. 4.28 Phase C+ simulation of a FAST based encoderless field-oriented controlled (FOC) drive, with motor identification: example shows motor under no load, after identification has been done)

- `Rs_recalb_on` button ON, in which case the Rs value is estimated using the 'off line' procedure applied during the standard identification procedure PRIOR to returning control to the user.
- `Rs_recalb_on` button OFF, in which case control is directly given to the user, to reduce any delay between converter action and drive operation.

During user operation, 'online' stator resistance measurement is possible by enabling the `Rs_onl_on` button. This will then display the online stator resistance value in the corresponding numerical display shown in Fig. 4.28, provided that BOTH `Rs_recalb_on` AND `ID_on` buttons are OFF. The FAST algorithm will use the online stator resistance value, when the `Rs_onl_active` button is ON. Note that a check should be made that the 'online' and 'offline' Rs values are in reasonable agreement, prior to activating said button. A screen-shot of phase C+ operation as given in Fig. 4.28 shows operation after conclusion of motor identification. An enlarged view of the VisSim scope module given in Fig. 4.29 shows the per unit currents $i_{n\alpha}$, i^{n}_{β} after completion of identification. Multiplication by the full scale current value of 20 A shows that a current vector of $\approx$ 0.2 A amplitude is used to generate the torque needed to overcome the friction of the drive. A substantial noise component is also present on these waveforms which is caused by the PWM switching process.

Five sliders are used, the purpose of which was discussed in the previous section. A post-scaling module is again used to convert the per unit measured DC bus voltage to actual voltage, as shown with a numeric display.

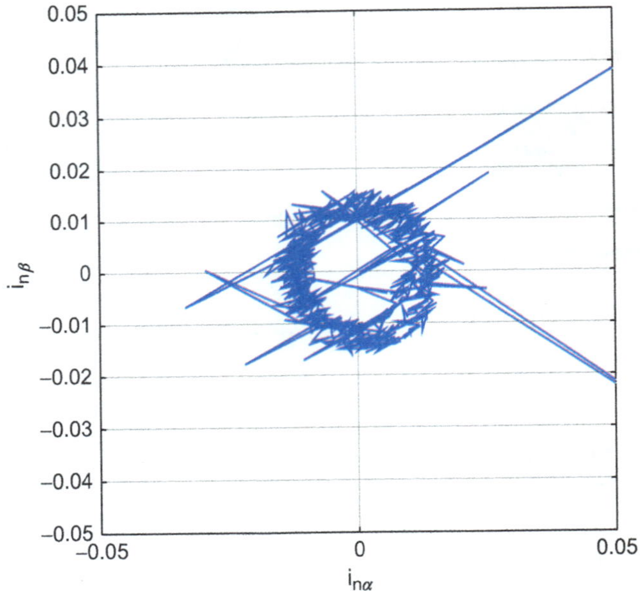

Fig. 4.29 Phase C+ simulation: Scope display showing per unit α, β currents after completion of motor identification (machine running with friction load only)

Note that a terminal-motor connection diagram is also shown in Fig. 4.28, which in this case is simply present to set a consistent rotational direction for the drive. A low-pass filter with a 1 Hz corner frequency, is used to filter the torque signal for display on a torque meter (see Fig. 4.28). Two 'Monitor Buffer' modules are used to display two selected (using a button connected to the `chan_sel` input) diagnostic signals. For this lab the diagnostic scope is set in `XY Plot` mode (right click on scope module to change this setting), hence the locus of the (per unit) current vector is shown.

Once motor identification has been activated (by setting the motor ID button to ON and switching on the converter) the gain/bandwidth estimates (`Kp`, `wi`), estimated machine parameters (`Rs`, `Ls`) sequentially appear on the numerical displays (see Fig. 4.28).

In terms of safety, the reader is again reminded that prior to activating this type of lab component it is prudent to execute the following 'Pre-Drive' check list:

- Dialog boxes used in C+ mode, match those of phase C: the run version is compiled with the dialog box entries specified under phase C. The dialog boxes shown in phase C+ are used by the 'Pre/Post' scaling modules.

(continued)

- Ensure that the sample time used is correct and latest (and correct) `.out` file has been downloaded to the MCU (Right mouse click on MCU module to show dialog box of these variables).
- Confirm that the user input values are set to either zero, or 'acceptable' values, which will not cause a current trip of the converter.
- Confirm that the converter 'switch' is set to OFF and the power supply is on (DC bus voltage present).
- Confirm that the motor is connected firmly and properly.

After completion of the Pre-Drive checklist, activate the program and confirm that the supply voltage source level shown in the digital display is 24 V. If not stop the program and restart. With the correct DC voltage level established, turn on the converter (using the ON button) and monitor the motor shaft and diagnostic scope, which in this example shown (Fig. 4.28) is set to show the per unit currents.

The experimental identification sequence for the PM motor as given in Fig. 4.30 corresponds with the operational sequence derived earlier with lab 2:2 phase B (see Fig. 4.21). Also shown in Fig. 4.30 are the estimator states present during parameter identification. These are linked to a timing diagram provided in Fig. 4.31, which shows a dialog box with the relevant section that controls the duration of the relevant estimator states. This figure is part of InstaSPIN dialog box (right click on the module shown in Fig. 4.27), which shows the timing cycles relevant (identified by estimator state) to PM identification. Actual time is found by taking the number of cycles shown and dividing by the estimator frequency in use. Hence with an estimator frequency of 15 kHz (as used in this laboratory) the actual time of, for example item (2), will be $^{75000}/_{15000} = 5$ s. A detailed discussion on the estimator states present during motor identification has already taken place in lab 2:2 phase B, hence only a brief summary of the sequence (using Figs. 4.30 and 4.31) and estimator states, is given below.

- State 2: Identification of the current controller gain and bandwidth. Frequency and amplitude (set in motor ID dialog box) are 300 Hz and 0.5 A (the current value used in this sequence is half the value used in the Rs estimation phase). The duration is set to 75000 cycles, hence 5 s, given the estimator frequency in use.
- State 3: 'offline' stator resistance measurement, DC current level (set in motor ID dialog box) is set tot 1.0 A. Note that the measured phase current may be unequal to the set value, as it depends on which phase is measured by the current probe. This sequence has three sub-cycles and the total time of this event is equal to 15000 cycles, hence 10 s (note that a scroll down in the `TTMotorware Block Properties` dialog box is required to show this number).

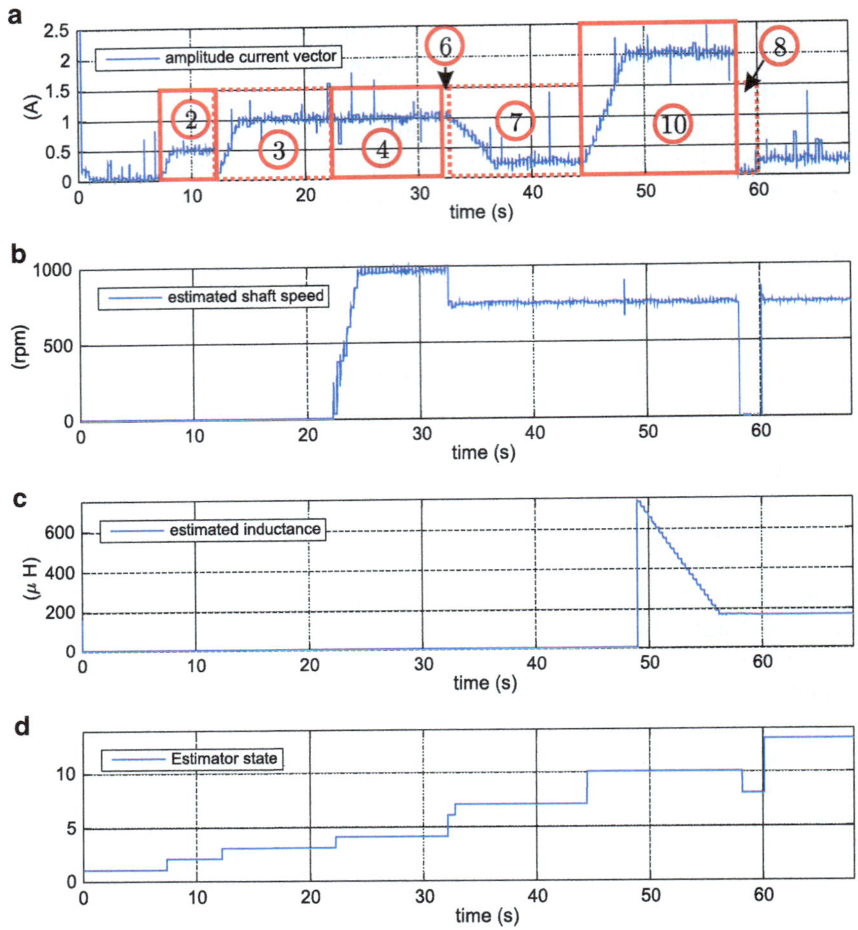

Fig. 4.30 Motor ID sequence for PM: operational data

- State 4: Open-loop speed control, with a current vector amplitude set to 1.0 A, which is half (default) of the flux amplitude set in the motor ID dialog box. The number of cycles shown in the dialog box is 15000 cycles, which corresponds to 10 s, given the estimator frequency in use. An additional time set to `USER_MOTOR_FLUX_EST_FREQ_Hz/USER_MAX_ACCEL_EST_Hzps` (equal to 2.5 s in this case) must be added to arrive at the total sequence time. Note that the estimated speed shown during this sub-cycle is not valid as the flux estimation process has not been completed. The actual speed is in this case 750 rpm as shown in Fig. 4.30.
- State 6: Open-loop flux estimation. Duration is set to 3000 cycles, hence 0.2 s.

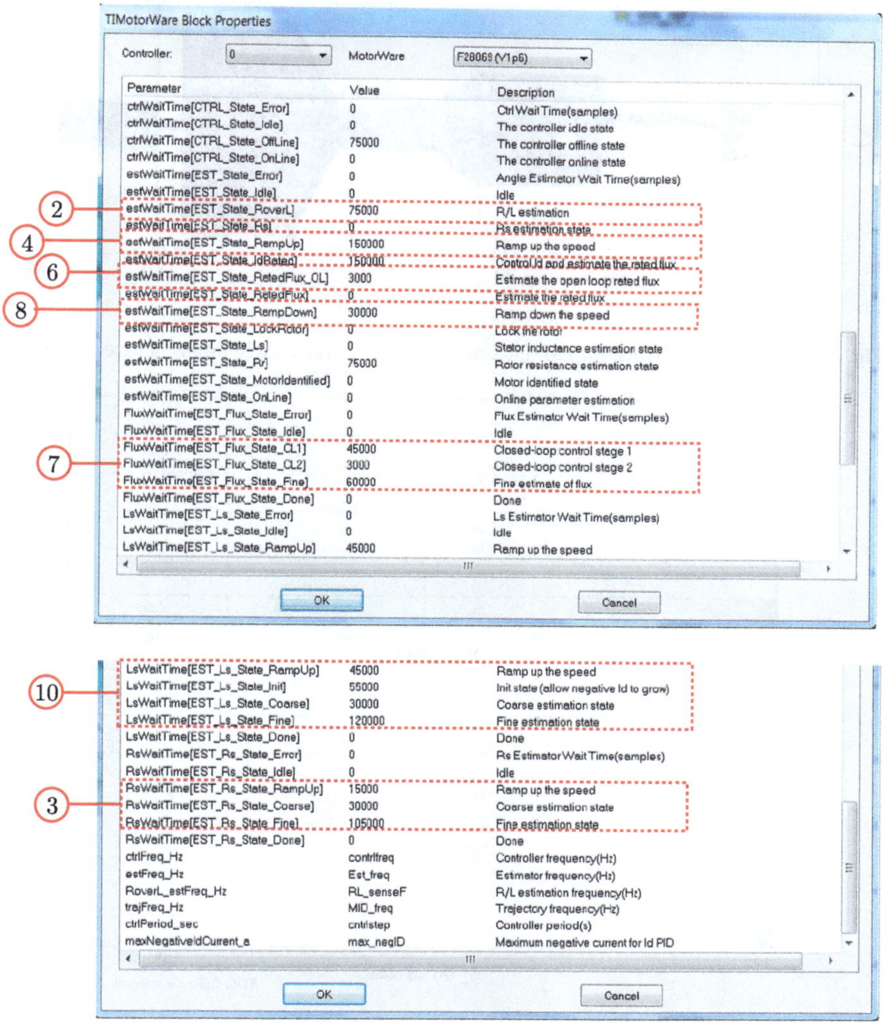

Fig. 4.31 Motor ID sequence for PM: timing data

- State 7: Closed-loop speed control. Current amplitude is determined by the load present, which should be friction only (no external load). This sequence has three sub-cycles and the total time of this event is equal to 108000 cycles, hence 7.2 s.
- State 10: Measurement of the stator inductance. Current amplitude of 2.0 A (set in motor ID dialog box), speed of 750 rpm, which is determined by selected frequency 50 Hz, in this case. This sequence has four sub-cycles and the total time of this event is equal to 250000 cycles, hence 16.6 s. Note that the inductance estimation process has been purposely added to Fig. 4.30, as the reader needs to make sure that a steady-state inductance value has been reached BEFORE this

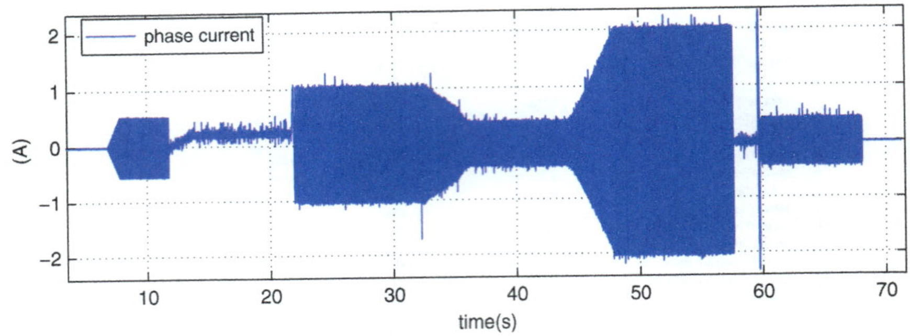

Fig. 4.32 Motor ID sequence for PM: measured with a DC-true current probe and scope

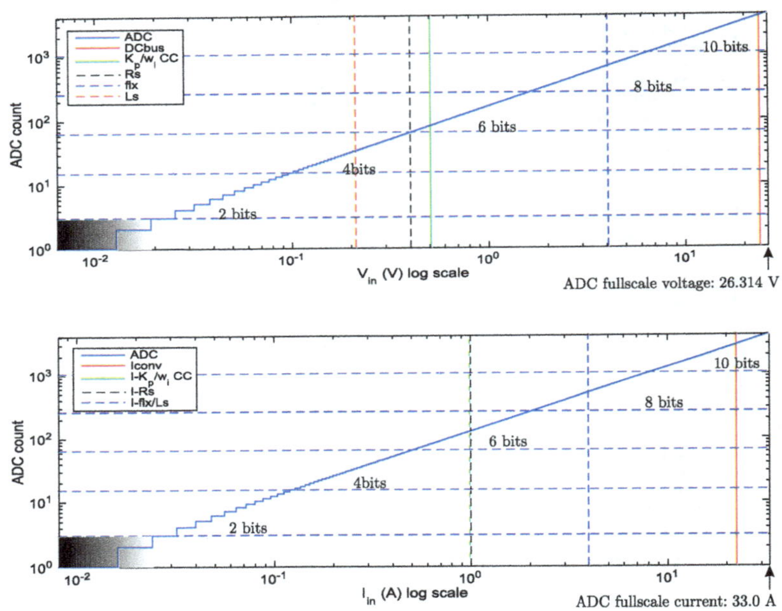

Fig. 4.33 Motor identification 'goodness' plots

cycle is terminated. Hence choosing a cycle time that is less than the time needed to reach a steady-state inductance value will lead to an erroneous result.

- State 8: Current ramp down, where motor speed will reduce to zero. The duration is set to 30000 cycles, hence 2 s, given the estimator frequency in use.

Prior to discussing the so called motor 'goodness plots' the phase current as measured with a Tektronix DC-true current probe and oscilloscope is provided in Fig. 4.32 for reference purposes. The results given in Fig. 4.32 clearly show the various stages of parameter identification as prior discussed using Figs. 4.30 and 4.22. Note that the phase current value shown during estimator state 2 (Resistance measurement) may vary from the assigned current vector value, as discussed previously. Furthermore, the converter is activated at $t = 0$ s, while user operation resumes at $t \approx 60$ s.

The motor identification routine, as described above, will provide accurate and meaningful results, provided that prudent choices are made with respect to the ADC full-scale voltage/current values and motor identification dialog box parameters. Furthermore, said variables should be selected, with reference to the motor and converter current/voltage ratings. In addition, particular attention should be given to the 12 bit ADC converter, because quantization errors can lead to erroneous results, if insufficient attention is given to the issues mentioned above. To assist the reader with this matter a set of so called 'Motor Identification goodness' plots, shown in Fig. 4.33 have been developed.

> Critical for achieving good sensorless operation is a continued awareness by the reader of the number of ADC bits that are in use by the algorithm when measuring the filtered converter voltages and currents. In this context the selection of the ADC full-scale values is important.

These 'goodness' plots show the voltage/current levels, which occur during motor identification in the context of the voltage/current ADC transfer function 'ADC' as shown in Fig. 4.33. More specifically, the goodness plots show the ADC bit utilization, which is, as was mentioned above, critical to achieving good sensorless performance. In addition, the ratings of the converter and motor can be added (provided they are within the voltage/current range used). The voltage 'ADC transfer' function, given in Fig. 4.33 (top plot) shows the 'ADC count', which is the decimal output value that corresponds to the input voltage range $0 \rightarrow$ ADC-Fullscale voltage, where the ADC-Fullscale voltage) (see Eq. (3.3a)) value is equal to 26.314 V for the BOOSTXL-DRV8301 module. Also shown (in both plots) are a set of ADC binary output levels, which have been added to show the number of bits in use for a given ADC count value. The least significant bit LSB_v for the voltage transfer function is equal to $\text{ADC-Fullscale voltage}/4095 = 6.42\,\text{mV}$. This value is used to set the horizontal logarithmic scale, which ranges from $LSB_v \rightarrow$ ADC-Fullscalevoltage. The vertical logarithmic scale ranges from ADC count $1 \rightarrow 4095$ (given that zero cannot be displayed on a log scale).

The lower plot given in Fig. 4.33 also shows an 'ADC' function, which is the current transfer function, with input current range of $LSB_i \rightarrow$ *ADC full scale current*, where the *ADC full scale current* (see Eq. (3.3b)) value is equal to 33.0 A (peak to peak current) for a BOOSTXL-DRV8301 module. Furthermore, the least significant bit LSB_i for the current transfer function is equal to $\textit{ADC full scale current}/4095 = 8.0\,\text{mA}$. The vertical axis of both plots show the ADC count from $1 \rightarrow 4095$. Note that the quantization error becomes significant, when input voltage/current levels are of the same order of magnitude as LSB_v/LSB_i respectively. It is therefore prudent to ensure that voltage/currents levels used during motor identification remain well above the

LSB values. Hence the introduction of 'gray' shaded regions in Fig. 4.33, where the ADC resolution is such that the region should be avoided by a prudent choice of currents and frequency during the motor identification phase. Consequently, the figures show how well the ADC converters are utilized, for a chosen set of motor ID input variables and ADC full-scale settings (set by the board design). Note that this 'post-processing' type of analysis, makes use of the measured motor parameters. As such these plots show whether or not the identification procedure has been carried out with appropriate parameter settings (as set in the motor identification dialog box). Furthermore, these plots can provide some guidelines as to what input parameters may need to be changed to improve identification, given the machine and converter board under consideration.

A brief discussion on the method used to calculate the voltage/currents shown in Fig. 4.33 is given below, for each of the legend plot variables. Note that all AC voltage measurements are shown in the voltage plot as **peak to peak** values.

- `DCbus`: Rated DC voltage $u_{\mathrm{DC}} = 24\,\mathrm{V}$ of the converter.
- `Iconv`: Rated converter current is 8.0 A RMS, which corresponds to a peak to peak current value of `Iconv`= 22.6 A, as shown in the lower plot.
- `K_p/w_i CC`: voltage level used during estimation of the current controller gain and bandwidth. With the machine at standstill, the impedance is readily found using $Z_{\mathrm{cc}} = \sqrt{X_{\mathrm{cc}}^2 + R_{\mathrm{s}}^2}$, where $X_{\mathrm{cc}} = 2\,\pi f_{\mathrm{cc}} L_{\mathrm{s}}$, X_{cc} and R_{s} represent the reactance and resistance respectively. The variable $f_{\mathrm{cc}} = 300\,\mathrm{Hz}$ is the frequency (Hz) used during this estimation phase. The peak voltage is thus found using $u_{\mathrm{cc}} = i_{\mathrm{cc}} Z_{\mathrm{cc}}$, where $i_{\mathrm{cc}} = 0.5\,\mathrm{A}$ is the peak current value used during estimation. The value shown in the top plot is $2\,u_{\mathrm{cc}} = 0.51\,\mathrm{V}$ and the corresponding peak to peak current `I-K_p/w_i cc` $= i_{\mathrm{cc}} = 1.0\,\mathrm{A}$ is given in the lower plot.
- `Rs`: The DC voltage during estimation of the resistance is equal to $u_{\mathrm{Rs}} = R_{\mathrm{s}}\, i_{\mathrm{Rs}}$, where i_{Rs} is the DC current value used during identification. The DC value , $u_{\mathrm{Rs}} = 0.4\,\mathrm{V}$, was calculated using $i_{\mathrm{Rs}} = 1.0\,\mathrm{A}$. Shown in the lower plot is DC current `I-Rs=`$i_{\mathrm{Rs}} = 1.0\,\mathrm{A}$ for this measurement.
- `flx/Ls`: The peak voltages related to PM flux estimation, can be found using $u_{\mathrm{flx}} = 2\pi f_{\mathrm{flx}}\, \psi_{\mathrm{PM}}$, where $f_{\mathrm{flx}} = 50\,\mathrm{Hz}$ is the frequency (Hz) used during estimation. A peak to peak value $2\,u_{\mathrm{flx}} = 4.08\,\mathrm{V}$ is shown in the top plot, while the corresponding peak to peak current `I-flx/Ls:`$2\,i_{\mathrm{flx/Ls}} = 4.0\,\mathrm{A}$ is given in the lower plot. The peak to peak voltage across the inductance L_{s} is equal to $u_{\mathrm{Ls}} = 2\pi f_{\mathrm{flx}}\, L_{\mathrm{s}}\, i_{\mathrm{flx/Ls}} = 0.21\,\mathrm{V}$.

Note that the RMS rated voltage and current for the LVSERVOMTR PM motor are 60 V (line to line) and 7.1 A respectively. Care should be taken that the motor ID currents/voltage used remain within the converter and machine limits.

Overall observation of the voltage/current levels used during estimation learns that they are well outside the LSB voltage/current values of the ADC converter and within the converter rated values. Utilization of the current ADC is satisfactory for all measurements, given that the lowest current level (used during current controller gain/bandwidth estimation) utilizes at least 6 bits. Increasing current levels during estimation should be done with care, to avoid overheating of the machine. In

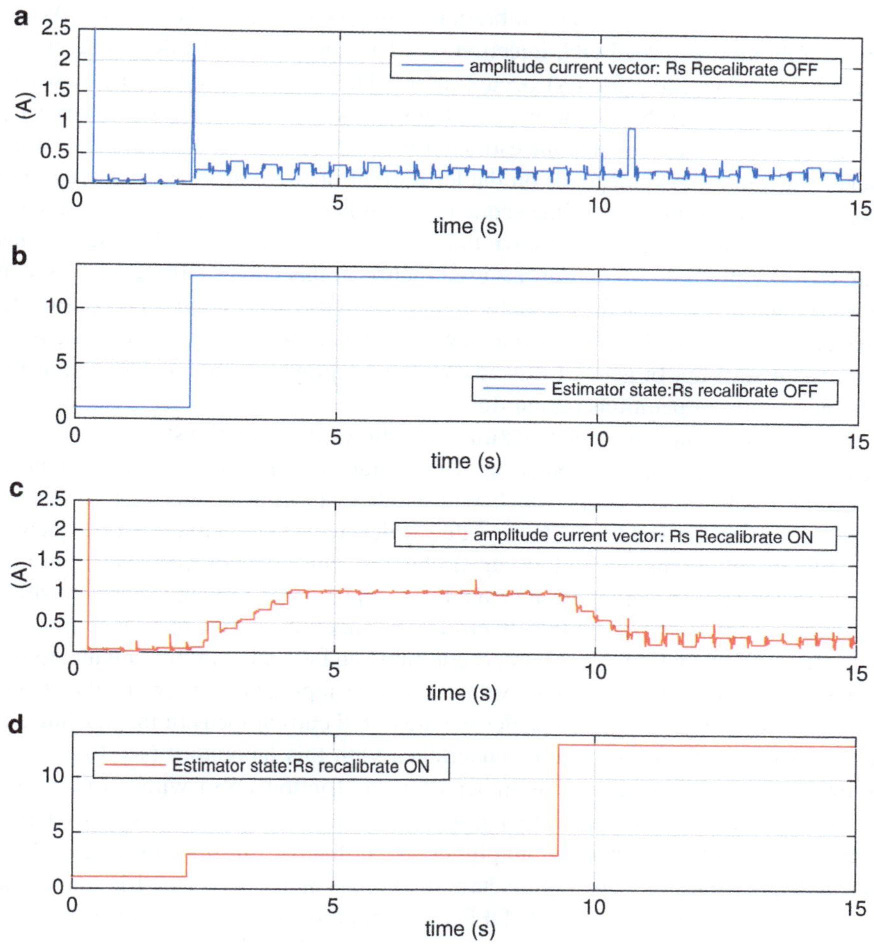

Fig. 4.34 Stator resistance estimation sequence, with recalibration OFF and ON

addition, the PWM frequency could be further increased given that low inductance motors will exhibit a high current ripple, which complicates the sampling of said variables.

In the sequel to this section on motor identification, two additional features will be discussed, which relate to measurement of the stator resistance prior to normal operation and so called 'online' estimation. When the `Rs_recal_on` button is ON, the stator resistance will be identified, immediately after the converter is turned on. The sequence of events following converter activation, with Rs recalibration active is shown in Fig. 4.34. The Rs recalibration sequence for the PM motor as given in Fig. 4.34 shows the current vector amplitude as function of time together with the Estimator state. Observation of Fig. 4.34 learns that upon converter activation ADC offset measurement takes place, which is followed by user control

(Estimator state 13) when Rs recalibration is OFF (see subplots (a) and (b)). When Rs recalibration is turned ON, injection of a DC current (same value as set in the motor ID dialog entry for Rs) occurs after ADC offset calibration as shown in subplots (c) and (d). Note that the actual amplitude of the phase current may vary between $0 \rightarrow 1.0$ A given that the estimator sets the $i_{\mathrm{d}}^{\mathrm{ref}}$ value to 1.0 A (in this case), hence the value shown depends on the location of the current vector relative to the phase being measured. The time sequence for this event is governed by estimator state 3 of the timing sequence shown in Fig. 4.31. However, of the three sub-cycles shown, only the last two entries are used, hence in this case 135000 cycles, which corresponds to 9 s, given the estimator frequency in use. Recalibration is considered mandatory in particular when using low inductance machines in order to ensure successful start-up. In general it is advisable to deploy recalibration prior to drive operation if user operations permit this.

So called 'online' resistance estimation, allows the user to estimate the stator resistance during 'normal' operation. This feature is particular useful if extended operation under load occurs which will result in heating up the motor. This in turn will affect the stator resistance value which will then increase. Consequently an amount of de-tuning will occur if the resistance is not corrected. On-line stator resistance estimation ensures that the resistance value remains correct, which therefore counters any de-tuning effects. Furthermore, the online resistance value, is a function of the motor temperature, accordingly an indication of motor temperature can be deduced from the online Rs value. A simple approach to observing the effects of online Rs estimation is to consider the measured current locus of the machine, as given in Fig. 4.35. These results, undertaken with this laboratory (see Fig. 4.28), show the per unit current locus under no-load (friction only) with a direct axis current $i_{\mathrm{d}}^{\mathrm{ref}}$ value of -1.0 A. Note that the current locus amplitude is slightly larger than 0.05 (which is the per unit amplitude given that the full scale current is 20 A) due to the presence of a quadrature current component caused by the friction torque of the machines. With 'online' Rs OFF, the current locus will be a concentric circle relative to the center of the x-y diagram, as shown in Fig. 4.35. However, when the online Rs function is ON, the current plot will be offset relative to the center of the x-y plot. Furthermore, said locus circle will rotate at a speed dictated by the Rs online estimation frequency f_{mod}, which by default is set to $f_{\mathrm{mod}} = f_{\mathrm{est}}/10^5$, where f_{est} is the estimator frequency in use. Consequently, with $f_{\mathrm{est}} = 15$ kHz, as used in the lab, the modulation frequency will be $f_{\mathrm{mod}} = 15$ mHz, which corresponds to an electrical rotation speed of approximately 0.9 rpm. The eccentricity of the locus circle is determined by the variable `OnlineRs__est_current(A)` set in the motor ID variable box, which is set to 0.25 A. In Fig. 4.35 the radius of this offset circle has purposefully been enlarged for the sake of readability.

Finally, when operating the drive with the motor ID function disabled, the machine parameters as present in the dialog box will be used for drive operation. However, it is recommended to operate the drive at least once under motor ID mode (before disabling this function), in order to update the machine parameters in the dialog box (notably the stator resistance) and most importantly, determine the current controller gain/bandwidth settings.

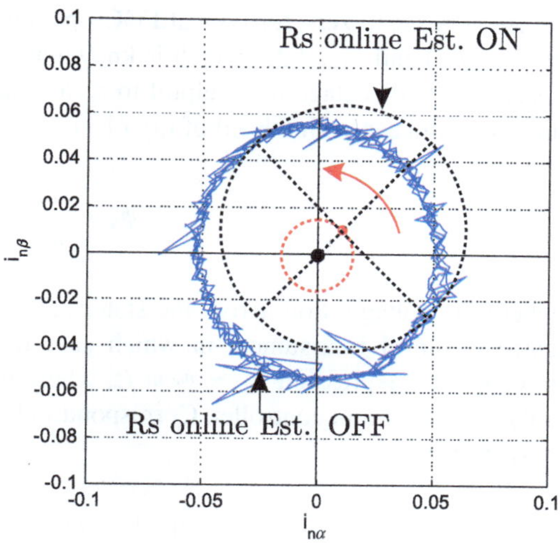

Fig. 4.35 Current locus plot with online stator resistance estimation OFF/ON

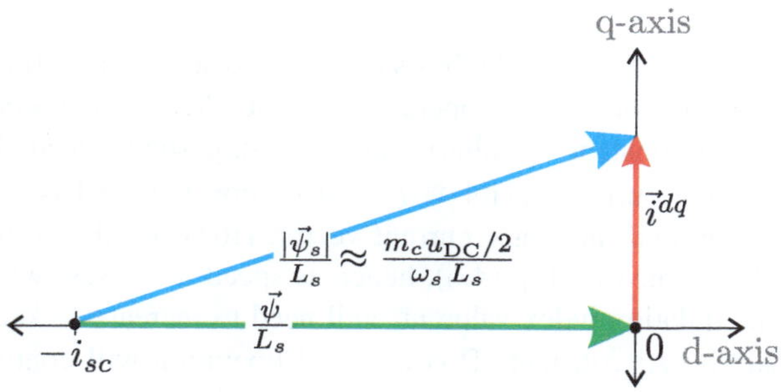

Fig. 4.36 Current vectors in synchronous reference frame, without field weakening

4.4 Laboratory 2:3: PM FOC Sensorless Control with Field Weakening

Prior to discussing how field weakening can be implemented, a brief overview will be given on the purpose of this control technique. For this analysis it is helpful to reconsider Eq. (4.3), which may also be written as

$$\frac{\vec{\psi}_s}{L_s} = \underbrace{\frac{\psi}{L_s}}_{i_{sc}} + \vec{i}^{dq} \tag{4.4}$$

This expression shows the relationship between a current vector associated with the stator flux $\vec{\psi}_s$, PM flux $\vec{\psi}^{dq} = \psi$ and stator current $\vec{i}^{dq}$ expressed in a synchronous coordinate reference frame, as shown in Fig. 4.36. The stator current vector $\vec{i}^{dq}$ has been purposefully aligned with the q-axis, to maximize the torque to current ratio, hence i_d is set to zero for typical FOC operation. A term $i_{sc} = \psi/L_s$ is introduced in Eq. (4.4) and Fig. 4.36, which is known as the 'PM short-circuit current' of the machine. At this stage it is helpful to recall that the stator flux, is also defined by Eq. (4.2) and can with the aid of Eq. (4.1) be written as

$$\vec{\psi}_s \approx \frac{\vec{u}_s}{j\omega_s} \tag{4.5}$$

where the voltage drop across the stator resistance $R_s \vec{i}^{dq}$ is ignored. Furthermore, current control is assumed, in which case the motor voltage amplitude will be governed by expression $u_s = m_c u_{DC}/2$, where m_c is the absolute modulation index value of the current controller. Correspondingly, the amplitude of the current vector may be written as

$$\frac{|\vec{\psi}_s|}{L_s} \approx \frac{m_c u_{DC}/2}{\omega_s L_s} \tag{4.6}$$

which, is shown in Fig. 4.36. On the basis of Fig. 4.36 some interesting conclusions can be drawn when considering operation of the drive under constant torque condition (hence a constant i_q value) and increasing shaft speed. For example, the amplitude of the current vector $\vec{\psi}_s/L_s$ will be constant and is defined by the short-circuit current and the stator current vector. However, the amplitude of this vector is also determined by Eq. (4.6), hence as speed increases (which means ω_s increases) the modulation index value m_c will need to increase to keep the current vector amplitude $|\vec{\psi}_s/L_s|$ constant. This mode of operation will continue, until the current controller modulation index reaches its maximum value m_c^{max} (set in the 'Operations' dialog box). The shaft speed at which this occurs can with the aid of Fig. 4.36 and use of Pythagoras's theorem, be written as

$$\omega_s^{max} = \left(\frac{m_c^{max}\, u_{DC}}{2\, L_s}\right) \frac{1}{\sqrt{i_{sc}^2 + i_q^2}} \tag{4.7}$$

which corresponds to a mechanical shaft speed of

$$n_m^{max} = \frac{30\, \omega_s^{max}}{\pi p} \tag{4.8}$$

where p is number of pole pairs. Application of Eqs. (4.7), (4.8) to the LVSERVOMTR PM motor in use (parameter as measured in the previous lab) gives a maximum shaft speed of $n_m^{max} = 4436$ rpm, with the conditions: $i_q = 2.5$ A, $m_c^{max} = 1$ and DC bus voltage of $u_{DC} = 24$ V, as used for the LAUNCHXL-F28069M, BOOSTXL-DRV8301 module. Furthermore, the short-circuit current

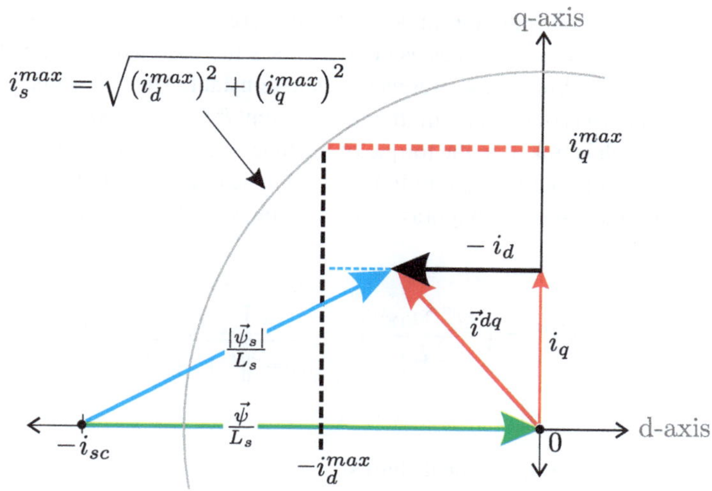

Fig. 4.37 Current vectors in synchronous reference frame, with field weakening

and stator inductance of the PM motor are equal to $i_{sc} = 37.9$ A and $L_s = 0.169$ mH respectively.

If operation above the maximum operating speed indicated above is undertaken, then the current vector amplitude $|\vec{\psi}_s/L_s|$ must reduce as the modulation index is now constant $m_c = m_c^{max}$, in which case the current i_q (and therefore the torque output) will reduce.

Operation above the max operating speed n_m^{max} with the condition $m_c = m_c^{max}$ is however, possible by introducing a 'field weakening' controller, which has as input the absolute modulation index variable m_c (as derived from the synchronous current controller module) and output the reference i_d^{ref} variable. The controller algorithm is as follows:

$$\text{if } (\text{mc}^{\text{max}} - \text{m}_\text{c}) \leq \epsilon \text{ then : reduce } \text{i}_\text{d}^\text{ref} \tag{4.9a}$$

$$\text{if } (\text{mc}^{\text{max}} - \text{m}_\text{c}) > \epsilon \text{ then : } \text{i}_\text{d}^\text{ref} = 0 \tag{4.9b}$$

where $m_c = \sqrt{m_d^2 + m_q^2}$. This algorithm basically adds a direct axis current component to the quadrature reference (as defined, for example, by the speed controller) to shorten the current vector amplitude $|\vec{\psi}_s/L_s|$ as may be observed from Fig. 4.37. Hence the field weakening controller will maintain the modulation index m_c of the synchronous current controller close to the m_c^{max} value by reducing the i_d^{ref} reference current value as speed increases. The action of the controller is therefore to generate a magnetic flux component that opposes that of the permanent magnet, which is why this method of control is referred to as 'field weakening'. During field weakening, the current amplitude $|\vec{i}^{dq}|$ will increase as may be observed from Fig. 4.37, hence efficiency reduces as copper losses increase. Once a specified i_d^{max} value is reached, the direct axis reference is held constant. The i_d^{max} value is usually calculated on the basis of the maximum motor current i_s^{max} and maximum quadrature current i_q^{max} (which is linked to the torque of the machine) as shown in Fig. 4.37. The maximum achievable shaft speed (rad/s) which can be reached with field weakening active, can with the aid of Pythagoras's theorem and Fig. 4.37 be written as

$$\omega_{sF}^{max} = \left(\frac{m_c^{max}\, u_{DC}}{2L_s}\right)\frac{1}{\sqrt{\left(i_{sc} - i_d^{max}\right)^2 + i_q^2}} \tag{4.10}$$

which corresponds to a mechanical shaft speed of

$$n_{mF}^{max} = \frac{30\,\omega_{sF}^{max}}{\pi p} \tag{4.11}$$

Application of Eqs. (4.10), (4.11) to the LVSERVOMTR PM motor in use and using an i_d^{max} value of 3.0 A gives n_{mF}^{max}=4814 rpm, which is a 378 rpm increase in operating speed, when compared to the value (without field weakening) calculated using Eq. (4.7). The speed increase is, in this case, modest because the short circuit current of the machine is relatively large with respect to the assumed maximum current of the machine, in this case set (for safety reasons) to i_s^{max} = 3.9 A. In practice, either the ratings of converter or the motor will determine the i_s^{max} value that can be used.

4.4.1 *Lab 2:3: Phase C*

Prior to being able to work with an operational drive, an `.out` file must be generated of a 'Drive Controller with FAST and Field Weakening' module, as shown in Fig. 4.38, which represents the drive under consideration. Hence details of the drive structure and corresponding dialog box parameter assignments will be discussed in this subsection.

The following information is relevant for this laboratory component:

- Reference program : `lab2-3_LaunchXLphCv3.vsm`.
- Description: Sensorless control of a PM motor using an IstaSPIN module, with field weakening capability.
- Equipment/Software: Texas Instruments LAUNCHXL-F28069M, with BOOSTXL-DRV8301 module ('aft' position) and VisSim simulation program.

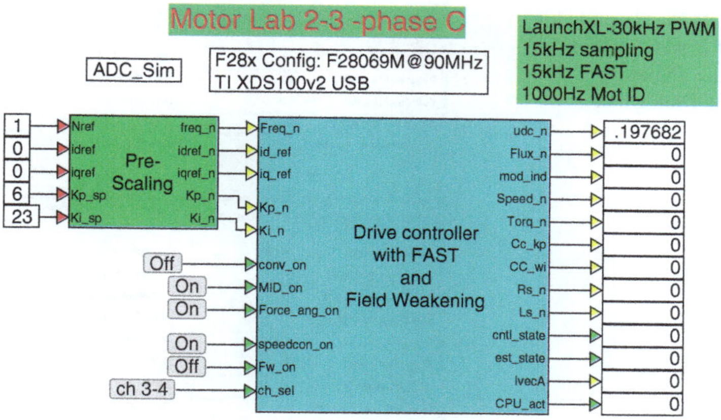

Fig. 4.38 Phase C simulation of a FAST based encoderless (FOC) drive, with field weakening capability

- Outcomes: generate the `.out` file needed to represent the drive structure under investigation.

As mentioned above, a single boost pack located in the 'aft' position (furthest away from the USB connector) is used for single motor operation, in which case the following jumper and dip-switch positions on the LAUNCHXL-F28069M module are required:

- Jumpers `JP1` and `JP2` OPEN
- Jumpers `JP4` and `JP5` CLOSED
- Jumpers `JP3`, `JP6` and `JP7` CLOSED
- Dip-switches `SW1` to `SW3` ON

Inputs to the Controller are five variables, which provide the user with the ability to set the reference shaft speed, direct/quadrature reference and speed gains. The `MID_on` button activates motor parameter estimation when the converter is activated. Note this will occur only after running the C+ file. A button `Force_ang_on` activates the 'force angle' function of the InstaSPIN algorithm, which is particularly useful for starting PM machines with a high cogging torque. Speed or torque control mode is controlled by the button `speedcon_on`. A `Conv_on'` button and `Ch_sel` function are used to respectively activate the converter and select which diagnostic channel combination is to be displayed in phase C+. Field weakening control is provided by the `Fw_on` button, which activates the field weakening controller. Outputs, as discussed previously, of the controller module are, `udc_n (V)`: per unit DC bus voltage, `Flux_n`: PM motor flux and `Speed_n`: per unit shaft speed. Additional outputs are, `Torq_n`: shaft torque, motor parameters: `Rs_n`, `Ls_n`, current control gain/bandwidth `CC_kp`, `CC_wi` and controller/estimator states `cntl_state`, `est_state`. For convenience sake, two output variables `mod_ind` and `IvecA` have been added which represent the

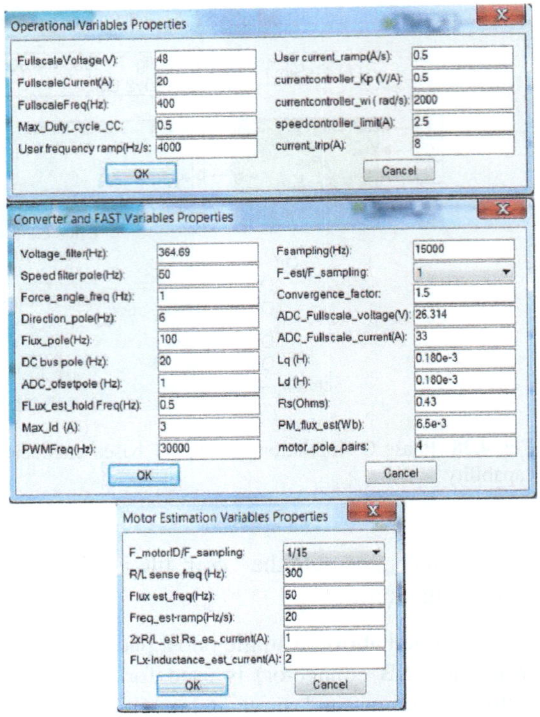

Fig. 4.39 Phase C simulation of a FAST based encoderless FOCdrive, with field weakening: Dialog box entries

amplitude of the modulation index and the current vector amplitude respectively. The variable `mod_ind`, when multiplied by half the DC bus voltage, yields the reference voltage amplitude generated by the FOC controller. As in the previous lab an output `CPU_act` is available to accurately show percentage CPU usage. Moving one level lower into the 'Drive Controller with FAST and Field Weakening' module brings up the by now familiar set of compound blocks centered around the 'main module control' module and a set of dialog boxes. The parameters used in said dialog boxes are presented in Fig. 4.39.

The 'Operational Variables' dialog box, contains all the parameters needed for scaling, rate limit values for user inputs and gain/limit settings for current/speed control. In this dialog box, the maximum duty cycle amplitude `Max_Duty_cycle_CC` has been purposely set to 0.5 to demonstrate field weakening operation, at a lower operating speed as to avoid current measurement problems at high modulation values due of the use of shunts (as discussed in Sect. 2.1.6). A `Max_Id` parameter shown in the 'Converter and FAST variable' dialog box, sets the maximum direct axis reference current that can be used by the field weakening controller, i.e. the current $i_{\mathrm{d}}^{\mathrm{ref}}$ is then limited to $-3.0\,\mathrm{A}$ in this example. The meaning of the variables and values assigned in the 'Converter and FAST Variables' and 'Motor Estimation variables' dialog boxes are as discussed in the previous two laboratories (see Figs. 4.10 and 4.25). Moving one level lower into the 'main motor control' module, brings up a set of modules given in Fig. 4.40,

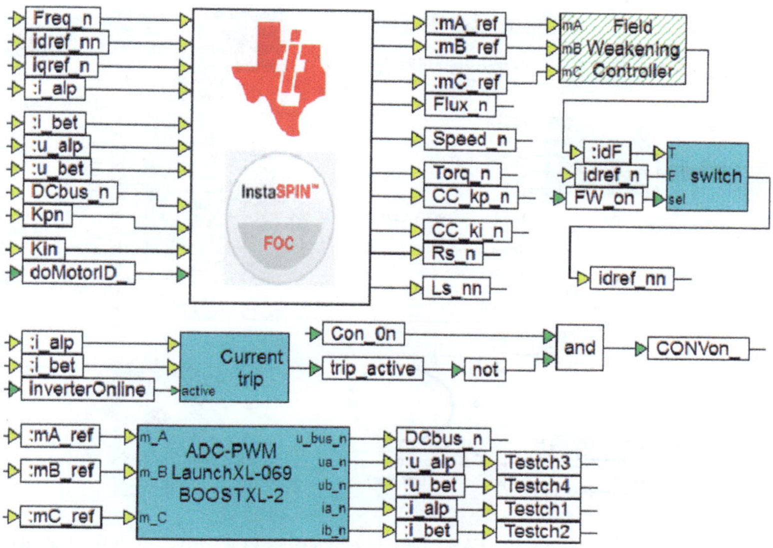

Fig. 4.40 Phase C simulation of a FAST based encoderless FOC drive, with field weakening: One level into the 'Main Motor Control' module

which have been identified in the previous laboratory. However, in this laboratory a field weakening controller has been introduced which complies with the algorithm represented by Eq. (4.9). Output of this module is the per unit reference direct axis current `idF`. A switch with a logic control input that is controlled by the field weakening button `Fw_on`, determines if the direct axis reference current used by the FOC controller is set by the user or the field weakening controller.

4.4.2 *Lab 2:3: Phase C+*

Phase C+, is the operational component of the laboratory and is basically a run version of the `.out` file compiled and downloaded to the MCU in phase C (see previous subsection).

The following information is relevant for this laboratory component:

- Reference program [11]: `lab2-3_LaunchXLphCv3_d.vsm`.
- Description: FOC sensorless control of a PM machine with field weakening capability.
- Equipment/Software: Texas Instruments LAUNCHXL-F28069M, with BOOSTXL-DRV8301 module ('aft' position), Texas Instruments LVSERVOMTR PM motor and VisSim simulation program.
- Outcomes: to use the InstaSPIN algorithm for PM motor parameter identification purposes, sensorless FOC operation and understand the role of field weakening.

Note that the required jumper and dip-switch settings for the LAUNCHXL-F28069M module are given in phase C.

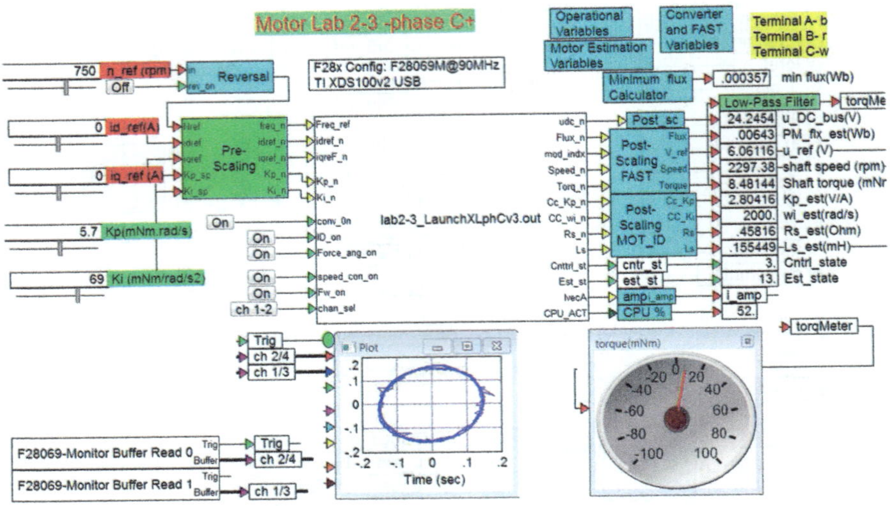

Fig. 4.41 Phase C+ simulation of a FAST based encoderless FOC drive, with field weakening

The run version as shown in Fig. 4.41 uses a VisSim run module, which executes the `.out` file, shown in said module. Five sliders are used, the purpose of which was discussed in the previous section. A post-scaling module is again used to convert the per unit measured DC bus voltage to actual voltage, as shown with a numeric display.

Four additional 'Post scaling' modules are used to generate the following variables:

- `Pm_flx_est (Wb)`: estimated magnetic flux (in Wb) of the machine in use
- `u_ref (V)`: reference voltage amplitude generated by the FOC current controller.
- `shaft speed (rpm)`: shaft speed of the motor as calculated by the FAST algorithm.
- `shaft torque (mNm)`: shaft torque (in milli-Nm) of the machine in use. A low-pass filter with a 1 Hz corner frequency, is used to filter this signal for display in the torque meter (see Fig. 4.12).
- `Rs_est (Ohm)`: estimated stator resistance.
- `Ls_est (mH)`: estimated stator inductance.
- `Cntrl_state`: controller status.
- `Est_state`: estimator status.
- `IvecA`: per unit measured current vector amplitude.

Two 'Monitor Buffer' modules are used to display two selected (using an on/off button connected to the `chan_sel` input) diagnostic signals. An enlarged view of the VisSim scope module given in Fig. 4.42 shows the per unit currents i^{n}_{α}, i^{n}_{β} after completion of identification. Multiplication by the full scale current value of 20 A give actual current values.

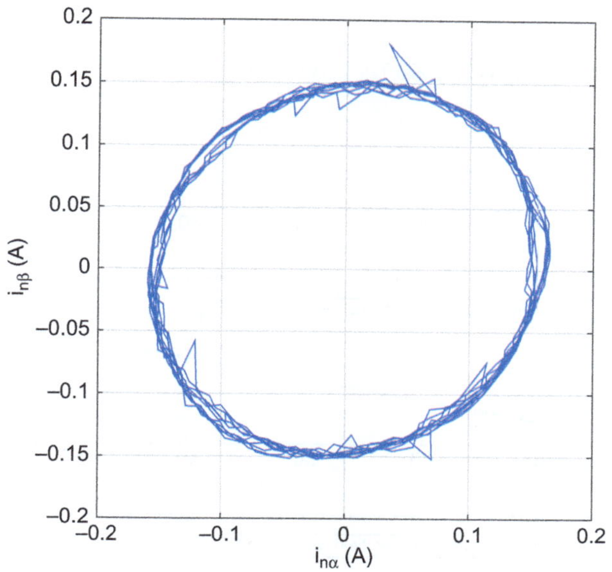

Fig. 4.42 Phase C+ simulation: Scope display showing per unit α, β currents after completion of motor identification (machine running with friction load only) with field weakening ON

Prior to activating this type of lab component, the reader needs to be aware of the fact that he or she is about to activate a complex electrical system, with live voltages. Hence it is prudent, **always**, to execute the following 'Pre-Drive' check list:

- Dialog boxes used in C+ mode, match those of phase C: the run version is compiled with the dialog boxes entry specified under phase C. The dialog boxes shown in phase C+ are used by the 'Pre/Post' scaling modules.
- Ensure that the sample time used is correct and latest (and correct) `.out` file has been downloaded to the MCU (Right mouse click on MCU module to show dialog box of these variables).
- Confirm that the user input values are set to either zero, or 'acceptable' values, which will not cause a current trip of the converter.
- Confirm that the converter 'switch' is set to OFF and the power supply is on (DC bus voltage present).
- Confirm that the motor is connected firmly and properly.

After completion of the Pre-Drive checklist, activate the program and confirm that the supply voltage source level shown in the digital display is 24 V. If not stop the program and restart. With the correct DC voltage level established, turn on the converter (using the ON button and monitor the motor shaft and diagnostic scope (set in 'x-y' mode), which in this example (Fig. 4.42) is set to show the per unit current components i_{α}^{n}, i_{β}^{n}.

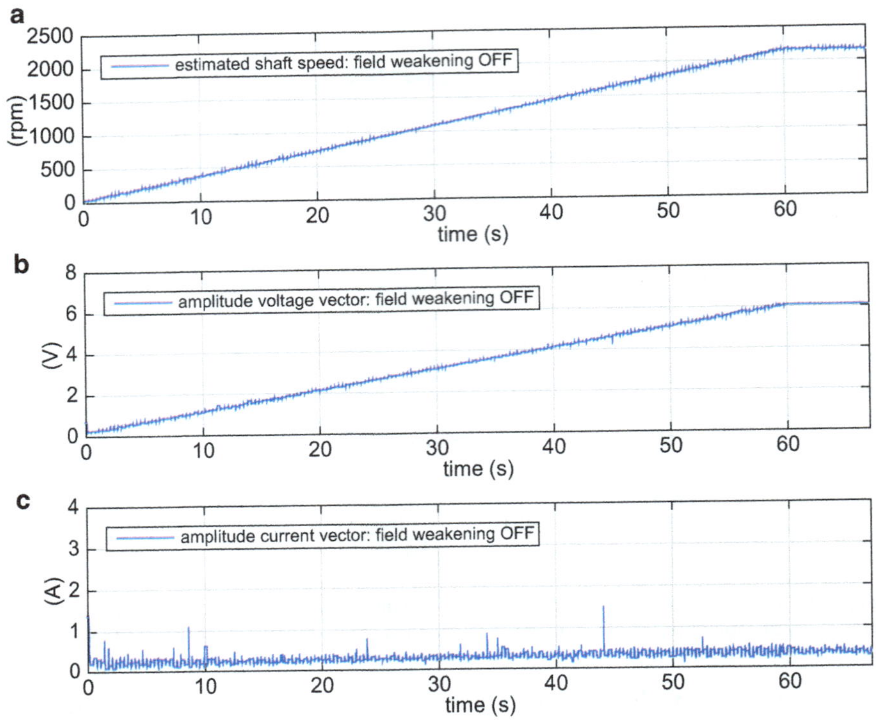

Fig. 4.43 Phase C+ experimental result: estimated shaft speed, voltage/current amplitude without field weakening. Maximum shaft speed achievable is 2150 rpm

Observation of Fig. 4.42 learns that the machine is currently operating under closed-loop speed control, with no external load, except friction, as is evident from the torque meter value. Furthermore, drive operation under field weakening is active, as is evident by the ON button, `Fw_on` and the amplitude of the currents shown in the scope. For this laboratory the maximum negative $i_\mathrm{d}^\mathrm{ref}$ was set to 3.0 A, and the per unit current amplitude shown, confirms drive operation under maximum field weakening conditions.

It is instructive to evaluate drive operation by imposing a speed ramp for the drive which accelerated the reference shaft speed from $0 \rightarrow 2500$ rpm in a time interval $0 \rightarrow 70$ s, after motor identification has been carried out. During this experiment the current vector amplitude, estimated shaft speed and the voltage vector generator by the current controller are to be monitored/displayed. The results of this analysis for the case WITHOUT field weakening active is shown in Fig. 4.43. Clearly observable in Fig. 4.43 is that the speed and voltage amplitude increase as anticipated, because the EMF increases. The machine is operating under no-load (friction only) hence the current, which is the i_q value required to generate the torque needed to overcome friction, remains constant. With increasing speed the modulation index is increased

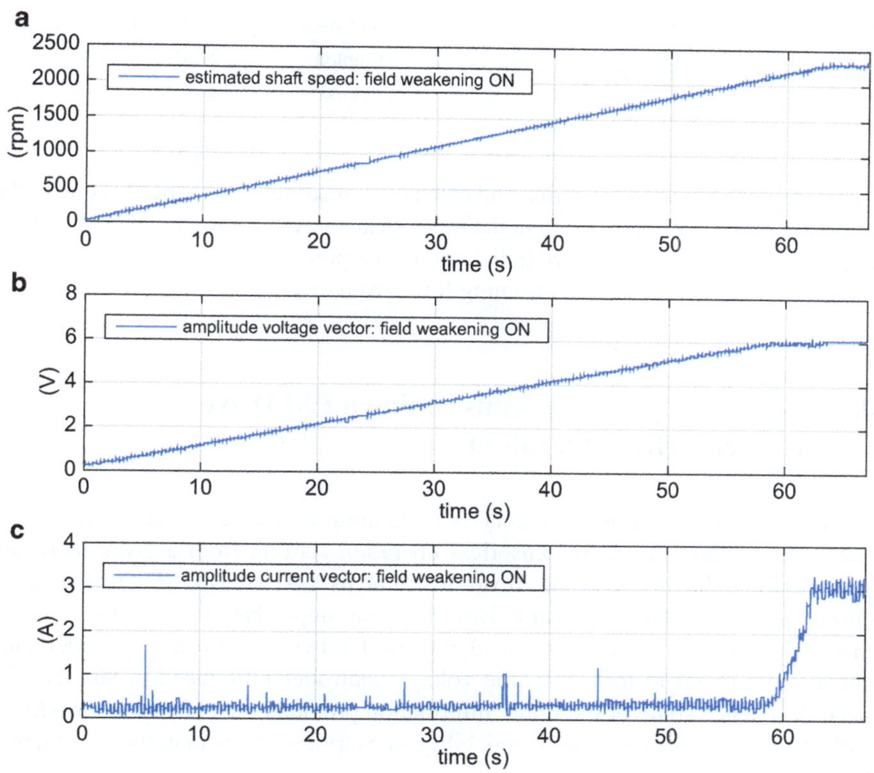

Fig. 4.44 Phase C+ experimental result: estimated shaft speed, voltage/current amplitude with field weakening. Maximum shaft speed achievable is 2297 rpm

by the current controller until the user set limit value of $m = 0.5$ is reached in which case the output voltage remains constant as may be observed from Fig. 4.43. Consequently the shaft speed remains constant thereafter.

When the experiment is repeated with field weakening active (button `Fw_on` ON) the result as shown in Fig. 4.44 appear. Observation of Fig. 4.44 learns that the field weakening control starts to provide a negative direct axis reference current once the modulation index generated by the current control approaches the user set reference value. Consequently the drive shaft speed will increase for as long as negative i_d current can be generated. Once the maximum direct axis reference value is reached speed will remain constant thereafter as may be observed.

The maximum speed achievable, with a maximum modulation index value of 0.5 (as specified in the 'Operations dialog' box), can be compared with field weakening enabled/disabled, see Table 4.1. Also given in Table 4.1 is the maximum theoretical shaft speed achieved under the specified conditions, as calculated using Eqs. (4.7), (4.10). For the theoretical analysis a quadrature current of $i_q = 0.3$ A was used, which corresponds (approximately) to the friction torque present in the drive.

Table 4.1 Field weakening results: shaft speed (rpm), with a maximum modulation value $m_c^{\max} = 0.5$

Field weakening	Measured	Calculated
Disabled	2150	2218
Enabled	2297	2407

The difference between measured and calculated maximum shaft speed is due to the omission of the stator resistance in the theoretical analysis. However, the theoretical analysis does provide a relative indication of the potential speed increase that may be achievable by using field weakening for a given drive.

4.5 Laboratory 2:4: Commissioning a PM Drive for Sensorless Operation

Primary objective of this laboratory is to familiarize the reader with all aspects needed to set up a drive for sensorless operation starting from a given hardware configuration. In this case the Texas Instruments LAUNCHXL-F28069M kit is chosen as the 'unknown' board. However, the steps that will be discussed in this laboratory are virtually identical for any hardware platform the reader may encounter. Central to this lab is the voltage controller introduced in lab 1:1 (see Sect. 3.1) hence some of the initial information provided here has been introduced earlier. However, for the sake of readability an overview of this material (as required for this lab) is given here.

For this laboratory the Texas Instruments LAUNCHXL-F28069M, with BOOSTXL-DRV8301 module setup, as shown in Fig. 4.45, is used together with the dual LVSERVOMTR/LVACIMTR motor combination. The BOOSTXL-DRV8301 (top PCB in Fig. 4.45 located in the aft (furthest away from the USB)) position is the converter module which houses the DRV 8301 module that contains the MOSFET power electronics devices, voltage/current sensing and protection circuitry. Interfaced with this module is the LAUNCHXL-F28069M module that contains the F28069M MCU and JTAG emulation with the corresponding USB interface. Observation of Fig. 4.45 learns that the LAUNCHXL-F28069M setup has a small footprint which (in this case) translates to a relative inexpensive drive setup. The setup given, which also shows the encoder cable connected to the module, is used for the 'single' machine (power to one machine) laboratories in this book. Using the 'aft' boost-pack location is preferred as it provides the reader with easy access to the LEDS on the LAUNCHXL-F28069M module (these are obscured when a 'forward' boost-pack is added). Note that a 24 V DC power supply must also be acquired for this laboratory. Furthermore, Fig. 4.45 shows a oscilloscope and DC-true current probe, both of these are 'optional'. However, the presence of this type of equipment certainly facilitates the drive learning experience and speeds up the debugging process.

Fig. 4.45 Typical experimental setup

Table 4.2 LVSERVOMTR PM motor: M-2310P-LN-04K motor data

Parameter	Value	Units
Inductance: L_{sLL}	0.4	mH
Resistance: R_{sLL}	0.72	Ω
EMF constant: K_e	4.64	$V_{peak}/krpm$
Pole pairs: p	4	–

Prior to considering the Converter/MCU in more detail the PM motor will be considered where use is made of the manufacturer data to derive the required information for the laboratory. The data shown in Table 4.2 is assumed for this motor. The inductance and resistance values shown in Table 4.2 are 'line to line' values, i.e. as measured between two phases of the machine. Accordingly, the stator resistance is equal to $R_s = {}^{R_{sLL}}/_2 = 0.362\ \Omega$. Similarly, the stator inductance is equal to $L_s = {}^{L_{sLL}}/_2 = 0.2$ mH. Furthermore, the EMF constant K_e has been provided (via the data sheet) and allows the calculation of the PM flux using

$$\psi(\mathrm{Wb}) = \frac{\mathrm{K_e}}{2000\,\pi\,\sqrt{3}} \tag{4.12}$$

which in this case leads to a value of $\psi = 0.42\,\mathrm{mWb}$ for the machine in question. Given the stator inductance of the machine, it is prudent to choose a PWM frequency of 30 kHz (which is higher than the assumed 15 kHz controller sampling frequency) in order to reduce current ripple.

4.5.1 *Lab 2:4: Voltage Controller, Phase C*

Prior to using current control in any application, it is helpful to consider drive operation under voltage control (see Sect. 3.1). Using this approach, the voltage and current measurements can be examined as was shown in lab 1:1. Emphasis in this laboratory will be on configuring the ADC-PWM module, triggering and allocating the ADC channels. The following information is relevant for this laboratory component:

- Reference program [11]: `lab2-4_LaunchXLphCv2.vsm`.
- Description: PM drive commissioning example.
- Equipment/Software: Texas Instruments LAUNCHXL-F28069M, with BOOSTXL-DRV8301 module ('aft' position), and VisSim simulation program.
- Outcomes: to gain familiarity of all aspects of operation needed to commission a drive.

As mentioned above, a single boost pack located in the 'aft' position (furthest away from the USB connector) is used for single motor operation, in which case the following jumper and dip-switch positions on the LAUNCHXL-F28069M module are required:

- Jumpers `JP1` and `JP2` OPEN
- Jumpers `JP4` and `JP5` CLOSED
- Jumpers `JP3`, `JP6` and `JP7` CLOSED
- Dip-switches `SW1` to `SW3` ON

The phase C, development phase, given in Fig. 4.46 has been adapted for this application. Inputs are the electrical frequency and direct/quadrature voltage reference variables. Moving one level down into the 'Open loop voltage controller' reveals a set of modules and dialog boxes as indicated in Fig. 4.47. This figure also shows a dialog box which appears upon a 'right click' on the 'main control module'. The critical entry in this case is the 'execute on interrupt' feature which must be set as shown. This implies that the module in question will be executed on an interrupt signal set by the ADC unit. More specifically, the interrupt is in

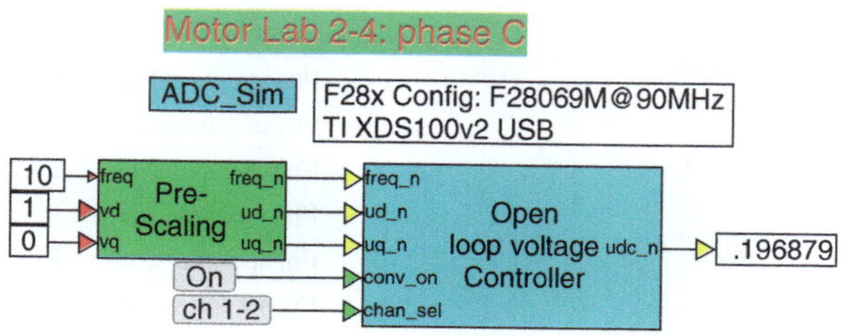

Fig. 4.46 Development Phase C: voltage controller set up for the LAUNCHXL-F28069M module

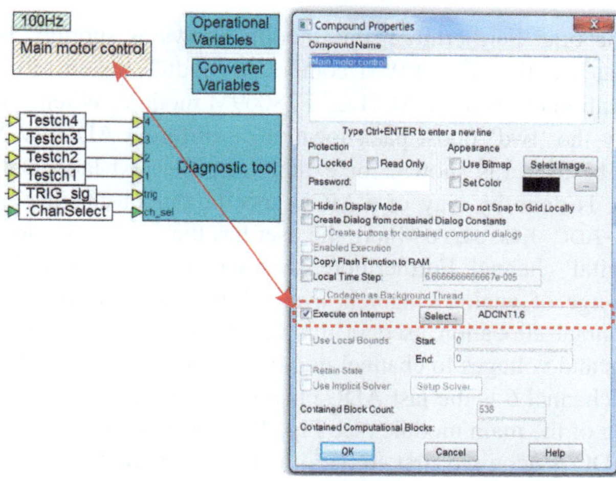

Fig. 4.47 Development Phase C, one level into 'Open loop voltage controller'

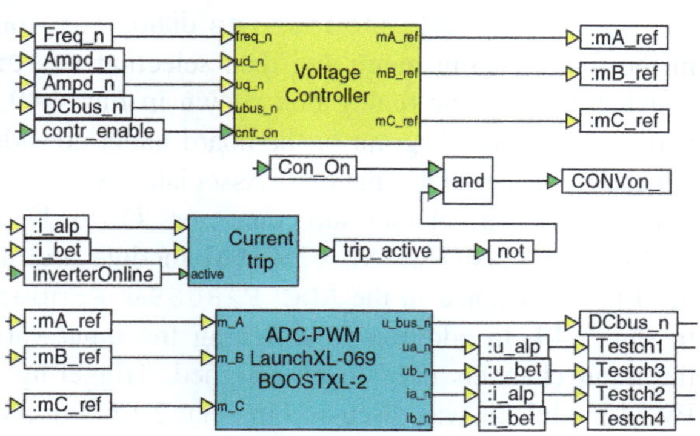

Fig. 4.48 Development Phase C: one level into 'main motor control' module

this case to `ADCINT1:6`, which implies that the signal is generated by ADC input channel 6. The ADC input channels are allocated within the ADC-PWM module, which appears when moving one level lower into the 'main control module' as shown in Fig. 4.48. The modules shown are as discussed in Sect. 3.1. However, in this case configuring the ADC-PWM module is of interest, as was mentioned earlier. This type of module is board specific, which is why the board type is

Table 4.3 Allocation of input variables to ADC input channels

Input variable	ADC input channel
Phase current i_A	0
Phase current i_B	1
Phase current i_C	2
Phase voltage u_A	3
Phase voltage u_B	4
Phase voltage u_C	5
Bus voltage u_{DC}	6

included in the title. Furthermore, this module shows as subtitle 'BOOSTXL-2' to indicate that it is the ADC-PWM compound module linked to the boost pack in the 'aft' location on the LAUNCHXL-F28069M module. When using two boost packs (or only the 'fwd' boost pack location) a different ADC-PWM compound module must be used (as shown, for example in the next lab). Data acquisition using VisSim, is done by way of a MCU specific 'input channel' module, such as `F28069M-ADC RESULT6` where the user has the option of selecting either an 'analog or digital' channel. Furthermore, the user must allocate a channel number (within the range of analog channels available) as mentioned above. Table 4.3 shows the channel allocation used in this laboratory. Note that the allocation of phase currents and voltages to channel numbers is up to this stage done arbitrarily. Furthermore, channel 6 is the last ADC channel to be sampled, which is precisely why triggering of the main module (see Fig. 4.47) occurs with said channel. Hence data by the ADC is acquired first so that it can then be used by the 'main control' module. A critical commissioning step is to allocate the channel numbers introduced in Table 4.3 to actual MCU variables. This requires access to the (for the MCU in use) `ADC F28069M Properties` dialog box, found by selecting `Embedded` in the main VisSim menu and then selecting `F280x` and finally `ADC Config` which brings up the dialog box shown in Fig. 4.49. Also given in this figure is part of the circuit diagram of the board under consideration, which shows the pin allocation of the ADC channels. Associated with the current/voltage variables and pin numbers (one of four sub connectors J5, J6, J7, J8 that link to the 'aft' boost pack) of the LAUNCHXL-F28069M module are alphanumeric variables which need to be assigned in the `ADC F28069M Properties` dialog box, as shown in Fig. 4.49. In addition to assigning the input variables, the trigger signal for the input channels need to be assigned. Triggering of the ADC is linked to the PWM module as was discussed in Sect. 2.1.6. For all the input channels the 'Start Of Conversion' signal `ePWM4-SOCA` is used which implies that the pulse width modulation module 4 generates the trigger for the input ADC channels in use. Note also that the `ADC F28069M Properties` dialog box has a dialog box `Interrupt on conversion start`, which interrupts the MCU before an ADC conversion has taken place. This is important for the triggering of the 'main control' module (see Fig. 4.47) as said module must be executed AFTER all the ADC conversions (including channel 6) have been completed. Note that in practice,

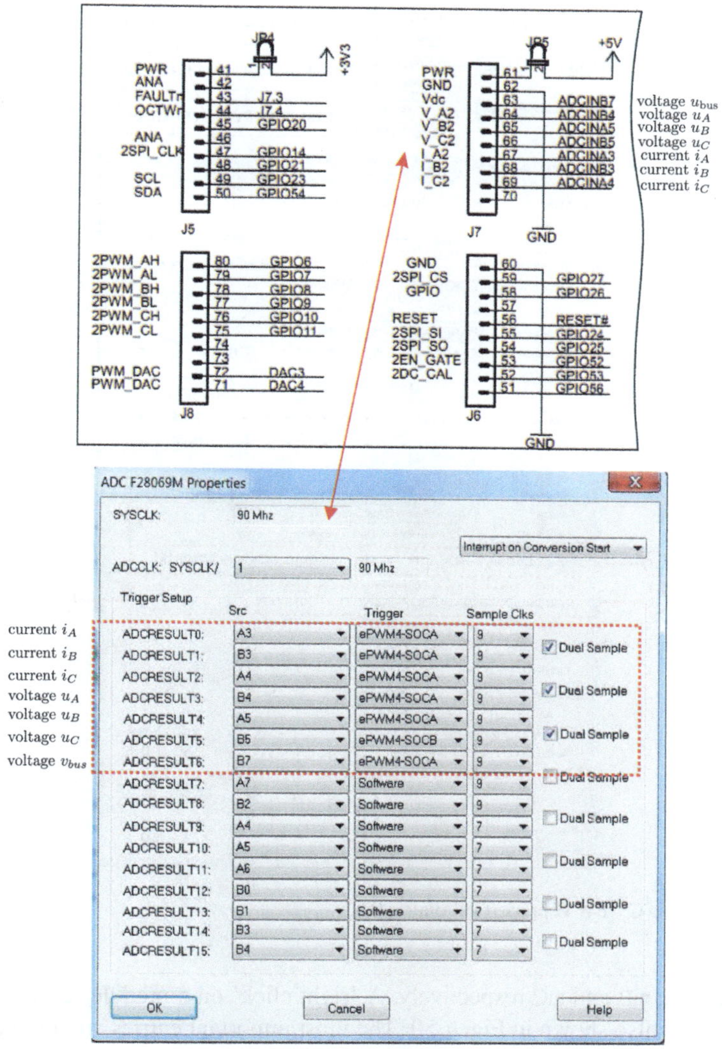

Fig. 4.49 Development Phase C: 'ADC properties' dialog box and link to circuit diagram

sufficient time is available to also sample channel 6, hence the 'before' option will suffice. A further observation of the `ADC F28069M Properties` dialog box shows that the option `Dual sample` has been activated for the first 6 channels. This implies that so called 'dual' sampling is used, which significantly speeds up the sampling process. This feature can however, only be used if the hardware has been designed for this task, i.e. two channels such as `A3`, `B3`, or `A0`, `B0`, etc. must be assigned to either currents or voltages to be sampled.

Moving one level lower into the 'ADC-PWM' module shown in Fig. 4.47 reveals three PWM modules (see Fig. 4.50), which have as inputs the modulation index

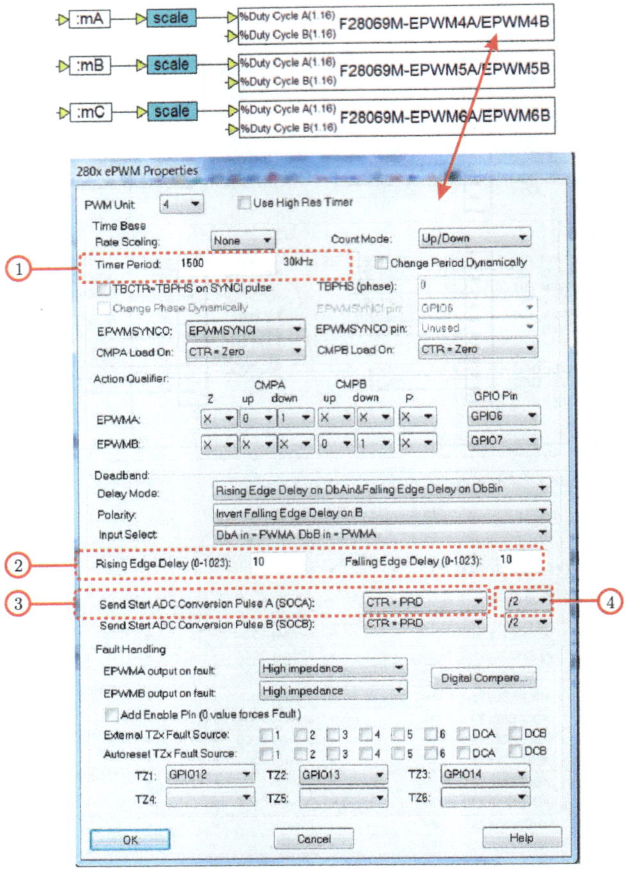

Fig. 4.50 Phase C: 'PWM properties' dialog box

variables `mA`, `mB` and `mC` respectively. A 'right click' on a module reveals a dialog box which is also shown in Fig. 4.50. The most important entries are numbered and need to be configured as follows:

1. `Timer Period`: represent the number of cycles of the PWM staircase as discussed in Sect. 2.1.6. The number to be entered here is defined by the clock frequency divided by the timer period times 2. In this laboratory a 30 kHz PWM frequency is required, which with a system clock frequency of 90 MHz, corresponds to a timer period value of 1500 as shown.
2. `Rising Edge Delay`, `Falling Edge Delay`: These variables represent the 'dead time' setting used to avoid converter leg 'shoot through' (for example, top switch turning ON, while the bottom switch is still in the process of turning OFF). Hence the delay is there to give the converter switches time to turn on and off. The 'dead-time' is represented in terms of the number of clock cycles:

in this case a value of 10 is chosen (given the devices in use), which corresponds to $t_{\text{dead}} = {}^{10}/_{(90\cdot 10^6)} = 111\,\text{ns}$.

3. `send start ADC Conversion Pulse A (SOCA)`: this entry is critical, in that it determines when the conversion pulse should be given with respect to the PWM cycle, as shown in Fig. 2.23. In this case the option `PRD` is chosen because triggering of the ADC should occur at the top of the PWM staircase, as shown in Fig. 2.23. Note that in Fig. 4.49 ADC triggering refers to `ePWM4-SOCA`, which implies that ePMW module 4 generates the SOC signal for all ADC channels.
4. `2`: this entry specifies if the ADC pulse should be given after every first, second or third `PRD` event. In this case a 30 kHz PWM frequency has been selected and the ADC sample frequency in use is 15 kHz (which is also the rate at which the controller module will be executed, with a sampling frequency set in the main VisSim module) hence the PWM to ADC ratio must be set to $^1/_2$. This implies that a ADC SOC pulse will be given on every second PRD event in this case.

Note that only a subset of the ADC and PWM dialog entries have been discussed, given their importance. The reader is however, advised to refer to the relevant Texas Instruments application notes for further details on these modules.

In the sequel to this section attention is given to the computation of the ADC full scale voltage/current parameters and the low-pass filter corner frequency. These parameters can be found by careful examination of the schematic diagram of the converter under consideration. Part of the relevant (to this topic) circuit diagram sections of the Texas Instruments BOOSTXL-DRV8301 module, as given in Fig. 4.51 will be used to derive the required data. Central to this figure is the lower switch of one converter leg of the three-phase converter, with the associated shunt resistor `R34` used to measure the phase current. An 'op-amp' based current signal conditioning circuit is used to generate the 'current' variable i_A, which is the input voltage to the ADC unit shown in Fig. 4.48. The ADC full scale current is the peak to peak current value that corresponds to a peak to peak ADC input voltage of 3.3 V. Analysis of the current conditioning circuit shows that the ADC full scale current is found using

$$ADC\,full\,scale\,current\,(A_{\text{pp}}) = 3.3\left(\frac{R44}{R34\cdot R43}\right) \tag{4.13}$$

where $R43, R44$ and $R34$ are the relevant resistors of the current gain circuit and shunt resistance respectively. Substitution of the resistance values (as shown in the diagram) into Eq. (4.13) yields an ADC full scale current value of 33.0 A. Note that the current signal conditioning circuit inverts the shunt voltage, which is convenient, given that a positive current is represented by an outgoing current from the converter terminal `A`.

Computation of the ADC full scale voltage requires evaluation of the ADC phase and DC bus voltage conditioning circuits. For both circuits, the same voltage attenuation is used as may be observed from Fig. 4.51. For the phase voltage circuit

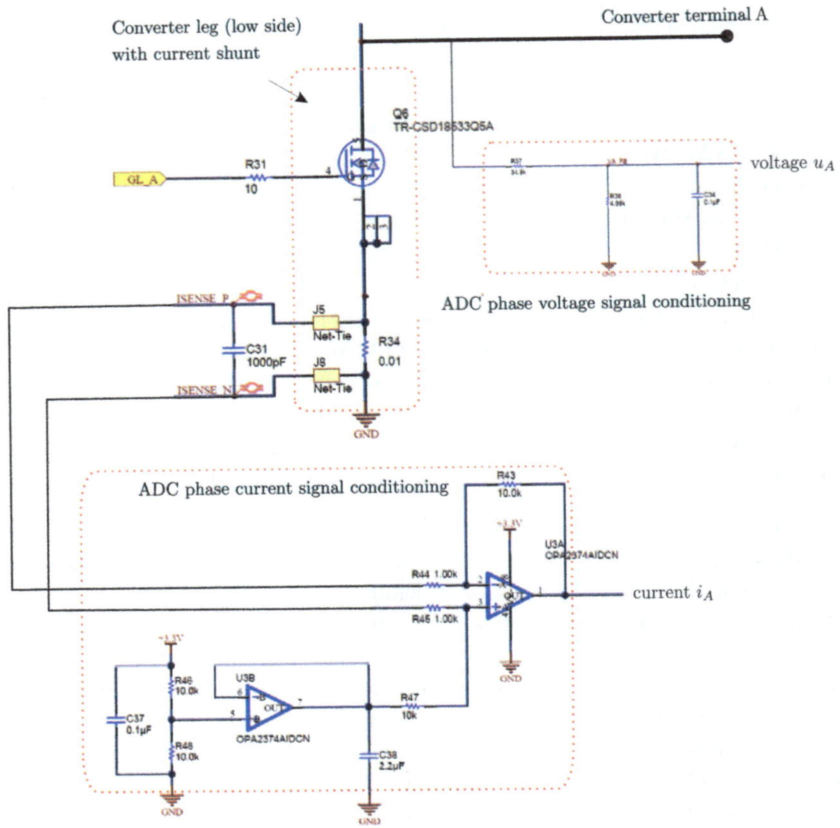

Fig. 4.51 Phase C: Part of BOOSTXL-DRV8301 module circuit diagram showing the low power side of one converter leg, with associated current and voltage signal conditioning [7]

the attenuation is defined as $R38/(R37+R38)$. The ADC full scale voltage is the peak to peak phase voltage that corresponds to a peak to peak ADC input voltage of 3.30 V. Analysis of the phase voltage conditioning circuit, shows that the ADC full scale voltage is found using

$$ADC\,full\,scale\,voltage\,(V_{pp}) = 3.3\left(\frac{R37 + R38}{R38}\right) \tag{4.14}$$

where $R37$ and $R38$ are the relevant resistors of the voltage gain circuit. Substitution of the resistance values (as shown in the diagram) into Eq. (4.14) yields an ADC full scale voltage value of 26.314 V. In conclusion, the reader is reminded of the fact that the phase voltage signal conditioning circuit is also a low-pass filter, of which the corner frequency may be found using

$$LPF\,corner\,freq\,(Hz) = \left(\frac{1}{2\pi}\right)\left(\frac{R37 + R38}{R37 \cdot R38 \cdot C34}\right) \tag{4.15}$$

Substitution of the resistance and capacitance values (as shown in the diagram) into Eq. (4.15) yields a low-pass corner frequency value of 364.69 Hz. This parameter is, among others, required for the InstaSPIN based laboratories.

4.5.2 *Lab 2:4: Phase C+*

Phase C+, is the operational component of the laboratory and is basically a run version of the `.out` file compiled and downloaded to the MCU in phase C (see previous subsection).

The following information is relevant for this laboratory component:

- Reference program [11]: `lab2-4_LaunchXLphCv2_d.vsm`.
- Description: Description: PM drive commissioning example.
- Equipment/Software: Texas Instruments LAUNCHXL-F28069M, with BOOSTXL-DRV8301 module ('aft' position), Texas Instruments LVSERVOMTR PM motor connected to Texas Instruments LVACIMTR machine (optional) and VisSim simulation program.
- Outcomes: to confirm basic drive operation by observing the phase voltage and current waveforms.

Note that the required jumper and dip-switch settings for the LAUNCHXL-F28069M module are given in phase C.

The run version as shown in Fig. 4.52 uses a VisSim run module, which executes the `.out` file, shown in said module. Three sliders are used which allow the user to set the frequency and direct/quadrature voltage value. Output of this module is the measured DC bus voltage. Two 'Monitor Buffer' modules are used to display two selected (using a button connected to the `ch_sel` input) diagnostic signals. An enlarged view of the VisSim scope module given in Fig. 4.53 shows the per unit currents u_{α}^{n}, i_{α}^{n}. Multiplication by the full scale current value of 20 A and full scale voltage of 48 V for the respective waveforms gives the actual current and voltage values.

The dialog box entries for this drive are shown for completeness in Fig. 4.54. Note that the ADC full scale shown are those which have been calculated in the previous development phase.

Prior to activating this type of lab component, the reader needs to be aware of the fact that he or she is about to activate a complex electrical system, with live voltages. Hence it is prudent, **always**, to execute the following 'Pre-Drive' check list:

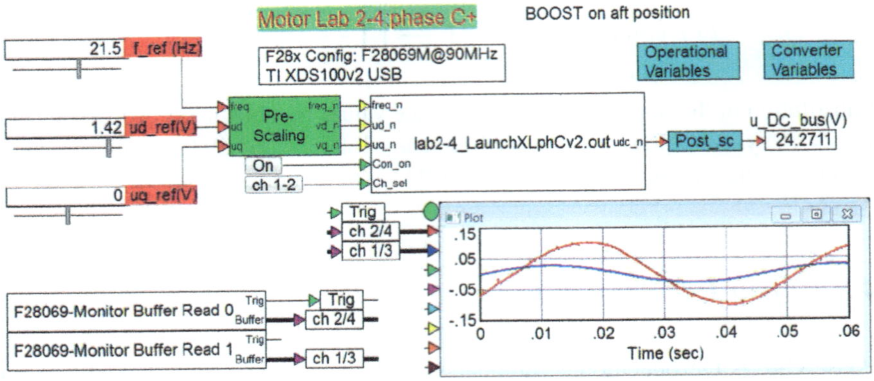

Fig. 4.52 Phase C+: Embedded voltage controller

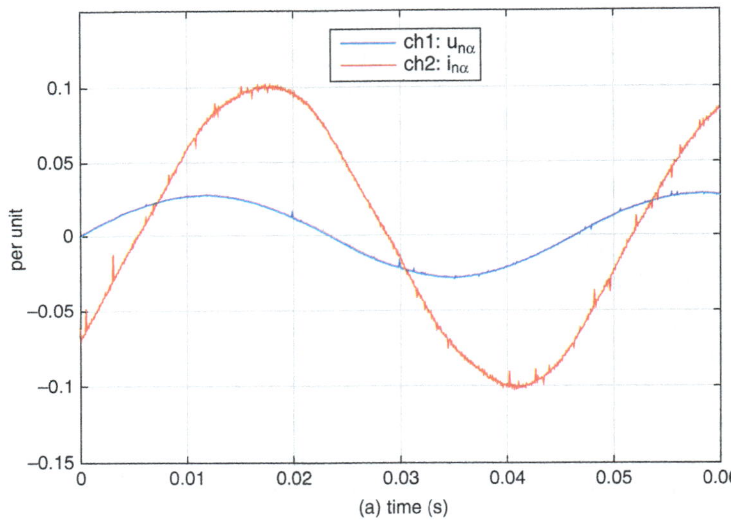

Fig. 4.53 Phase C+ simulation: Scope display showing per unit α current and voltage

- Dialog boxes used in C+ mode, match those of phase C: the run version is compiled with the dialog box entries specified under phase C. The dialog boxes shown in phase C+ are used by the 'Pre/Post' scaling modules.
- Ensure that the sample time used is correct and latest (and correct) `.out` file has been downloaded to the MCU (Right mouse click on MCU module to show dialog box of these variables).
- Confirm that the user input values are set to either zero, or 'acceptable' values, which will not cause a current trip of the converter.

(continued)

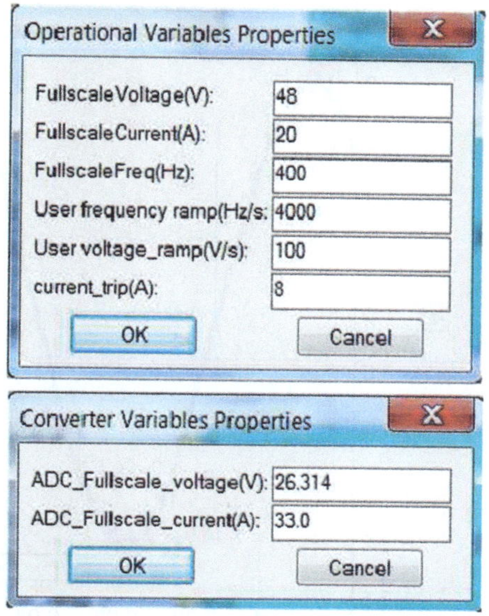

Fig. 4.54 Phase C+: dialog boxes for LAUNCHXL-F28069M commissioning drive

- Confirm that the converter 'switch' is set to OFF and the power supply is on (DC bus voltage present).
- Confirm that the motor is connected firmly and properly.

After completion of the Pre-Drive checklist, activate the program and confirm that the supply voltage source level shown in the digital display is approximately 24 V. If not stop the program and restart. With the correct DC voltage level present, turn on the converter (using the ON button) and monitor the motor shaft and diagnostic scope, which in this example shown (Fig. 4.53) is set to show the per unit voltage component u_{α}^{n} and current i_{α}^{n}. Several important observations can be made on the basis of the information provided in Fig. 4.53 namely;

- The reference frequency has been set to 21.5 Hz hence the period time of the waveforms should be 46.51 ms.
- A reference voltage amplitude of 1.42 V has been set by the slider `ud_ref` hence the amplitude of the per unit u_{α}^{n} should be equal to $1.42/u_{\mathrm{fs}} = 0.0296$, where u_{fs} is the full scale voltage value set to 48 V. In reality the voltage amplitude will be less as dead-time effects and the voltage drop across the converter switches will affect the amplitude shown. The per unit voltage shown is in this case equal to 0.0279 (see Fig. 4.53).
- Independent verification of the current amplitude can be done by making use of a DC-true current probe and oscilloscope, which leads to the result given in Fig. 4.55. Observation of the experimental current waveform reveals a current

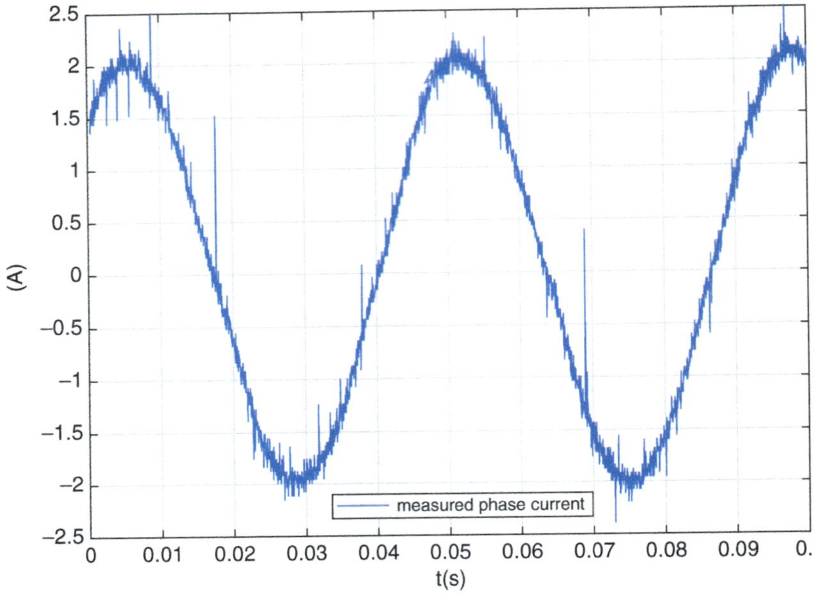

Fig. 4.55 Phase current measured using a DC-true current probe and oscilloscope

amplitude of ≈ 2.0 A which corresponds to a per unit value of $^{2.0}/_{i_{\text{fs}}} = 0.1$, where i_{fs} is the full scale current value set to 20 A. Examination of the scope module in Fig. 4.53 confirms that the per unit current is indeed correct.

- The phase current waveform should lag the voltage phase waveform given that the motor has inductance and resistance elements.

Finally, observe that the motor is rotating correctly. With a reference frequency of 21.5 Hz the speed of the 8 pole motor will be 322.5 rpm, which is difficult to measure without suitable equipment. However, if the reference frequency is set to 4 Hz the rotational speed will be 60 rpm, i.e. one revolution per second, which can be visually determined. This measurement will verify that the assumed pole pair number for the machine is indeed correct.

4.6 Laboratory 2:5: Dual Control of Two PM Machines

The final laboratory of this chapter considers dual operation of two permanent magnet machines. For one PM machine a sensored field-oriented controller is used, as discussed in lab 1:3 (see Sect. 3.3), which is configured for torque control. This PM machine is mechanically connected to the second PM machine, which is connected to a converter operating with a FOC Sensorless speed controller as discussed in lab 2:2 (see Sect. 4.3).

Fig. 4.56 Dual PM/PM drive setup with LAUNCHXL-F28069M board

The purpose of this laboratory is to explore dual machine operation, where, for example, one machine is acting as a load and the other as a motor. Care should be taken in this case to ensure that both machines are NOT operating as motors simultaneously (which could occur when 'torque control' is selected for both drives), given that a speed runway situation could then occur. In this laboratory use is made of a LAUNCHXL-F28069M board, with two BOOSTXL-DRV8301 modules, which are each connected to a Texas Instruments LVSERVOMTR PM machine as shown in Fig. 4.56. Clearly observable are the two boost packs of which the 'aft' unit is directly (without a Molex connector) attached to the lower PM machine (of the two) shown in this figure. Also shown are the DELTA ELEKTRONIKA SM400 -AR-4 power supply and DC-true probe (attached to one phase of the sensorless machine) used for this laboratory. Development phases C and C+ are shown in this laboratory. The encoder of the sensored PM machine (upper PM machine shown in Fig. 4.56) is used to provide shaft angle information for the sensored drive. Note that the encoder of the PM machine operating under FOC sensorless control is not connected.

4.6.1 Lab 2:5: Phase C

For dual operation the controller is configured to activate both boost packs simultaneously so that ADC offset correction for the (common) ADC module can be undertaken before both controllers become operational. The following information is relevant for this laboratory component:

- Reference program [11]: `lab2-5_LaunchXLphCv3.vsm`.
- Description: Dual control of a sensored PM and sensorless PM machine.
- Equipment/Software: LAUNCHXL-F28069M, BOOSTXL-DRV8301 module ('aft' position) connected to Texas Instruments LVSERVOMTR PM machine, BOOSTXL-DRV8301 module ('forward' position) connected to LVSERVOMTR PM machine, DC power supply and VisSim simulation program.
- Outcomes: Develop a complete drive algorithm, which handles dual machine control. Compile and download an `.out` file for use in phase C+.

As mentioned above, two boost packs are used in this case, both of which are connected to LVSERVOMTR PM machines. Both machines are present in the Texas Instruments 2MTR-DYNO kit. The module in the 'forward' position (closest to the USB connector) will be used together with a sensored PM machine operating under FOC torque control. The resultant sensored drive serves as a dynamometer for the sensorless drive. The 24 V DC power supply to be used for this laboratory must be connected to both BOOSTXL-DRV8301 modules in this case. The following jumper and dip-switch settings on the LAUNCHXL-F28069M module are required:

- Jumpers `JP1` and `JP2` OPEN
- Jumpers `JP4` and `JP5` OPEN
- Jumpers `JP3`, `JP6` and `JP7` CLOSED
- Dip-switches `SW1` to `SW3` ON

Furthermore, the 'J4' encoder connector of the PM machine cable used for sensored FOC (which has a Molex connector) should be attached to encoder input `QEP_A` of the LAUNCHXL-F28069M module.

A phase C development stage, cannot be used to run a drive, but its primary task is to assemble all the modules needed for drive operation with a given controller configuration. The drive setup as given in Fig. 4.57, shows the 'Dual PM Drive controller with FAST and sensored' module, which must be compiled to generate an `.out` file. Inputs to the controller module (via the upper pre-scale module) are five variables (shown as 'constants for simplicity), which allow the user to set the speed `Freq_n`, direct/quadrature current reference values and speed controller gains `Kp_n`, `Ki_n` for the FAST controller. The lower pre-scale module has as input the quadrature reference current for the sensored machine and generates the per unit variable `iq1_n`. Five logic inputs are used of which `conv1-2_on`, activates BOTH power packs together and initializes the common ADC module. Three inputs: `MID_on`, `Force_ang_on` and `speedcon_on` are used to activate motor identification, force angle operation and select speed/torque control for the sensorless drive as discussed in laboratory 2:2 (see Sect. 4.3). A `chan_sel` button is again used to select the diagnostic channels to be displayed. Two outputs of the controller module are used to show per unit DC bus-voltages `udc_n1` and `udc_n2` of the forward and aft booster packs respectively. Three outputs `fluxn_2`, `speed_n2` and `Torque_n2` represent the per unit estimated flux, speed and torque of the sensorless controlled machine. In addition, four outputs are allocated

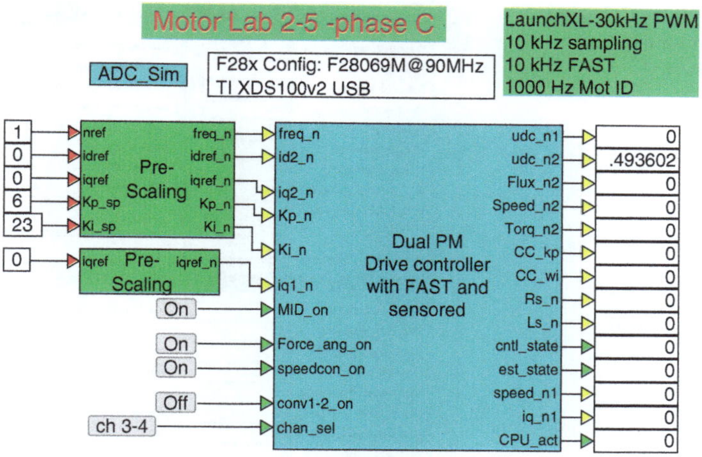

Fig. 4.57 Phase C simulation of a dual PM/PM machine drive

Fig. 4.58 One level down into the controller module

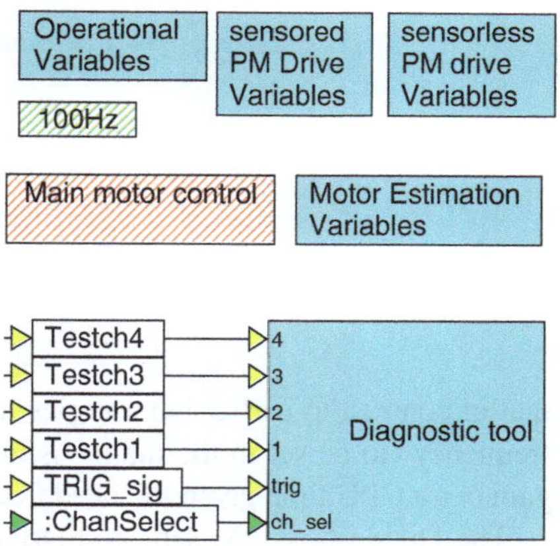

to the sensorless drive and show the required per unit current gain/bandwidth `CC_Kp`, `CC_wi` and estimated stator resistance/inductance parameters `Rs_n`, `Ls_n` respectively. Furthermore, two outputs `cntl_state`, `est_state` are used to show the status of the controller and estimator respectively. The three remaining as yet unnamed outputs are the per unit encoder shaft speed `speed_n1`, quadrature current `iq_n1` of the sensored FOC controller and a variable `CPU_act` used to show the MCU activity during phase C+ operation.

Moving one level down into the controller module leads to the set of modules/dialog boxes shown in Fig. 4.58. The four dialog boxes given in Fig. 4.59, contain all the user settings required for this laboratory as explained in previously laboratories. The 'Operational Variables' dialog box is used to assign common (to both controllers) full scale values for the voltage, current, and frequency. In addition, the ADC full-scale current/voltage values and sampling frequency (as a ratio relative to the ADC sample frequency) for both speed loops are also assigned here. Note

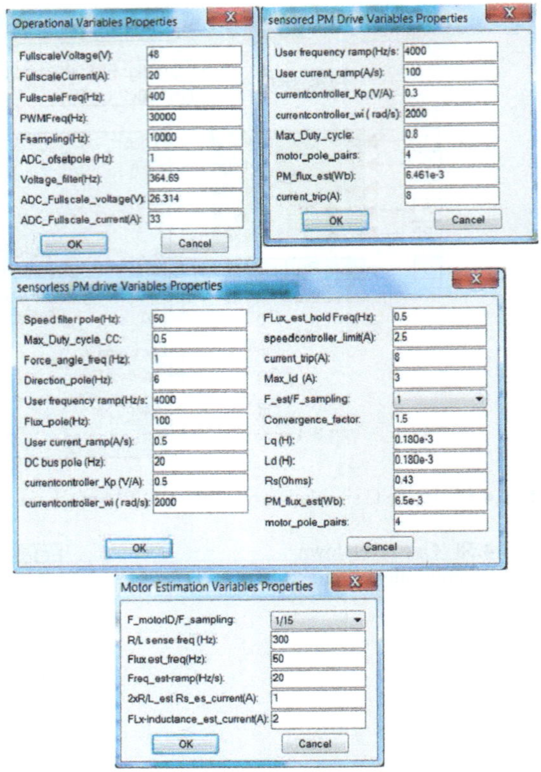

Fig. 4.59 Phase C; Dialog boxes used for the dual machine drive

that in contrast to earlier laboratories in this chapter the PWM and ADC sampling frequency (to be set in the main VisSim menu) are reduced to 10 kHz, in order to gain extra CPU time given the need to execute two controller algorithms.

The Dialog box 'PM sensored Drive Variables', assigns the parameters needed for the sensored PM controller. Hence, current controller gains, maximum modulation index and PM flux amplitude (assumed known in this case) must be provided by the user. The remaining two dialog boxes: 'Converter and FAST Variables' and 'Motor Estimation Variables' are used by the sensorless drive and define the parameters needed for 'normal operation' as well as motor identification. Observation of Fig. 4.59 learns that FAST estimation is set to operate at 10 kHz. For both drives the maximum duty cycle value has purposely been set to 0.8 because current measurement is done with the aid of shunt resistors. This implies that modulation values in excess of 0.8 can impair current measurement, if no measures are taken (as is the case here) to counter this issue. So called 'current reconstruction' [9] measures can be deployed to solve this problem, which allows drive operation with a modulation index greater than 0.8. Furthermore, the rate of change of control variables and pole pair number (equal for both machines) is set in the corresponding dialog boxes for both controllers.

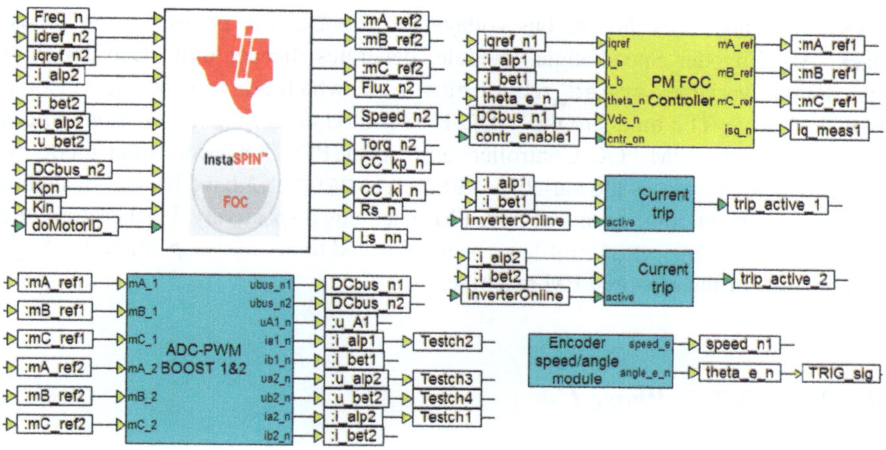

Fig. 4.60 One level down into the 'Main motor control' module

A 'Diagnostic tool' module is again used to buffer two variables, which can subsequently displayed in a graph when operating in phase C+. A two channel multiplexer is used to allow the user to either display variables `Testch1`, `Testch2` or `Testch3`, `Testch4` by using a `chan_sel` button discussed above. A '100 Hz' module acts as a 'heart beat' and flashes a 'red' LED on the LAUNCHXL-F28069M module. A 'blue' LED remains ON if a current trip condition occurs on either boost pack, in which case BOTH converters are shut down. Furthermore, this module executes all the drive background tasks at a sampling rate of 100 Hz.

Moving one level lower into the 'Main motor Control' module reveals a set of modules, as shown in Fig. 4.60. Of these, the 'PM FOC controller' and 'InstaSPIN FOC' modules are used to implement sensored FOC control and sensorless FOC control, as discussed in laboratories 1:3 and 2:2 respectively. The 'ADC-PWM BOOST 1&2' unit shown, generates the α, β current variables for both converters, as well as the per unit DC bus voltages. Note that the ADC module samples six currents and six voltages, where use is made of so called 'dual sampling', where two signals are sampled simultaneously. Furthermore, a PWM frequency of 30 kHz is used, as was mentioned above, together with an ADC sampling frequency of 10 kHz which corresponds to an ADC sampling time of 0.1 ms, that is set in the VisSim 'System Properties' pull down menu. When the button `conv1-2_on` is activated, both boost packs are initialized after which the `inverterOnline` flag is set and the ADC offsets of the 'ADC-PWM BOOST 1&2' unit are determined. After this sequence a `control_enable` is set which sequentially activates both controllers. Relevant (to this laboratory) outputs of the ADC unit for the sensored drive are the DC bus voltage `DCbus_n1`, and the per unit currents i_{α}^{n}, i_{β}^{n}, represented by the variables `i_alp1`,`i_bet1` respectively. The sensorless drive requires the α, β currents/voltage components as represented by the variables `i_alp2`, `i_bet2` and `u_alp2`, `u_bet2` respectively. In addition, the

FAST drive also uses the DC bus voltage variable `DCbus_n2` of the 'aft' boost pack. An 'Encoder Speed/angle' module, generates the per unit electrical shaft speed/angle variables `speed_n1`, `theta_e_n`, which are used by the sensored FOC PM drive. The inputs to the 'ADC-PWM' module are the modulation indices generated by the 'PM FOC Controller' and 'InstaSPIN FOC' controller modules. For diagnostic purposes the variables `TESTch1`, `TESTch2` have been allocated to per unit α current components of both PM machines respectively. Furthermore, the variables `TESTch3`, `TESTch4` have been allocated to per unit α, β filtered voltage components of the sensored PM machine.

4.6.2 Lab 2:5: Phase C+

Phase C+, is the drive operational component of the laboratory and is basically a run version of the `.out` file compiled and downloaded to the MCU in phase C (see previous subsection).

The following information is relevant for this laboratory component:

- Reference program [11]: `lab2-5_LaunchXLphCv3_d.vsm`.
- Description: Dual operation of a sensorless FOC PM and sensored FOC PM drive
- Equipment/Software: LAUNCHXL-F28069M, BOOSTXL-DRV8301 module ('aft' position) connected to Texas Instruments LVSERVOMTR PM machine, BOOSTXL-DRV8301 module ('forward' position) connected to LVSERVOMTR PM machine, DC power supply and VisSim simulation program.
- Outcomes: To investigate dual machine drive operation, using a sensored and sensorless FOC controller.

Note that the required jumper and dip-switch settings for the LAUNCHXL-F28069M module are given in phase C. Furthermore, the 'J4' encoder connector of the sensored PM machine cable should be attached to encoder input `QEP_A` of the LAUNCHXL-F28069M module.

The run version shown in Fig. 4.61 uses a VisSim run module, which executes the `.out` file, shown in said module. An enlarged view of the VisSim scope module given in Fig. 4.62 shows the per unit currents $i_{n\alpha1}$, $i_{n\alpha2}$ after completion of identification for both machines. Multiplication by the full scale current value of 20 A gives actual current values. Prior to activating the drive the following 'Pre-Drive' check list should be executed:

- Dialog boxes used in C+ model must match those of C (and vice-versa): the run version is compiled with the dialog box entries specified under phase C. The dialog boxes shown in C+ are used by the 'Pre/Post' scaling modules.

(continued)

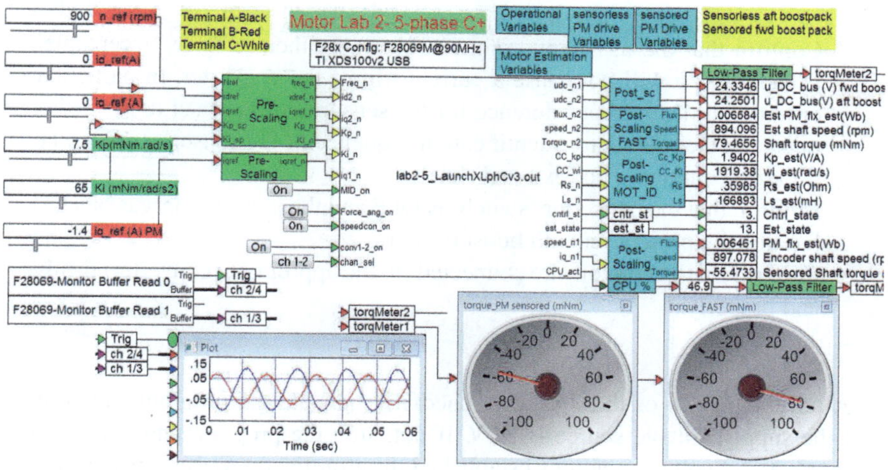

Fig. 4.61 Phase C+ dual operation of a PM/PM drive

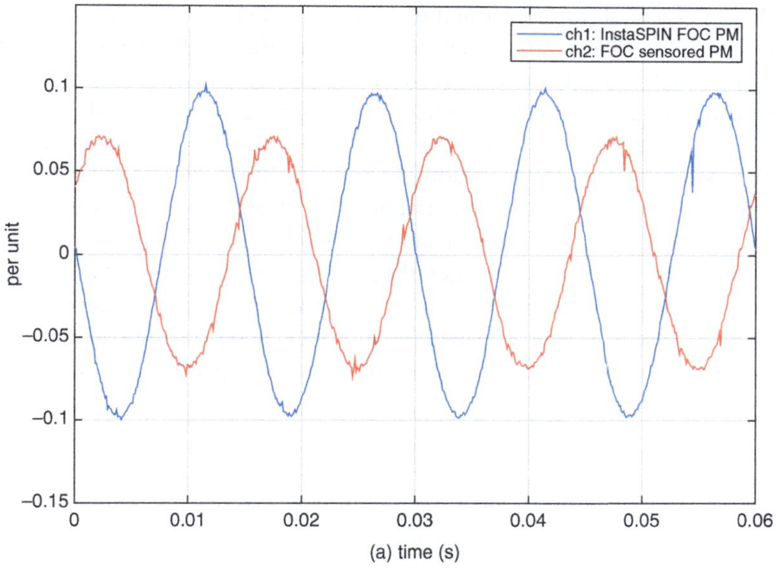

Fig. 4.62 Phase C+ simulation: Scope display showing per unit α currents for sensorless and sensored machine

- Ensure that the sample time used is correct and latest (and correct) `.out` file has been downloaded to the MCU (Right mouse click on MCU module to show dialog box of these variables).

(continued)

- Confirm that the user input values are set to either zero, or 'acceptable' values, which will not cause a current trip of the converter. In particular ensure that the current reference for the sensored drive is set to zero (this to ensure that the motor identification sequence of the sensorless machine is carried out under no (external) load).
- Confirm that the converter 'switch' is OFF and the power supply is on (DC bus voltage present on both boost packs).
- Confirm that the motors are connected (to the appropriate converter) firmly and properly.

After completion of the Pre-Drive checklist, activate the program and confirm that the supply voltage source is 24 V. If not, stop the program and restart. With the correct DC voltage readings established, turn on the converters (using the ON button, for both converters) and monitor drive operation. After motor identification of the sensorless machine, monitor the diagnostic scope, which in this example (Fig. 4.62) is set to show the per unit α current components of both machines. The screen shot given in Fig. 4.61 shows that the sensorless PM machine is operating under speed control, while the sensored PM machine, is operating under FOC torque control and acting as a load, i.e. operating in generator mode with a torque of ≈ -60 mNm. Note that the torque generated by the sensorless FOC drive is equal to ≈ 80 mNm. The sum of two readings represents the friction torque of the dual drive, which is ≈ 20 mNm. A `CPU` variable with attached scope module provides, on a scale from $0 \rightarrow 100\,\%$, an indication of CPU activity.

This laboratory session concludes the chapter on sensorless PM control. The reader is reminded of the fact that only the most relevant information related to these laboratories has been provided. A thorough understanding of the material presented requires an effort on the part of the reader to experimentally work through the laboratories shown.

Chapter 5
Laboratory Sessions: Module 3

A series of laboratory modules will be discussed in this chapter, which have been designed to lead the reader through the development phases needed to understand and work with a sensorless field-oriented IM drive. Sensorless control implementation is based on the use of the InstaSPIN algorithm, which will be configured for different modes of operation. This implies application of the algorithm as a 'software encoder' as well as full motor parameter identification. In addition, a special inductance machine algorithm referred to as 'PowerWarp' will be discussed, which allows energy efficient operation under partial load conditions.

Like the previous chapter, implementation phases B, C and C+ are discussed, which will include the VisSim 'Processor In the Loop' (PIL) software tool used for communication with the FAST algorithm. Dual drive operation, where the PM sensored drive acts as a load/torque transducer for a sensorless controlled IM drive will provide the reader with the opportunity to explore motor/generator operation. Simultaneously, the reader will be guided through the hardware and software steps needed to achieve induction machine based sensorless control using the InstaSPIN-FOC approach.

5.1 Introduction to Sensorless Control for Induction Machines

Sensorless or more accurately, encoderless (without use of a physical shaft encoder which measures either angle or speed) field oriented control of asynchronous drives

---

Electronic supplementary material: The online version of this chapter (doi: 10.1007/978-3-319-20043-9_5) contains supplementary material, which is available to authorized users.

D.W.J. Pulle et al., *Applied Control of Electrical Drives*,
Power Systems, DOI 10.1007/978-3-319-20043-9_5

remains desirable for reasons which are similar to those discussed in the previous chapter for PM based drives namely:

- Cost reduction: the ability to remove the encoder can be highly beneficial in terms of reducing the overall cost of a drive application.
- Smaller drive footprint: removing the encoder from the motor means that less space is required for the application and space saving up to 30 % is possible.
- Reliability: the use of encoder inevitably requires the use of a thin sensor cable (in addition to the, often bulky, three-phase motor leads) to connect the encoder to the drive. Such a sensor cable is easily damaged, which can lead to lengthy (costly) interruptions in drive operation. Elimination of the encoder (plus cable) improves reliability and simplifies motor installation.
- Redundancy: in applications where an encoder is required a sensorless 'back-up' solution may still be desirable, to ensure that a drive remains functional when an encoder related fault occurs.
- Dual shaft operation: for applications where both ends of the motor shaft are connected to a load, the use of a shaft encoder is not possible, hence encoderless operation is required.

Given the above, there has been extensive research undertaken to find a sensorless algorithm, that makes use of the electrical voltage and/or phase currents to generate an estimate for the rotor flux angle, i.e. its orientation relative to the stator frame. Some insight, into the problems associated with such algorithms can be given by considering so called 'Direct Field Oriented Control' (DFO) with voltage and current transducers' [4]. This DFO algorithm is briefly outlined below, where use is made of Fig. 5.1. For rotor flux oriented control the IM flux ψ is aligned with the d-axis of a synchronous reference frame. The current vector $\vec{i}_s$ has two components: i_d, i_q, which regulate the flux and torque of the machine respectively (see Sect. 2.1.3), Field oriented control (FOC), requires access to an accurate estimate of the rotor flux angle θ hence any algorithm must be able to satisfy this requirement. The DFO approach estimates the voltage EMF vector $\vec{e}_s$, defined as

$$\vec{e}_s = \vec{u}_s - R_s \vec{i}_s \tag{5.1}$$

which according to Eq. (5.1) requires measurement of the motor current $\vec{i}_s$ and voltage $\vec{u}_s$ vectors. In addition, the stator resistance R_s must be known and

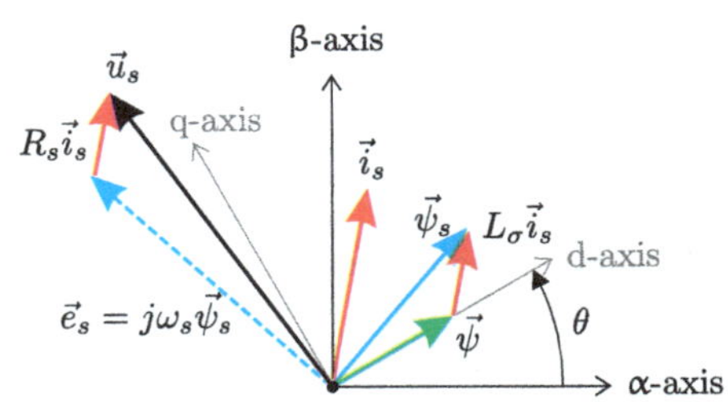

Fig. 5.1 Space vector diagram for IM motor

subsequently tracked, given that its value changes as the motor heats up. In a converter the voltage vector $\vec{u}_s$ is typically made up of two active vectors and a 'zero' vector during each sampling interval T_s, as was discussed in Sect. 2.1.6. This implies that finding the average voltage per sample (with a very high degree of accuracy) of the converter is not a simple task. Once the EMF vector is identified, the corresponding stator flux vector $\vec{\psi}_s$ can be found using

$$\vec{\psi}_s = \frac{\vec{e}_s}{j\omega_s} \tag{5.2}$$

where ω_s is the stator frequency of the EMF vector, which is linked to the shaft speed via the slip of the machine. Hence the immediate problem with this approach is that estimation of this flux vector using Eq. (5.2) becomes difficult, if not impossible, at low stator frequencies. The magnitude and orientation of the IM rotor flux vector $\vec{\psi} = \psi$ can be found using

$$\vec{\psi} = \vec{\psi}_s - L_\sigma \vec{i}_s \tag{5.3}$$

where L_σ represents the leakage inductance of the machine (as represented in a 'four parameter model', see Fig. 2.12), hence its value must be known for correct identification of the rotor flux vector $\vec{\psi}$. In general, a 'good' sensorless algorithm for induction machines must be able to meet the following guidelines:

- Tracking of the rotor flux angle θ: ability to calculate this angle over a wide speed range, including operation at ultra-low stator frequency.
- Fast response: ability to track near-instantaneous changes in $\omega_s = {}^{d\theta}/_{dt}$.
- Robust: ability to provide an accurate rotor angle estimate in the face of motor parameter variations, notably the stator resistance R_s.
- Flux tracking: ability to track the magnitude of the rotor flux over a wide speed range in order to estimate the shaft torque of the machine.
- Motor parameters: ability to accurately identify the parameters needed for accurate rotor angle estimation.

The sensorless algorithm InstaSPIN-FOC developed by Texas Instruments [9] as discussed in the previous chapter (see Fig. 4.2) meets the guidelines set out above. Most importantly it is stable at $\omega_s = 0$. Central to these laboratories is the aim to familiarize the reader with the parameters and use of the InstaSPIN module as relevant to induction machine drives. In this chapter use is again made of the same two VisSim InstaSPIN modules discussed in the previous chapter, which handle internally all the communication with the TMS320F2806xF MCU that houses the sensorless algorithm. However, in this case these modules will be configured to develop a fully operational FOC sensorless IM based drive with motor parameter estimation.

5.2 Laboratory 3:1: FOC Sensorless Control of a IM Machine

Primary objective of the laboratory is to familiarize the reader with the InstaSPIN parameters required to achieve sensorless operation with KNOWN motor parameters, i.e. it is assumed that these have been measured (as indeed they have) prior to this laboratory. A field-oriented controlled drive as discussed in Sect. 3.3 is adapted to sensorless operation by making use of the FAST module as shown in Fig. 4.2. The latter produces a shaft angle estimate, which will replace the shaft angle input from the encoder. However the encoder is still used in this drive example (for phases C and C+ only) to allow the user to operate under either sensorless or sensored operation. Primary objective of the laboratory is to familiarize the reader with sensorless drive operation with KNOWN motor parameters, i.e. it is assumed that these have been measured (as indeed they have) prior to this laboratory.

5.2.1 Lab 3:1: Phase B

The purpose of this introductory laboratory is to demonstrate operation of the FAST based sensorless drive with the aid of a set of vector plots which show estimated and actual (machine model) rotor flux vectors, whilst operating under speed or torque control. The basic FOC drive is identical to that discussed for FOC sensored control (see lab 1:6 in Sect. 3.6.1). The key difference is that rotor angle, rotor flux, shaft speed and torque data are now provided indirectly by a FAST module located within the 'Drive Controller' shown in Fig. 5.2. The term 'indirectly' refers to the fact that said FAST module contains a set of function calls which communicate with the TMS320F28069M unit located on the LAUNCHXL board. This 'Processor In the Loop' (PIL) approach, as discussed earlier for the PM drive, makes it possible to evaluate drive performance under simulated conditions using the actual FAST algorithm located in the hardware.

The following information is relevant for this laboratory component:

- Reference programs [11]: `lab3-1_LaunchXLphBv2.vsm` and `FAST_IM_v2.vsm`.
- Description: Sensorless control of an IM motor using FAST module as an observer.
- Equipment/Software: Texas Instruments LAUNCHXL-F28069M (no boost packs) and VisSim simulation program.
- Outcomes: Show the various vector plots that are present in the drive when using a fixed point representation of the controller when operating under field oriented control using the FAST sensorless algorithm.

As indicated above no boost packs are used in the laboratory, nor is an external 24 V power supply required, given that power to the MCU is directly provided via

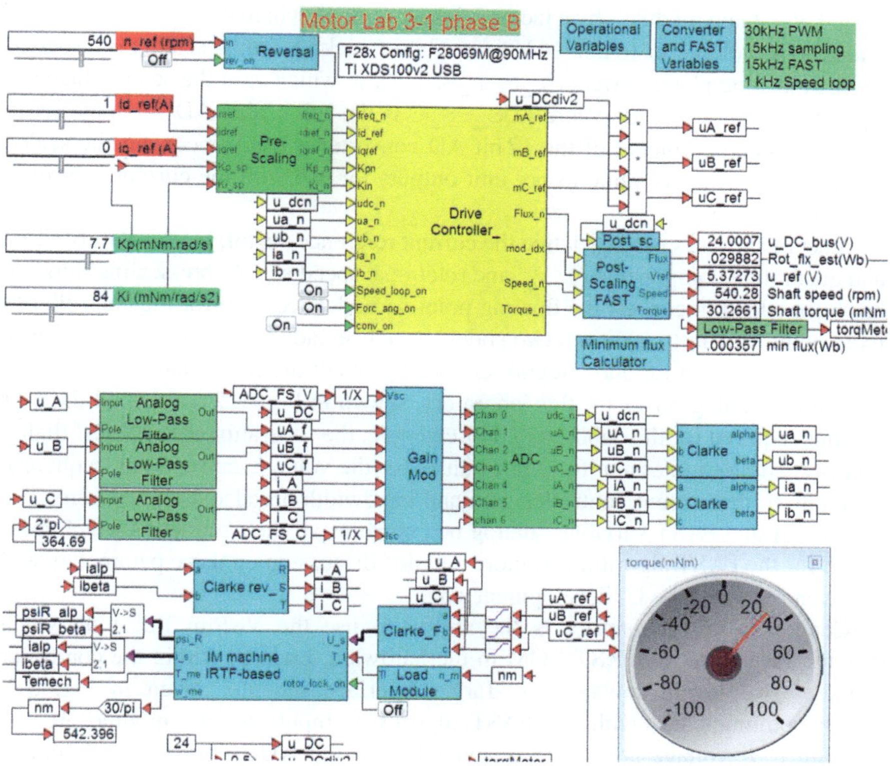

Fig. 5.2 Phase B simulation of a FAST based encoderless field-oriented controlled (FOC) drive

the USB cable connected to the laptop. The machine parameters used in the lab are those of the LVACIMTR IM. In this case the following jumper and dip-switch positions on the LAUNCHXL-F28069M module are required:

- Jumpers `JP1` and `JP2` CLOSED
- Jumpers `JP4` and `JP5` OPEN
- Jumpers `JP3`,`JP6` and `JP7` CLOSED
- Dip-switches `SW1` to `SW3` ON

Note that there are two reference files required for this lab: the first is the main simulation model which contains a PIL 'interface module' that makes use of a second file `FAST_IM_v2.vsm`. The outputs of the 'Drive Controller' shown in Fig. 5.2 are the three reference modulation indices `mA_ref`, `mB_ref`, `mC_ref`. These variables are multiplied by the term $u_{DC}/2$ and connected to the 'converter', which in this case is simply represented by three limiters, that limit the voltage to the motor to $\pm u_{DC}/2$. The phase voltages `u_A`, `u_B`, `u_C` of the machine are used by the machine model and by the FAST algorithm. For the latter case the converter outputs are filtered using 'analog low-pass filter' modules that replace the electrical circuits which are present in the actual hardware. The filtered voltages are then scaled with the aid of a 'Gain module', by a factor 3.3/ADC_FS_V, where ADC_FS_V is the full scaled ADC voltage set to 26.314 V for the boost pack. A similar scaling is also used for the machine phase currents i_A, i_B, i_C, in which case the scaling factor is equal to 3.3/ADC_FS_C, where ADC_FS_C is the full scaled ADC voltage set to 33.0 A. A detailed model of the 12 bit AD converter is used to convert the scaled voltage/current inputs value to per unit outputs for use with the current controller and FAST module.

A set of sliders is used to define the current reference vector, in terms of its direct and quadrature components $i_{\mathrm{d}}^{\mathrm{ref}}$, $i_{\mathrm{q}}^{\mathrm{ref}}$ and rotational speed $n_{\mathrm{m}}^{\mathrm{ref}}$. A 'pre-scaling' module has been added to convert the floating point variables required for the controller to fixed point format. In addition, two sliders have been added which allow the user to set the proportional Kp and integral Ki speed gains of the controller.

The two dialog boxes for this laboratory, shown in Fig. 5.2, contain all the user settings required for this laboratory. For example, the 'operational Variables' dialog box is again used to assign full scale values for the voltage, current and frequency. In addition, the current controller gain and bandwidth are also assigned here. The 'Converter and FAST variables' dialog box contains all the parameters which must be set for the FAST algorithm. A more detailed discussion on these parameters will be given in the 'phase C' development stage.

Central to this laboratory is the ability to use the VisSim PIL tool which provides access to the FAST ROM in the hardware. Located within the controller module (two levels down) is a 'Target Interface module' given in Fig. 5.3, which communicates with the FAST algorithm. Inputs to this module are the per unit α, β voltage and currents generated by the ADC unit, DC bus voltage VdcN and a set of control inputs. Key outputs are the per unit inverse DC bus voltage invB (used by the current controller), rotor flux estimate Flux, rotor angle estimate Angle and Speed/Torque estimates. A set of LED_in/LED_out variables are used to activate a 'blue' LED on the LAUNCHXL board when PIL activity takes place. Of central importance in the dialog box given in Fig. 5.3 is the Target Execution File in this case set to FAST_IM_v2.out. When the main simulation program is activated (via the 'green' arrow button), this .out file will be downloaded into the MCU prior to the start of the actual simulation. Once downloaded the rate of data exchange between MCU and VisSim simulation will be dictated by the Sample Rate (Hz) entry (set to 200 Hz in this case) provided the Syn Target to This Block option has been activated. Note that the simulation frequency (set in the system properties) is set to 15 kHz given that this is the sampling frequency specified. However, actual run time during PIL operation is dictated by the Sample Rate (Hz) entry mentioned above.

Generation of the FASTv2.out file is carried out by compilation of the module 'FAST EST IM' shown in Fig. 5.4, which is present in the file FAST_IM_v2.vsm. This module contains all the functions that are needed to access the FAST ROM module. More details on the content of the 'FAST EST IM' module will be given in development phase C.

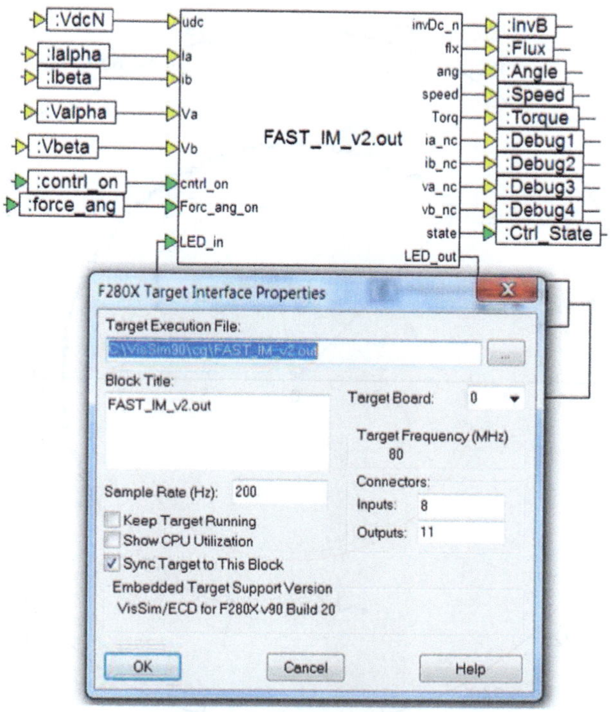

Fig. 5.3 Target interface module with corresponding dialog box entry

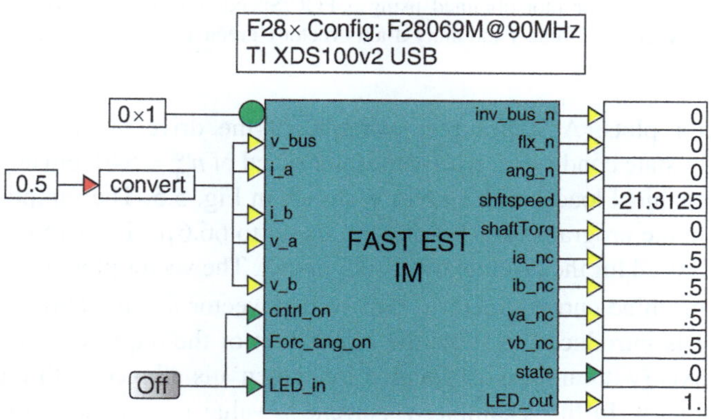

Fig. 5.4 FAST estimation module

Starting this simulation is done via the usual 'green' button (VisSim main menu) in which case the `.out` file for the PIL process is downloaded. Then the ADC offsets are determined within the AD unit, which takes ≈ 1.4 s, (simulation time) AFTER that period the PIL should become active (flashing 'blue' LED on LaunchXL board) and drive operation should become apparent from observation

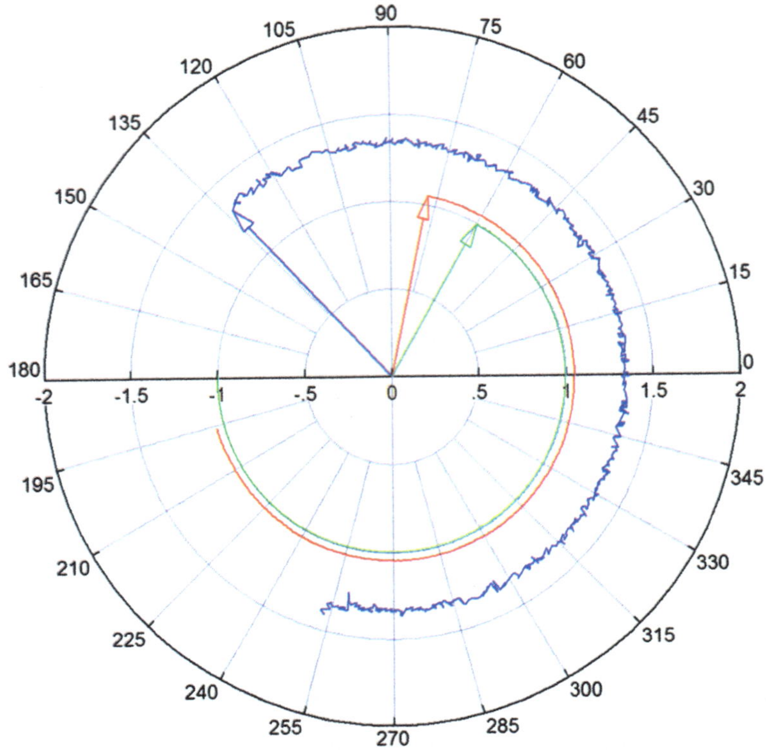

Fig. 5.5 Phase B: Vector plot obtained using a FOC/Speed sensorless control with current $\vec{i}_s$ ('red'), voltage vector $\vec{u}_s$ ('blue') and unity d-axis vector ('green')

of the vector plots. A vector plot example of the drive in question operating under steady-state conditions, with a constant speed of $n_m^{ref} = 540$ rpm and arbitrarily chosen mechanical load of 30.2 mNm is shown in Fig. 5.5. The sampling time T_s, as defined in the program pull-down menu, is set to 66.6 μs, i.e. a 15 kHz sampling frequency is used for the discrete controller model. The vector plot given in Fig. 5.5, shows the machine current vector $\vec{i}_s$ and voltage vector $\vec{u}_s$. In addition, a unity 'd-axis' vector is introduced to show the orientation of the d,q reference frame. The positive d-axis (with amplitude 1 and color 'green') is aligned with the estimated rotor flux vector. In this example, a current i_q value is generated given that the machine is producing torque to match the load torque applied. Consequently the current vector should have an i_q and $i_d = 1.0$ A component, as is indeed the case. The voltage vector is made up of a back EMF that is orthogonal (leading) the rotor flux vector and a resistive voltage component, which is aligned with the current vector. The third component of the voltage vector is due to the inductive (leakage inductance) component of the machine and its value will depend (among others) on the electrical frequency and will be orthogonal (leading) to the current vector.

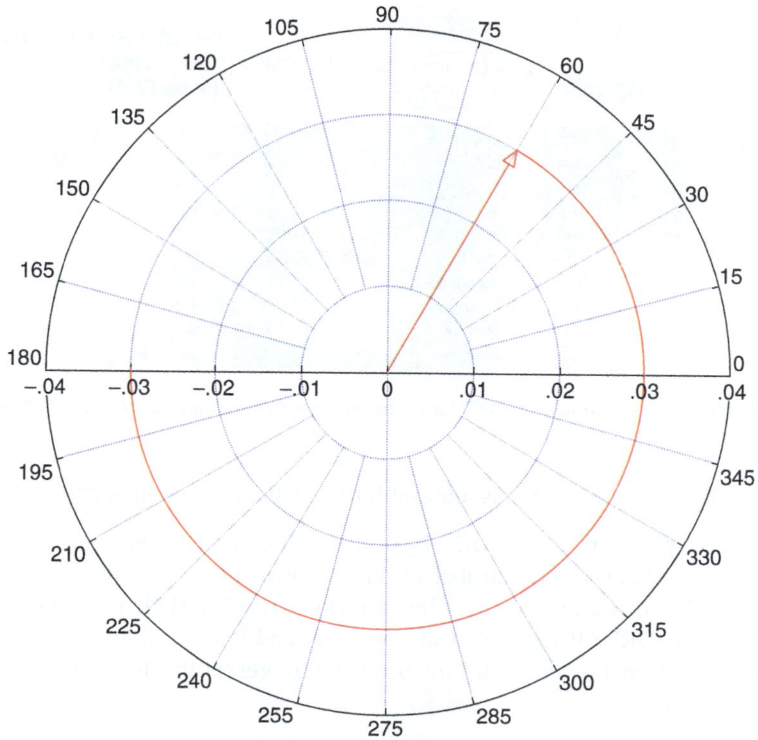

Fig. 5.6 Phase B: Rotor flux vectors generated by the machine model $\vec{\psi}_R$ ('red'), and FAST estimator $\vec{\psi}_R^{est}$ ('blue')

An indication of the flux tracking capabilities of the FAST algorithm may be deduced from the vector plot given in Fig. 5.6. The theoretical amplitude of the flux vector is given by $\psi_R = L_M\, i_d$, which in this case is equal to 30 mWb, as is indeed the case. Clearly observable from Fig. 5.6 is the high degree of alignment between the two vectors.

The reader is again encouraged to explore all modes of operation as to obtain a better understanding of the sensorless control approach presented here, before moving on to a more in depth discussion on the actual experimental implementation.

5.2.2 Lab 3:1: Phase C

Prior to being able to work with an operational drive, an `.out` file must be generated of a 'Current Controller with FAST' module, as shown in Fig. 5.7, which represents the drive under consideration. Hence details of the drive structure and corresponding dialog box parameter assignments will be discussed in this subsection.

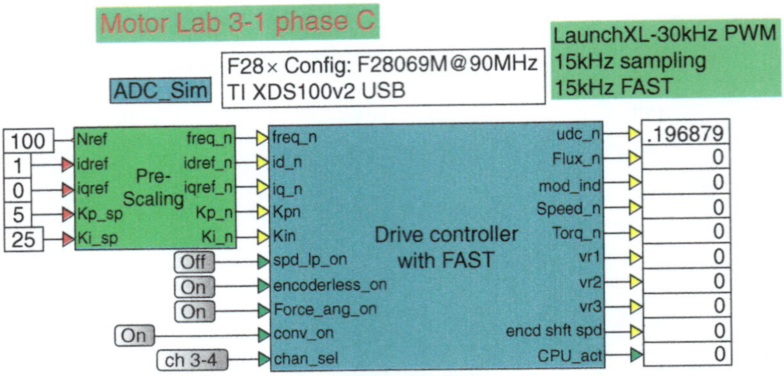

Fig. 5.7 Phase C simulation of a FAST based encoderless field-oriented controlled (FOC) drive

The following information is relevant for this laboratory component:

- Reference program [11] : `lab3-1_LaunchXLphCv2.vsm`.
- Description: Sensorless control of a IM motor using FAST module as an observer.
- Equipment/Software: Texas Instruments LAUNCHXL-F28069M, with BOOSTXL-DRV8301 module ('aft' position) and VisSim simulation program.
- Outcomes: Generate the `.out` file needed to represent the drive structure under investigation.

As mentioned above, a single boost pack located in the 'aft' position (furthest away from the USB connector) is used for single motor operation (PM machine is not used electrically), in which case the following jumper and dip-switch positions on the LAUNCHXL-F28069M module are required:

- Jumpers `JP1` and `JP2` OPEN
- Jumpers `JP4` and `JP5` CLOSED
- Jumpers `JP3`, `JP6` and `JP7` CLOSED
- Dip-switches `SW1` to `SW3` ON

Inputs to the Controller are five variables (shown for simplicity as 'constants'), which set the per unit reference shaft speed, direct and quadrature current amplitudes `id_n`, `iq_n` and speed gains `Kpn`, `Kin`. Note that the quadrature current input variable will only be functional when the speed control `spd-lp_on` function is disabled. The `encoderless_on` button activates the sensorless algorithm, when OFF, the drive operates under current control only, where the rotational speed of the current vector is set by the reference frequency and amplitude by the current reference sliders. A button `Force_ang_on` activates the 'force angle' function of the InstaSPIN algorithm, which is helpful when a high starting torque value is required at start-up. The buttons connected to the module inputs `conv_on'` and `chan_sel` are used to respectively activate the converter and select which diagnostic channel combination is to be displayed in phase C+. Outputs of the controller module, in per unit format, are: measured DC bus voltage

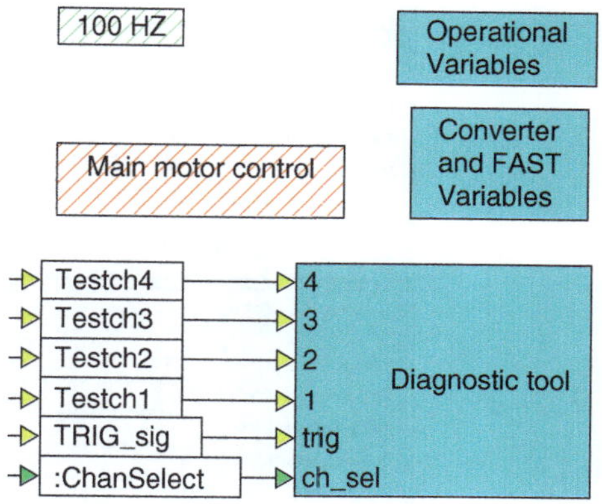

Fig. 5.8 Phase C simulation of a FAST based encoderless field-oriented controlled (FOC) drive: one level into the drive controller module

`udc_n`, `Flux_n` the estimated rotor flux (in Wb), `mod_ind` the modulation index amplitude from the current controller which will be used to calculate the reference output voltage. Additional outputs are: `speed_n` estimated shaft speed and shaft torque `Torq_n`. Three additional outputs `VR1`, `VR2`, `VR3` are used for diagnostic purposes, hence variables of interest can be allocated before compilation, so that they can be subsequently (in Phase C+) shown on a numerical display. In this case the stator resistance, leakage inductance and rotor resistance (four parameter model) values as used by FAST will be shown. Two outputs `encd shft_spd` and `CPU_act` have also been added, which generate the per unit shaft speed (from encoder) value and percentage MCU usage value respectively.

Moving one level lower into the 'Drive Controller with FAST' module reveals the set of modules shown in Fig. 5.8. Of these shown, the 'Diagnostic tool' module has been discussed in previous laboratories. A '100 Hz' module is used to execute background tasks which are non time critical. Of interest are the 'Operational' and 'Converter and FAST variable' dialog boxes, shown in Fig. 5.9. These contain all the data entries that must be provided by the user in order to achieve successful sensorless drive operation.

The 'Operational Variables' dialog box, contains all the parameters needed for scaling, rate limit values for user inputs and gain/limit settings for current/speed control. These are similar to those discussed for laboratory 1:6 (see Sect. 3.6). The meaning of the variables and values to be assigned on the 'left' column of the 'Converter and FAST Variables' dialog box are itemized as follows:

- `Voltage_filter (Hz)`: this is the corner frequency value (in Hz) of the low-pass filter used to measure the phase voltages, which can be found with the aid of Eq. (3.1b).
- `Speed_filter pole (Hz)`: this is the corner frequency (in Hz) of a low-pass filter that is used to filter the estimated shaft speed value. Typically its value is a factor 5 higher than the bandwidth of the speed-loop controller and less than 100 Hz.

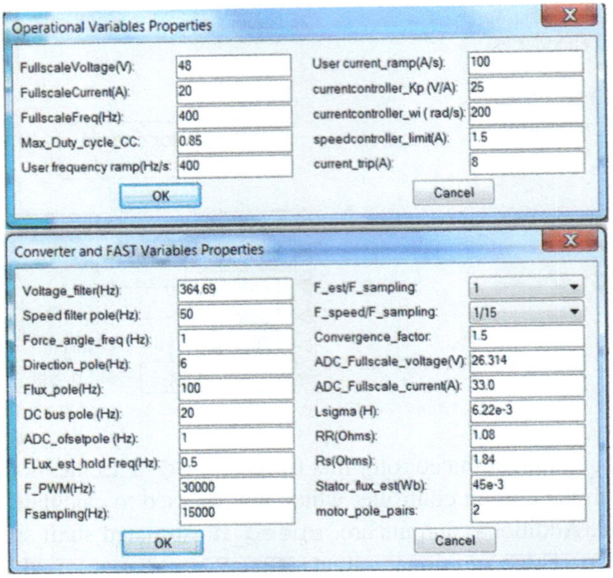

Fig. 5.9 Phase C simulation of a FAST based encoderless field-oriented controlled (FOC) drive: Dialog box entries

- `Force_angle_freq (Hz)`: the force angle function (for induction machine based drives) is used when a high start up torque is required. When speed reversals are required, the function should be turned off.
- `Direction_pole (Hz)`: motor direction is identified by the FAST algorithm, and this function is tied to a low-pass corner frequency, with a default value of 1 Hz.
- `Flux_pole (Hz)`: flux estimation in the FAST algorithm is tied to a low-pass corner frequency, with a default value of 100 Hz.
- `Flux_est_hold Freq (Hz)`: the absolute frequency where the estimator 'freezes' estimation, i.e. it holds the flux constant within the 0.5 Hz band set by this variable.
- `DC bus_pole (Hz)`: The inverse bus voltage (for use with the current controller) is calculated in the FAST algorithm and is tied to a low-pass corner frequency with a default value of 20 Hz.
- `ADC_offsetpole(Hz)`: corner frequency of a low-pass filter used to measure the ADC offsets.
- `F_PWM(Hz)`: converter PWM frequency set to 30 kHz inside the ADC-PWM module.
- `Fsampling(Hz)`: the sampling frequency used for the ADC, the value is shown here for convenience and reference purpose only. Its value is in fact set in the main program, via the pull-down menu 'system properties'. For this application it is set to 15 kHz.

The variables and values to be assigned on the ‘right’ column of the ‘Converter and FAST Variables’ dialog box are more self explanatory namely:

- `F_est/F_sampling`: ratio between sampling frequency used by the FAST estimator algorithm and ADC sampling frequency. The ratio is an integer value, hence the user can select the ratio from a dialog box pull-down menu. For example, a choice of `1/1` implies an estimator sampling frequency of 15000 Hz, given that the ADC sampling frequency is set to 15000 Hz.
- `Convergence_factor`: this variable governs the transient ability of the algorithm to ‘home in’ on the actual rotor angle, its default value is 1.5.
- `ADC_Fullscale_voltage(V)`: a value defined by the voltage signal conditioning circuit and can be calculated using Eq. (3.3a). The value shown is the (theoretical) peak to peak voltage needed to generate a 3.3 V peak to peak signal on the ADc input.
- `ADC_Fullscale_current(A)`: a value defined by the current signal conditioning circuit and can be calculated using Eq. (3.3b). The value shown is the (theoretical) peak to peak current needed to generate a 3.3 V peak to peak signal on the ADC input.
- `L_sigma(H)`: the leakage inductance (in H) of the machine, which is defined in relation to a four parameter model as discussed in Sect. 2.1.4. Its value must be known for this laboratory, via a data sheet, or measured.
- `R_R(Ohms)`: the rotor resistance (in Ω) of the machine, which is defined in relation to a four parameter model as discussed in Sect. 2.1.4. Its value must be known for this laboratory, via a data sheet, or measured.
- `R_s(Ohms)`: the stator resistance (in Ω) of the machine, which must be known for this laboratory, via a data sheet, or measured using the InstaSPIN motor identification algorithm, or with the aid of an Ohm meter.
- `Stator_flux_est(Wb)`: an estimate for the stator flux (in Wb). In this example its value is set to 45 mWb, which can be found by considering the rated RMS line to line voltage U_{LL} of the machine and the rated frequency f_s (Hz). On the basis of this data a flux estimate can be found using $\psi^{est} = U_{LL}\sqrt{2}/(2\pi f_s\sqrt{3})$.
- `motor_pole_pairs`: the number of motor poles in use. Its value must be derived from the data sheet or measured. For example, by using open-loop current control (encoderless operation OFF) experiment and adjusting the input electrical frequency in such a manner that the shaft (with no external load) rotates at 1 revolution per second (which can be observed visually). The pole pair number is then found from the ratio of user frequency and observed shaft speed. The machine in use has four poles (two pole pairs $p = 2$).

Moving one level lower into the ‘main motor control’ module, shows the set of modules given in Fig. 5.10. Of these shown, all with exception of the ‘Drive Controller’ have, in fact, been identified in earlier laboratories. Inputs to the controller module, are those identified and discussed in laboratory 1:6. However, for the sensorless controller considered here, both (per unit) voltage and current α, β variables are now required. Moving yet one level lower into the ‘Drive Controller’ module, reveals a set of modules given in Fig. 5.11. Hopefully familiar, are the

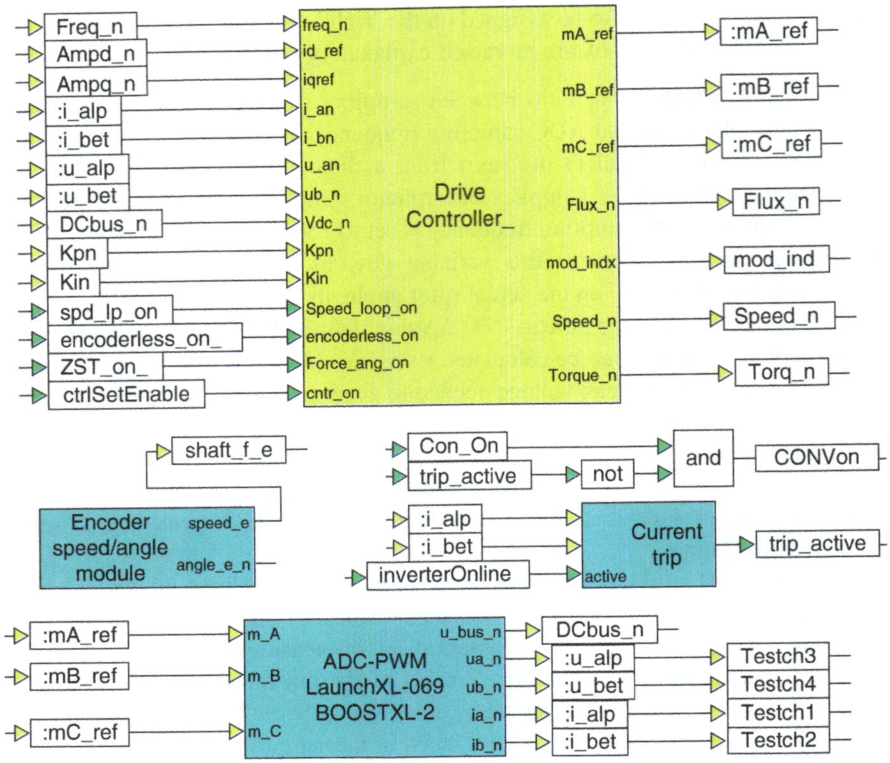

Fig. 5.10 Phase C simulation of a FAST based encoderless field-oriented controlled (FOC) drive: One level into the 'Main Motor Control' module

modules linked to current and speed control. Central to the figure is the InstaSPIN module, which in this case is configured to operate as a software encoder. Input variables to this module are the (per unit) α, β voltages/currents and DC bus voltage. Outputs of the current controller are the three phase modulation indices. The InstaSPIN module provides the following instantaneous (per unit) output variables:

- `INVDC`: inverse DC bus voltage.
- `ROTORFLX`: rotor flux estimate.
- `ang`: estimate for electrical rotor flux angle.
- `rotfreq`: filtered estimate for shaft frequency.
- `eTT` estimate for shaft torque.

Observation of Fig. 5.11 learns that use is made of the `encoderless_on` (user activated) function to activate a 'switch' module that ensures that the variables `Cos_th`, `Sin_th` (as used for Park transformations) are switched to the variables `FCos_th`, `FSin_th`, which are the Sin/Cos functions that make use of the FAST generated variable `ang`. When the function `encoderless_on` is off, speed control is disabled and the Sin/Cos functions for the speed controller are

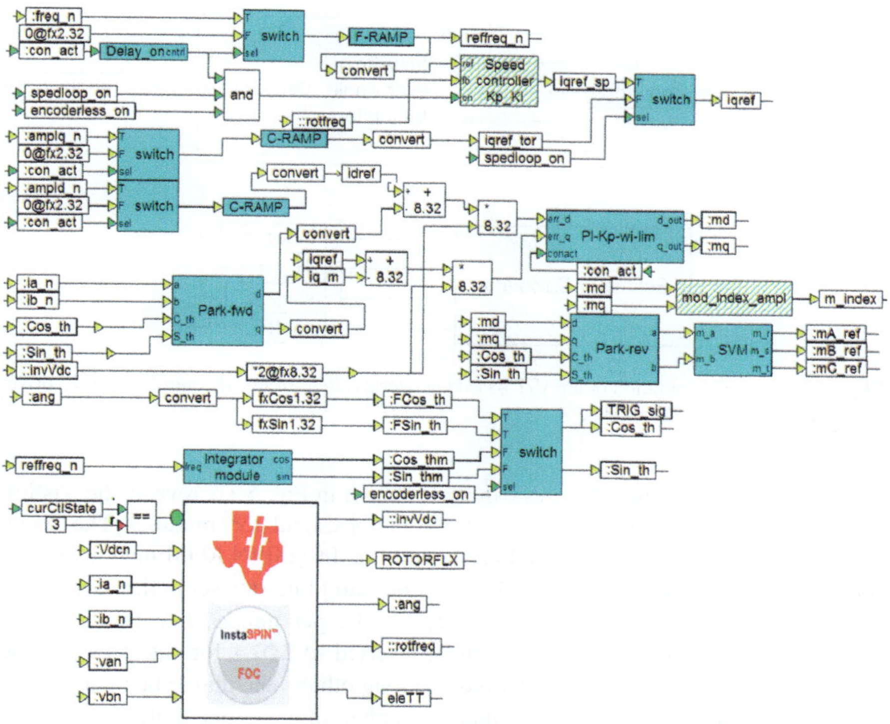

Fig. 5.11 Phase C simulation of a FAST based encoderless field-oriented controlled (FOC) drive: One level into the 'Drive Controller' module

generated via an integrator that is linked to the reference frequency `reffreq_n` variable. If the logic variable `speed_loop_on` is zero (speed loop not active) and `encoderless_on` is on, then the quadrature reference variable `iqref` will be connected to the variable `amplq_n`, which is set by the `i_q` slider. Under these conditions, the sensorless drive is operating under 'torque control' and care should be taken to MAINTAIN a load on the machine, as a speed runaway situation can occur (the machine will try and maintain a constant torque value, defined by the user set `i_q` value, independent of speed). A `mod_index_ampl` module calculates the amplitude of the d, q modulation index variables generated by the current controller module. This module and the `Speed controller Kp_Ki` module are hatched 'green', which signifies that they operate (in this case) with a sampling frequency of 1 kHz, which is 15 times lower than the other modules shown in this figure. This has been done to reduce CPU loading and improve accuracy of the speed-controller's integrator. The InstaSPIN module is triggered by the ADC start of conversion signal (SOC) which is generated by PWM module 4 (located inside the ADC-PWM module, see Fig. 5.10) and is conditional to the controller state `curCtlState` being active. This variable is in turn generated within the '100 Hz' module via the converter ON button.

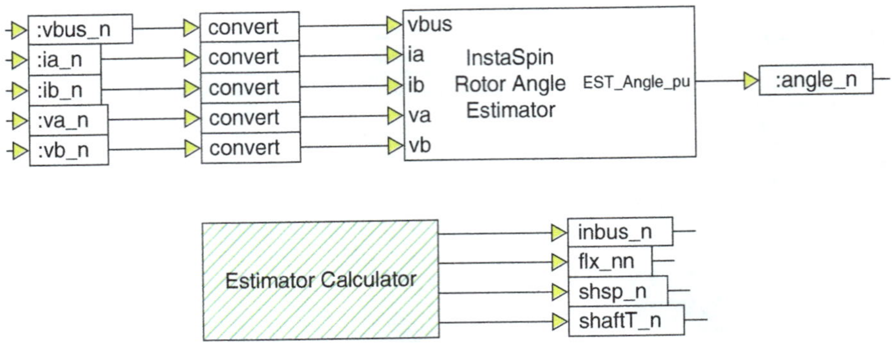

Fig. 5.12 Phase C simulation of a FAST based encoderless field-oriented controlled (FOC) drive: One level into the 'FAST' module

Moving one level into the FAST module shown in Fig. 5.12, reveals the VisSim 'InstaSPIN Rotor Angle Estimator' and 'Estimator Calculator' modules. The 'angle estimator' module communicates directly with the InstaSPIN ROM module, which returns the flux angle estimate `angle_n` at the sampling rate set in the user dialog box (see Fig. 5.9). Inputs to this module are the per unit α, β currents/filtered voltages and DC bus voltage, which are converted to `IQ24` format. This module has an extensive dialog box, which holds (among others) the dialog box parameters introduced in Fig. 5.9. Note also that a number of less time critical estimator tasks have been allocated to the '100 Hz' module given in Fig. 5.8. An 'Estimator calculator' compound module, also shown in Fig. 5.12, is used to derive the per unit inverse-bus voltage: `inbus_n`, rotor-flux: `flx_nn`, shaft-speed: `shsp_n` and shaft torque: `shaftT_n` values from the ROM. The use of a hatched surface, indicates that the sampling rate has been set independently of the main simulation frequency, which is typically the ADC sampling frequency. From a practical perspective, it is prudent to set the sampling frequency for the 'Estimator Calculator' to the same value used by the speed control module. Note that the modules shown in Fig. 5.12 are also those which appear in the 'FAST EST' module (see Fig. 5.4) used for PIL operation in phase B.

5.2.3 Lab 3:1: Phase C+

Phase C+, is the operational component of the laboratory and is basically a run version of the `.out` file compiled and downloaded to the MCU in phase C (see previous subsection). The following information is relevant for this laboratory component:

- Reference program [11]: `lab3-1_LaunchXLphCv2_d.vsm`.
- Description: FOC sensorless control of a IM machine.

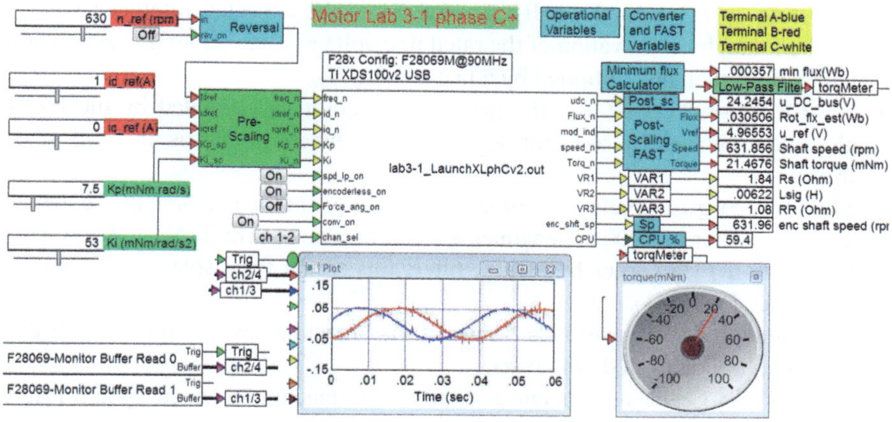

Fig. 5.13 Phase C+ simulation of a FAST based encoderless field-oriented controlled (FOC) drive

- Equipment/Software: Texas Instruments LAUNCHXL-F28069M, with BOOSTXL-DRV8301 module ('aft' position), Texas Instruments LVACIMTR IM machine motor connected to Texas Instruments LVSERVOMTR PM machine and VisSim simulation program.
- Outcomes: to gain familiarity with the FAST InstaSPIN algorithm when used as a 'software encoder'.

Note that the required jumper and dip-switch settings for the LAUNCHXL-F28069M module are given in phase C. Furthermore, the 'J4' encoder connector of the PM machine cable should be attached to encoder input `QEP_A` of the LAUNCHXL-F28069M module. Its purpose is purely to generate an encoder based shaft speed reading for this lab.

The run version shown in Fig. 5.13 uses a VisSim module, which executes the `.out` file compiled in the previous development phase. Five sliders are deployed, the purpose of which was discussed in the previous section. A post-scaling module is again introduced to convert the per unit measured DC bus voltage to actual voltage, as shown with a numeric display. An additional 'Post scaling FAST' module is added, for which a number of scaling factors are introduced to generate the required outputs based on inputs from the 'run' module. The computation of the scaling factor for the output variables shown is as follows:

- `u_DC_bus(V)`: measured DC bus voltage, scaled by the fullscale voltage `u_fs`.
- `Rot_flx_est(Wb)`: the estimated rotor flux (Wb) of the machine. Calculated using the input `Flux_n`, which (in this example) is not scaled.
- `u_ref(V)`): amplitude of the reference output voltage generated by the current controller. This variable makes use of the modulation index amplitude `mod_ind`, which is scaled by half the DC bus voltage u_{DC}. This variable is useful for determining the rated direct axis current of the machine. With the machine

operating at, for example, half the rated speed the expected reference voltage can be found by observation of the rated line voltage value (from the nameplate) which must then be multiplied by a factor $0.5\sqrt{2/3}$.

- `shaft speed(rpm))`: shaft speed of the motor as estimated by the FAST algorithm. Scaled by a factor `F_fs*60/p`, where `F_fs` is the full scale frequency.
- `shaft torque (mNm))`: estimated shaft torque of the machine in use, where a scaling factor of 1000 is generates a result in terms of (milli-Nm). A low-pass filter with a 1 Hz corner frequency, filters this signal for display on the torque meter (see Fig. 5.13).
- `Rs (Ohm)`: stator resistance value of the machine as used by the FAST algorithm (value from dialog box).
- `Lsig (H)`: Leakage inductance value of the machine as used by the FAST algorithm (value from dialog box).
- `RR (Ohm)`: rotor resistance value of the machine as used by the FAST algorithm (value from dialog box).
- `enc_shft_sp` shaft speed measured by the encoder located on the attached PM machine. Scaled by a factor `F_fs*60/p`.
- `CPU` a module attached to the `CPU_act` input of the run module that gives an accurate MCU usage reading. Note that a 'right click on the run module reveals a box 'show CPU utilization' which is NOT used (this feature has been superseded).

Two 'Monitor Buffer' modules display two selected (using an on/off button connected to the `chan_sel` input) diagnostic signals. An enlarged view of the VisSim scope module given in Fig. 5.14 shows the per unit currents i_{α}^{n}, i_{β}^{n}. Multiplication by the full scale current value of 20 A gives actual current values.

Prior to activating this type of lab component, the reader needs to be aware of the fact that he or she is about to activate a complex electrical system, with live voltages. Hence it is prudent, **always**, to execute the following 'Pre-Drive' check list:

- Dialog boxes used in C+ mode, match those of phase C: the run version is compiled with the dialog box entries specified under phase C. The dialog boxes shown in phase C+ are used by the 'Pre/Post' scaling modules.
- Ensure that the sample time used is correct and latest (and correct) `.out` file has been downloaded to the MCU (Right mouse click on MCU module to show dialog box of these variables).
- Confirm that the user input values are set to either zero, or 'acceptable' values, which will not cause a current trip of the converter.
- Confirm that the converter 'switch' is set to OFF and the power supply is on (DC bus voltage present).
- Confirm that the motor is connected firmly and properly.

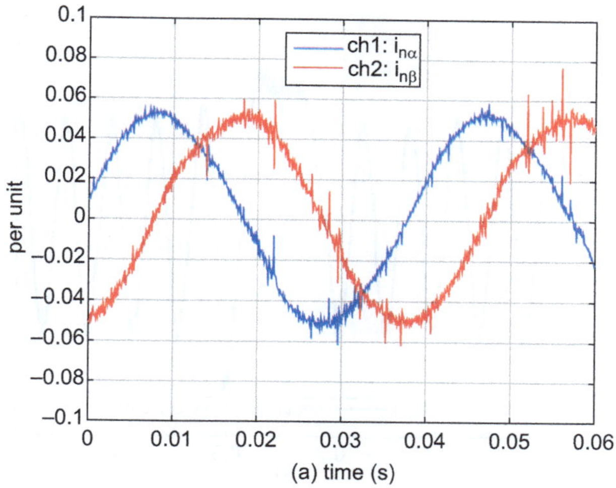

Fig. 5.14 Phase C+ simulation: Scope display showing per unit α, β for sensorless operation

After completion of the Pre-Drive checklist, activate the program and confirm that the supply voltage source level shown in the digital display is 24 V. If not stop the program and restart. With the correct DC voltage reading established, turn on the converter (using the ON button) and monitor the motor shaft and diagnostic scope, which in this example (Fig. 5.14) is set to show the per unit current components $i^{\mathrm{n}}_{\alpha}, i^{\mathrm{n}}_{\beta}$.

Observation of Figs. 5.13, 5.14 learns that the machine is currently operating under closed-loop speed control without an external load (friction only), as is evident from the torque meter value and the (per unit) i_{α}, i_{β} currents displayed in the diagnostic scope. Operation with $i_{\mathrm{d}} = 1.0\,\mathrm{A}$ is undertaken in this case, hence the maximum phase current value will be equal to $i_{\mathrm{max}} = \sqrt{(i_{\mathrm{d}})^2 + (i_{\mathrm{q}})^2}$, where i_{q} represents the quadrature reference current value (set by the speed controller). Sensorless operation is in use, as is evident by the ON button, `encoderless_on`.

The 'Minimum Flux Calculator' module (see Fig. 5.13) calculates the lowest flux level with which the estimator will function given the chosen dialog variables. To attempt FOC sensorless operation with an anticipated flux level that is likely to be lower than the minimum flux level shown, is not recommended and may fail. In that case the full scale voltage and/or estimator frequency should be adjusted. In addition to the above, an oscilloscope and DC-true current probe were used to measure the phase current during a speed reversal, as shown in Fig. 5.15. During the speed reversal the quadrature current is limited to $i_{\mathrm{q}} = 1.5\,\mathrm{A}$ (set in the dialog box) hence the maximum current that can occur during the reversal is $i_{\mathrm{max}} = \sqrt{(i_{\mathrm{d}})^2 + (i_{\mathrm{q}})^2}$, which is equal to 1.8 A given the direct axis current reference value i_{d}= 1.0 A in use.

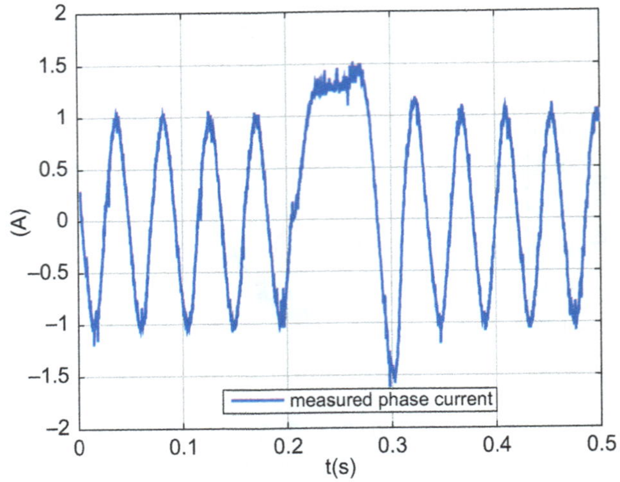

Fig. 5.15 Phase C+ experimental result: phase current waveform, during a speed reversal $n_m = 630 \rightarrow -630\,\text{rpm}$

5.3 Laboratory 3:2: IM FOC Sensorless Control with Motor Identification

Critical to any sensored field-oriented drive application is the ability to correctly estimate: stator resistance R_s, leakage inductance L_σ, rotor resistance R_R and rotor flux ψ, where the latter variable is used for torque estimation purposes. Note that a four parameter machine module is assumed, as discussed in Sect. 2.1.4. Furthermore, it is helpful for current controller tuning purposes, to have access to estimates for the (current controller) gain K_p and bandwidth ω_i. Primary objective of this laboratory is to familiarize the reader with the parameters, which need to be set to realize motor parameter identification with the InstaSPIN-FOC module given in Fig. 4.2. In addition, said module will be deployed to realize sensorless FOC operation, where use will be made of the ROM based current/speed controllers, instead of external (to the ROM) modules, introduced in the previous laboratory. In addition to the above, stator resistance recalibration will also be discussed. Finally, so called ‘goodness’ plots will be introduced at the end of this section, which provide guidance, in terms of the chosen motor identification parameters.

5.3.1 *Lab 3:2: Phase B*

The purpose of this laboratory is to demonstrate operation of the FAST based sensorless drive with motor identification. In particular the various stages of identification will be discussed. For this purpose the VisSim processor in the loop

(PIL) approach will again be used, which communicates with the TMS320F28069M unit located on the LAUNCHXL board. The following information is relevant for this laboratory component:

- Reference programs [11]: `lab3-2_LaunchXLphBv1.vsm` and `IM_InstaSPINv1.vsm`.
- Description: Sensorless control of a IM motor with motor parameter identification.
- Equipment/Software: Texas Instruments LAUNCHXL-F28069M and VisSim simulation program.
- Outcomes: To familiarize the reader with the InstaSPIN motor identification algorithm for IM drives.

No boost packs are used in the laboratory, nor is an external 24 V power supply required, given that power to the MCU is directly provided via the USB cable connected to the laptop. In this case the following jumper and dip-switch positions on the LAUNCHXL-F28069M module are required:

- Jumpers `JP1` and `JP2` CLOSED
- Jumpers `JP4` and `JP5` OPEN
- Jumpers `JP3`, `JP6` and `JP7` CLOSED
- Dip-switches `SW1` to `SW3` ON

Note that there are two reference programs required for this lab: the first is the main simulation model which contains a PIL 'interface module' which in turn makes use of a second program `IM_InstaSPINv1.vsm`. The outputs of the 'Drive Controller' shown in Fig. 5.16 are three reference modulation indices `mA_ref`, `mB_ref`, `mC_ref`, which must be multiplied by the term $u_{DC}/2$. The resultant reference voltages `uA_ref`, `uB_ref`, `uC_ref`, are connected to the 'converter', which in this case is simply represented by three limiters, that limit the voltage to the motor to $\pm u_{DC}/2$. The phase voltages `u_A`, `u_B`, `u_C` of the machine are used by the machine model and by the InstaSPIN algorithm. For the latter case the converter outputs are filtered using 'analog low-pass filter' modules that replace the electrical circuit models which are present in the actual hardware. The filtered voltages are then scaled with the aid of a 'Gain module', by a factor 3.3/`ADC_FS_V`, where `ADC_FS_V` is the full scale ADC voltage set to 26.314 V for the boost pack. A similar scaling is also used for the machine phase currents `i_A`, `i_B`, `i_C`, in which case the scaling factor is equal to 3.3/`ADC_FS_C`, where `ADC_FS_C` is the full scale ADC current value set to 33.0 A. A detailed model of the 12 bit AD converter is again used to convert the scaled voltage/current inputs value to per unit outputs for use with the current controller and InstaSPIN module.

A set of sliders define the current reference vector, in terms of its direct and quadrature components i_d^{ref}, i_q^{ref} and rotational speed n_m^{ref}. A 'pre-scaling' module has been added to convert the floating point variables required for the controller to fixed point format. In addition, two sliders have been added which allow the user to set the proportional K_p and integral K_i speed gains of the controller.

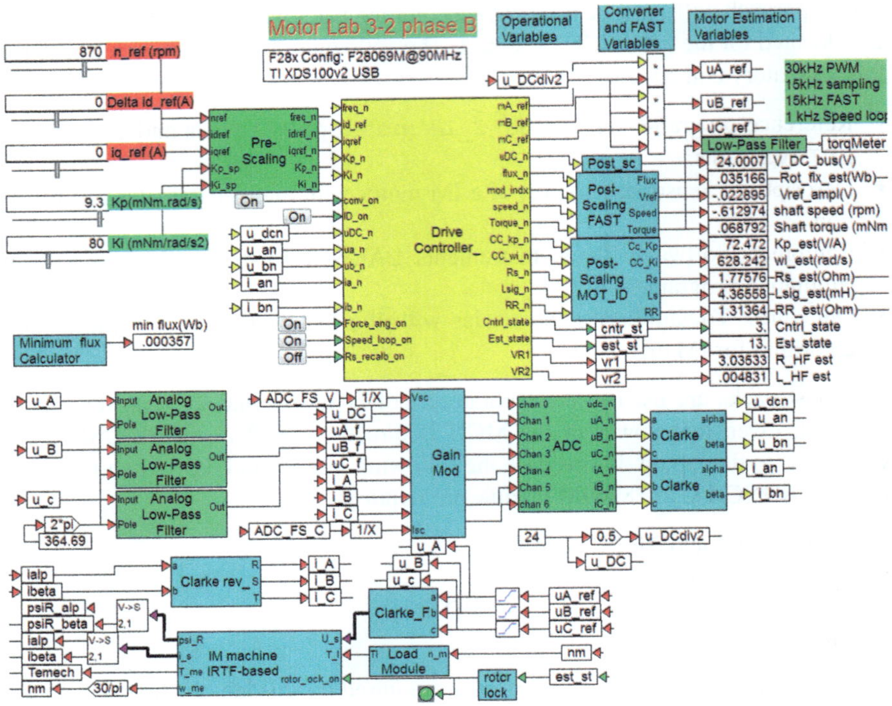

Fig. 5.16 Phase B simulation of a FOC IM drive with motor identification capability: screen shot taken after completion of parameter estimation

The three dialog boxes for this laboratory, shown in Fig. 5.16, contain all the user settings required for this laboratory. For example, the 'Operational Variables' dialog box is again used to assign full scale values for the voltage, current and frequency. In addition, the current controller gain and bandwidth are assigned here. Also defined in this dialog box is the rate of variable change for current and frequency, i.e. introduced as a safety measure to avoid erratic slider action from causing over-currents in the drive. The 'Converter and FAST variables' dialog box contains all the parameters which must be set for the InstaSPIN algorithm, which includes among others the full-scale ADC value and low-pass filter corner frequency. A third dialog box 'Motor Estimation Variables' contains the parameters that define the motor identification sequence. A more detailed discussion on these dialog boxes will be given in the 'phase C' development stage.

Central to this laboratory is the ability to use the VisSim PIL tool, which synchronizes the simulation with the MCU in the hardware. Located within the controller module (two levels down) is a 'Target Interface' module given in Fig. 5.17 that communicates with the InstaSPIN ROM. Inputs to this module are the per unit α, β voltage and currents generated by the ADC unit and a set of control inputs. Of these, the logic input `MID_on` activates the parameter identification process.

Further details on all inputs will be given in the next lab section. Key outputs are the modulation indices: mA_ref to mC_ref, DC bus voltage: DCbus_n, rotor flux estimate: Flux_n and (among others) the estimated leakage inductance: Lsig_n and stator resistance: Rs_n. Furthermore, estimates are provided for the current controller gain: CC_kp and bandwidth: CC_wi. To monitor operation during the identification sequence the controller state: cntrl_state and parameter estimation state: est_state are also generated by this module. The LED_in/LED_out variables activate a 'blue' LED on the LAUNCHXL board which flashes when PIL operation is active. Of central importance in the dialog box shown in Fig. 5.17 is the Target Execution File, which in this case is set to IM_InstaSPINv1.out. When the main simulation program is activated (via the 'green' arrow button), this .out file will be downloaded into the MCU prior to the start of the actual simulation. Once downloaded the rate of data exchange between MCU and VisSim simulation will be dictated by the Sample Rate (Hz) entry (set to 200 Hz in this case) provided that the Syn Target to this Block option has been activated. Note that the simulation frequency (set in the system properties) is set to 15 kHz given that this is the sampling frequency specified. However, actual run time during PIL operation is dictated by the Sample Rate (Hz) entry mentioned above.

The required IM_InstaSPINv1.out file is generated by separate compilation of the module 'InstaSPIN PIL' shown in Fig. 5.18 first, which is present in the program IM_InstaSPINv1.vsm. This module contains all the functions which are needed to access the InstaSPIN ROM module. More details on the content of the 'InstaSPIN PIL' module will be given in development phase C.

Starting this simulation is done via the usual 'green' button in which case the .out file for the PIL process is downloaded first. Then the ADC offsets are determined within the ADC unit, which takes ≈ 1.4 s, (simulation time) AFTER that period the PIL should become active (flashing 'blue' LED on LaunchXL board) and drive operation should become apparent by monitoring the controller and estimator states. Note that the use of the PIL for this type of analysis is time consuming as a time sequence of 80 s, as considered here, takes approximately ≈ 5 hours to complete in real time.

A scope module is used to collect key variables over the motor identification time sequence and these results will be discussed below. Figure 5.19 provides details on the operational aspects of identification and as such it shows the observable phase current, shaft speed (in this case generated by the motor model) and the estimator state. The second Fig. 5.20, also generated on the basis of VisSim scope results for the phase B simulation, shows the estimated parameters: stator resistance, leakage inductance and rotor resistance. These results are shown together with the actual (reference) parameters as used in the four parameter model of the LVACIMTR IM machine.

On the basis of these two figures an overview will be given of the motor identification cycle as identified by the estimator state. Events shown, assume activation of the converter at $t = 0$ s with motor identification active (button MID_on ON).

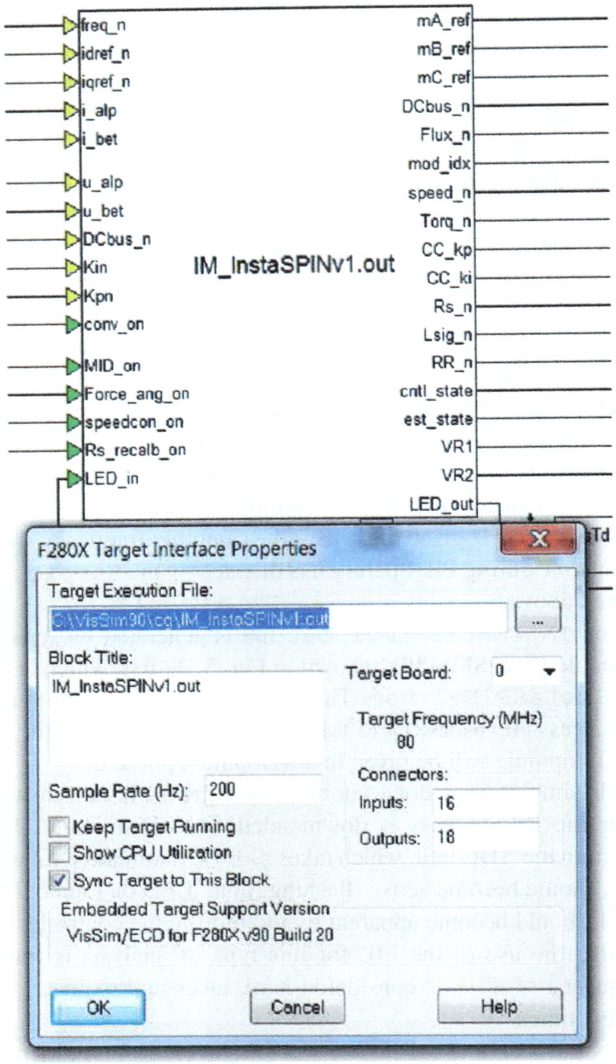

Fig. 5.17 Target interface module with corresponding dialog box entry

The button `Rs_recalb_on` (used for resistance estimation purposes outside the 'normal' identification sequence) is set to OFF. Furthermore, the identification cycle makes use of the dialog box settings specific to this example. A quadratic mechanical load of the machine is assumed in this example to represent the friction present in the drive. Note also that not all estimator state numbers appear below, as some refer to permanent magnet machine identification hence not relevant to this discussion.

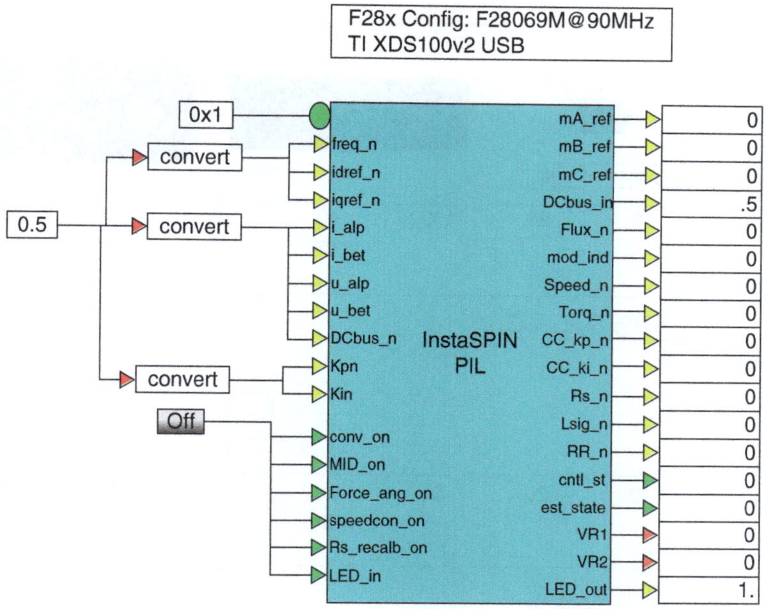

Fig. 5.18 InstaSPIN PIL module

The following estimator states are present during IM machine identification given below:

- State 1: referred to as 'idle' because the estimator is not in use. After converter start up: (button `conv_on` ON) an ADC offset measurement sequence takes place to remove any DC offsets in the ADC converter from the measurement process. Motor should **not** rotate during this phase.
- State 2: referred to as 'RoverL'. The phase current frequency and amplitude are set to 100 Hz and 0.5 A respectively in this case. Note that current amplitude used is half the value used in the next phase. During this measurement sequence the so called high frequency (HF) stator resistance R_{HF} and leakage inductance L_{HF} estimated values are found, which determine the current controller gain/bandwidth. Motor should **not** rotate during this phase. Values generated should provide a realistic approximation of the leakage inductance and resistance of the machine and these are then used to dimension the current controller, as discussed in Sect. 2.1.5. In some practical cases where it is not possible to lock the rotor during the motor identification stage, use can be made of the 'high frequency' leakage inductance value for InstaSPIN. The R_{HF} value represents an estimate for the sum of the stator and rotor resistance. Hence an estimate for the rotor resistance can be found (if a locked rotor test cannot be done) from this measurement and subtracting the stator resistance value. The latter can be found from the next (state 3) measurement cycle. Note that in a real machine skin-effect and iron losses can affect the R_{HF} value.

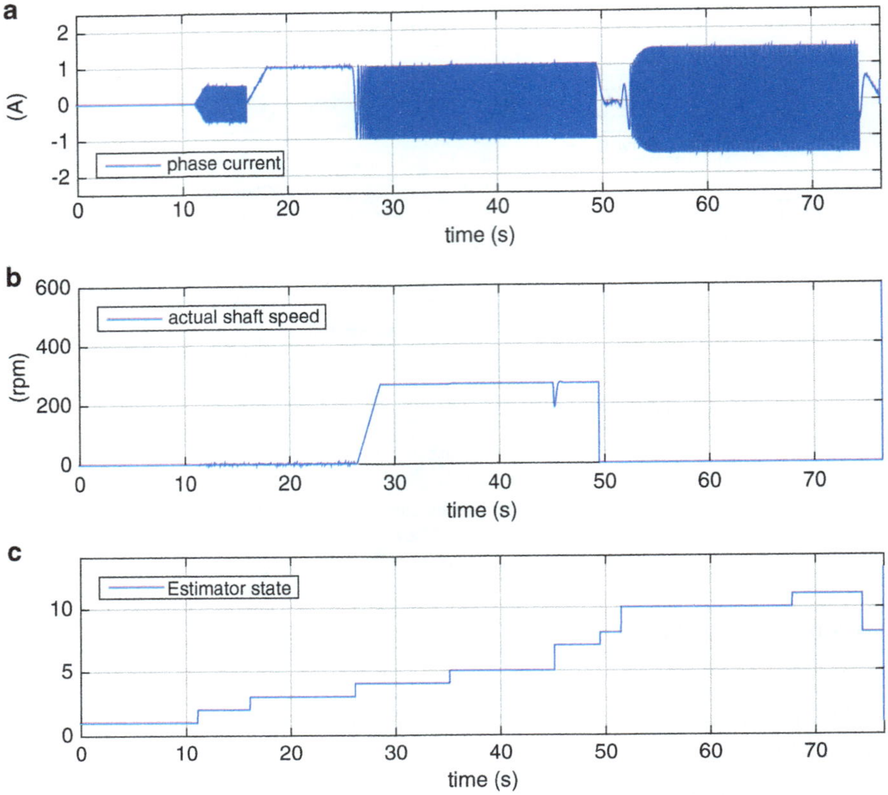

Fig. 5.19 Phase B simulation of a FOC IM drive with motor identification capability: operation cycle

- State 3: referred to as 'Rs'. DC current level is set tot 1.0 A. Note that the measured phase current may be unequal to the set value, as it depends on which phase is measured by the current probe, i.e. one phase can (for example) show 1.0 A, while the other two will then show 0.5 A each. Measurement of the stator resistance takes place as may be observed from Fig. 5.20. Motor should **not** rotate during this phase.
- State 4: referred to as ' Ramp up', which undertakes open-loop speed control, with a current vector amplitude set to 1.0. Motor should increase speed which matches the rotational frequency 10 Hz of the current vector as may be observed from Fig. 5.19. Hence, the synchronous shaft speed will be $^{10*60}/_{p} = 300$ rpm, where $p = 2$ is the pole pair number of the machine in use. Note that during open-loop current control, stable speed operation is often impaired by the lack of electrical damping. Hence the use of a friction load in this laboratory. In practice mechanical damping caused by friction is usually sufficient to ensure stable operation during this sequence.

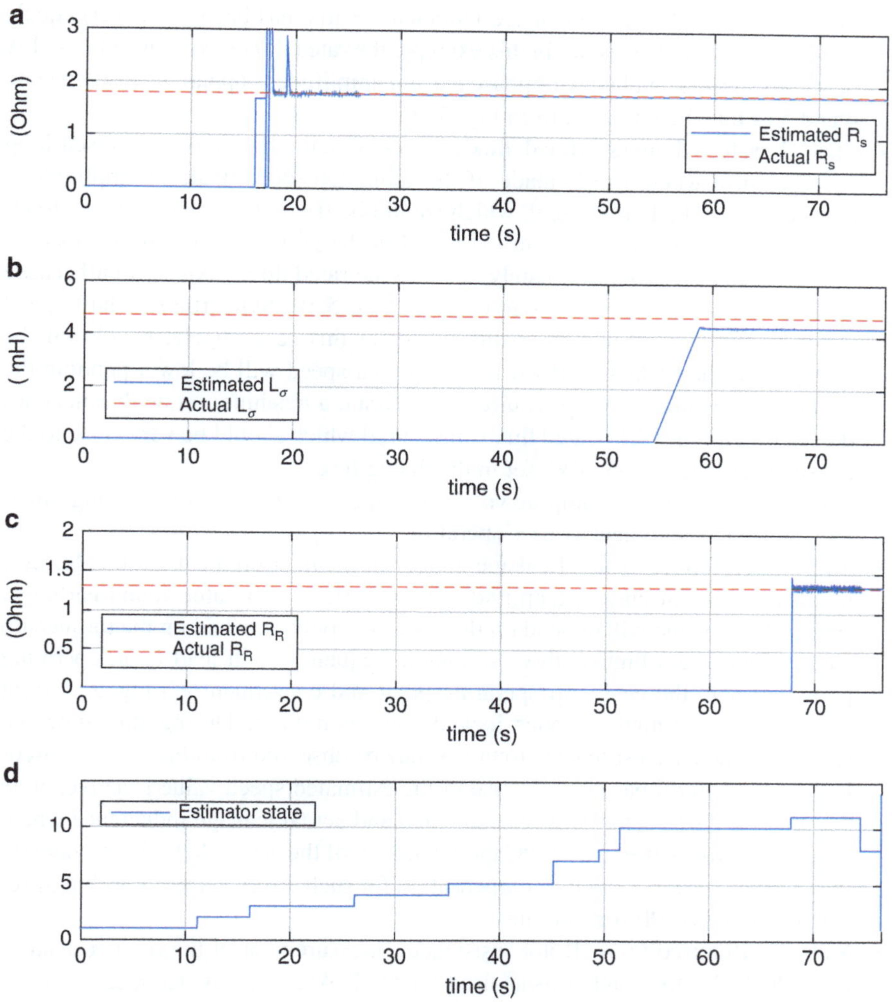

Fig. 5.20 Phase B simulation of a FOC IM drive with motor identification capability: parameter measurement

- State 5: referred to as 'Rated direct axis current', makes use of the user entered stator flux value set the 'Converter and FAST variables' dialog box. In this case value was set to 35 mWb. Operation is under open-loop speed control where the synchronous shaft speed is held at 300 rpm (which corresponds to the 'flux estimation' frequency value set in the motor ID dialog box of 10 Hz), hence the actual speed will be lower pending on the friction load present. The current step size `Idstepsize__est_current(A)` sets the incremental value used to determine the rated current for the chosen flux value. Consequently, the direct axis current will be ramped up until the estimator flux value (which is at present the stator flux value as no leakage inductance value has been set as yet) equals the user set stator flux value. In this example the rated direct axis current is ≈ 1 A hence no significant change of phase current amplitude is apparent between state 4 and 5 as may be observed from Fig. 5.19.

- State 7: referred to as 'Rated flux', which is carried out under closed-loop speed control where use is made of the estimated speed. Current amplitude is determined by the load present, which should be friction only (no external load) and the rated direct axis current (as found in the previous state 5). In this case the phase current is approximately equal to the rated direct axis current because friction is low, as may be observed from Fig. 5.19. Synchronous shaft speed is set by the flux estimation frequency of (in this case) 10 Hz, which will be 300 rpm for the LVACIMTR IM motor. Actual speed will be lower pending the friction torque present (slip required to generate a machine torque that matches the friction torque). The rated flux is measured which should be very close to the chosen user value of 35 mWb set in the dialog box.
- State 8: referred to as 'ramp down': motor speed to zero, before starting either inductance measurement or user operation.
- State 10: referred to as 'Leakage inductance': measurement of the leakage inductance. Initial current amplitude set by `rated Id` value found earlier in state 5. Final value will depend on the operating point reached on the torque/slip curve. Choosing a higher 'flux estimation frequency' will lead to an operating point closer to the peak slip operating point and consequently a higher current value during estimation under locked rotor condition. During this state the leakage inductance estimate is found as may be observed from Fig. 5.20. Closed-loop speed control based on the use of the estimated speed value is active. Note that there is a 6 % error between estimated and actual leakage inductance which is attributed to the use of an ADC model instead of the actual ADC. Consequently the sampling process and the compensation for such errors in the algorithm differ between real and PIL operation.
- State 11: Referred to as 'Rotor resistance': measurement of the rotor resistance under locked rotor conditions as used for the leakage inductance measurement. Actual shaft speed is zero, hence the estimated shaft speed must also be zero, which is achieved by correct estimation of the rotor resistance as may be observed from Fig. 5.20. Closed-loop speed control based on the use of the estimated speed value is active.
- State 13: referred to as 'user operation', where control is transferred to the user.

5.3.2 Lab 3:2: Phase C

Prior to being able to work with an operational drive, an `.out` file must be generated of a 'Drive Controller with FAST and Motor Identification' module, as shown in Fig. 5.21, which represents the drive under consideration. Hence details of the drive structure and corresponding dialog box parameter assignments will be discussed in this subsection.

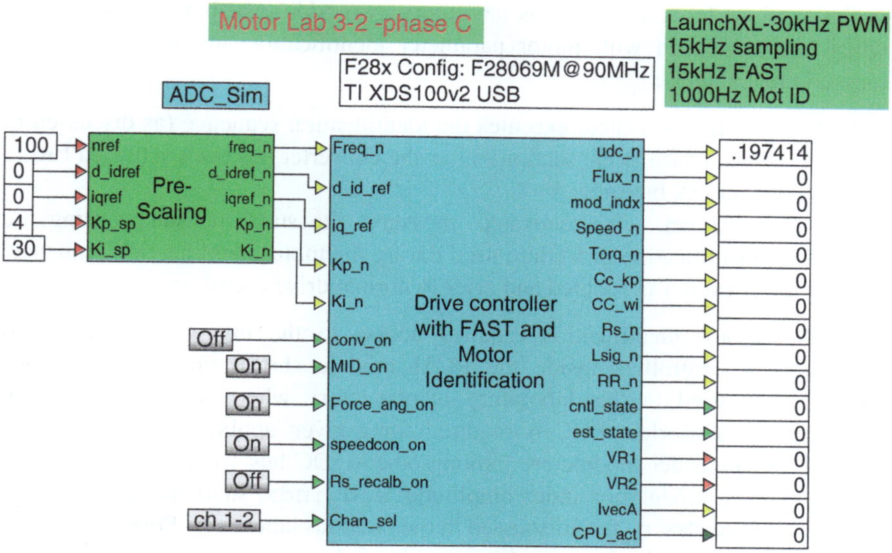

Fig. 5.21 Phase C simulation of a InstaSPIN FAST based encoderless field-oriented controlled (FOC) drive, with motor identification capability

The following information is relevant for this laboratory component:

- Reference program : `lab3-2_LaunchXLphCv3.vsm`
- Description: Motor Parameter Identification and sensorless control of a IM motor using the InstaSPIN module.
- Equipment/Software: Texas Instruments LAUNCHXL-F28069M, with BOOSTXL-DRV8301 module ('aft' position) and VisSim simulation program.
- Outcomes: Generate the `.out` file needed to represent the drive structure under investigation.

As mentioned above, a single boost pack located in the 'aft' position (furthest away from the USB connector) is used for single motor operation (PM machine is not used electrically), in which case the following jumper and dip-switch positions on the LAUNCHXL-F28069M module are required:

- Jumpers `JP1` and `JP2` OPEN
- Jumpers `JP4` and `JP5` CLOSED
- Jumpers `JP3`, `JP6` and `JP7` CLOSED
- Dip-switches `SW1` to `SW3` ON

Inputs to the Controller are five variables (shown as 'constants' for readability), which set the reference shaft speed, direct/quadrature reference currents, and speed controller gains. Four logic buttons are used to select speed/torque control, diagnostic channels used, force angle and converter ON/OFF control, as discussed in the previous laboratory. In this laboratory, two additional buttons are present, which are associated with motor parameter identification and stator resistance measurement namely:

- `MID_on`: when activated, executes the identification sequence (as discussed in the previous, phase B lab section), after the converter has been activated (using the converter ON button).
- `Rs_recalb_on`: when activated, measures the stator resistance using the 'offline' identification procedure used during the motor identification sequence, after the converter is enabled and prior to normal drive operation.

Note that motor identification requires access to the internal (ROM) based current/speed controllers inside the ROM module shown in Fig. 4.2. Hence the approach used in this laboratory is mandatory when motor identification together with sensorless FOC is required for a given application. Outputs (per unit) of the controller module are (among others), DC bus voltage: `udc_n`, rotor flux: `FLux_n`, modulation index amplitude: `mod_indx`, shaft speed: `speed_n` and shaft torque: `Torq_n` as discussed in the previous laboratory. Furthermore, the drive controller module also provides ten additional outputs namely;

- `CC_kp`: estimate for the proportional current controller gain `K_p`. Note that this variable can, according to Eq. (2.23), also be expressed as $K_p = L_\sigma/T_s$, where T_s is the ADC sampling time, set to 66.67 μs in this case.
- `CC_wi`: estimate for the current controller bandwidth.
- `Rs_n`: per unit estimate for the stator resistance.
- `Lsig_n`: estimate for the leakage inductance.
- `RR_n`: per unit estimate for the rotor resistance. `cntrl_state`: status of the controller. `est_state` status estimator.
- `VR1`: estimated for the 'high frequency' stator plus rotor resistance (value generated during estimation state 2, see below).
- `VR2`: estimated for the 'high frequency' leakage inductance (value of L_σ generated during estimation state 2, see below).
- `IvecA`: which represents the per unit amplitude of the measured current vector $\vec{i} = \sqrt{i_\alpha^2 + i_\beta^2}$. This variable is useful for evaluating motor identification in the next phase C+, as will be come apparent shortly.
- `CPU_act`: used to monitor % MCU activity.

Moving one level lower into the 'Drive Controller' module reveals the modules shown in Fig. 5.22. Of these shown in the figure, the '100 Hz' and 'Diagnostic tool' modules have been discussed in previous laboratories. The dialog box parameters for this laboratory are given in Fig. 5.23. The parameters associated with the 'Operation variables' and 'Converter and FAST Variables' dialog boxes, have been discussed in the previous laboratory. Of interest here is therefore, dialog box 'Motor Estimation Variables', where the following motor identification parameters must be assigned:

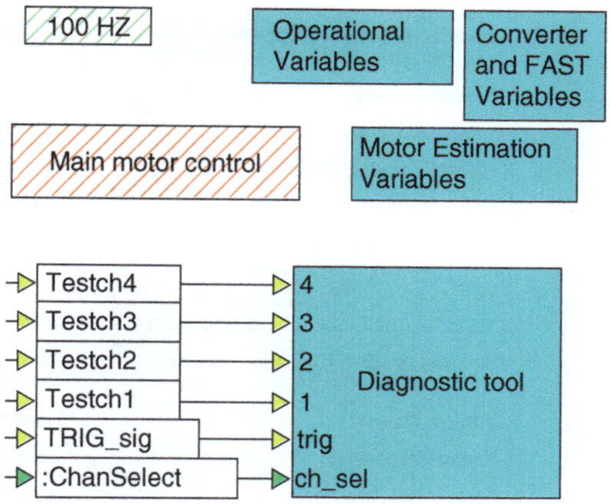

Fig. 5.22 Phase C simulation of a FAST based encoderless FOC drive, with motor identification: one level into the drive controller module

- `F_motorID/F_sampling`: ratio between sampling frequency used by the Identification algorithm and ADC sampling frequency. The ratio is an integer value, hence the user can select the ratio from a dialog box pull-down menu. For example, a choice of `1/15` implies an estimator sampling frequency of 1000 Hz, given that the ADC sampling frequency is set to 15000 Hz.
- `R/L sense freq (Hz)`: frequency used to estimate the current controller gain and bandwidth.
- `Flux est_freq (Hz)`: frequency used to estimate the direct axis rated current, rotor flux and leakage inductance. Its value should be less than the pull-out slip $\hat{s}$ of the machine.
- `Freq_est-ramp (Hz/s)`: rate of frequency change used for motor identification purposes.
- `2xR/L_est,Rs_es_current(A)`: current amplitude used for current controller gain/bandwidth and stator resistance estimation. Note that the value used for `R/L_est` (current controller gain estimation) is half the value shown in this dialog box entry, i.e. if stator resistance estimation is to be done at 1 A, then `R/L_est` will take place at 0.5 A.
- `idstepsize_est_current`: current step size used during estimation of the rated direct axis current value. The algorithm will identify the magnetizing current that corresponds to the rated stator flux entry set by the user. The relevant rated stator flux value is set in the 'Converter and FAST' dialog box (currently set to 35 mWb).

Moving one level lower into the 'main motor control' module, reveals a set of modules given in Fig. 5.24. Of these shown, all with exception of the 'InstaSPIN-FOC module' have, in fact, been identified in earlier laboratories. Key inputs to the controller module are the per unit α, β current/voltage variables: `ia`, `ib`, `va`, `vb` and DC bus voltage: `DCbus_n`, generated by the ADC-PWM module. User reference variables: `Freq_n`, `idrefd`, `iqrefq_n` and `Kpn`, `Kin` represent the scaled users sliders for shaft speed, direct/quadrature currents and speed controller

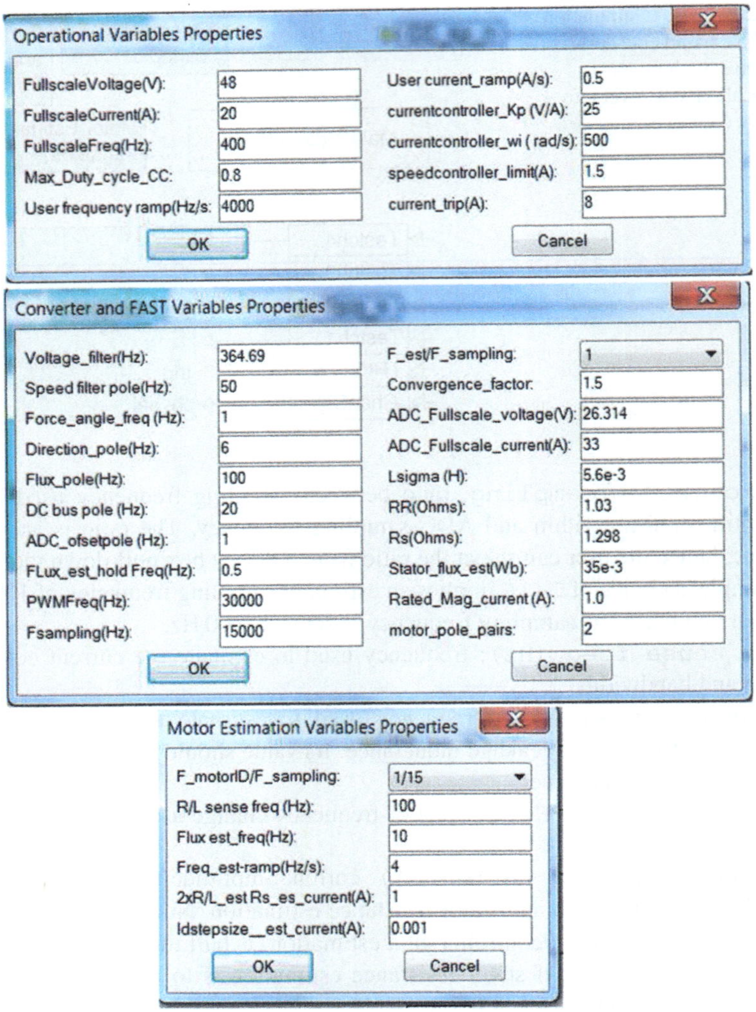

Fig. 5.23 Phase C simulation of a InstaSPIN encoderless FOC drive, with motor identification: Dialog box entries

gains. Note that the direct axis reference input is a **differential input**, i.e. its value will be either added or subtracted from the estimated rated direct axis current. Hence this value should be set to zero by default. Also shown in Fig. 5.24, is the `Mod_ind(abs)` module which calculates the modulation index amplitude generated by the current controller. It is hatched 'green' given that it is executed at a lower sampling rate (equal to 1 kHz in this case) than the other modules shown in the diagram. In this figure logic input `doMotorID` is used to activate the motor identification sequence. Outputs of the 'InstaSPIN-FOC' module are (among others) the modulation indices: `mA_ref`, `mB_ref`, `mC_ref` generated by

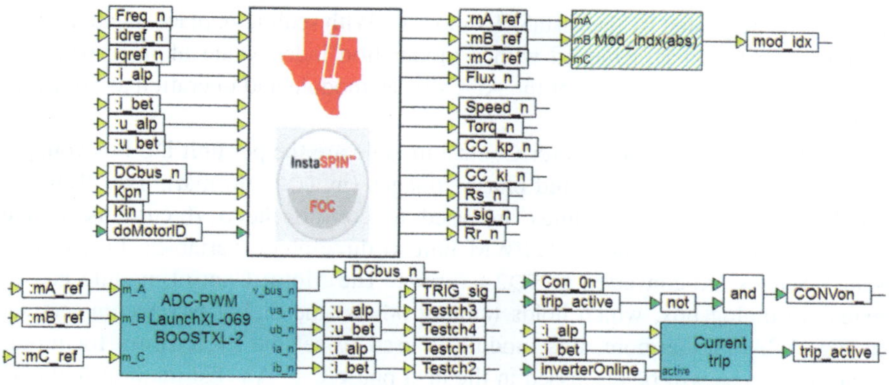

Fig. 5.24 Phase C simulation of a FAST based encoderless field-oriented controlled (FOC) drive, with motor identification: one level into the 'Main Motor Control' module

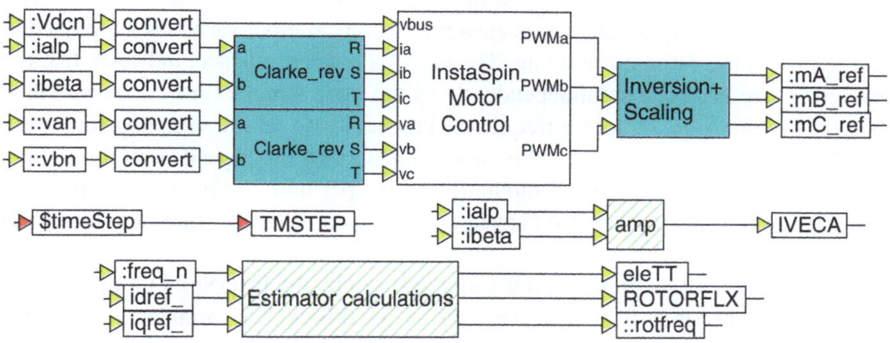

Fig. 5.25 Phase C simulation of a FAST based encoderless field-oriented controlled (FOC) drive, with motor identification: one level into the InstaSPIN-FOC module

the internal current controller. The remaining outputs are those which have been named earlier. For diagnostic purposes the test variables: `TESTch1`, `TESTch2` have been connected to per unit variables: i_α^n and i_β^n respectively. Furthermore, test variables: `TESTch3`, `TESTch4` have been connected to per unit variables: u_α^n and u_β^n respectively.

Moving one level into the InstaSPIN-FOC module, shown in Fig. 5.25, reveals the basic VisSim 'InstaSPIN Motor Control' and 'estimator calculation' modules. The 'InstaSPIN Motor Control' module communicates directly with the InstaSPIN ROM module, which returns the modulation indices used by the ADC/PWM unit. Note that an 'inversion+scaling' module is used to invert and scale the outputs of the Motor control unit. Inversion is required because the relationship between the modulation index polarity and converter phase leg top/bottom switches as used by InstaSPIN is different to that used here (see Fig. 2.23 for convention used in this book). Scaling by a factor $\sqrt{3}/2$ is required because the FAST module has a gain factor of $2/\sqrt{3} = 1.157$ between the outputs of the internal current controller and the outputs generated by the InstaSPIN module. With scaling in place, the boundary of linear operation is achieved with a modulation index value of $2/\sqrt{3}$, which is consistent with the operation of the ADC-PWM module and operation as shown in Fig. 2.21.

Inputs to the InstaSPIN Motor Control module are the per unit DC bus voltage, three phase currents and filtered phase voltages (in `IQ24` format). Two 'Reverse Clarke' transformation modules are used to convert the α, β currents/voltage variables provided by the ADC/PWM unit to three-phase variables. In addition, said variables are converted to `IQ24` format. The Motor Control module has an extensive dialog box, which holds (among others) the dialog box parameters set in Fig. 5.23. Furthermore, this module is used to set the cycle times for motor identification as will be discussed in the next phase C+. An 'Estimator calculator' compound module, also shown in Fig. 5.25, derives the rotor flux, shaft speed and shaft torque values from the ROM. In addition, the per unit reference frequency: `freq_n` and reference direct axis (which is a differential input variable in this case) current: `idref_n`/quadrature axis current: `iqref_n` input variables are passed to the ROM via the estimator module. The use of a hatched surface, indicates that the sampling rate has been set independently of the main simulation frequency, which is typically the ADC sampling frequency. Practical is to set the sampling frequency for the 'Estimator Calculator' to the same value used by the 'Speed control' module, i.e. 1000 Hz in our case. The computation of the per unit current vector amplitude (output variable `IvecA`) is also carried out at a lower sampling rate to reduce CPU processing time.

Note that not all the input/output variables used by the InstaSPIN module shown in Fig. 5.24 appear in Fig. 5.25. The reason for this is that variables such as, for example, `Kpn`, `Kin`, `Rs_n` etc. do not need to be determined at the estimator frequency sampling rate, instead they are executed in the '100 Hz' module, i.e. they are sampled at 100 Hz instead of 15 kHz to reduce CPU loading.

Finally, it is noted that the modules shown in Fig. 5.25 are precisely those which are also used in the PIL module discussed in phase B of this laboratory, see Fig. 5.18.

5.3.3 Lab 3:2: Phase C+

Phase C+, is the operational component of the laboratory and is basically a run version of the `.out` file compiled and downloaded to the MCU in phase C (see previous subsection).

The following information is relevant for this laboratory component:

- Reference program [11]: `lab3-2_LaunchXLphCv3_d.vsm`.
- Description: FOC sensorless control of a IM machine, with motor parameter identification.

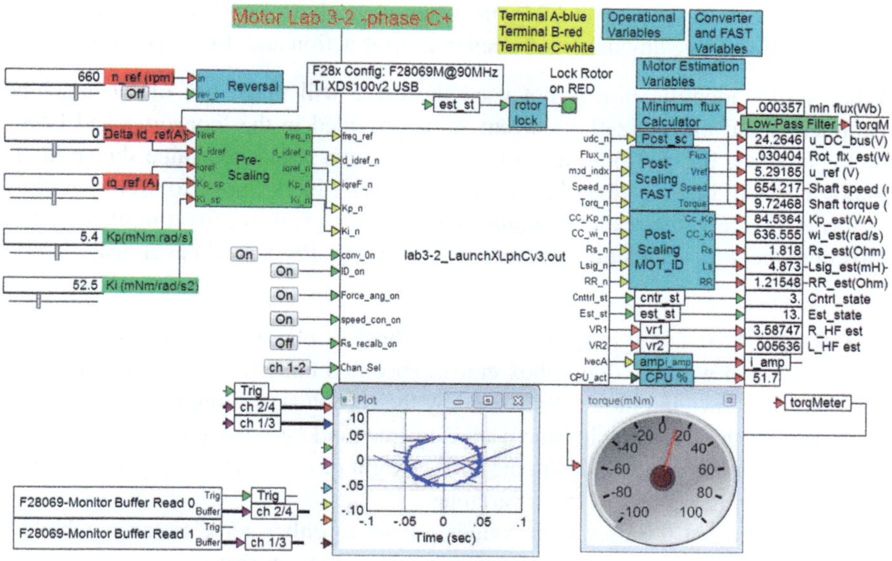

Fig. 5.26 Phase C+ simulation of a FAST based encoderless field-oriented controlled (FOC) drive, with motor identification: example shows motor under no load (friction only), after identification has been done)

- Equipment/Software: Texas Instruments LAUNCHXL-F28069M, with BOOSTXL-DRV8301 module ('aft' position), Texas Instruments LVACIMTR IM machine and VisSim simulation program.
- Outcomes: to use the InstaSPIN algorithm for IM motor parameter identification purposes and sensorless FOC operation.

Note that the required jumper and dip-switch settings for the LAUNCHXL-F28069M module are given in phase C.

The run version shown in Fig. 5.26, uses a VisSim run module, which executes the `.out` file, shown in said module. Several modes of operation are possible AFTER the converter has been enabled via the `Conv_on` button namely:

- Motor Identification mode: active, `MID_on` button ON, in which case the drive will execute the identification procedure when the converter is enabled. Upon completion (estimator status: 12) control is returned to the user and the parameters are shown on the numerical displays (see Fig. 5.26).
- Motor Identification mode: not active, `MID_on` button OFF, in which case drive operation will depend on the status of the `Rs_recal_on` button, hence there are two options:
  - `Rs_recal_on` button ON: in which case the Rs value is estimated using the 'offline' procedure applied during the standard identification procedure PRIOR to returning control to the user.

– `Rs_recal_on` button OFF: in which case control is directly given to the user, to reduce any delay between converter action and drive operation.

When motor identification is not used the drive will make use of the machine parameters and magnetization current value provided in the 'converter and FAST dialog box (see Fig. 5.23). In the latter case the stator resistance value shown in the dialog box will be used if no R_s recalibration is carried out prior to user operation. In terms of safety, the reader is again reminded that prior to activating this type of lab component it is prudent to execute the following 'Pre-Drive' check list:

- Dialog boxes used in C+ mode, match those of phase C: the run version is compiled with the dialog box entries specified under phase C. The dialog boxes shown in phase C+ are used by the 'Pre/Post' scaling modules.
- Ensure that the sample time used is correct and latest (and correct) `.out` file has been downloaded to the MCU (Right mouse click on MCU module to show dialog box of these variables).
- Confirm that the user input values are set to either zero, or 'acceptable' values, which will not cause a current trip of the converter.
- Confirm that the converter 'switch' is set to OFF and the power supply is on (DC bus voltage present).
- Confirm that the motor is connected firmly and properly.

After completion of the Pre-Drive checklist, activate the program and confirm that the supply voltage source reading shown in the digital display is 24 V. If not stop the program and restart. With the correct DC voltage reading established, turn on the converter (using the ON button) and monitor the motor shaft and diagnostic scope. An enlarged view of the VisSim scope module given in Fig. 5.27 shows the per unit currents i_α^n, i_β^n after completion of identification. Multiplication by the full scale current value of 20 A shows that a current vector of ≈ 1.0 A amplitude which corresponds to the magnetization current amplitude of the drive.

Five sliders are used, the purpose of which was discussed in the previous section. It is however, again emphasized that the slider `Delta _Id_ref` is a differential variable, i.e. the value set by the user, will be added or subtracted from the rated direct axis value found during estimation. A post-scaling module is again used to convert the per unit measured DC bus voltage to actual voltage, as shown with a numeric display. A low-pass filter with a 1 Hz corner frequency, is used to filter the torque signal for display on a torque meter (see Fig. 5.26). Two 'Monitor Buffer' modules display two selected (using an on/off button connected to the `chan_sel` input) diagnostic signals. In this example the per unit current variables i_α^n, i_β^n are depicted in an 'xy' plot, in which case the locus of the current vector appears. Under no load the current depicted is the magnetizing current of the machine. The corresponding magnetizing curve showing rotor flux as function of magnetizing current, can be derived by recording said variables as the slider `Delta_Id_ref` is changed over the range $-0.6 \rightarrow +0.5$ A, which leads to the results shown in

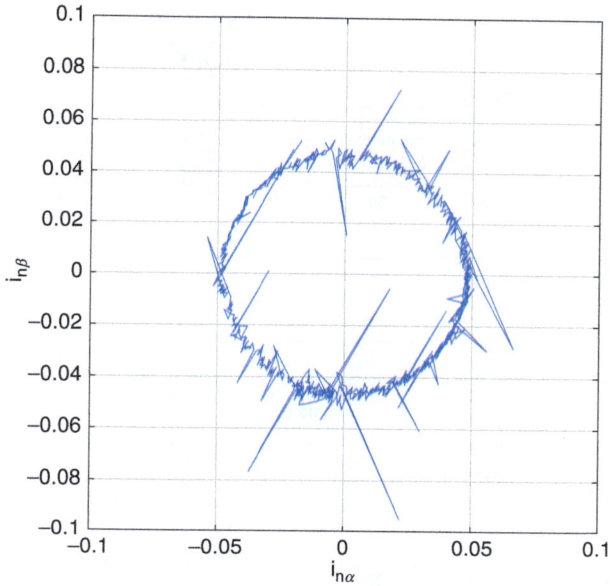

Fig. 5.27 Phase C+ simulation: Scope display showing per unit α, β currents after completion of motor identification (machine running with friction load only)

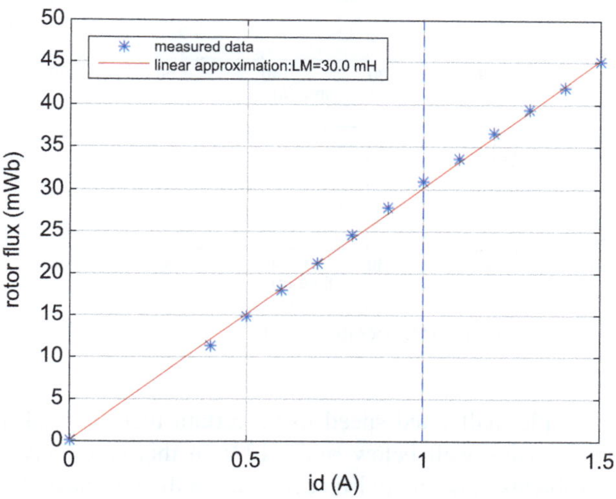

Fig. 5.28 Magnetization curve of the LVACIMTR IM machine

Fig. 5.28. Also added to this figure is a linear function which represents the magnetizing inductance of the machine, which is in this case approximately 30 mH. These results indicate that the machine shows virtually no magnetic saturation effects for the current range in use. Note that this type of measurement should be

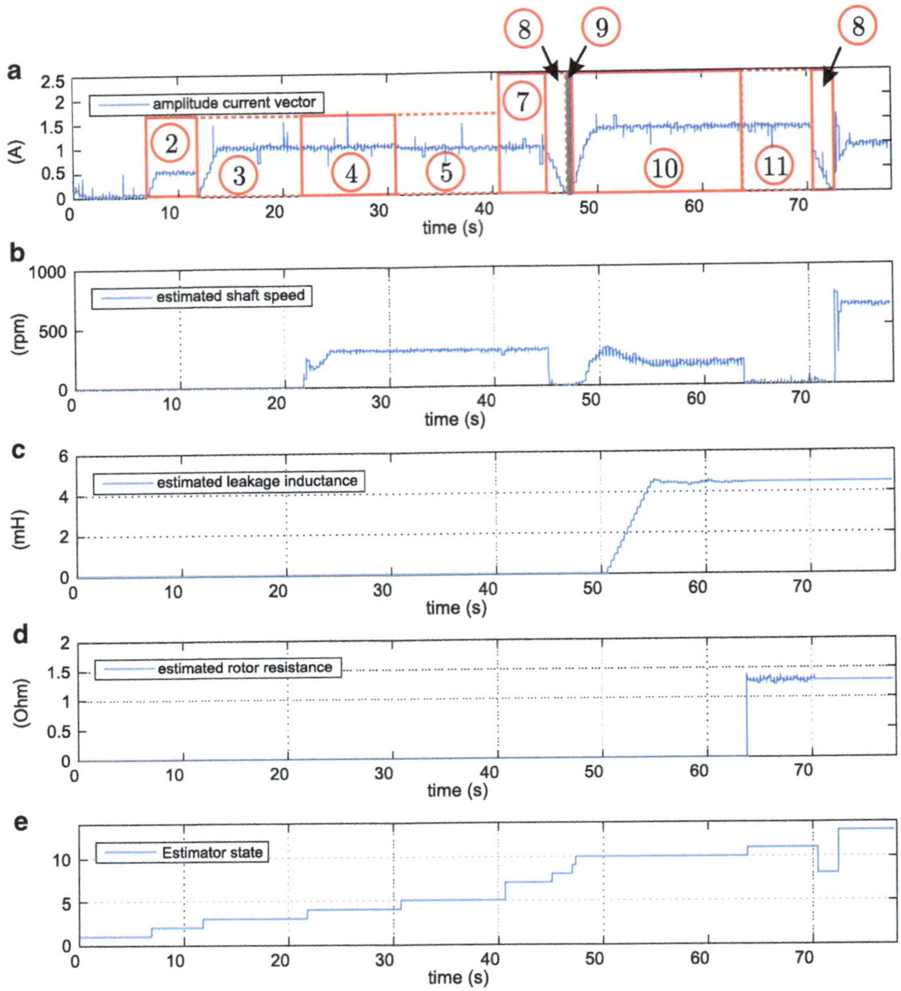

Fig. 5.29 Motor ID sequence for IM: operational data

done at, for example, half rated speed to be certain that the peak phase voltage of the machine remains well below $u_{DC}/2$ (12 V in this case), given that current measurement problems may otherwise occur, due to the presence of shunt resistors (see Sect. 2.1.6 for further details on this issue).

The experimental identification sequence for the IM motor as given in Fig. 5.29 corresponds with the operational sequence derived earlier with lab 3.2 phase B (see Fig. 5.19). Also shown in Fig. 5.29 are the estimator states present during parameter identification. These are linked to a timing diagram given in Fig. 5.30, which shows a dialog box with the relevant section that controls the duration of the relevant estimator states. This figure is part of InstaSPIN dialog box (right click on the module shown in Fig. 5.25), which shows the timing cycles relevant (identified

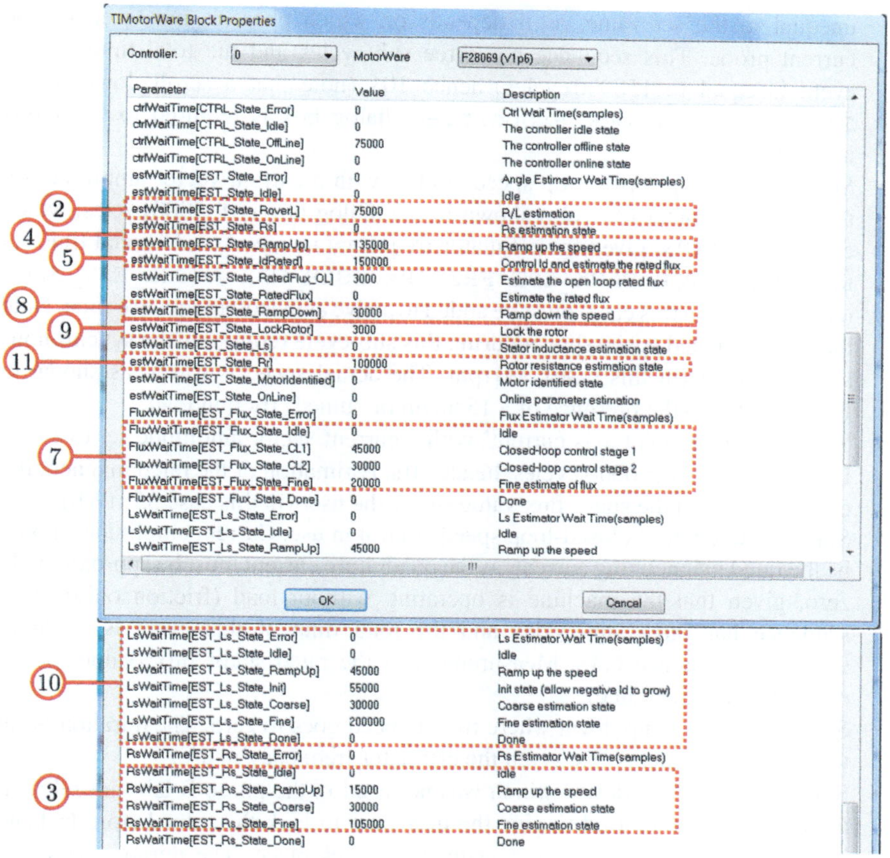

Fig. 5.30 Motor ID sequence for IM: timing data

by estimator state) to IM identification. Actual time is found by taking the number of cycles shown and dividing by the estimator frequency in use. Hence with an estimator frequency of 15 kHz (as used in this laboratory) the actual time of, for example item (2), will be $75000/15000 = 5$ s. A detailed discussion on the estimator states present during motor identification has already taken place in lab 3:2 phase B, hence only a brief summary of the sequence (using Figs. 5.29 and 5.30) and estimator state, is given below.

- State 2: identification of the current controller gain and bandwidth. Frequency and amplitude (set in motor ID dialog box) are 100 Hz and 0.5 A (the current value used in this sequence is half the value used in the Rs estimation phase). The duration is set to 75000 cycles, hence 5 s, given the estimator frequency in use.
- State 3: 'offline' stator resistance measurement, DC current level (set in motor ID dialog box) is set tot 1.0 A. Note that the measured phase current may be

unequal to the set value, as it depends on which phase is measured by the current probe. This sequence has three sub-cycles and the total time of this event is equal to 150000 cycles, hence 10 s (note that a scroll down in the `TTMotorware Block Properties` dialog box is required to show this number).

- State 4: 'ramp up' open-loop speed control with a current vector amplitude set to 1.0. The number of cycles shown in the dialog box is 135000 cycles, which corresponds to 9 s, given the estimator frequency in use. An additional time set to `USER_MOTOR_FLUX_EST_FREQ_Hz/USER_MAX_ACCEL_EST_Hzps` (equal to 2.5 s in this case) must be added to arrive at the total sequence time. Note that the estimated speed shown during this sub-cycle represents the synchronous shaft speed of (in this case) 300 rpm. The actual speed is lower, as shown in Fig. 5.29 due to the presence of a 15 mNm (assumed) friction load.
- State 5: 'Rated direct axis current' with a current vector amplitude set to 1.0 A. Duration is set to 150000 cycles, hence 10 s. Estimation of the rated magnetizing current based on the stator flux value set by the user (in this case 35 mWb).
- State 7: 'Rated flux' closed-loop speed control in use. Current amplitude, should be the rated magnetizing current, as the quadrature current must be approximately zero, given that the machine is operating without load (friction only). This sequence has three sub-cycles and the total time of this event is equal to 95000 cycles, hence 6.3 s. Measurement of the rated stator flux (value should match the user set value).
- State 8: current ramp down where motor speed goes to zero. The duration is set to 30000 cycles, hence 2 s, given the estimator frequency in use.
- State 6: 'Rotor lock down' which was not used in the phase B lab (not relevant there). This state is used to give the user time to lock the rotor before leakage inductance and rotor resistance estimation takes place. The duration is set to 3000 cycles, hence 0.2 s, given the estimator frequency in use.
- State 10: measurement of the leakage inductance. Current amplitude is set by the slip frequency encountered by the machine. Hence value depends on the measurement frequency set by the user, i.e. 10 Hz, in this case. If the torque levels are higher then wanted, reduce the slip frequency by selecting a low frequency for this cycle. Speed estimate shown is not valid. This sequence has four sub-cycles and the total time of this event is equal to 330000 cycles, hence 22 s. Note that the inductance estimation process has been purposely added to Fig. 5.29, as the reader needs to make sure that a steady-state inductance value has been reached BEFORE the next (rotor resistance) cycle is started. Hence choosing a cycle time that is less than the time needed to reach a steady-state inductance value will lead to a erroneous result.
- State 11: measurement of the rotor resistance. Current amplitude is set by the slip frequency encountered by the machine. Hence value depends on the estimator frequency set by the user, i.e. 10 Hz, in this case. Speed estimate shown is not valid until a steady-state rotor resistance value has been found. This sequence of this event is equal to 100000 cycles, hence 6.6 s. Note that the resistance estimation process has been purposely added to Fig. 5.29, as the reader needs

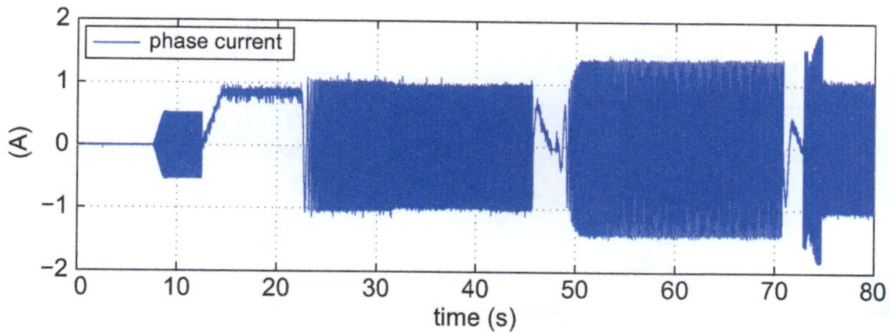

Fig. 5.31 Motor ID sequence for IM: measured with a DC-true current probe and scope

to make sure that a steady-state resistance value has been reached BEFORE the cycle is terminated. Hence choosing a cycle time that is less than the time needed to reach a steady-state value will lead to an erroneous result.

- State 13: user control, which occurs after ramp-down (state 8). In this case speed control is active under no load (friction only), with the machine operating with the rated magnetizing current $\approx 1.0\,\mathrm{A}$.

Prior to discussing the so called motor 'goodness plots' the phase current as measured with a Tektronix DC-true current probe and oscilloscope is provided in Fig. 5.31 for reference purposes. The results given in Fig. 5.31 clearly show the various stages of parameter identification as prior discussed using Figs. 4.30 and 4.22. Note that the phase current value shown during estimator state 2 (Resistance measurement) may vary from the assigned current vector value, as discussed previously. Furthermore, the converter is activated at $t = 0\,\mathrm{s}$, while user operation resumes at $t \approx 70\,\mathrm{s}$.

The motor identification routine, as discussed above, will provide accurate and meaningful results, provided that prudent choices are made with respect to the ADC full-scale voltage/current values and motor identification dialog box parameters. Furthermore, said variables should be selected, with reference to the motor and converter current/voltage ratings. In addition, particular attention should be given to the 12 bit ADC converter, because quantization errors can lead to erroneous results, if insufficient attention is given to the issues mentioned above. The 'Motor Identification goodness' plots, as introduced in the previous chapter for PM machines are also used here to provide the user with an indication on the 'quality' of the results achieved. The plots for the motor and motor ID settings in question are shown in Fig. 5.32.

These 'goodness' plots show the voltage/current levels, which occur during motor identification in the context of the voltage/current ADC transfer function 'ADC' as shown in Fig. 5.32. In addition, the ratings of the converter and motor can be added (provided they are within the voltage/current range used). The voltage 'ADC transfer' function, given in Fig. 5.32 (top plot) shows the 'ADC count', which is the decimal output value that corresponds to the input voltage range $0 \rightarrow$ *ADC full scale voltage*, where the *ADC full scale voltage* (see Eq. (3.3a)) value

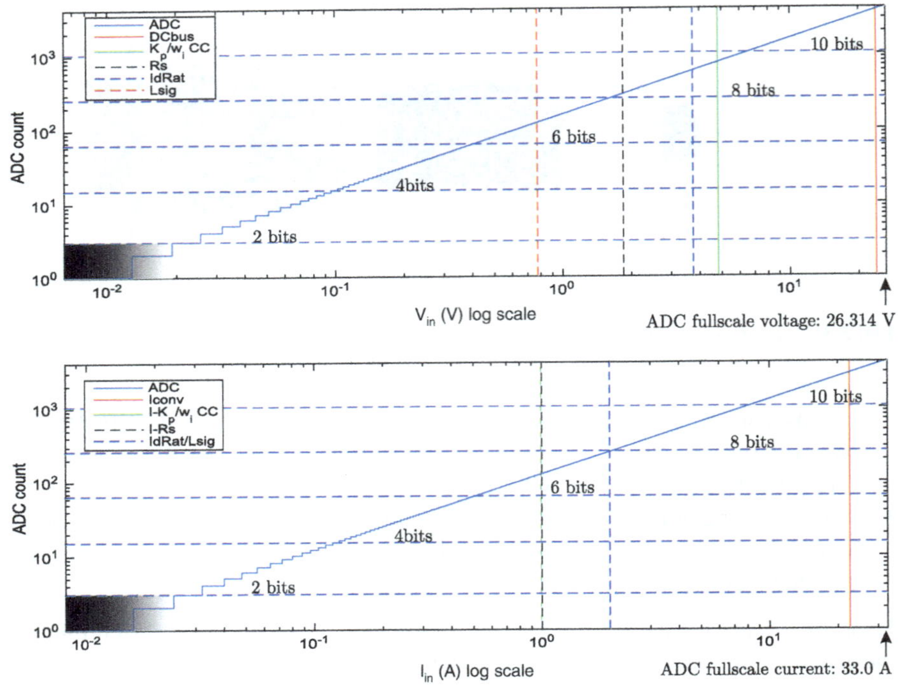

Fig. 5.32 'Motor Identification goodness' plots

is equal to 26.314 V (peak to peak) for the BOOSTXL-DRV8301 module. Also shown (in both plots) are a set of ADC binary output levels, which have been added to show the number of bits in use for a given ADC count value. The least significant bit LSB_v for the voltage transfer function is equal to $^{ADC\,full\,scale\,voltage}/_{4095} = 6.42\,\text{mV}$. This value is used to set the horizontal logarithmic scale, which ranges from $LSB_v \rightarrow$ *ADC full scale voltage*. The vertical logarithmic scale ranges from ADC count $1 \rightarrow 4095$ (given that zero cannot be displayed on a log scale).

The lower plot given in Fig. 5.32 also shows an 'ADC' function, which is the current transfer function, with input current range of $LSB_i \rightarrow$ *ADC full scale current*, where the *ADC full scale current* (see Eq. (3.3b)) value is equal to 33.0 A (peak to peak) current for a BOOSTXL-DRV8301 module. Furthermore, the least significant bit LSB_i for the current transfer function is equal to $^{ADC\,full\,scale\,current}/_{4095} = 8.0\,\text{mA}$. The vertical axis of both plots show the ADC count from $1 \rightarrow 4095$. Note that the quantization error becomes significant, when input voltage/current levels are of the same order of magnitude as LSB_v/LSB_i respectively. Hence it is prudent to ensure that voltage/currents levels used during motor identification remain well above the LSB values. Hence the introduction of 'gray' shaded regions in Fig. 5.32, where the ADC resolution is such that the region should be avoided by a prudent choice of currents and frequency during the motor identification phase. Consequently, the figures show how well the ADC converters are utilized, for a chosen set of motor ID input variables and ADC full-scale settings (set by the board design). Note that this 'post-processing' type of analysis, makes use of the measured motor parameters. As such these plots show whether or not the identification procedure has been carried out with appropriate parameter settings (as set in the motor identification dialog box). Furthermore, these plots can provide some guidelines as to what input parameters/variables may need to be changed to improve identification, given the machine under consideration. A brief discussion on the method used to calculate the voltage/currents shown in Fig. 5.32 is given below, for each of the legend variables. Note that all AC voltage measurements are shown in the voltage plot as **peak to peak** values.

- `DCbus`: rated DC voltage $u_{\mathrm{DC}} = 24\,\mathrm{V}$ of the converter. The rated converter current is 8.0 A RMS, which corresponds to a peak to peak current value of `Iconv`=22.6 A, as shown in the lower plot.
- `K_p/w_i CC`: voltage level used during estimation of the current controller gain and bandwidth. With the machine at standstill, the impedance is readily found using $Z_{\mathrm{cc}} = \sqrt{X_{\mathrm{cc}}^2 + R^2}$, with $X_{\mathrm{cc}} = 2\,\pi f_{\mathrm{cc}} L_{\sigma}$. The variables X_{cc} and R represent the reactance (associated with the leakage inductance) and total resistance $R = R_{\mathrm{s}} + R_{\mathrm{R}}$ respectively. The variable $f_{cc} = 100$ Hz is the frequency (Hz) used during this estimation phase. The peak voltage is thus found using $u_{\mathrm{cc}} = i_{\mathrm{cc}}\, Z_{\mathrm{cc}}$, where $i_{\mathrm{cc}} = 0.5$ A is the peak current value used during estimation. The value shown in the top plot is $2\,u_{\mathrm{cc}} = 4.86\,\mathrm{V}$ and the corresponding peak to peak current `I-K_p/w_i cc` $= i_{\mathrm{cc}} = 1.0$ A is given in the lower plot.
- `R_s`: DC voltage during estimation of the DC resistance is equal to $u_{\mathrm{Rs}} = R_{\mathrm{s}}\, i_{\mathrm{Rs}}$, where i_{Rs} is the DC current value used during identification, which is set to the same value as used above. The DC voltage value $u_{\mathrm{Rs}} = 1.84$ V, was calculated using $i_{\mathrm{Rs}} = 1.0$ A. Shown in the lower plot is the DC current `I-Rs`: $i_{\mathrm{Rs}} = 1.0$ A for this measurement.
- `IdRat/Lsig`: peak voltage during Id rated $i_{\mathrm{d}}^{\mathrm{rat}}$ estimation, can be found using $u_{\mathrm{flx}} = 2\pi\, f_{\mathrm{flx}}\, L_{\mathrm{M}}\, i_{\mathrm{d}}^{\mathrm{rat}}$, where $f_{\mathrm{flx}} = 10$ Hz is the frequency Hz used during estimation and L_{M} the magnetizing inductance. A peak to peak value $2\,u_{\mathrm{flx}} = 3.76$ V is shown in the top plot, while the corresponding peak to peak current `IdRat`=2.0 A is given in the lower plot. The voltage across the leakage inductance is equal to $u_{L_\sigma} = 2\pi\, f_{\mathrm{flx}}\, L_{\sigma}\, i_{\mathrm{d}}^{\mathrm{rat}}$ which corresponds to a peak to peak voltage of 0.77 V which is also shown by legend `Lsig` in the top plot of Fig. 5.32.

Note that the rated peak voltage/current for the LVACIMTR IM motor are 12 V and 1.17 A RMS respectively. Hence the corresponding peak to peak values are 24 V and 3.32 A respectively, which implies that the current/voltage values used during motor identification are within the rated values of the machine and the converter.

Overall observation of the voltage/current levels used during estimation learns that they are well outside the LSB voltage/current values of the ADC converter and within the converter rated values. For example, the lowest voltage level (used during leakage estimation) still utilizes approximately 6 bits of the ADC. Utilization of the current ADC's is also satisfactory, given that the lowest current level (used during current controller gain/bandwidth estimation) utilizes at least 6 bits.

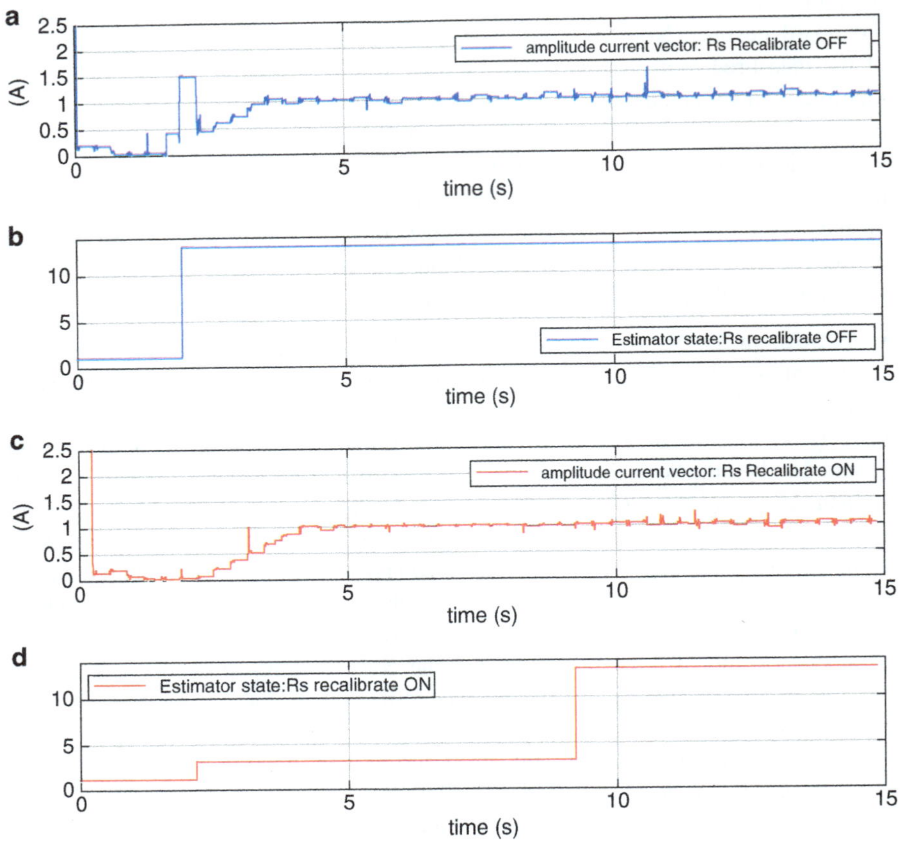

Fig. 5.33 Stator resistance estimation sequence, with recalibration OFF and ON, machine at standstill

In the sequel to this section on motor identification, one additional feature will be discussed, which relates to measurement of the stator resistance prior to normal operation. When the `Rs_recalb_on` button is ON, the stator resistance will be identified, immediately after the converter is turned on and before user operation starts. The R_s recalibration sequence for the IM motor as given in Fig. 5.33 shows the current vector amplitude as function of time together with the Estimator state. Observation of Fig. 5.33 learns that upon converter activation ADC offset measurement takes place, which is followed by user control (Estimator state 13) when Rs recalibration is OFF(see subplots (a) and (b)). In this example the drive is operating under no external load (friction only) conditions and speed control. The magnetizing current is 1.0 A, hence the current vector amplitude will be the same value approximately (a small quadrature current component will be present to generated the torque required to overcome friction). When Rs recalibration is turned ON, injection of a DC current (same value as set in the motor ID dialog entry for Rs) occurs after ADC offset calibration as shown in subplots (c) and (d). In this example motor identification is turned OFF hence the drive will make use of the user entered leakage inductance and stator/rotor resistance values. If no R_s recalibration is done the user defined value will be used, otherwise the estimated R_s will be used. If motor ID is activated, then full parameter identification will take place first where the stator resistance, leakage inductance and rotor resistance are derived. If Rs recalibration is ON the stator resistance will be remeasured prior to normal operation. The time sequence for recalibration is governed by estimator state 3 of the timing sequence shown in Fig. 5.30. However, of the three sub-cycles shown, only the last two entries are used, hence in this case 135000 cycles, which corresponds to 9 s, given the estimator frequency in use. In general it is advisable to deploy stator resistance recalibration prior to drive operation if user operations permit this.

Finally, note that the use of 'online' stator resistance estimation is in principle possible, but not recommended as the technique discussed for PM drives in the previous chapter requires modulation of the direct axis current. For induction machine this leads to modulation of the rotor flux which will cause a torque ripple component.

5.4 Laboratory 3:3: Efficient FOC Drive Operation of Induction Machines

In contrast to permanent magnet machines, the opportunity exists with induction machine based drives to change the rotor flux ψ magnitude by altering the direct axis current i_d. The so called magnetizing characteristic of the machine, as given in Fig. 5.28 shows the relationship between rotor flux and current i_d for the LVACIMTR IM motor in use. This function can be conveniently approximated by the following expression (provided saturation is not significant) namely

$$\psi = L_M \, i_d \tag{5.4}$$

where L_M represents the magnetizing inductance in a four parameter model (see Fig. 2.12). Varying the current i_d and therefore the rotor flux is useful in partial load conditions, i.e. where the steady-state load torque is well below the rated torque. Under these conditions reducing the current i_d by typically 50 % will effectively reduce the energy dissipated due to copper losses by 75 %. In addition, iron losses will also reduce because the flux level is reduced. Consequently, having access to an algorithm such as PowerWarp that can minimize energy losses during partial load operation is useful.

In order to explain the use of the PowerWarp concept, it is helpful to consider the steady-state shaft torque equation of the machine under field oriented control namely

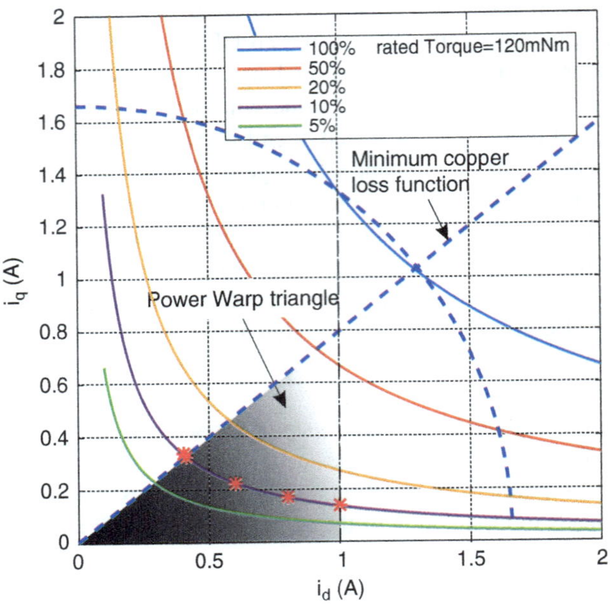

Fig. 5.34 Efficient partial load operation of induction machine using PowerWarp

$$T_{\mathrm{m}} = \frac{3}{2} p\, i_{\mathrm{q}}\, \psi \tag{5.5}$$

where p and i_{q} represent the pole pair number and quadrature current of the machine respectively. Substitution of Eq. (5.4) into Eq. (5.5) gives

$$T_{\mathrm{m}} = \frac{3}{2} p\, L_{\mathrm{M}}\, i_{\mathrm{d}}\, i_{\mathrm{q}} \tag{5.6}$$

which shows the relation between i_{q} and i_{d} as function of the shaft torque, which under steady-state conditions (as considered here) is also the load torque. A graphical representation of Eq. (5.6) for the two pole pair ($p = 2$) LVACIMTR IM motor is given in Fig. 5.34 for a set of load torque values ranging from 100 % rated shaft torque (which is equal to 120 mNm for this machine) to 5 % rated torque. Also shown in this figure is the 1.66 A rated (peak) current circle (for one operating quadrant) of the machine which represents the boundary of the area within operation can take place without damage to the machine. Normally drive operation occurs under rated direct axis current conditions, in this case with $i_{\mathrm{d}}^{\mathrm{rat}} = 1.0$ A. This implies operating with the highest (rated) possible flux level in the machine and implies operating along the vertical line $i_{\mathrm{d}}^{\mathrm{rat}} = 1.0$ A as the load torque changes. This is most practical from the perspective of requiring the lowest quadrature current to realize a given load torque (an approach used for PM machines) but is also the most energy inefficient method of operation. From an energy efficiency perspective it is desirable to operate the machine on the 'minimum copper loss function' of the machine, indicated in Fig. 5.34. This curve, generated by the PowerWarp algorithm, identifies which combination of (steady-state) direct and quadrature current values should be used for a given load torque condition to minimize energy (copper losses) usage of the drive. For example, when operating with a load torque of 10 % rated torque (which is the friction torque of the induction machine, when not connected to any other load) a set of possible operating points are available that fall within the so called 'Power Warp triangle' identified by the 'gray' shaded area in Fig. 5.34. Operation without the PowerWarp algorithm implies operating under rated direct axis current and rated flux conditions ie, with $i_\mathrm{d} = 1.0\,\mathrm{A}, i_\mathrm{q} \approx 0.15\,\mathrm{A}$. When PowerWarp is activated the operating point will move along the 10 % load curve (assuming that the load torque remains unchanged) to an operating point on the 'minimum copper loss function', which in this case corresponds to operation at 40 % rated flux with $i_\mathrm{d} \approx 0.4\,\mathrm{A}, i_\mathrm{q} \approx 0.33\,\mathrm{A}$. Consequently, the amplitude of the current vector (defined as $|\vec{i}| = \sqrt{i_\mathrm{d}^2 + i_\mathrm{q}^2}$) is approximately halved, which implies that the copper losses are now at 25 % of their previous value (when operating at rated flux). Note that the effectiveness of the algorithm will reduce as load torque is increased, and becomes ineffective when the top of the 'Power Warp triangle', operating point $i_\mathrm{d} = 1.0\,\mathrm{A}, i_\mathrm{q} \approx 0.8\,\mathrm{A}$ is reached. Furthermore, the algorithm must be able to increase the direct axis current when a higher load torque is required. However, the time constant associated with increasing the torque using this approach is limited by the time constant $L_\mathrm{M}/R_\mathrm{R}$ of the machine, which is typically in the order of 30 ms for the machine in question. Hence the price to pay for improved energy efficiency is a reduction in transient torque performance.

5.4.1 Lab 3:3: Phase C

Prior to being able to work with an operational drive, an `.out` file must be generated of a 'Drive Controller with FAST Motor ID and PowerWarp' module, as shown in Fig. 5.35, which represents the drive under consideration. Hence details of the drive structure and corresponding dialog box parameter assignments will be discussed in this subsection.

The following information is relevant for this laboratory component:

- Reference program [11]: `lab3-3_LaunchXLphCv3.vsm`.
- Description: Sensorless control of a IM motor with InstaSPIN-FOC with motor identification and energy efficiency algorithm PowerWarp.
- Equipment/Software: Texas Instruments LAUNCHXL-F28069M, with BOOSTXL-DRV8301 module ('aft' position) and VisSim simulation program.
- Outcomes: Generate the `.out` file needed to represent the drive structure under investigation.

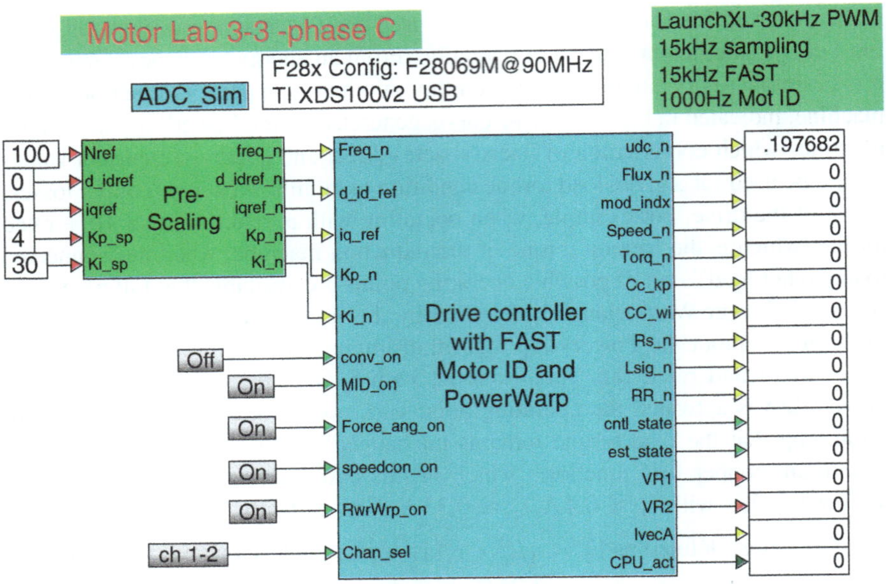

Fig. 5.35 Phase C simulation of a InstaSPIN based encoderless FOC drive, with PowerWarp

As mentioned above, a single boost pack located in the 'aft' position (furthest away from the USB connector) is used for single motor operation (PM machine is not used electrically), in which case the following jumper and dip-switch positions on the LAUNCHXL-F28069M module are required:

- Jumpers `JP1` and `JP2` OPEN
- Jumpers `JP4` and `JP5` CLOSED
- Jumpers `JP3`, `JP6` and `JP7` CLOSED
- Dip-switches `SW1` to `SW3` ON

Inputs to the Controller are five variables, which provide the user with the ability to set the reference shaft speed, differential direct axis/quadrature axis reference currents and speed gains. The `MID_on` activates motor parameter estimation when the converter is activated. Note this will occur only after running the C+ file A button `Force_ang_on` activates the 'force angle' function of the InstaSPIN algorithm. Activating this function ensures faster synchronization with the rotor flux vector during start up. However, this function should not be used when speed reversals are required. Speed or torque control mode is controlled by the button `speedcon_on`. A `Conv_on` button and `Chan_sel` function are used to respectively activate the converter and select which diagnostic channel combination is to be displayed in phase C+. Energy efficient operation using PowerWarp is activated by the button `PwrWrp_on`. Outputs (per unit) of the controller module are, `u_dc_n`: DC bus voltage, `Flux_n`: IM motor rotor flux, `Speed_n`: shaft speed. Additional outputs are, `Torq_n`: shaft torque, current controller gains/bandwidth: `CC_kp`, `CC_wi`, motor parameters: `Rs_n`, `Lsig`, `RR_n`, controller `cntrl_state` and estimator `est_state`. For convenience sake, output variables `mod_indx` and `IvecA` have been added, which represent the amplitude of the modulation index and current vector amplitude respectively. Furthermore, three output variables have been added of which two `VR1`, `VR2` are used to display user defined internal variables. The third `CPU_act` is used to provide an indication of percentage CPU utilization when running in phase C+. Moving one level lower into the 'Drive Controller with FAST Motor ID and PowerWarp' module brings up the by now familiar 'main control','diagnostic tool', 'Operational', 'Converter and FAST variable' and 'Motor Estimation variables' dialog boxes. The parameters used in said dialog boxes are presented in Fig. 5.36.

The dialog boxes shown are identical to those discussed in the previous section, with exception of the `I_ramp_PwrWrp(A/s)` variable shown in the 'Converter and FAST variables' dialog box. This variable controls the rate of current decay AFTER PowerWarp has been activated, hence it allows the user to control the transition time from rated $i_{\mathrm{d}}^{\mathrm{rat}}$ to the i_{d} value on the 'minimum copper loss function' (see Fig. 5.34). The rate of current increase due to encountering a higher load torque or after deactivating the PowerWarp function is controlled by the `User current_ramp(A/s)` variable in the 'Operational variables' dialog box. Actual activation of the PowerWarp algorithm is undertaken in the '100 Hz' module (located one level into the 'Drive control with FAST and motor identification' module).

5.4.2 *Lab 3:3: Phase C+*

Phase C+, is the operational component of the laboratory and is basically a run version of the `.out` file compiled and downloaded to the MCU in phase C (see previous subsection).

The following information is relevant for this laboratory component:

- Reference program [11]: `lab3-3_LaunchXLphCv3_d.vsm`.
- Description: FOC sensorless control of a IM machine using Powerwarp.
- Equipment/Software: Texas Instruments LAUNCHXL-F28069M, with BOOSTXL-DRV8301 module. ('aft' position), Texas Instruments LVACIMTR IM machine and VisSim simulation program.
- Outcomes: to gain familiarity with the FAST InstaSPIN algorithm when used with PowerWarp to realize energy efficient operation.

Note that the required jumper and dip-switch settings for the LAUNCHXL-F28069M module are given in phase C.

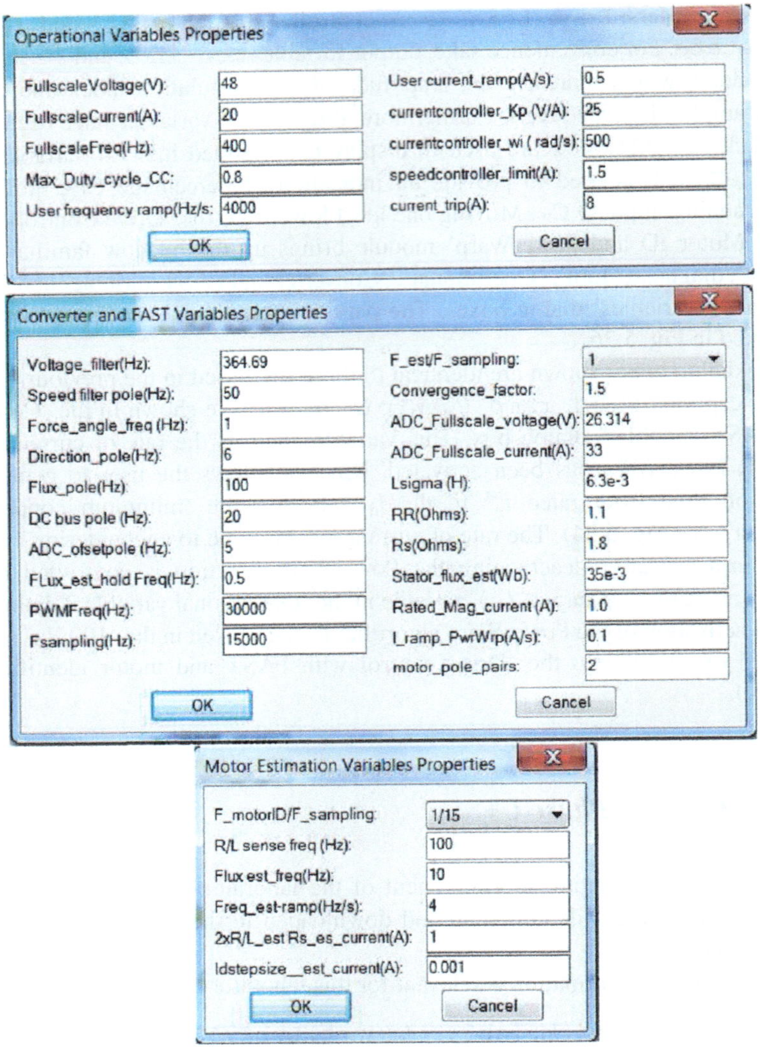

Fig. 5.36 Phase C simulation of a FAST based encoderless FOC drive, with field-weakening: Dialog box entries

The run version shown in Fig. 5.37 uses a VisSim run module, which executes the `.out` file, shown in said module. Five sliders are used, the purpose of which was discussed in the previous section. A post-scaling module is again used to convert the per unit measured DC bus voltage to actual voltage, as shown with a numeric display.

Three additional 'Post scaling' modules are used to generate the following variables:

- `Rot_flx_est (Wb)`: estimated magnetic rotor flux (in Wb) of the machine in use.

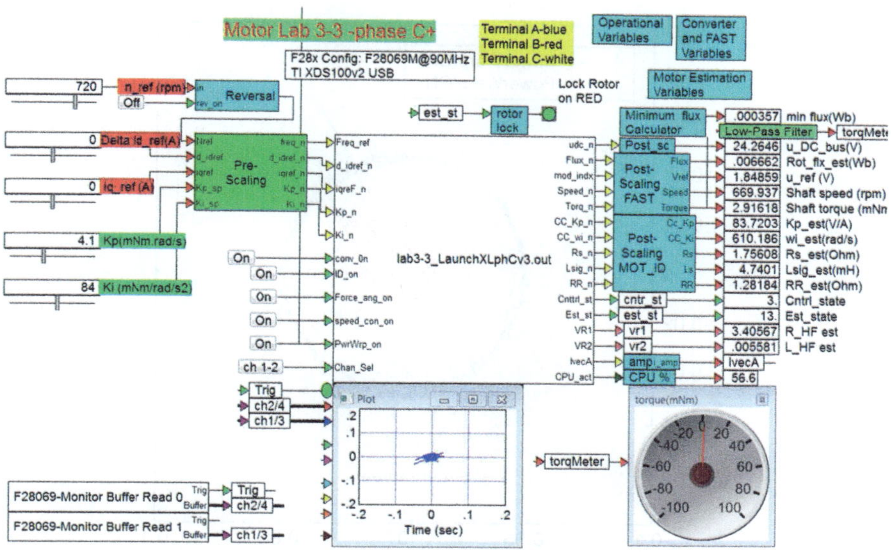

Fig. 5.37 Phase C+ simulation of a FAST based encoderless FOC drive, with PowerWarp

- `u_ref (V))`: amplitude of the reference voltage generated by the current controller.
- `shaft speed (rpm)`: estimated shaft speed.
- `shaft torque (mNm)`: estimated shaft torque (in milli-Nm) of the machine in use. A low-pass filter with a 1 Hz corner frequency, is used to filter this signal for display in the torque meter (see Fig. 5.37).
- `Kp_est (Ohm)`: estimated current controller gain (V/A).
- `wi_est (Ohm)`: estimated current controller bandwidth (rad/s).
- `Rs_est (Ohm)`: estimated stator resistance.
- `Lsig_est (mH)`: estimated leakage inductance.
- `RR_est (Ohm)`: estimated rotor resistance.
- `Cntrl_state`: controller status.
- `Est_state`: estimator status.
- `u_ref(V)`: reference voltage amplitude generated by the FOC current controller.
- `VR1`: diagnostic output for user, set to show the 'high frequency' stator resistance estimate.
- `VR2`: diagnostic output for user, set to show the 'high frequency' leakage inductance estimate.
- `IvecA`: diagnostic output used to plot the current vector amplitude.
- `CPU_act`: percentage utilization of the MCU.

Two 'Monitor Buffer' modules are used to display two selected (using an button connected to the `chan_sel` input) diagnostic signals. An enlarged view of the

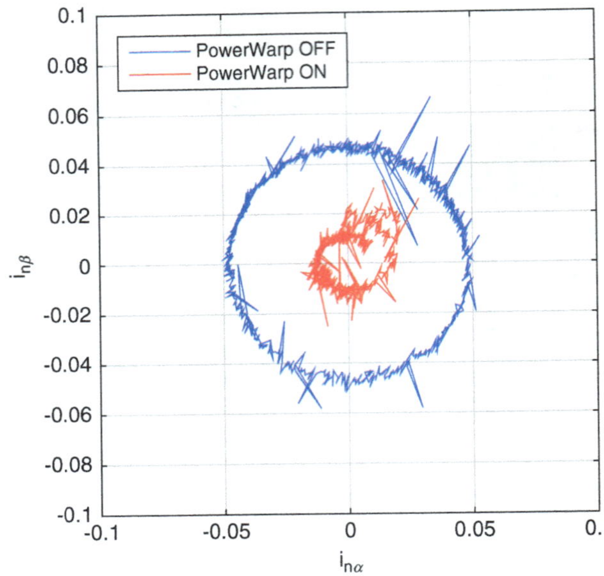

Fig. 5.38 Phase C+ simulation: Scope display showing per unit α, β currents operating with PowerWarp OFF/ON (machine running with friction load only)

VisSim scope module given in Fig. 5.38 shows the per unit currents i_{α}^{n}, i_{β}^{n} after completion of identification. Multiplication by the full scale current value of 20 A gives actual values.

Prior to activating this type of lab component, the reader needs to be aware of the fact that he or she is about to activate a complex electrical system, with live voltages. Hence it is prudent, **always**, to execute the following 'Pre-Drive' check list:

- Dialog boxes used in C+ mode, match those of phase C: the run version is compiled with the dialog box entries specified under phase C. The dialog boxes shown in phase C+ are used by the 'Pre/Post' scaling modules.
- Ensure that the sample time used is correct and latest (and correct) `.out` file has been downloaded to the MCU (Right mouse click on MCU module to show dialog box of these variables).
- Confirm that the user input values are set to either zero, or 'acceptable' values, which will not cause a current trip of the converter.
- Confirm that the converter 'switch' is set to OFF and the power supply is on (DC bus voltage present).
- Confirm that the motor is connected firmly and properly.

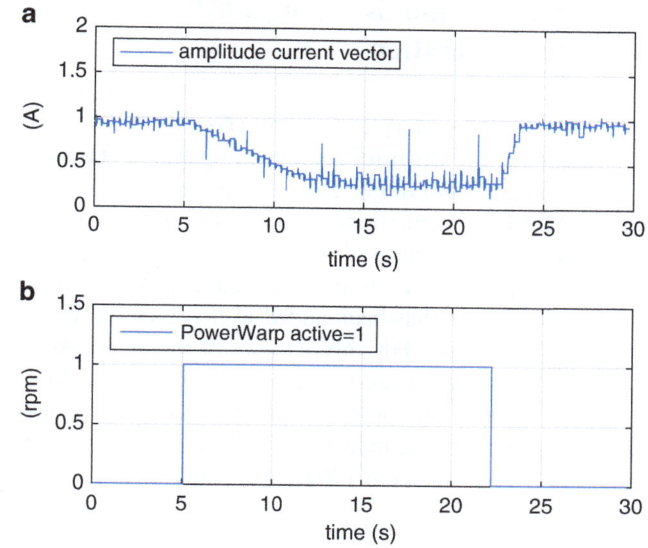

Fig. 5.39 Current vector amplitude during a PowerWarp OFF/ON sequence

After completion of the Pre-Drive checklist, activate the program and confirm that the supply voltage source reading shown in the digital display is 24 V. If not stop the program and restart. With the correct DC voltage reading established, turn on the converter (using the ON button, shown) and monitor the motor shaft and diagnostic scope, which in this example shown (Fig. 5.38) is set to show the per unit current components i_α, i_β with the scope set in 'xy' mode. Note that the actual current values are found multiplying the per unit value with the full-scale current value set to 20 A in this case.

Observation of Fig. 5.37 learns that the machine is currently operating under closed-loop speed control, with no external load, except friction, as is evident from the torque meter value. Furthermore, energy efficient drive operation using PowerWarp is active, as is evident by the ON button, `PwrWrp_on` and the amplitude of the current locus shown in Fig. 5.37. In addition to the above, the current vector amplitude `IvecA` was plotted during a PowerWarp turn ON/OFF cycle, as shown in Fig. 5.39.

Observation of this result learns that the current amplitude is reduced as energy efficient operation is invoked as discussed with the aid of Fig. 5.34. The rate of current decay after PowerWarp is turned ON is then controlled by the `I_ramp_PwrWrp(A/s)` variable, as discussed above. When PowerWarp is switched OFF the current amplitude returns to the original value, in which case the rate of current change is governed by the setting `user_current_ amp` present in the 'converter and FAST' dialog box mentioned earlier.

5.5 Laboratory 3:4: Commissioning a IM Drive for Sensorless Operation

Primary objective of this laboratory is to familiarize the reader with all aspects needed to set up an induction drive for sensorless operation starting from a given hardware configuration as shown in Fig. 5.40. The approach taken is similar to that discussed for the PM drive commissioning laboratory. In this case the DRV8301-HC-EVM-Rev D board is chosen as the 'unknown' board, which also uses the DRV8301 converter chip. However, unlike the Launchpad, an additional 3 phase converter with discrete devices has been added to this board to enhance its phase current capabilities (Table 5.1). For this laboratory the F28069M control card, will be used together with the Texas Instruments LVACIMTR IM machine. The latter will be used as the 'test' motor, which has the in this table known/measured parameters given in Table 5.1. The inductance and resistance value indicated were measured using an L-C-R meter set to 100 Hz and Multimeter respectively. In both

Fig. 5.40 Commissioning IM drive setup

Table 5.1 Texas instruments LVACIMTR IM machine: measured motor parameters

Parameter	Value	Units
Inductance: L_{sLL}	9.6	mH
Resistance: R_{sLL}	3.7	Ω
Pole pairs: p	2	–

cases the meter was connected between two phases of the machine. On the basis of these measurements the stator resistance was found to be $R_s = {}^{R_{sLL}}/_2 = 1.85\,\Omega$. Furthermore, the leakage inductance was found to be $L_\sigma = {}^{L_{sLL}}/_2 = 4.8\,\text{mH}$. For this drive the DC supply is set to $u_{DC} = 24\,\text{V}$. The rated RMS phase current for this machine is 0.91 A, hence the peak phase currents to be used in this laboratory should typically not exceed $\approx 1.3\,\text{A}$. In terms of choosing a PWM frequency, a value typically twice (or more) the controller sampling frequency is preferred in order to reduce current ripple, which will improve the current measurements, given that shunts are used in this converter. However, a higher PWM frequency will also increase the converter switching losses, hence a trade-off (between switching losses and current ripple) is required. In this laboratory the PWM frequency will be set to 30 kHz, whilst the ADC sampling frequency is taken to be 15 kHz, hence a factor 2 lower.

5.5.1 Lab 3:4: Voltage Controller, Phase C

Prior to using current control in any application, it is helpful to consider drive operation under voltage control first. Using this approach, the voltage and current measurements are examined. Furthermore, the phase relationship between voltage and current can be established, which is important prior to using current control. The following information is relevant for this laboratory component:

- Reference program [11]: `lab3-4_8301phCv2.vsm`.
- Description: IM drive commissioning example.
- Equipment/Software: Texas Instruments DRV8301-HC-EVM, Rev D board, with F28069M ISO control card and VisSim simulation program.
- Outcomes: to gain familiarity of all aspects of operation needed to commission a drive.

The voltage controller as discussed in Sect. 3.1 is used for this analysis. However, emphasis in this laboratory will be on configuring the ADC-PWM module, triggering and allocating the ADC channels for this new board. The phase C, development phase, given in Fig. 5.41 has been adapted for this application. Inputs are the electrical frequency and direct/quadrature voltage reference variables. Two control buttons are present to enable the drive and select the diagnostic channels to be viewed on the scope. Moving one level down into the 'Open loop voltage controller' reveals a set of modules and dialog boxes as indicated in Fig. 5.42. This figure also shows a dialog box which appears upon a 'right click' on the 'main control module'. The critical entry in this case is the 'execute on interrupt' feature which must be set as shown. This implies that the module in question will be executed on an interrupt signal set by the ADC unit. More specifically, the interrupt is in this case `ADCINT1:7`, which implies that the signal is generated by ADC input channel 7. The ADC input channels are allocated within the ADC-PWM module, which appears when moving one level lower into the 'main control module' shown in Fig. 5.43. The modules shown are as discussed in Sect. 3.1. Note

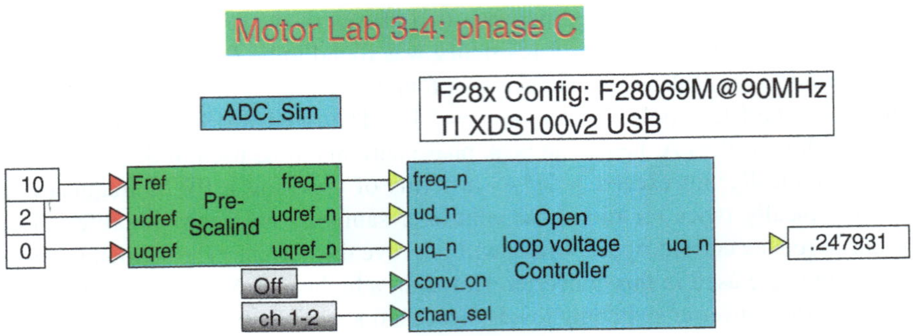

Fig. 5.41 Development Phase C: voltage controller set up

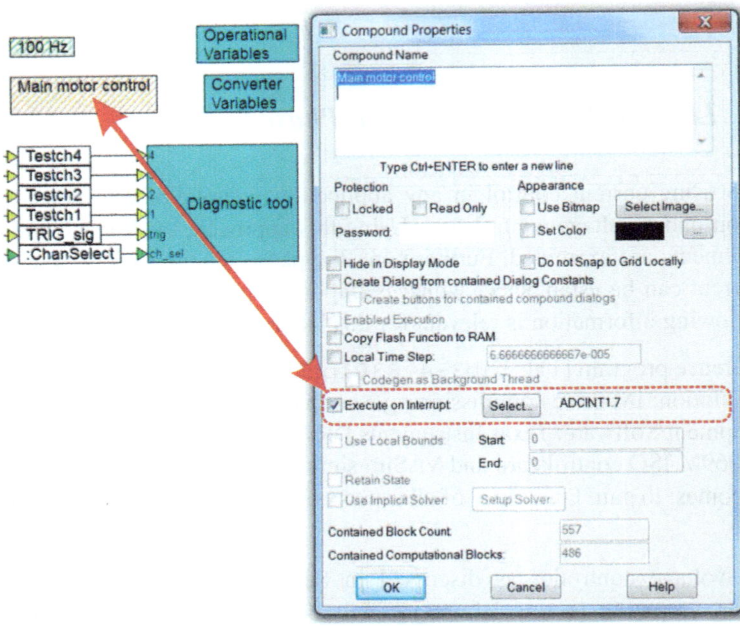

Fig. 5.42 Development Phase C, one level into 'Open-loop voltage controller'

that in this laboratory the per unit currents `i_alp`, `i_bet`, which are defined in a α, β reference frame are in `IQ24` instead of the `IQ30` format used in previous laboratories. The reason for this is that the full-scale current value is set to 10 A which implies that a scaling factor $^{\mathrm{ADC_{fc}}}/_{i_{\mathrm{fs}}}$ as used in the ADC-PWM module will be greater than 2, hence outside the `IQ30` range. By reformatting the current outputs to `IQ24` the allowable scaling factor is extended to 128. Note also that the ADC/PWM module is board specific, which is why the board type is included in the title. Data acquisition using VisSim, is done by way of a MCU specific 'input channel' module, such as `F28069M Input Channel` in which the user

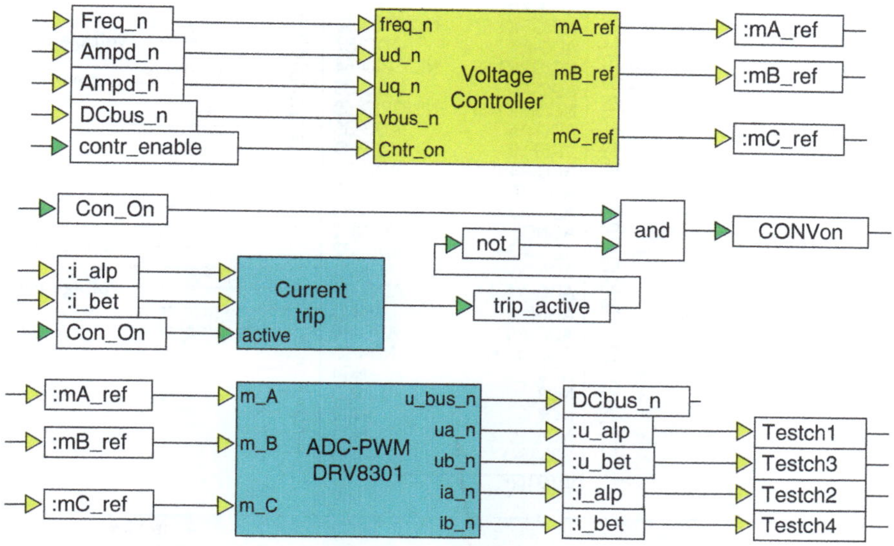

Fig. 5.43 Development Phase C: one level into 'main motor control' module

Table 5.2 Allocation of input variables to ADC input channels

Input variable	ADC input channel
Phase current i_A	1
Phase current i_B	2
Phase current i_C	3
Phase voltage u_A	4
Phase voltage u_B	5
Phase voltage u_C	6
Bus voltage u_{DC}	7

has the option of selecting either an 'analog or digital' channel. Furthermore, the user must allocate a channel number (within the range of analog channels available) as mentioned above. Table 5.2 shows the channel allocation used in this laboratory. Note that the allocation of phase currents and voltages is up to this stage done arbitrarily. Observation of said table shows that channel 0 is not used, as it is the first ADC conversion result and may contain corrupted data, hence it is discarded here. A critical commissioning step is to allocate the channel numbers introduced in Table 5.2 to actual MCU variables. This requires access to the `ADC F28069M Properties` dialog box, found by selecting `Embedded` in the main VisSim menu and then selecting `F280x` and finally `ADCconfig` which brings up the module shown in Fig. 5.44. Also given in this figure is part of the circuit diagram of the board under consideration, which shows the pin allocation of the current/voltage variables. The DRV8301-HC kit makes use of (in this case) the F2806x ISO, rev 2 control card which slots into the `DIMM100` connector shown in Fig. 5.44. Associated with the current/voltage variables and pin numbers of the

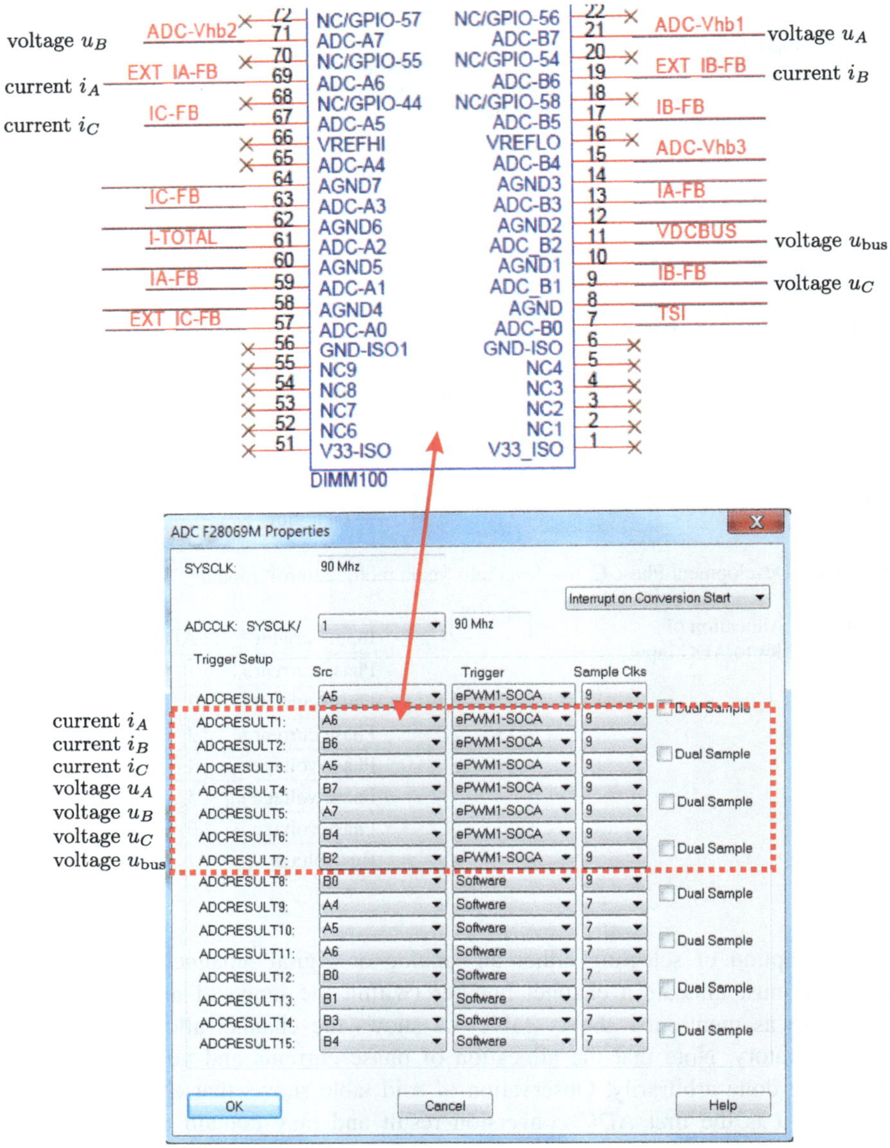

Fig. 5.44 Development Phase C: 'ADC properties' dialog box and link to circuit diagram

`DIMM100` connector are alphanumeric variables which need to be assigned in the `ADC F28069M Properties` dialog box, as shown. In addition to assigning the input variables the trigger signal for the input channels need to be assigned. Triggering of the ADC is linked to the PWM module as was discussed in Sect. 2.1.6. For all the input channels the 'Start Of Conversion' signal `ePWM1-SOCA` is used which implies that the pulse width modulation module 1 generates the trigger signal for the input ADC channels in use.

Moving one level lower into the 'ADC-PWM' module shown in Fig. 5.43 reveals three PWM modules (see Fig. 5.45),which have as inputs the modulation index variables `mA`, `mB` and `mC` respectively. A 'right click' on a module reveals a dialog box which is also shown in Fig. 5.45. The most important entries are numbered and need to be configured as follows:

1. `Timer Period`: represents the number of cycles of the PWM staircase as discussed in Sect. 2.1.6. The number to be entered here is defined by the clock frequency divided by the timer period times 2. In this laboratory a 30 kHz PWM frequency is required, which with a system clock frequency of 90 MHz, corresponds to a timer period value of 1500 as shown.
2. `Rising Edge Delay, Falling Edge Delay`: these variables represent the 'dead-time' setting used to avoid converter leg 'shoot through' (for example, top switch turning ON, while the bottom switch is still in the process of turning OFF). Hence the delay is there to give the converter switches time to turn on and off. The 'dead-time' is represented in terms of the number of clock cycles: in this case a value of 10 is chosen (given the devices in use), which corresponds to $t_{dead} = {}^{10}/_{(90 \cdot 10^6)} = 111$ ns.
3. `send start ADC Conversion Pulse A (SOCA)`: this entry is critical, in that it determines when the conversion pulse should be given with respect to the PWM cycle, as shown in Fig. 2.23. In this case the option `PRD` is chosen because triggering of the ADC should occur at the top of the PWM staircase, as shown in Fig. 2.22. Note that in Fig. 5.44 ADC triggering refers to `ePWM1-SOCA`, which implies that ePMW module 1 generates the SOC (Start Of Conversion) signal for all ADC channels.
4. `2`: this entry specifies if the ADC pulse should be given after every first, second or third PWM clock cycle. In this case a 30 kHz PWM frequency has been selected and the ADC sample frequency in use is 15 kHz (which is also the rate at which the controller module will be executed, with a sampling frequency set in the main VisSim module) hence the PWM to ADC ratio must be set to ${}^{1}/_{2}$.

Note that only a subset of the ADC and PWM dialog entries have been discussed, given their importance. The reader is however, advised to refer to the relevant Texas Instruments application notes for further details on these modules.

In the sequel to this section attention is given to the computation of the ADC full scale voltage/current parameters and the low-pass filter corner frequency. These parameters can be found by careful examination of the schematic diagram of the converter under consideration. Part of the relevant (to this topic) circuit diagram sections of the DRV8301-HC board, as given in Fig. 5.46 will be used to derive the required data. Central to this figure is one leg of the three-phase converter, where the associated shunt resistor `R80` is used to measure the phase current. An 'op-amp' based current signal conditioning circuit is used to generate the 'current' variable i_A, which is the input voltage to the ADC unit shown in Fig. 5.44. The ADC full scale current is the peak to peak current value that corresponds to a peak to peak ADC input voltage of 3.3 V. Analysis of the current conditioning circuit shows that the ADC full scale current is found using

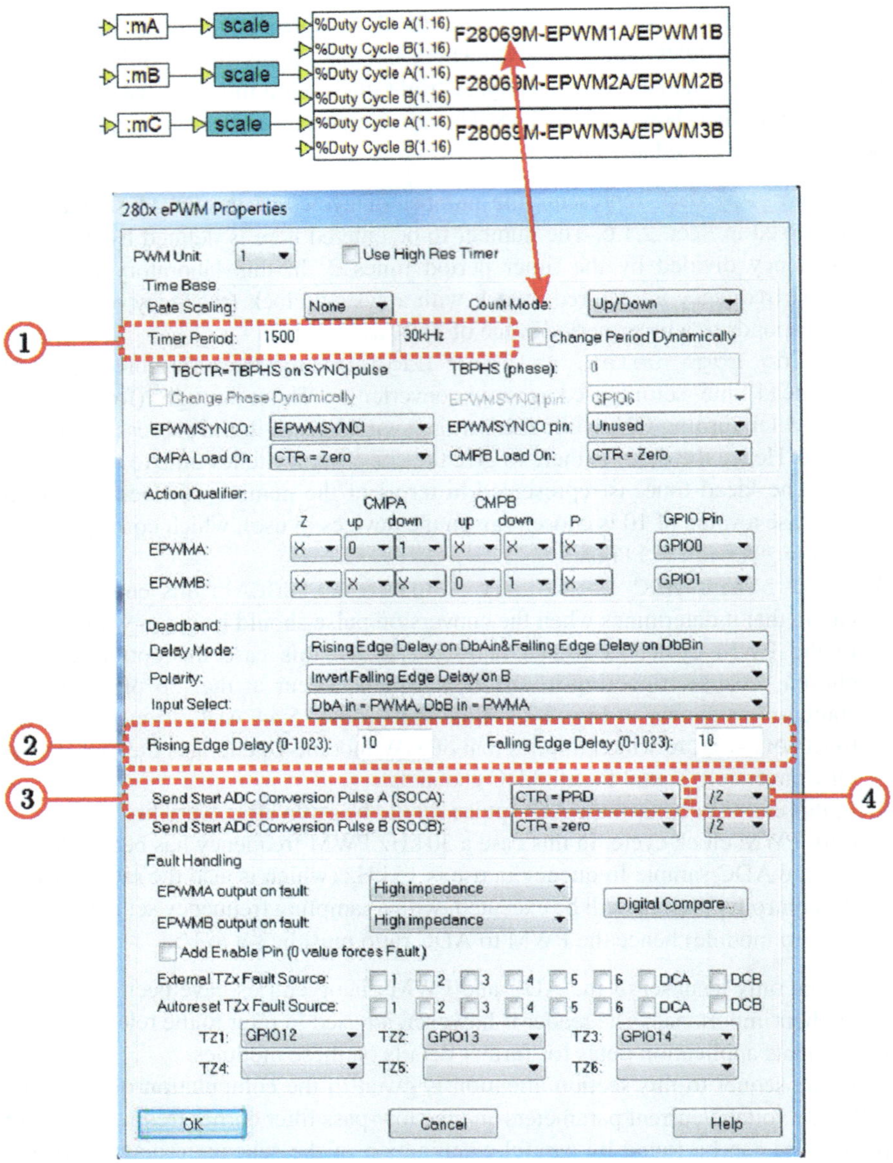

Fig. 5.45 Phase C: 'PWM properties' dialog box

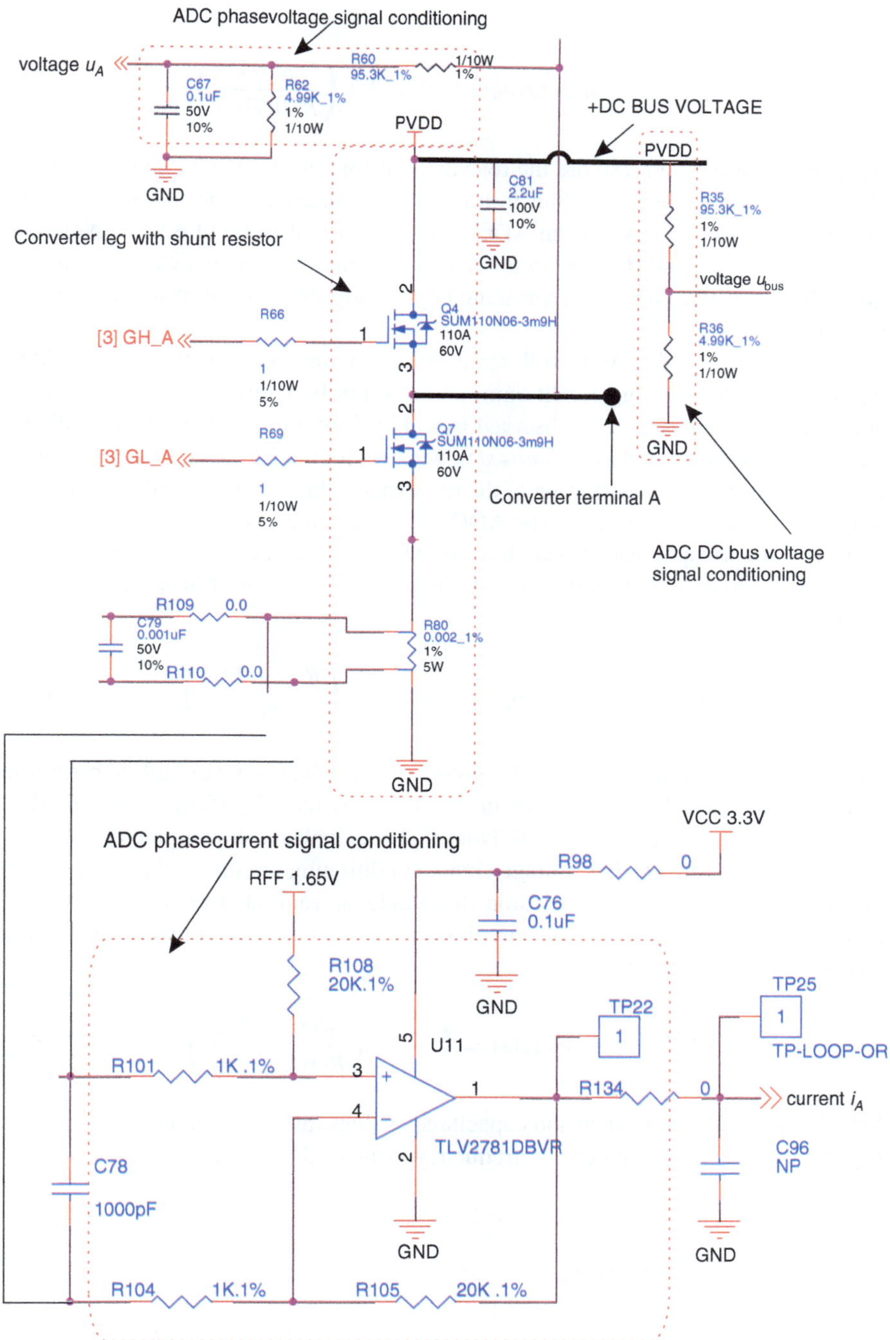

Fig. 5.46 Phase C: Part of DRV8301-HC circuit diagram showing one converter leg, with associated current and voltage signal conditioning

$$ADC\,full\,scale\,current\,(\mathrm{A_{pp}}) = 3.3\left(\frac{R104}{R80\,R105}\right) \tag{5.7}$$

where $R104, R105$ and $R80$ are the relevant resistors of the current gain circuit and shunt resistance respectively. Substitution of the resistance values (as shown in the diagram) into Eq. (5.7) yields an ADC full scale current value of 82.5 A. Note that the current signal conditioning circuit inverts the shunt voltage, which is convenient, given that a positive current is represented by an outgoing current from the converter terminal A.

Computation of the ADC full scale voltage requires evaluation of the ADC phase and DC bus voltage conditioning circuits. For both circuits, the same voltage attenuation is used as may be observed from Fig. 5.46. For the phase voltage circuit the attenuation is defined as $R62/(R60+R62)$, whereas for the DC bus circuit attenuation is set to $R36/(R35+R36)$. Substitution of the resistance values, learns that the attenuation of both circuits is indeed equal. The ADC full scale voltage is the peak to peak phase voltage that corresponds to a peak to peak ADC input voltage of 3.3 V. Analysis of the phase voltage conditioning circuit, shows that the ADC full scale voltage is found using

$$ADC\,full\,scale\,voltage\,(\mathrm{V_{pp}}) = 3.3\left(\frac{R60 + R62}{R62}\right) \tag{5.8}$$

where $R60$ and $R62$ are the relevant resistors of the voltage gain circuit. Substitution of the resistance values (as shown in the diagram) into Eq. (5.8) yields an ADC full scale voltage value of 66.32 V. Note that the analysis shown, could also have been done using the DC bus voltage signal conditioning circuit, as that would have given the same result. In conclusion, the reader is reminded of the fact that the phase voltage signal conditioning circuit is also a low-pass filter, of which the corner frequency may be found using

$$LPF\,corner\,freq\,(\mathrm{Hz}) = \left(\frac{1}{2\pi}\right)\left(\frac{R60 + R62}{R60\,R62\,C67}\right) \tag{5.9}$$

Substitution of the resistance and capacitance values (as shown in the diagram) into Eq. (5.9) yields a low-pass corner frequency value of 335.6 Hz.

5.5.2 *Lab 3:4: Phase C+*

Phase C+, is the operational component of the laboratory and is basically a run version of the `.out` file compiled and downloaded to the MCU in phase C (see previous subsection).

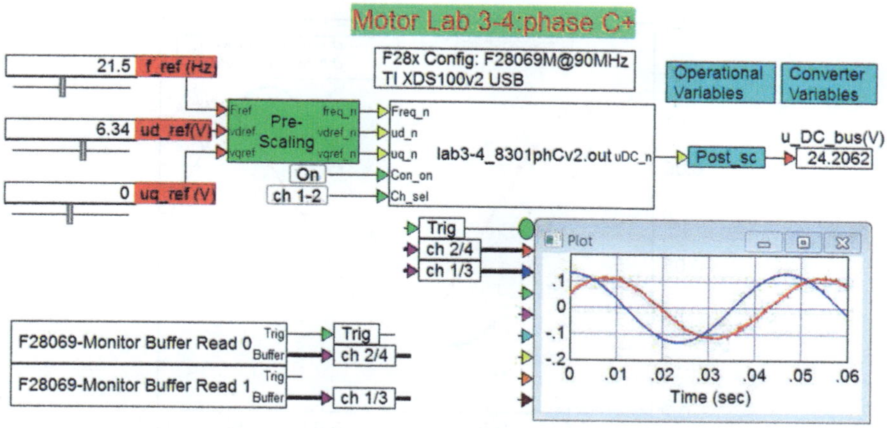

Fig. 5.47 Phase C+: Embedded voltage controller

The following information is relevant for this laboratory component:

- Reference program (on CD): `lab3-4_8301phCv2_d.vsm`.
- Description: Description: IM drive commissioning example.
- Equipment/Software: Texas Instruments DRV8301-HC-EVM, Rev D board with F28069M ISO control card, Texas Instruments LVACIMTR IM machine and VisSim simulation program.
- Outcomes: to confirm basic drive operation by observing the voltage and current waveforms.

The run version shown in Fig. 5.47 uses a VisSim run module, which executes the `.out` file, shown in said module. Three sliders are used to control the stator frequency and direct/quadrature axis reference voltages. Two buttons are used to enable the drive and select the diagnostic channels to be viewed on the scope. A post-scaling module is again used to convert the per unit measured DC bus voltage to actual voltage, as shown with a numeric display.

Two 'Monitor Buffer' modules are used to display two selected diagnostic signals. An enlarged view of the VisSim scope module given in Fig. 5.48 shows the per unit current i_α^n and voltage u_α^n. Multiplication by the full scale current value of 10 A and 48 V for the per unit current and voltage respectively, gives the full scale values. The dialog box entries for this drive are shown for completeness in Fig. 5.49. Note that the ADC full scale shown are those which have been calculated in the previous development phase. Prior to activating this type of lab component, the reader needs to be aware of the fact that he or she is about to activate a complex electrical system, with live voltages. Hence it is prudent, **always**, to execute the following 'Pre-Drive' check list:

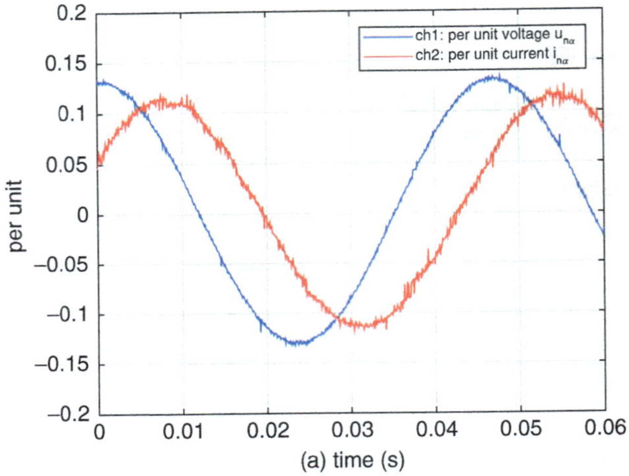

Fig. 5.48 Phase C+ simulation: Scope display showing per unit α current and voltage (machine running with friction load only)

Fig. 5.49 Phase C+: dialog boxes for IM drive commissioning laboratory

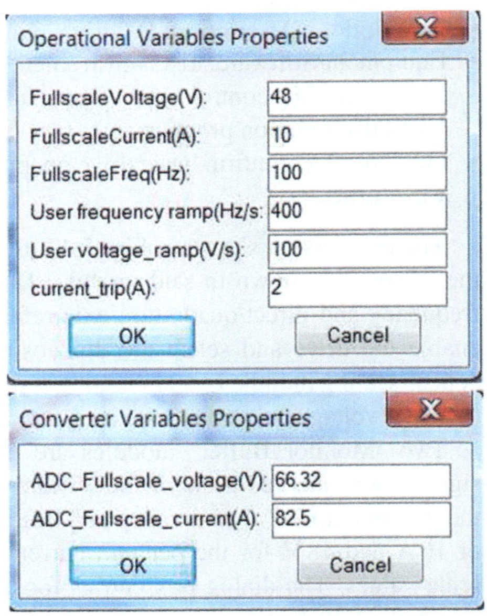

- Dialog boxes used in C+ mode, match those of phase C: the run version is compiled with the dialog box entries specified under phase C. The dialog boxes shown in phase C+ are used by the 'Pre/Post' scaling modules.
- Ensure that the sample time used is correct and latest (and correct) `.out` file has been downloaded to the MCU (Right mouse click on MCU module to show dialog box of these variables).
- Confirm that the user input values are set to either zero, or 'acceptable' values, which will not cause a current trip of the converter.
- Confirm that the converter 'switch' is set to OFF and the power supply is on (DC bus voltage present).
- Confirm that the motor is connected firmly and properly.

After completion of the Pre-Drive checklist, activate the program and confirm that the supply voltage source reading shown in the digital display is approximately 24 V. If not stop the program and restart. With the correct DC voltage level present, turn on the converter (using the ON button) and monitor the motor shaft and diagnostic scope. Several important observations can be made on the basis of the information provided in Fig. 5.48 namely;

- The reference frequency has been set to 21.5 Hz hence the period time of the waveforms should be 46.5 ms.
- A reference voltage amplitude of 6.34 V has been set by the slider `ud_ref` hence the amplitude of the per unit u_α should be approximately (as dead-time effects and the voltage drop across the converter switches will affect the amplitude shown) equal to $^{6.34}/_{u_{\rm fs}} = 0.13$, where $u_{\rm fs}$ is the full scale voltage value set to 48 V.
- Verification of the current amplitude can be done by making use of a DC-true current probe and oscilloscope, which leads to the result given in Fig. 5.50. Observation of the experimental current waveform reveals a current amplitude of 1.1 A, which corresponds to a per unit value of $^{1.1}/_{i_{\rm fs}} = 0.11$, where $i_{\rm fs}$ is the full scale current value set to 10 A. Examination of the scope results in Fig. 5.48 confirms that the per unit current is indeed correct.
- The phase relationship between the voltage and current waveform should be such that the latter is lagging, given the presence of a resistive/inductive type load.

Finally, observe that the motor is rotating correctly. With a reference frequency of 21.5 Hz the speed of the 4 pole motor will be approximately 645 rpm (no external load connected hence slip is approximately zero, as there will be friction due to the bearings), which is difficult to measure without suitable equipment. However, if the reference frequency is set 2 Hz the rotational speed will be ≈ 60 rpm, i.e. one revolution per second, which can be visually determined. This measurement will verify that the assumed pole pair number for the machine is indeed correct.

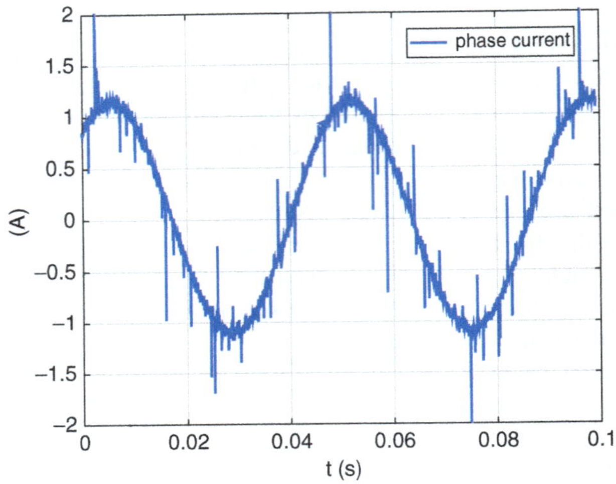

Fig. 5.50 Phase C+: Experimental result: phase current

5.6 Laboratory 3:5: Dual Control of IM and PM Machine

The final laboratory of this chapter considers dual operation of an induction machine connected to a PM machine. The latter machine is used in conjunction with a sensored field-oriented controller, as discussed in lab 1:3 (see Sect. 3.3), which will be operated under torque control. This PM machine is mechanically connected to an induction machine, which is electrically connected to a converter operating with a FOC Sensorless controller as discussed in lab 3:2 (see Sect. 5.3).

The purpose of this laboratory is to explore dual machine operation, where, for example, one machine is acting as a load and the other as a motor. Care should be taken in this case to ensure that both machines are not operating as motors simultaneously, given that a speed runway situation could then occur. In this laboratory use is made of a LAUNCHXL-F28069M board, with two BOOSTXL-DRV8301 modules, which are each connected to a Texas Instruments LVACIMTR IM and LVSERVOMTR PM machine as shown in Fig. 5.51. Clearly observable are the two boost packs of which the aft unit is attached to the IM machine shown in this figure. Also shown are the DELTA ELEKTRONIKA SM400-AR-4 power supply and DC-true probe (attached to one phase of the sensorless machine) used for this laboratory. The encoder of the PM machine is used to provide actual shaft speed and shaft angle information required for the sensored drive. Development phases C and C+ are shown in this laboratory.

Fig. 5.51 Dual IM/PM drive setup with LAUNCHXL-F28069M board

5.6.1 *Lab 3:5: Phase C*

For dual operation the controller is configured to activate both boost packs simultaneously so that ADC offset correction for the (common) ADC module can be undertaken before both controllers become operational. The following information is relevant for this laboratory component:

- Reference program [11]: `lab3-5_LaunchXLphCv2.vsm`.
- Description: Dual control of a sensored PM and sensorless IM machine.
- Equipment/Software: LAUNCHXL-F28069M, BOOSTXL-DRV8301 module ('aft' position), BOOSTXL-DRV8301 module ('forward' position), DC power supply and VisSim simulation program.
- Outcomes: Develop a complete drive algorithm, which handles dual machine control. Compile and download and `.out` file for use in phase C+.

The 24 V DC power supply to be used for this laboratory must be connected to both BOOSTXL-DRV8301 modules in this case. Furthermore, the following jumper and dip-switch settings on the LAUNCHXL-F28069M module are required:

- Jumpers `JP1` and `JP2` OPEN
- Jumpers `JP4` and `JP5` OPEN
- Jumpers `JP3`, `JP6` and `JP7` CLOSED
- Dip-switches `SW1` to `SW3` ON

For this laboratory the 'J4' encoder connector of the PM machine (which has a Molex connector) cable should be attached to encoder input `QEP_A` of the LAUNCHXL-F28069M module.

A phase C development stage, cannot be used to run a drive, but its primary task is to assemble all the modules needed for drive operation with a given controller configuration. The drive setup as given in Fig. 5.52, shows the 'Dual Drive controller with FAST-IM and sensored PM' module, which must be compiled to generate an `.out` file. Inputs to the controller module (via the pre-scale modules) are six variables (shown as 'constants' for simplicity), which allow the user to set the quadrature reference current for both controllers and the 'differential' direct axis current for the IM controller. The term 'differential' refers to the fact that the FAST controller identifies the required magnetizing current $i_d^{\rm rat}$ for the machine (based on a user defined stator flux reference value), hence the direct axis reference slider is only used to change the $i_{\rm d}$ value relative to this value. In addition, two inputs have been

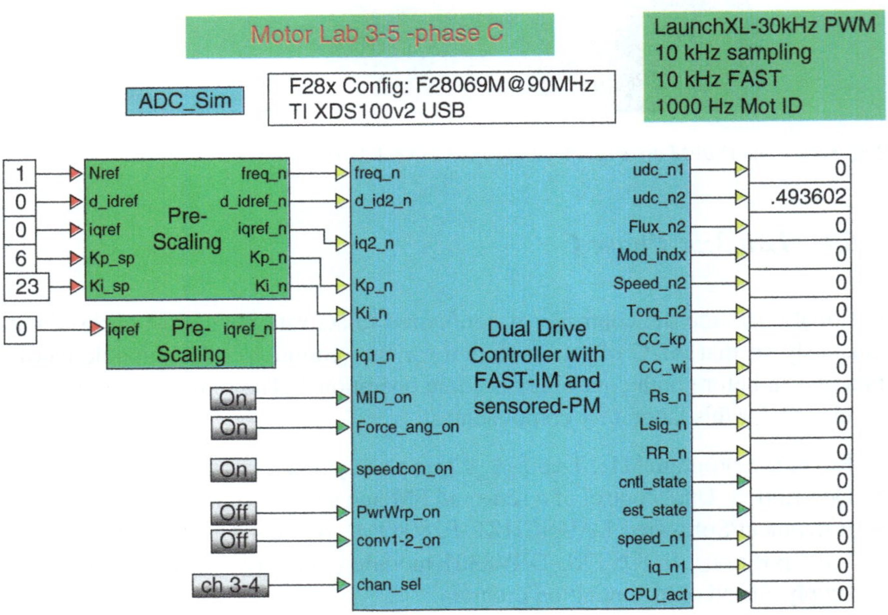

Fig. 5.52 Phase C simulation of a dual IM/PM machine drive

assigned to set the speed gain `Kp_n` and speed bandwidth `Ki_n` for the induction machine drive. Six logic inputs are present, of which `conv1-2_on` is used to activate both boost packs sequentially and undertake ADC offset compensation, after which both controllers are enabled. Three logic inputs: `speedcon_on`, `Force_ang_on` and `MID_on` are used to select speed/torque control, force angle operation and motor identification for the sensorless drive as discussed in laboratory 3:2 (see Sect. 5.3). The buttons connected to the `chan_sel` and `PwrWrp_on` inputs are used to select the diagnostic channels to be displayed and enable PowerWarp operation respectively. Of the sixteen outputs on the Controller module, two are used to show per unit DC bus-voltages `udc_n1`/`udc_n2` of the 'fwd' and 'aft' boost packs respectively. Four outputs are allocated for IM drive operational purposes namely: estimated rotor flux `flux_n2`, current controller modulation index amplitude `Mod_indx`, estimated shaft speed `Speed_n2` and estimated shaft torque `Torque_n2`. An additional five outputs are allocated for IM motor parameter identification purposes namely: current controller gain `CC_kp`, current controller bandwidth `CC_wi`, stator resistance `Rs_n`, leakage inductance `Lsig_n` and rotor resistance `RR_n`. Two outputs `cntrl_st` and `est_state` have been added to monitor operation of the controller and estimator respectively. The remaining three as yet unnamed outputs provide the encoder shaft speed `speed_n1`, quadrature current of the PM machine (used to estimate the torque, given that the PM flux level is known) and `CPU_act` which is used to provide a percentage indication of CPU activity during phase C+ operation.

Moving one level down into the Controller module leads to the set of modules/dialog boxes shown in Fig. 5.53. A 'Diagnostic tool' module is again used to buffer two variables, which can subsequently be displayed in a graph when operating in phase C+. A two channel multiplexer allows the user to either display variables `Testch1`, `Testch2` or `Testch3`, `Testch4` by using a button connected to the `chan_sel` input as discussed above. A '100 Hz' module acts as a 'heart beat' and flashes a 'red' LED on the LAUNCHXL-F28069M module. A 'blue' LED also on

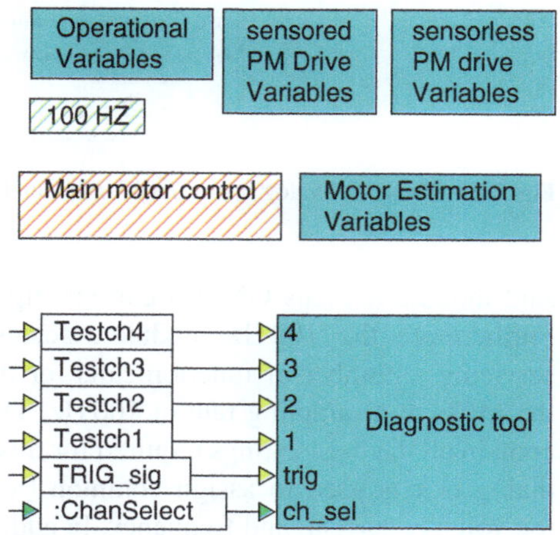

Fig. 5.53 One level down into the controller module

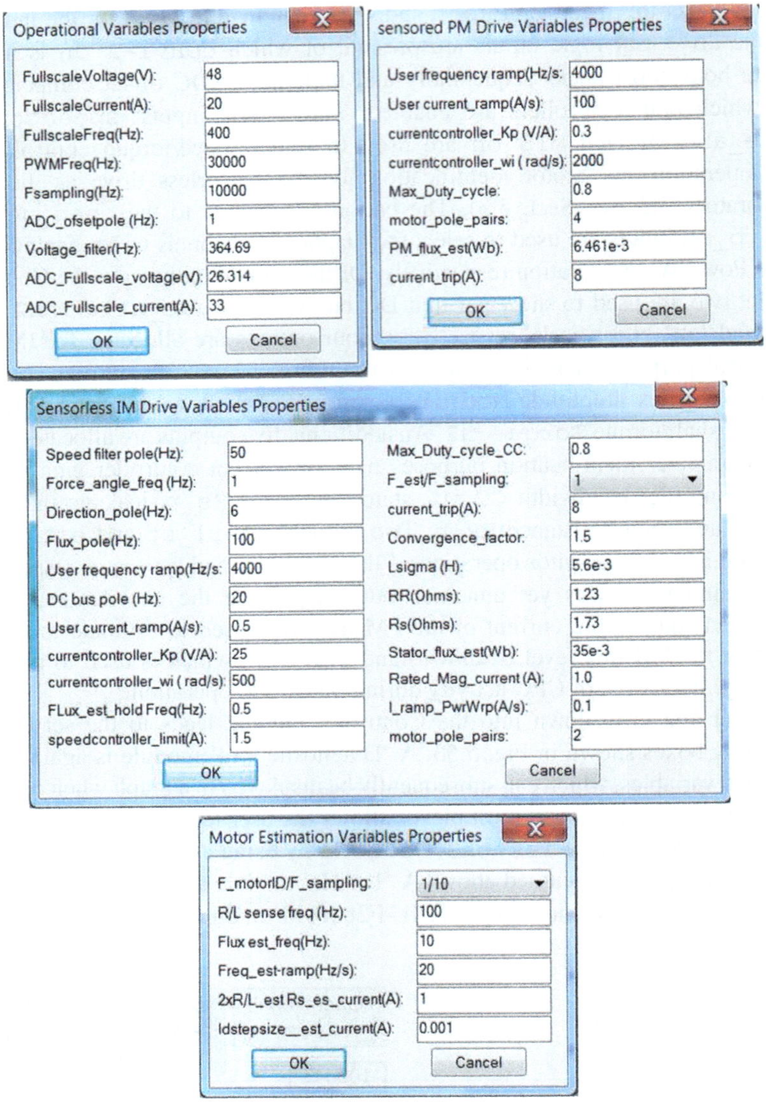

Fig. 5.54 Phase C; Dialog boxes used for the dual machine drive

said module remains ON if a current trip condition occurs on either boost pack. Furthermore, the '100 Hz' module executes all the drive background tasks such as servicing a number of function calls for the InstaSPIN algorithm in the ROM of the MCU at a sampling rate of 100 Hz. The four dialog boxes given in Fig. 5.54, contain all the user settings required for this laboratory. The 'Operational Variables' dialog box is used to assign common (to both controllers) full scale values for the voltage, current, and frequency. In addition, the ADC full-scale current/voltage values, PWM/sampling frequencies and corner frequency of the low-pass filters on the boost packs are assigned here. Note that in contrast to earlier laboratories in this chapter the PWM and ADC sampling frequency (which must also be set in the main VisSim menu) are set to 30 kHz and 10 kHz respectively. Hence a lower sampling frequency is used (when compared to single motor labs which use 15 kHz), in order to gain extra CPU time given the need to execute two controller algorithms.

The Dialog box 'sensored PM Drive Variables', assigns the parameters needed for the sensored PM controller. Hence, current controller gains, maximum modulation index and PM flux amplitude (assumed known in this case) which must be provided by the user. The two dialog boxes: 'Sensorless IM Drive Variables' and 'Motor Estimation Variables' are used by the sensorless drive and define the parameters needed for 'standard operation' as well as motor identification. Observation of Fig. 5.54 learns that the FAST estimation algorithm is set to operate at 10 kHz. For both drives the maximum duty cycle value has purposely been set to 0.8 because current measurement is done with the aid shunt resistors. This implies that modulation values in excess of 0.8 can impair current measurement, if no measures are taken (as is the case here) to counter this issue. So called 'current reconstruction' [9] measures can be deployed to solve this problem, which allows drive operation with a modulation index greater than 0.8. Furthermore, the rate of change of control variables and pole pair number (NOT equal for both machines) is set in the corresponding dialog boxes for both controllers.

Moving one level lower into the 'Main motor Control' module reveals a set of modules, as shown in Fig. 5.55. Of these, the 'PM FOC controller' and 'InstaSPIN FOC' modules are used to implement sensored FOC control and sensorless FOC control, as discussed in laboratories 1:3 and 3:2 respectively. The 'ADC-PWM' unit shown, generates the α, β voltage/current variables bus voltages for both boost packs. Furthermore, a PWM frequency of 30 kHz is used, together with a sampling frequency of 10 kHz, which corresponds to an ADC sampling time of 0.1 ms, that is set in the VisSim 'System Properties' pull down menu. The variable `control_enable1` activates the 'PM FOC controller' after the converters have been turned on (using the `conv1-2_on` button) and internal ADC/PWM initialization actions have been performed. A similar variable `ctrlSetEnable` (located in the '100 Hz' module) activates the sensorless controller. Relevant (to this laboratory) inputs of this unit for the sensored drive are the DC bus voltage `DCbus_n1`, and the per unit currents i_{α}^{n}, i_{β}^{n}, represented by the variables `ia1`, `ib1` respectively. The sensorless drive requires the α, β currents/ low-pass filtered voltage components as represented by the variables `i_alp2`, `i_bet2` and `u_alp2`, `u_bet2` respectively. In addition, the FAST drive also uses the DC bus voltage variable `DCbus_n2` of the aft boost pack. A module `Mod_indx(abs)` connected to the modulation outputs of the 'InstaSPIN-FOC' module, calculates the amplitude of the modulation index vector. In phase C+ this variable is multiplied by half the DC voltage which then gives the peak reference voltage amplitude generated by the 'InstaSPIN-FOC' current controller (used for diagnostic purposes only).

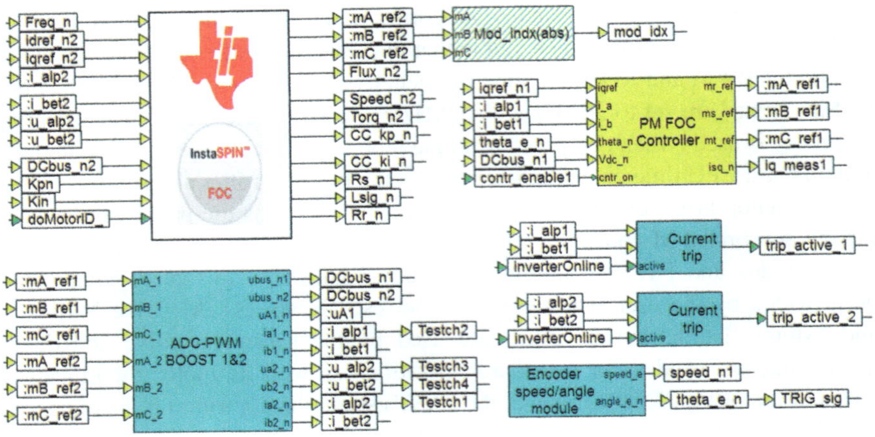

Fig. 5.55 One level down into the 'Main motor control' module

An 'Encoder speed/angle' module, generates the per unit electrical shaft speed/angle variables `speed_n1`, `theta_e_n`, which are used by the sensored FOC PM drive. Furthermore, the encoder speed variable is made available to the user during phase C+ operation, as to allow a comparison with the estimated shaft speed. The inputs to the 'ADC-PWM' module are the modulation indices generated by the 'current/speed' and 'InstaSPIN-FOC' controller modules. Note that in this case the ADC module samples six converter currents and voltages, in addition to two DC bus voltages, where use is made of so called 'dual sampling', which implies that two signals are sampled simultaneously to reduce the time needed to acquire all the data. For diagnostic purposes the variables `TESTch1`, `TESTch2` have been allocated to per unit α current components of both PM and IM machines respectively. Likewise, the variables `TESTch3`, `TESTch4` have been allocated to per unit β current components of both machines.

5.6.2 *Lab 3:5: Phase C+*

Phase C+, is the drive operational component of the laboratory and is basically a run version of the `.out` file compiled and downloaded to the MCU in phase C (see previous subsection).

The following information is relevant for this laboratory component:

- Reference program [11]: `lab3-5_LaunchXLphCv2_d.vsm`.
- Description: Dual operation of a sensorless FOC IM and sensored FOC PM drive.
- Equipment/Software: LAUNCHXL-F28069M, BOOSTXL-DRV8301 module ('aft' position) connected to Texas Instruments LVACIMTR IM machine, BOOSTXL-DRV8301 module ('forward' position) connected to LVSERVOMTR PM machine, DC power supply and VisSim simulation program.

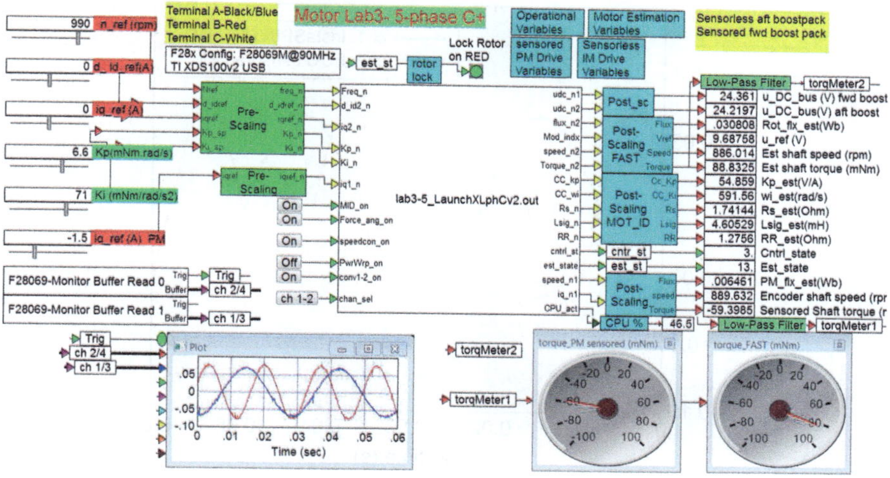

Fig. 5.56 Phase C+ dual operation of a IM/PM drive

- Outcomes: To investigate dual machine drive operation, using a FOC sensored PM machine drive and sensorless FOC induction machine drive.

Note that the required jumper and dip-switch settings for the LAUNCHXL-F28069M module are given in phase C. Furthermore, the 'J4' encoder connector of the sensored PM machine cable should be attached to encoder input `QEP_A` of the LAUNCHXL-F28069M module.

The run version shown in Fig. 5.56 uses a VisSim run module, which executes the `.out` file, shown in said module. An enlarged view of the VisSim scope module given in Fig. 5.57 shows the per unit currents i_{α}^{n} of both machines after completion of identification. Multiplication by the full scale current value of 20 A gives actual values.

Prior to activating the drive the following 'Pre-Drive' check list should be executed:

- Dialog boxes used in C+ model must match those of C (and vice-versa): the run version is compiled with the dialog box entries specified under phase C. The dialog boxes shown in C+ are used by the 'Pre/Post' scaling modules.
- Ensure that the sample time used is correct and latest (and correct) `.out` file has been downloaded to the MCU (Right mouse click on MCU module to show dialog box of these variables).

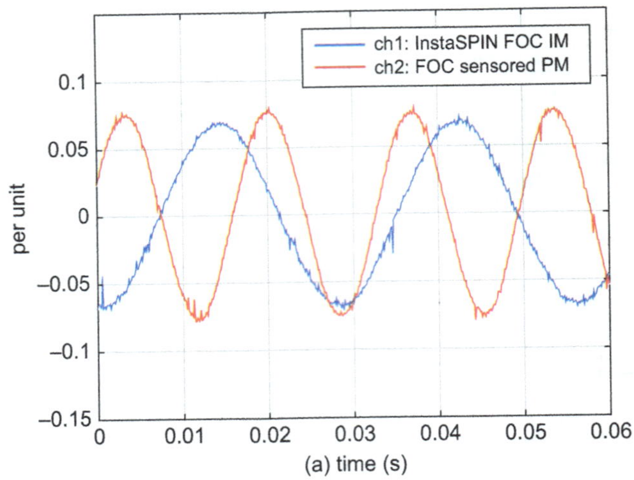

Fig. 5.57 Phase C+ simulation: Scope display showing per unit α currents of the sensorless IM and sensored PM machine

- Confirm that the user input values are set to either zero, or 'acceptable' values, which will not cause a current trip of the converter. In particular ensure that the current reference for the sensored drive is set to zero (this to ensure that the motor identification sequence of the sensorless machine is carried out under no (external) load).
- Confirm that the converter 'switch' is OFF and the power supply is on (DC bus voltage present on both boost-packs).
- Confirm that the motors are connected (to the appropriate converter) firmly and properly.

After completion of the Pre-Drive checklist, activate the program and confirm that the supply voltage reading is 24 V on both boost packs. If not stop the program and restart. With the correct DC voltage levels present, turn on the converters (using the ON button, for both converters) and monitor drive operation. After motor identification of the sensorless machine, monitor the diagnostic scope, which in this example (Fig. 5.56) is set to show the per unit α current components of both machines. The screen shot given in Fig. 5.56 confirms that the sensorless IM machine is operating under speed control, while the sensored PM machine, is operating under FOC torque control and acting as a load, i.e. operating in generator mode with a torque of ≈ -60 mNm. Note that the torque generated by the sensorless FOC drive is equal to ≈ 80 mNm. The sum of these two readings represents the friction torque of the dual drive, which is ≈ 20 mNm at the instant this screen shot was taken. A `CPU_act` variable provides a percentage indication of CPU activity.

This laboratory session concludes the chapter on sensorless IM control. The reader is reminded of the fact that only the most relevant information related to these laboratories has been provided. A thorough understanding of the material presented requires an effort on the part of the reader to experimentally work through the laboratories shown.

Chapter 6
VisSim Case Studies

In this chapter three VisSim based industrial sensorless drive examples will be discussed which are 'FAST' prepared. This implies that the required voltage/current measurement structure has been put in place in these converters to allow sensorless operation using InstaSPIN technology. The aim is to show the reader the developmental steps needed to arrive at a fully operational drive. In all three cases the starting point is a given industrial drive concept, of which relevant circuit aspects will be shown as needed to achieve commissioning and subsequently, sensorless operation of either a permanent magnet or induction machine. It is hoped that the content of this chapter will assist the reader in making the transition from the laboratory examples given in the previous two chapters to his or her specific sensorless drive implementation. Each case consists of two subsections, which describe the commissioning and sensorless implementation process. These steps have been discussed in previous chapters (for different drives), hence only the key steps needed to achieve the desired tasks will be discussed. The reader should refer to the previous two chapters for a more in depth analysis of the topics discussed in this 'case study' chapter. Most importantly, the examples shown in this chapter underline specific topics, which go beyond the material presented in the previous chapters. Consequently, the solutions provided are tailored to the applications to be discussed.

6.1 Case Study V1: Helicopter Drive

The purpose of this project is to evaluate a FOC sensorless converter that can be used to power the helicopter introduced in Chap. 2 (see Fig. 2.1). This type of

---

Electronic supplementary material: The online version of this chapter (doi: 10.1007/978-3-319-20043-9_6) contains supplementary material, which is available to authorized users.

D.W.J. Pulle et al., *Applied Control of Electrical Drives*,
Power Systems, DOI 10.1007/978-3-319-20043-9_6

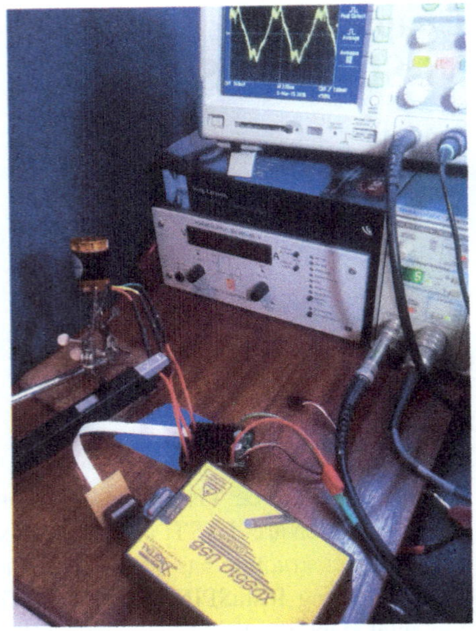

Fig. 6.1 FOC sensorless controller for an electric model helicopter

Table 6.1 Walkera: WK-WS-28-007A measured line to line parameters

Parameter	Value	Units
Inductance: L_{sLL}	2.37	μH
Resistance: R_{sLL}	0.062	Ω
Pole pairs: p	3	–

application is popular due to the advent of drone technology which requires access to compact, high performance and efficient electrical drive technology. Shown in Fig. 6.1 is an electric helicopter motor, which is connected to an InstaSPIN FOC sensorless controller. A current probe, also shown in this figure, is used to measure the phase current. A mini JTAG connection on the controller interfaces with the Spectrum Digital XDS510 USB JTAG, which in turn is connected to the laptop that runs the VisSim software. A 12 V battery (not shown) is used to provide power for the drive. In addition to the above, the FOC controller is provided with a serial interface that can, in principle, be connected to a telemetry unit (not used in the present setup). Using this approach the pilot (on the ground) could, for example, observe the torque generated on the rotor blades as well as the main rotor shaft speed on the transmitter display. In addition, an indication of motor temperature could be provided, which is based on an 'online' estimate of the motor stator resistance. Prior to connecting the machine to the drive some provisional measurements were carried out with the aid of a L-C-R meter as given in Table 6.1. The inductance and resistance value indicated were measured with the L-C-R meter (set to 1 kHz), that was connected between two phases of the machine. According to these measurements, the stator resistance and inductance were found to be $R_s = {}^{R_{sLL}}/_{2} = 0.032\,\Omega$ and $L_s = {}^{L_{sLL}}/_{2} = 1.2\,\mu$H respectively. Furthermore, the

DC supply for the helicopter drive is $u_{DC} = 12$ V, while the anticipated maximum operating speed is approximately $n_m = 48000$ rpm, which corresponds to a main rotor blade speed of 4160 rpm (given the main gear/motor shaft ratio of 150:13 present in the helicopter shown in Fig. 2.1). On the basis of the DC bus voltage and motor speed, an initial indication of the motor flux can be found using $\psi \approx {}^{u_{DC}}/_{(2\pi f_e)}$, where f_e is the electrical shaft frequency, which is equal to $f_e = {}^{n_m\,60}/_p$ Hz. On the basis of the data available, the maximum electrical frequency and initial flux estimate for the machine are $f_e^{max} = 2400$ Hz and $\psi \approx 0.8$ mWb, hence a low inductance, low flux and low inertia application, all of which point to a challenging engineering problem.

Given the low motor time constant (order of 37 μs), it is prudent to choose a PWM frequency of 45 kHz (which is higher than the chosen 15 kHz controller sampling frequency) in order to reduce current ripple. On the other hand, selecting an even higher converter switching frequency will increase switching losses, which can lead to thermal problems, hence a compromise is required. Note also that it is good practice to ensure that the ADC sampling frequency is at least a factor five higher than the highest anticipated operating frequency of this drive. Hence in this case an ADC sampling frequency of 15 kHz is desirable. Choosing an even higher sampling frequency may lead to MCU timing issues, i.e. insufficient time during a sampling interval to execute the required tasks.

6.1.1 *Case Study V1a: Helicopter Drive: Commissioning, Phase C*

Prior to using current control in any application, it is helpful to consider drive operation under voltage control first. Using this approach, the voltage and current measurements can be examined. Furthermore, the phase relationship between voltage and current can be established, which is important prior to using current control. The following information is relevant for this case study:

- Reference program : `Case1a_027phCv8.vsm`.
- Description: Commissioning of helicopter drive.
- Equipment/Software: Helicopter board from 'A2Spin,LLC' [12] with Texas Instruments F28027F MCU, Spectrum Digital XDS510 USB Emulator and VisSim simulation program.
- Outcomes: generate the `.out` file needed to represent the drive structure and calculate the full scale ADC values as well as LPF frequency of the voltage filter.

Note that the use of an F28027F MCU and its Flash memory implies the need to use an alternative program for downloading the generated `.out` file as VisSim *cannot* be used for this purpose (only downloads to RAM are possible). The voltage controller as discussed in Sect. 3.1 is used for this analysis. However, emphasis in this section will be on configuring the ADC-PWM module, triggering, allocating

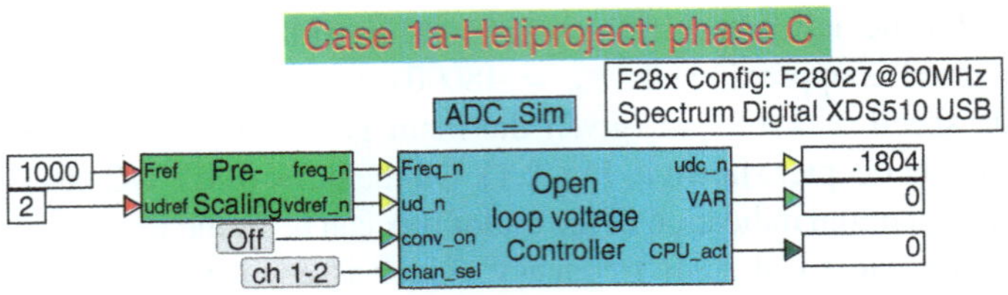

Fig. 6.2 Development Phase C: voltage controller set up for helicopter drive

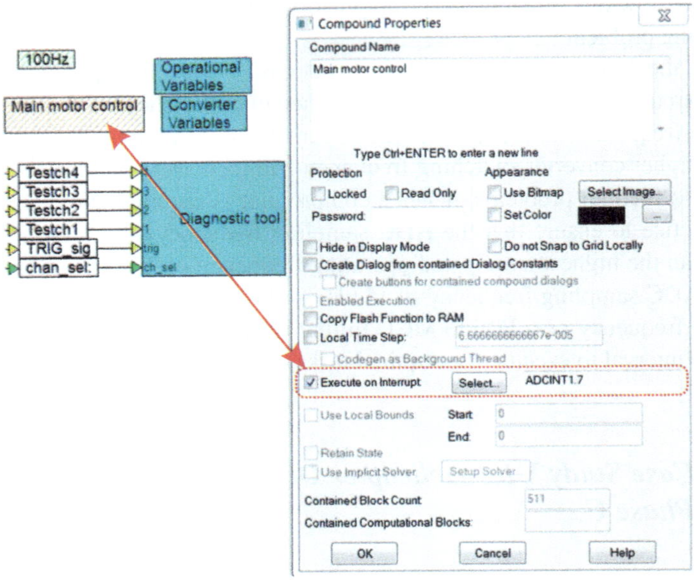

Fig. 6.3 Development Phase C, one level into 'Open loop voltage controller'

ADC channels and confirming that the current/voltage waveforms are correct in terms of amplitude and phase relationships. The phase C development stage shown in Fig. 6.2, has been adapted for this application. Inputs are the electrical frequency and direct axis voltage reference variable. Furthermore, a channel select input `chan_sel` has been added to toggle between diagnostic scope channels 1, 2 and 3, 4 in phase C+. Moving one level down into the 'Open loop voltage controller' reveals a set of modules and dialog boxes as indicated in Fig. 6.3. This figure also shows a dialog box which appears upon a 'right click' on the 'main control module'. The critical entry in this case is the 'execute on interrupt' feature which must be set as shown. This implies that the module in question will be executed on an interrupt signal set by the ADC unit. More specifically, the interrupt is in this case `ADCINT1:7`, which implies that the signal is generated by ADC input channel 7. The ADC input channels are allocated within the ADC-PWM module, which appears when moving one level lower into the 'main control module' as shown in Fig. 6.4. The modules shown are as discussed in Sect. 3.1. However, in

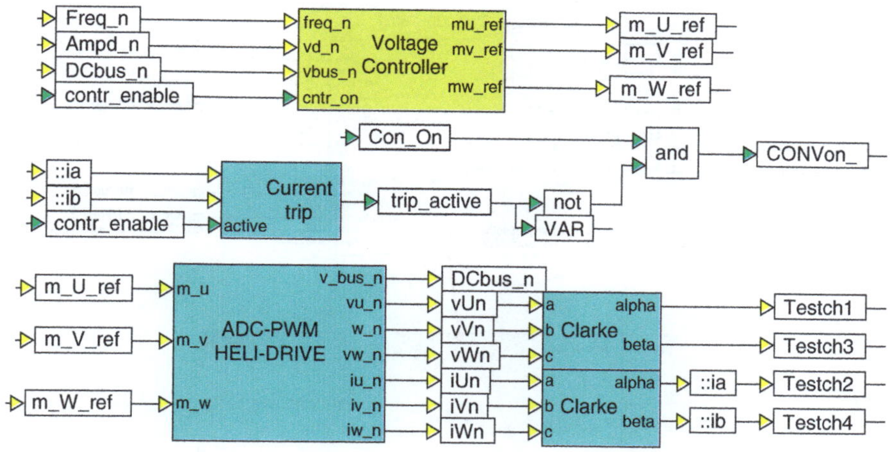

Fig. 6.4 Development Phase C: one level into 'main motor control' module

Table 6.2 Allocation of input variables to ADC input channels

Input variable	ADC input channel
Phase current i_U	2
Phase current i_V	3
Phase voltage u_U	4
Phase voltage u_V	5
Phase voltage u_W	6
Bus voltage u_{DC}	7

this case configuring the ADC-PWM module is of interest. This type of module is board specific, which is why the board type is included in the title. Moreover, this module outputs the phase current/voltages instead of the Clarke converted variables used in previous chapters. The InstaSPIN module can also accept the phase variables and in this case these are used directly to reduce MCU time, i.e. to avoid a forward/reverse Clarke transformation. Data acquisition via VisSim, is done by way of a MCU specific 'input channel' module, such as `F28027F Input Channel` in which the user has the option of selecting an 'analog or digital' channel. Furthermore, the user must allocate a channel number (within the range of analog channels available) as mentioned above. Table 6.2 shows the channel allocation used in this laboratory. Note that the allocation of phase currents and voltages to input channels is up to this stage done arbitrarily. Observation of said table shows that channels (also referred to as 'conversion results') 0 and 1 are not used, as these are the first two ADC analog channels and may contain corrupted data [13], hence they are discarded here. Note also that current i_W is not measured because no current transducer has been allocated, given space limitations on the board. Instead, the current i_W is calculated from the two measured currents using $i_W = -(i_U + i_V)$, which takes advantage of the fact that the sum of the three phase currents must be zero. A critical commissioning step is to allocate the channel

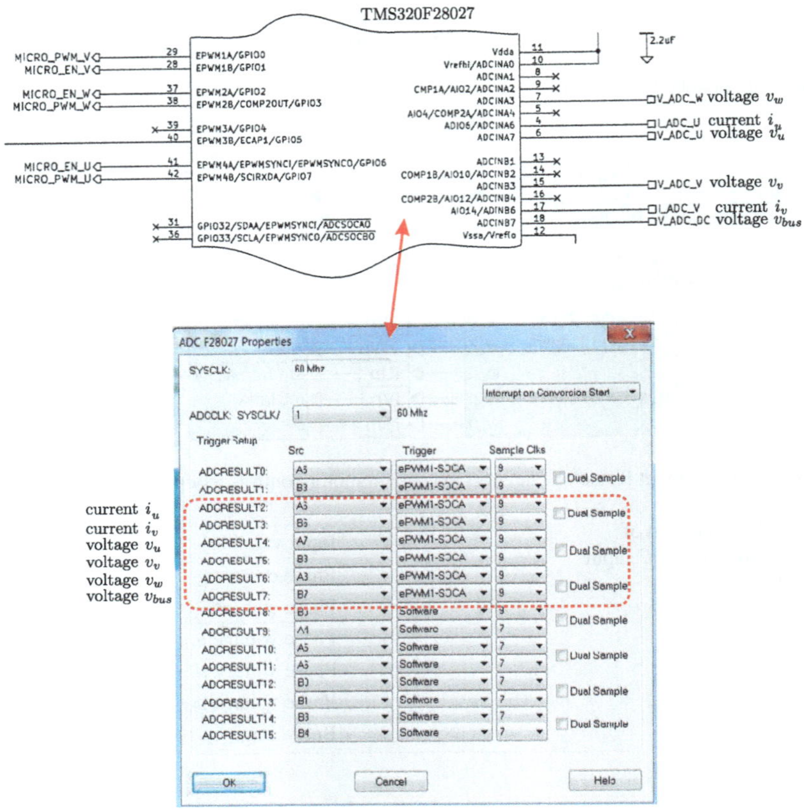

Fig. 6.5 Development Phase C: 'ADC properties' dialog box and link to circuit diagram

numbers introduced in Table 6.2 to actual MCU variables. This requires access to the `ADC F28027F Properties` dialog box, found by selecting `Embedded` in the main VisSim menu and then selecting `F280x` and finally `ADCconfig` which brings up the module shown in Fig. 6.5. Also given in this figure is part of the circuit diagram of the board under consideration, which shows the pin allocation of the current/voltage variables. The helicopter board makes use of the TMS32028027F processor shown (partly) in Fig. 6.5. Associated with the current/voltage variables and MCU pin numbers are alphanumeric variables which need to be assigned in the `ADC F28027F Properties` dialog box, as shown. In addition to assigning the input variables, the trigger signal for the input channels needs to be allocated. Triggering of the ADC is linked to the PWM module as was discussed in Sect. 2.1.6. For all the input channels the 'Start Of Conversion' signal `ePWM1-SOCA` is used, which implies that the pulse width modulation module 1 generates the trigger for the input ADC channels in use.

Moving one level lower into the 'ADC-PWM' module shown in Fig. 6.4 reveals three PWM modules (see Fig. 6.6), which have as inputs the modulation index

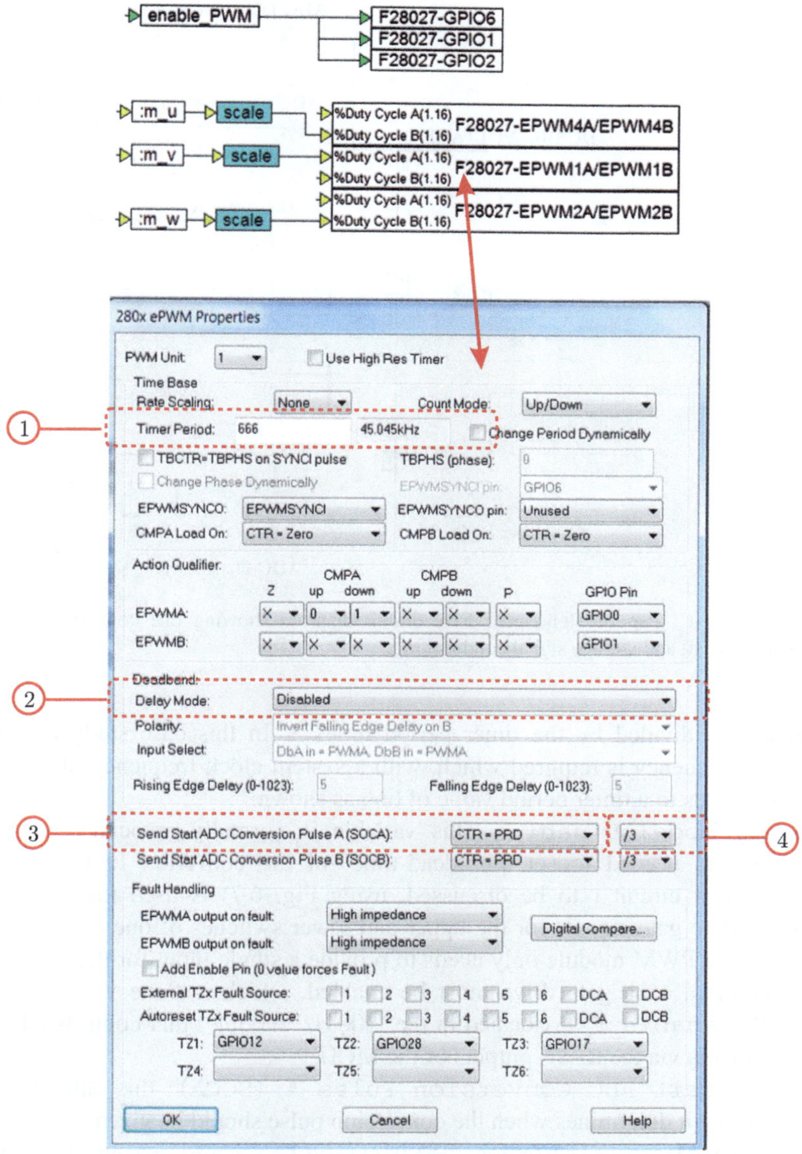

Fig. 6.6 Phase C: 'PWM properties' dialog box

variables m_u, m_v and m_w respectively. A 'right mouse click' on a module reveals a dialog box, which is also shown in Fig. 6.6. The most important entries are numbered and need to be configured as follows:

1. `Timer Period`: represent the number of cycles of the PWM staircase as discussed in Sect. 2.1.6. The number to be entered here is defined by the clock

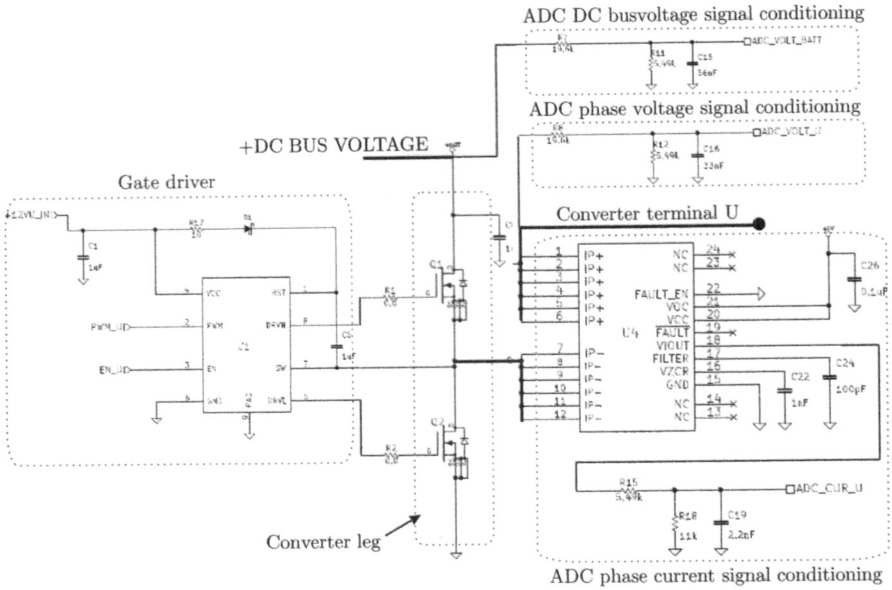

Fig. 6.7 Phase C: Part of Helicopter drive circuit diagram showing one converter leg, with associated current and voltage signal conditioning

frequency divided by the timer period times 2. In this case study a 45 kHz PWM frequency is required which, with a system clock frequency of 60 MHz, corresponds to a timer period value of 666 as shown.

2. `Delay mode: Disabled` This variable is normally associated with the parameters needed to set the 'dead-time' of the converter. In this drive a gate driver circuit (to be discussed, using Fig. 6.7) is used to provide the appropriate gate signals for the upper and lower switches of one converter leg, hence one PWM module only needs to provide a single input for the gate drive. In addition, each gate drive must be enabled, which is done via a software variable `enable_PWM` (located in the '100 Hz' module) that controls all three gate drivers via the `GPIO` outputs of the MCU.
3. `send start ADC Conversion Pulse A (SOCA)`: this entry is critical, in that it determines when the conversion pulse should be given with respect to the PWM cycle. The option `PRD` is chosen because triggering of the ADC should occur at the top of the PWM staircase, as shown in Fig. 2.23. Note that in Fig. 6.5 ADC triggering refers to `ePWM1-SOCA`, which implies that EPMW module 1 generates the SOC signal for all ADC channels.
4. `3`: this entry specifies if the ADC pulse should be given after every first, second or third PWM clock cycle. In this case a 45 kHz PWM frequency has been selected and the ADC sample frequency in use is 15 kHz (which is also the rate at which the controller module will be executed, with a sampling frequency set in the main VisSim module) hence the PWM to ADC ratio must be set to $^1/_3$.

Note that only a subset of the ADC and PWM dialog entries have been discussed, given their importance. The reader is however, advised to refer to the relevant Texas Instruments application notes [13] for further details on these modules.

Computation of the ADC full scale voltage/current parameters and the low-pass filter corner frequency are critical to successful sensorless operation. Accordingly these parameters must be either given or derived by careful examination of the schematic diagram of the converter under consideration. Part of the relevant (to this topic) circuit diagram [12] sections of the helicopter drive board, as given in Fig. 6.7, will be used to derive the required data.

Central to this figure is one leg of the three-phase converter, which is connected to a 'Gate driver' circuit that controls the two power electronic devices. Inputs to the gate drive module are the variable `PWM_U` which determines whether the top or bottom switch is activated and a variable `EN_B` which enables said unit. The latter variable is controlled via a `GPIO` output (see Fig. 6.6), which in turn is controlled by the `enable_PWM` signal that is set by the controller when converter activation is required. A Hall-effect current sensor `U4` is used to measure the phase current. This device provides galvanic isolation and generates an output voltage which in turn is linked to an attenuation network that consists of resistors `R15` and `R18`. The output of this network is the variable `ADC_cur_U`, which represents the inverted phase current 'U'. The variable `ADC_cur_U` is the input voltage to the ADC unit shown in Fig. 6.5. The ADC full scale current is the (peak to peak) current value that corresponds to a (peak to peak) ADC input voltage of 3.3 V. Analysis of the current signal conditioning circuit shows that the ADC full scale current is found using

$$ADC\,full\,scale\,current\,(\mathrm{A_{pp}}) = 3.3\left(\frac{(R15 + R18)}{R18\,K_{\mathrm{HeC}}}\right) \tag{6.1}$$

where $R15$ and $R18$ are the relevant resistors of the current gain circuit. The variable K_{HeC} is the transconductance gain of the Hall-effect sensor, which according to manufacturer specifications is equal to 28 $^{\mathrm{mV}}/_{\mathrm{A}}$ for the device in question. Substitution of the resistance values (as shown in the diagram) and transconductance value into Eq. (6.1) yields an ADC full scale current value of 176.6 A. Note that the current signal conditioning circuit inverts the output voltage `ADC_cur_U`, which must be corrected by software, given that a positive voltage is represented by an outgoing current from the converter terminal `U`.

Computation of the ADC full scale voltage requires evaluation of the ADC phase and DC bus voltage conditioning circuits. For both circuits, the same voltage attenuation is used as may be observed from Fig. 6.7. For the phase voltage circuit the attenuation is defined as $^{R12}/_{(R12+R8)}$, whereas for the DC bus circuit attenuation is set to $^{R11}/_{(R11+R7)}$. Substitution of the resistance values, learns that the attenuation of both circuits is indeed equal. The ADC full scale voltage is the peak to peak phase voltage that corresponds to a peak to peak ADC input voltage of 3.3 V. Analysis of the phase voltage conditioning circuit, shows that the ADC full scale voltage is found using

$$\mathit{ADC\,full\,scale\,voltage}\,(\mathrm{V_{pp}}) = 3.3\left(\frac{R8 + R12}{R12}\right) \tag{6.2}$$

where $R8$ and $R12$ are the relevant resistors of the voltage gain circuit. Substitution of the resistance values (as shown in the diagram) into Eq. (6.2) yields an ADC full scale voltage value of 15.08 V. Note that the analysis shown, could also have been done using the DC bus voltage signal conditioning circuit, as that would have given the same result. In conclusion, the reader is reminded of the fact that the phase voltage signal conditioning circuit is also a low-pass filter, of which the corner frequency may be found using

$$\text{LPF corner freq (Hz)} = \left(\frac{1}{2\pi}\right)\left(\frac{R8 + R12}{R8\,R12\,C16}\right) \tag{6.3}$$

Substitution of the resistance and capacitance values (as shown in the diagram) into Eq. (6.3) yields a low-pass corner frequency value of 662.6 Hz.

6.1.2 Case Study V1a: Helicopter Drive: Commissioning, Phase C+

Phase C+, is the operational component of the laboratory and is basically a run version of the `.out` file compiled in phase C (see previous subsection). The `F28027` MCU has less RAM than the `F28069M` used in the LaunchXL-069M module (as used in the previous chapters). Consequently, the `.out` file generated in phase C has to be loaded to the Flash memory of the CPU. This *cannot* be done directly via VisSim, hence use needs to be made of an alternative approach. In this case use is made of the Texas Instruments Code Composer (CCS) environment to be discussed more extensively in the next chapter.

The following information is relevant for this laboratory component:

- Reference program: `Case1a_027phCv8_d.vsm`.
- Description: Commissioning of helicopter drive.
- Equipment/Software: Helicopter board from 'A2Spin,LLC' [12], Texas Instruments F28027F MCU, Spectrum Digital XDS510 USB Emulator, Walkera WK-WS-28-007A PM motor and VisSim simulation program.
- Outcomes: to confirm basic drive operation by observing the voltage and current waveforms.

The run version as shown in Fig. 6.8 uses a VisSim run module, which executes the `.out` file, shown in said module. Two sliders are used, the purpose of which was discussed in the previous section. A post-scaling module is used to convert the per unit measured DC bus voltage to actual voltage, as shown with a numeric display. Furthermore, a variable `VAR` has been added, which serves to display internal variables, as required. In this case the variable is set to show the 'trip active'

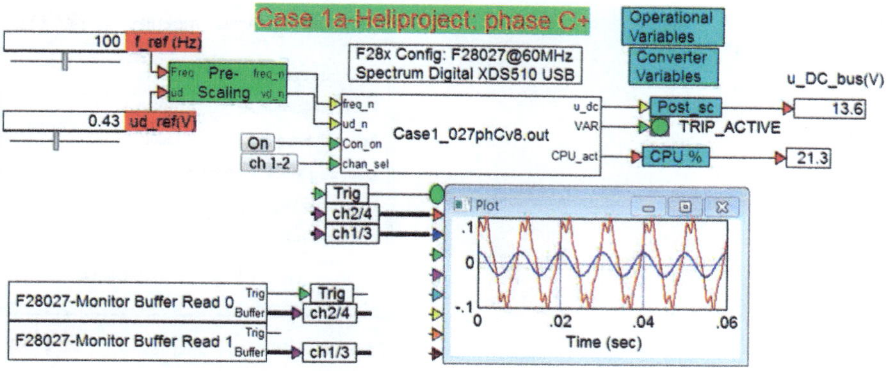

Fig. 6.8 Phase C+: Embedded voltage controller

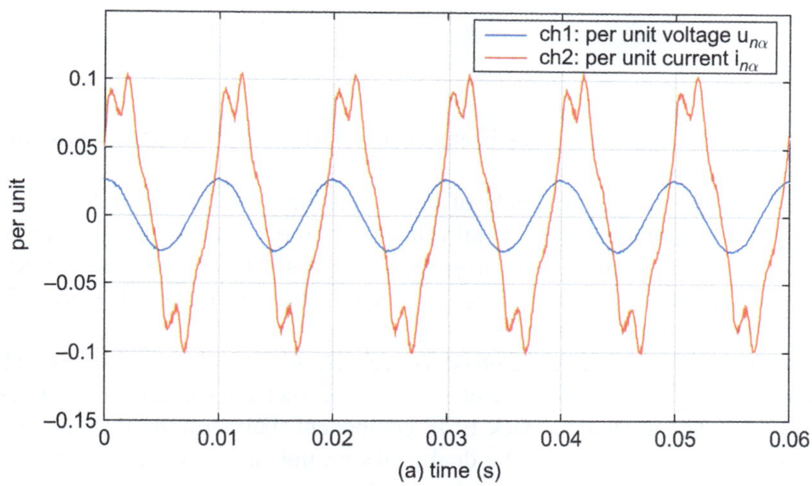

Fig. 6.9 Phase C+ simulation: Scope display showing per unit α current and voltage (Helicopter motor running with friction load only)

signal which is logic '1' when a current trip occurs. A third output `CPU_act` has been added which provides an accurate indication of the CPU activity.

Two 'Monitor Buffer' modules are used to display two selected diagnostic signals. An enlarged view of the VisSim scope module given in Fig. 6.9 shows the per unit current i_{n_α} and voltage u_{n_α}. Multiplication by the full scale current value of 100 A and 15 V for the per unit current and voltage respectively, gives the full scale values. The dialog box entries for this drive are shown for completeness in Fig. 6.10. Note that the ADC full scale shown are those which have been calculated in the previous development phase. Several important observations can be made on the basis of the information provided in Figs. 6.8, 6.9 namely:

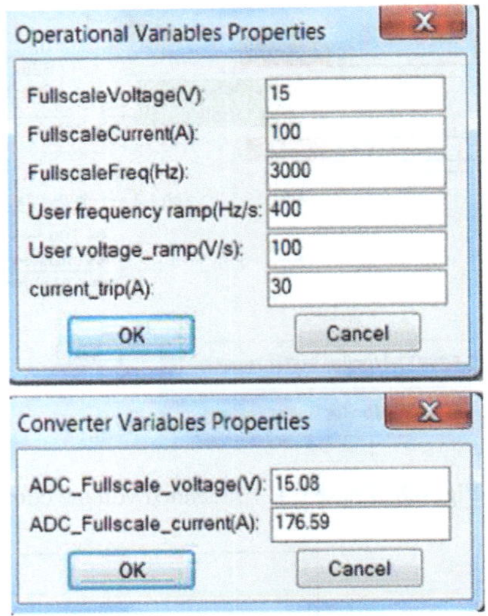

Fig. 6.10 Phase C+: dialog boxes for helicopter drive

- The reference frequency has been set to 100 Hz hence the period time of the waveforms should be 10 ms.
- A reference voltage amplitude of 0.43 V has been set by the slider `ud_ref` hence the amplitude of the per unit voltage waveform u_α should be approximately (as dead-time effects and the drop across the converter switches will affect the amplitude shown) equal to $^{0.43}/_{u_{fs}} = 0.0286$, where u_{fs} is the full scale voltage value which was set to 15 V.
- The current waveform has a phase lag relative to the voltage waveform which is expected given the presence of an inductive load and (in this case) a rotating machine, which causes a back EMF component. Furthermore, the EMF is not sinusoidal as is evident by the dual peaks on the current waveform. Note that the phase relationship between the β voltage/current waveforms must also be checked.

Finally, observe that the motor is rotating correctly. With a reference frequency of 100 Hz the speed of the 6 pole motor will be 2000 rpm, which is difficult to measure without suitable equipment. However, if the reference frequency is set to 3 Hz the rotational speed will be 60 rpm, i.e. one revolution per second, which can be visually determined. This measurement will verify that the assumed pole pair number for the machine is indeed correct.

6.1.3 Case Study V1b: Helicopter Drive: Sensorless Operation, Phase C

In this part of the helicopter case study, attention is given to the Phase C module which will allow motor identification and speed control of the drive. For this purpose

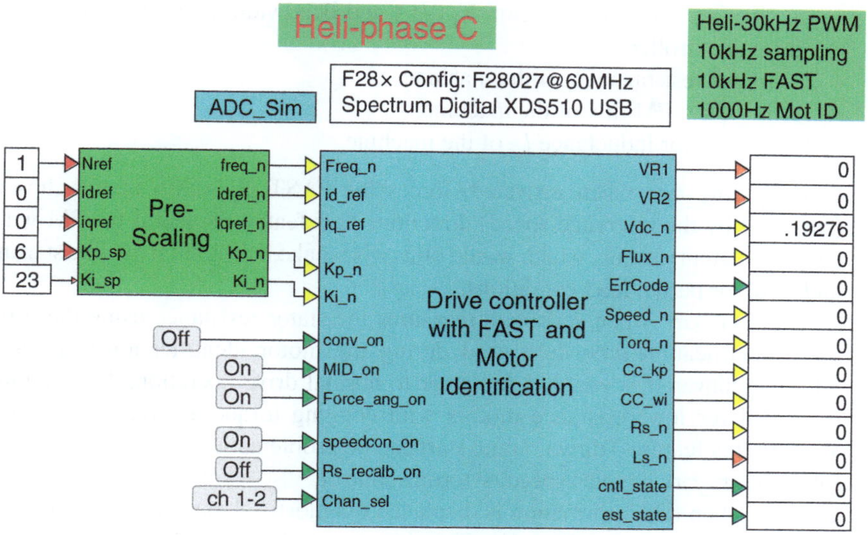

Fig. 6.11 Phase C simulation of a InstaSPIN-FOC helicopter drive with motor identification capability

the approach used in PM motor identification laboratory 2:2 (see Sect. 4.3) will be used. More specifically, said laboratory will be adapted to suit the specific requirements of this drive. Note that the PWM and sampling frequency used here differ from those in the previous section. Central to this development phase is the 'Drive Controller with FAST and Motor Identification' module, as shown in Fig. 6.11, which represents the drive under consideration. Hence details of the drive structure and corresponding dialog box parameter assignments will be discussed in this subsection. The following information is relevant for this laboratory component:

- Reference program: `Case1b_027phCv3.vsm`.
- Description: Motor parameter identification and sensorless control of helicopter drive using the InstaSPIN module.
- Equipment/Software: Helicopter board from 'A2Spin,LLC' [12] with Texas Instruments F28027F MCU, Spectrum Digital XDS510 USB Emulator and VisSim simulation program.
- Outcomes: Generate the `.out` file needed to represent the drive structure under investigation.

Inputs to the Controller are five variables, which set the reference shaft speed, direct/quadrature reference currents, and speed controller gains. In this laboratory, six buttons have been allocated to the controller namely:

- `conv_on`: activates the drive.
- `MID_on`: when activated, executes the identification sequence, after the converter has been activated (using the converter ON button) that identifies:

- `CC_kp/wi`: proportional gain K_p V/A and bandwidth ω_i rad/s used for the current controller.
- `Rs`: stator resistance R_s of the machine.
- `FLux_est`: PM flux of the machine.
- `L_est`: stator inductance L_s of the machine.

- `Force_ang_on`: when activated, forces the FAST algorithm in the chosen (specified by the reference speed) direction. This feature is useful for PM with cogging (detent) torque, which would otherwise make it difficult to realize startup under full or partial load conditions.
- `Rs_recal_on`: when activated, measures the stator resistance using the 'off-line' identification procedure used during the motor identification sequence, after the converter is enabled and prior to normal drive operation. This feature is critical for low resistance motors with cogging torque as considered here. Without this feature a direct online start can be problematic.
- `Speedcon_on`: enables speed or torque control.
- `ch1-2`: when this information is displayed diagnostic channels 1 and 2 will be shown in the scope in phase C+. This is a toggle button and can also display `ch3-4`, in which case channels 3 and 4 will be shown on the scope.

The module is provided with the following outputs (starting from top to bottom):

- `VR1`: a diagnostic variable which has been allocated to show the so called 'high frequency' stator resistance estimate R_{HF}. This value is identified during estimation state 2 of the motor identification cycle.
- `VR2`: a diagnostic variable which has been allocated to show the so called 'high frequency' stator inductance estimate L_{HF}. This value is also identified during estimation state 2 of the motor identification cycle.
- `Vdc_n`: per unit measured DC bus voltage.
- `ErrCode`: a diagnostic variable in this case connected to the 'current trip', hence will be '1' if a software current conditions occurs (current greater than the trip value set by the user).
- `Speed_n`: per unit estimated shaft speed.
- `Torque`: estimated torque (Nm).
- `CC_kp`: estimate for the proportional current controller gain `K_p`. This value is calculated using $K_p = L_{HF}/T_s$, where T_s is the ADC sampling time, set to 100 μs in this case. L_{HF} is the 'high frequency' inductance value, mentioned above.
- `CC_wi`: estimate for the current controller bandwidth (rad/s). This value is calculated using $\omega_i = R_{HF}/L_{HF}$, where R_{HF} is the 'high frequency' resistance value, mentioned above.
- `Rs_n`: per unit estimate for the stator resistance. Multiplication of this variable by a gain factor (u_{fs}/i_{fs}) gives an estimate for the stator resistance `Rs` (Ω).
- `Ls_n`: estimate for the stator inductance in (H)
- `cntrl_state`: status of the controller.
- `est_state`: status estimator.

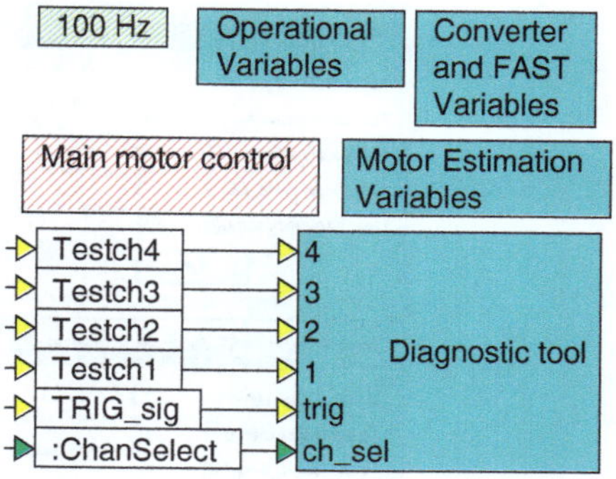

Fig. 6.12 Phase C simulation of a FAST based encoderless FOC drive, with motor identification: one level into the drive controller module

An ADC sampling frequency of $f_{\text{ADC}} = 10\,\text{kHz}$ has been selected for this laboratory, as was mentioned above. The current/speed controllers have also been set to operate at the ADC sampling frequency.

> The FAST and Motor Identification algorithms must be set to operate at sampling frequencies of $10000/i\,\text{kHz}$ and $10000/j\,\text{kHz}$ respectively, where i, j are integer values.

For this lab the values used are $i = 1$ and $j = 10$ which implies that the sampling frequency for the FAST and Motor ID algorithm have been set to 10000 Hz and 1000 Hz respectively. These frequencies are set in the appropriate dialog boxes as will be discussed below.

Moving one level lower into the 'Drive Controller' module reveals the modules shown in Fig. 6.12. Of these shown in the figure, the '100 Hz' and 'Diagnostic tool' modules have been discussed in previous laboratories. The dialog box parameters for this laboratory are given in Fig. 6.13. The parameters associated with the 'Operation variables' and 'Converter and FAST Variables' dialog boxes, have been discussed in the previous laboratory. Of interest here is therefore, the dialog box 'Motor Estimation Variables', where the following motor identification parameters must be assigned:

- `F_motorID/F_sampling`: ratio between sampling frequency used by the Identification algorithm and ADC sampling frequency. The ratio is an integer value, hence the user can select the ratio from a dialog box pull-down menu. For example, a choice of `1/10` implies an estimator sampling frequency of 1000 Hz, given that the ADC sampling frequency is set to 10000 Hz.
- `R/L sense freq (Hz)`: frequency used to estimate the current controller gain and bandwidth.
- `Flux est_freq (Hz)`: frequency used to estimate the PM flux and stator inductance.

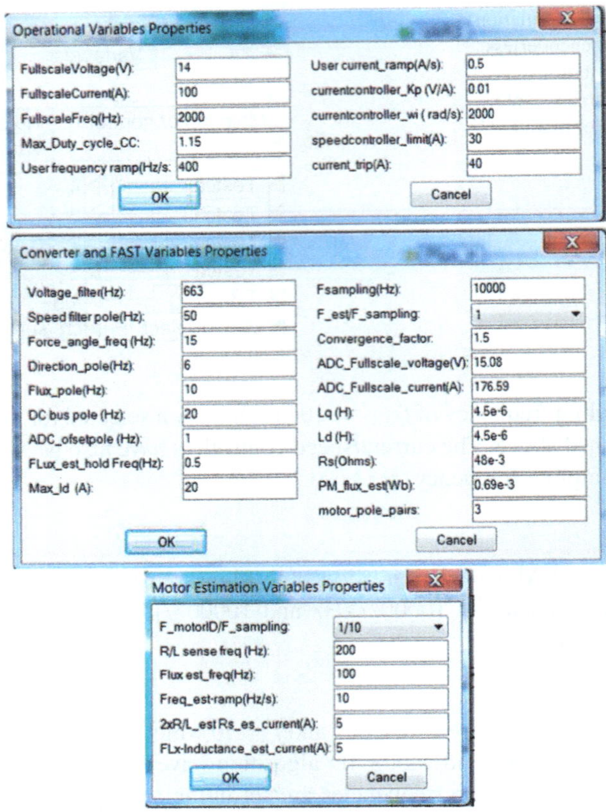

Fig. 6.13 Phase C simulation of a FAST based encoderless FOC drive, with motor identification: Dialog box entries

- `Freq_est_ramp (Hz/s)`: rate of frequency change used for motor identification purposes.
- `Current_est_ramp (A/s)`: rate of current change used for motor identification purposes.
- `2xR/L_est,Rs_es_current (A)`: current amplitude used for current controller gain/bandwidth and stator resistance estimation.
- `FLx-Inductance_est_current (A)`: current amplitude used for PM flux and stator inductance estimation.

Moving one level lower into the 'main motor control' module, reveals a set of modules given in Fig. 6.14. Key inputs to the InstaSPIN controller module are the (per unit) phase current/voltage variables: `iUn`, `iVn`, `iWn`, `vUn`, `vVn`,`vWn` and DC bus voltage variable `DCbus_n`, generated by the ADC-PWM module. User reference variables: `Freq_n`, `Ampd`, `Ampq_n` and `Kpn`, `Kin` represent the user inputs for shaft-speed, direct/quadrature currents and speed controller gains. Logic input `doMotorID_` is used to activate the motor identification sequence.

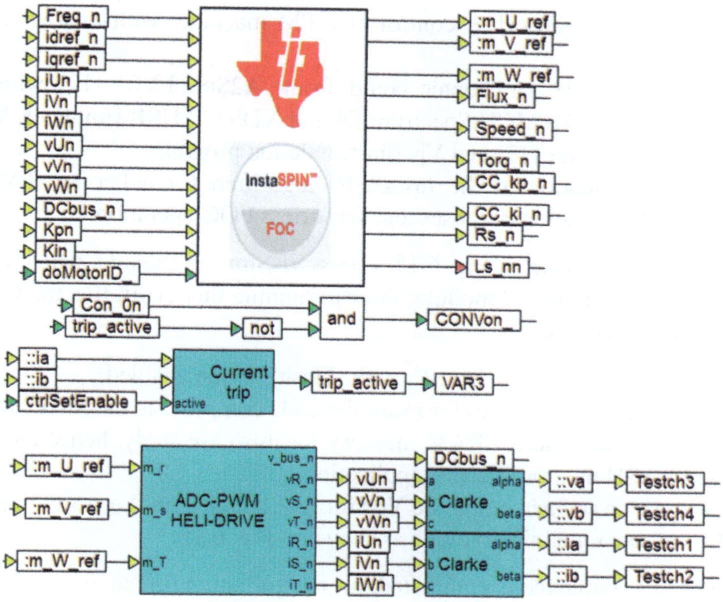

Fig. 6.14 Phase C simulation of a FAST based encoderless field-oriented controlled (FOC) drive, with motor Identification: one level into the 'Main Motor Control' module

Outputs of the 'InstaSPIN-FOC module' are (among others) the modulation indices: `m_U_ref`, `m_V_ref`, `m_W_ref` generated by the internal current controller. The remaining outputs are those which have been named above. Moving one level into the InstaSPIN-FOC module, leads to the basic VisSim 'InstaSPIN Motor Control' and 'estimator calculation' modules as discussed in previous chapters. For diagnostic purposes the test variables: `TESTch1`, `TESTch2` have been connected to per unit variables: i_α and i_β respectively.

6.1.4 *Case Study V1b: Helicopter Drive: Sensorless Operation, Phase C+*

Phase C+, is the operational component of the laboratory and is basically a run version of the `.out` file compiled in phase C (see previous subsection). As with the previous laboratory, download of the generated `.out` file must be done via an alternative (to VisSim) approach given the need to use the Flash memory of the MCU (insufficient RAM to store the program).

The following information is relevant for this laboratory component:

- Reference program : `Case1b_027phCv3_d.vsm`.

- Description: FOC sensorless control of a PM machine, with motor parameter identification.
- Equipment/Software: Helicopter board from 'A2Spin,LLC' [12] with Texas Instruments F28027F MCU,Spectrum Digital XDS510 USB Emulator, Walkera V450BD5 Helicopter [17] and VisSim simulation program.
- Outcomes: application of the InstaSPIN algorithm for helicopter PM motor parameter identification purposes and sensorless FOC operation.

The run version shown in Fig. 6.15, uses a VisSim run module, which executes the `.out` file, shown in said module. Prior to running this `.out` file, the following steps should be undertaken:

- Compile the `.out` file in phase C, with 'Flash' option enabled.
- Download the `.out` file via (for example) code composer to the MCU. Note that this MCU has insufficient RAM memory for this case study, hence the need to use the FLASH memory for this application.
- Prior to start up (by activating the VisSim 'go' button), set the `conv_on` button to OFF and the remaining buttons as shown in Fig. 6.15.

When the above sequence is completed, hit the go button (green arrow in VisSim) and monitor the DC bus voltage. After verification that the value shown is correct, turn on the drive (`conv_on` button ON), in which case the motor identification cycle should start as may be verified by monitoring the status of the variables `Cntrl_st`and `Est_st` shown in Fig. 6.16.

A brief outline of the identification sequence, based on the 'Estimator state' is given below, where use is made of Fig. 6.16 :

- Estimator state 1: estimator idle, ADC offset measurement sequence occurs after the converter has been activated (converter button ON and motor ID on). Controller state cycles from 'idle' via 'offline' to 'online'.
- Estimator state 2: Identification of the current controller gain and bandwidth, Frequency and amplitude (set in motor ID dialog box) are 200 Hz and 5 A (the current value used in this sequence is half the value used in the Rs estimation phase, i.e. 2.5 A during this phase). Motor should NOT rotate whilst a rotating current vector is applied to the machine. At the end of this sequence the so called 'high frequency' stator resistance R_{HF} (Ω) and inductance L_{HF} (H) are returned. In this laboratory these variables are shown by diagnostic variables `VR1` and `VR2` respectively. If this stage is executed correctly these results provide an initial estimate for the motor parameters R_s and L_s.
- Estimator state 3: Stator resistance measurement, DC current level (set in motor ID dialog box) is set tot 5 A. A stationary current vector is applied with an amplitude of 5 A. At the end of this sequence the stator resistance value is returned.
- Estimator state 4: Open-loop speed control, with a current vector amplitude set to 2.5 A, which is half (default) of the flux amplitude set in the motor ID dialog box and ramp up amplitude of 0.5 A/s (set in the 'operational dialog box'). The rotational speed of the vector will ramp up at a rate of 10 Hz/s to 100 Hz. Motor speed (rpm) should increase until it stabilizes at a value that matches the

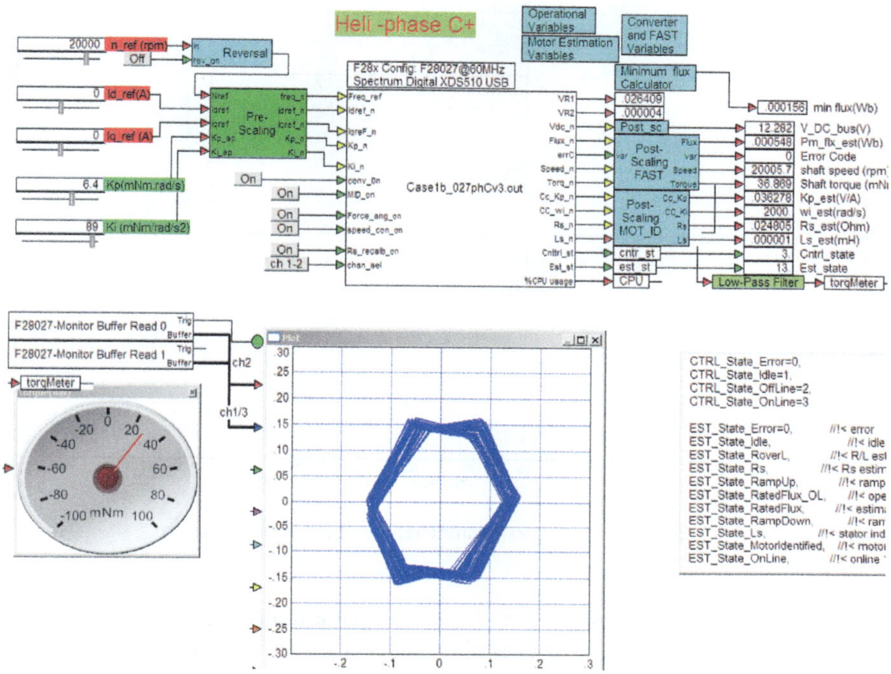

Fig. 6.15 Phase C+ simulation of a FAST based encoderless field-oriented controlled (FOC) drive, with motor identification: example shows helicopter operation at 20000 rpm, after identification has been done

frequency set in the motor ID dialog box. Hence, in this case, with 100 Hz, being the selected frequency the rotational speed will be $^{100*60}/_{p} = 2000$ rpm, where $p = 3$ is the pole pair number of the machine in use.

- Estimator state 6: Transitioning from open to closed-loop speed control.
- Estimator state 7: estimation of the rated PM flux under closed-loop speed control. Current amplitude, is determined by the load present, which should be friction only (no external load). Speed set by flux estimation frequency of (in this case) 100 Hz. At the end of the sequence the PM flux amplitude is identified.
- Estimator state 10: measurement of the stator inductance. Direct axis current amplitude of 5.0 A (set in motor ID dialog box), speed of 2000 rpm, which is set by the user selected frequency 100 Hz. Drive operation is under closed-loop speed control. At the end of this cycle the inductance value is (normally) returned to the user.
- Estimator state 8: Ramp down motor speed to zero
- Estimator state 3: Re-estimation of stator resistance.
- Estimator state 13: Drive online and control returned to the user.

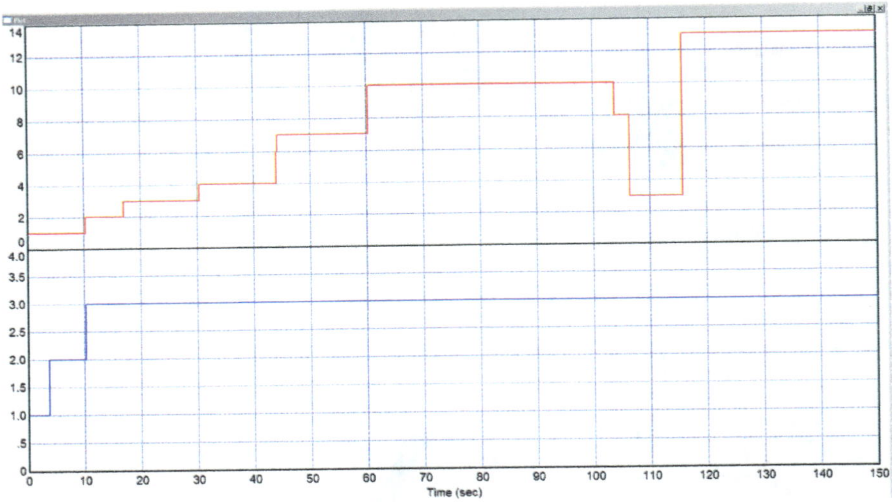

Fig. 6.16 Motor ID sequence showing estimator ('red') and controller states

Note that the inductance estimation phase 10 did not return a correct value in this case, given that the motor was operating under a load. In industrial applications where this occurs the reader has the following options;

- redo the motor identification with a motor under no load, by using for example a second machine (of the same type), which is the approach taken here, which led to an inductance value of 0.44 μH, which compares favorably with the high frequency estimate (shown by variable VR1) of 0.3 μH. This value must then be used by the controller.
- use the high frequency inductance estimate obtained during estimation phase 1.

Upon completion of the identification procedure (and control was returned to the user) the reference speed was gradually increased to 20000 rpm at which point the phase current was measured with a DC-true probe and scope as shown in Fig. 6.17. These results show operation with a peak current of $\approx$ 15 A, which is also the quadrature current generated by the speed controller, given that the direct axis reference current was set to zero. Hence the shaft torque generated would be equal to $T_m = \frac{3}{2}\, i_{\mathrm{q}}\, \psi_{\mathrm{PM}}\, p$ which is equal to $T_m = 29\,\mathrm{mNm}$. This value compares favorably with the average value shown in the torque display module (see Fig. 6.15). The diagnostic plot given in Fig. 6.15 shows the per unit $i_{n\alpha}, i_{n\beta}$ vector plot for the helicopter under partial load. As mentioned earlier, a higher PWM frequency of 45 kHz is advisable to reduce the current ripple and also to satisfy the guideline set out in the Excel sheet to be discussed at the end of this section.

The motor identification routine, as mentioned above, will provide accurate and meaningful results, provided that prudent choices are made with respect to the ADC full scale voltage/current values and motor identification dialog box parameters. For this purpose so called 'Motor Identification goodness' plots have been introduced in previous chapters which provide a 'post processing' analysis of the motor identification process. The 'goodness plots' for this drive shown in Fig. 6.18 are briefly discussed below.

The Voltage 'goodness' diagram (see Fig. 6.18 (top plot)) shows the voltage 'ADC transfer' function, which is the decimal output value that corresponds to the input voltage range $0 \rightarrow$ *ADC full scale voltage*, where the *ADC full scale voltage*) (see Eq. (3.3a)) value is equal to 15.08 V for this application. The least significant bit $\mathrm{LSB_v}$ for the voltage transfer function is equal to $^{ADC\,full\,scale\,voltage}/_{4095} = 3.68\,\mathrm{mV}$. This value is used to set the horizontal logarithmic scale, which ranges from $^{\mathrm{LSB_v}}/_{4} \rightarrow$ *ADC fullscalevoltage*.

The lower plot given in Fig. 6.18 shows the 'ADC current transfer' function, with an input current range of $^{\mathrm{LSB_i}}/_{4} \rightarrow$ *ADC fullscale current*, where the *ADC fullscale current* is equal to 176.59 A (peak to peak current) for this board. Furthermore, the least significant bit $\mathrm{LSB_i}$ for the current transfer function is equal to $^{ADC\,fullscale\,current}/_{4095} = 43.1\,\mathrm{mA}$. The vertical axis of both plots show the ADC count from $1 \rightarrow 4095$. Note that the quantization error becomes significant, when input voltage/current levels are of the same order of magnitude as $\mathrm{LSB_v}/\mathrm{LSB_i}$ respectively. Consequently, the figures show how well the ADC converters are utilized, for a chosen set of motor ID input variables. Observe that this 'post-processing' type of analysis, makes use of the measured motor parameters. As such these plots show whether or not the identification procedure has been carried out with appropriate parameter settings (as set in the motor identification dialog box), as will become apparent shortly. Furthermore, these plots can provide guidelines as to what input parameters may need to be changed to improve identification, given the machine under consideration.

A brief discussion on the method used to calculate the voltage/currents shown in Fig. 6.18 is given below, for each of the legend variables.

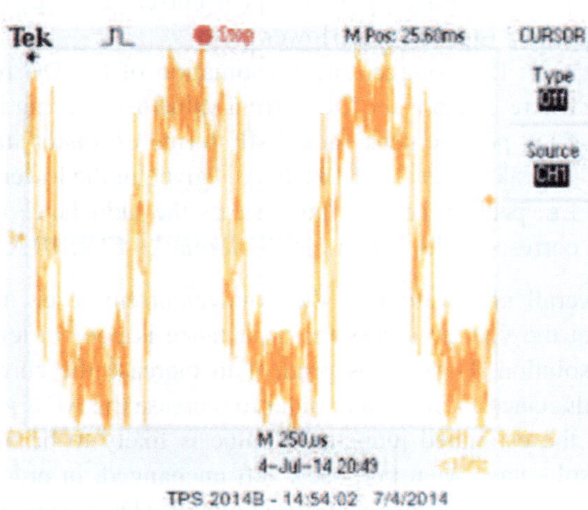

Fig. 6.17 Phase C+: Experimental result: helicopter drive operating at 20000 rpm, phase current 5.0 A/div

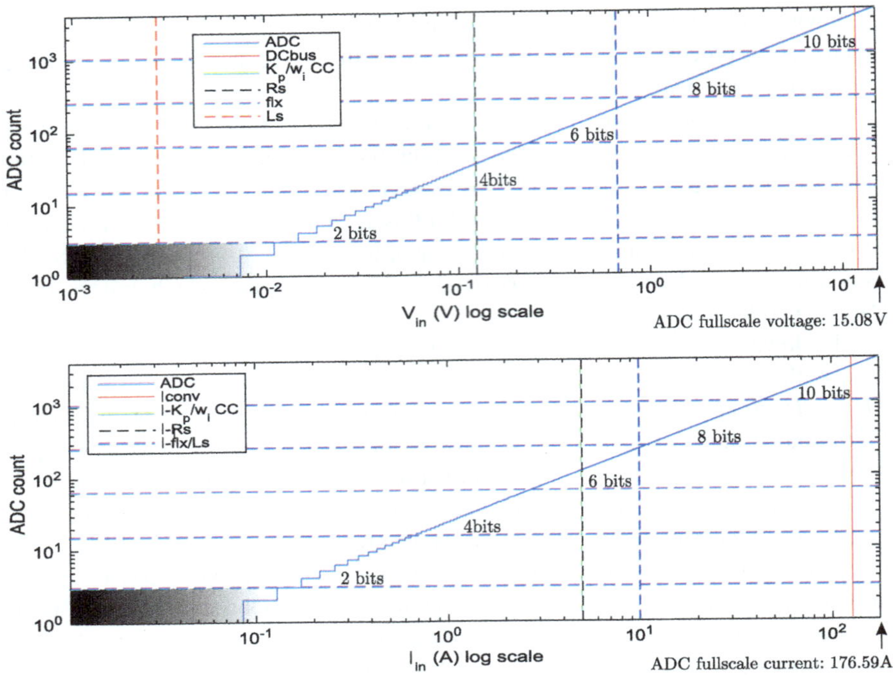

Fig. 6.18 'Motor Identification goodness' plots

- `DCbus`: rated DC voltage $u_{DC} = 12\,V$ of the converter.
- `Iconv`: rated converter current is 45 A RMS, which corresponds to a peak to peak current value of 127 A, shown in the lower plot.
- `K_p/w_i CC`: peak to peak phase voltage that occurs during estimator state 2. The corresponding peak to peak current `I-K_p/w_i cc` used during estimator state 2 is given in the lower plot.
- `R_s`: DC voltage during estimation of the DC resistance in estimator state 3, where `I-Rs` is the DC current value used during identification.
- `flx`: peak to peak back EMF during estimator state 7. The corresponding peak to peak current `I-flx`: 10 A is given in the lower plot.
- `Ls`: peak to peak voltage across the inductance during estimator state 10. The corresponding peak to peak current `I-flx`: 10 A is given in the lower plot.

Overall observation of the voltage/current levels used during estimation learns that the voltage across the inductance is poor in terms of available ADC voltage resolution. Hence it is prudent to increase the current or frequency used during inductance estimation in order to increase the ADC voltage measurement resolution as the estimated inductance value is likely to be unreliable. In this example the results have been purposely left unchanged, in order to illustrate how 'goodness' plots can be used to identify potential ADC resolution issues. In the sequel to this case study attention is given to the issue of selecting key parameters such as those

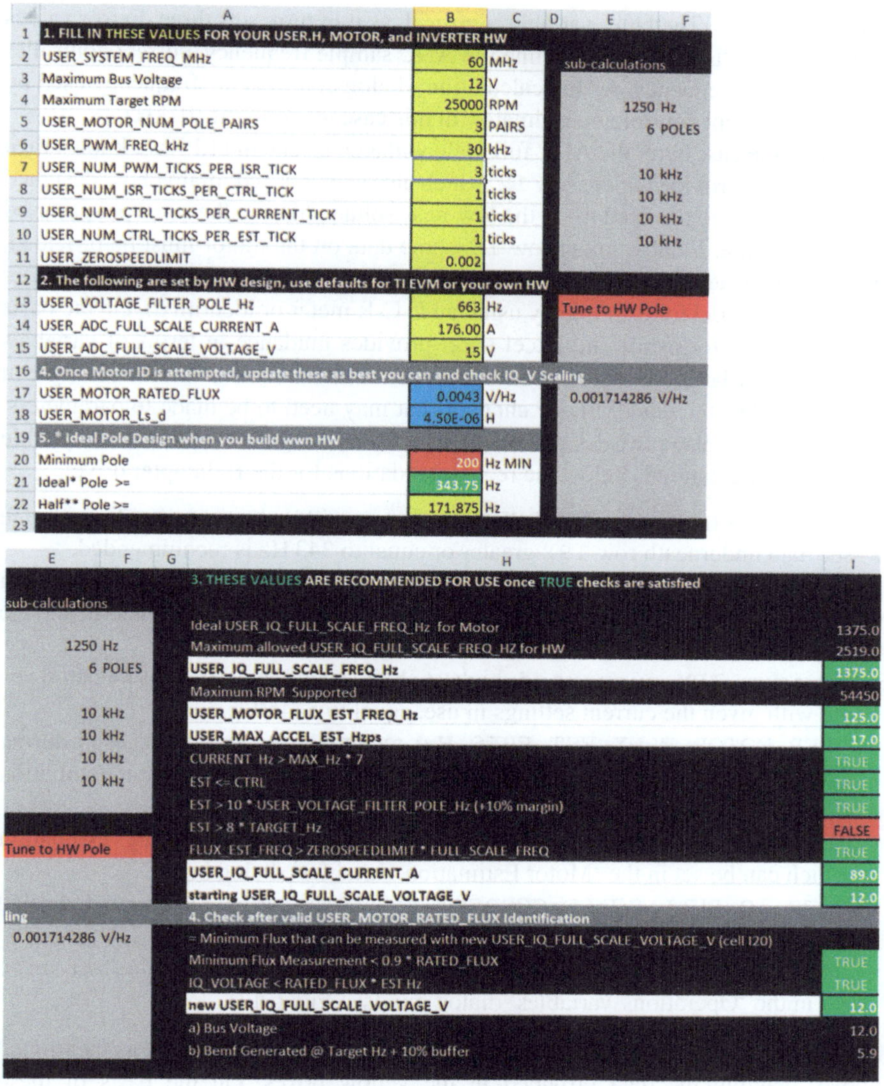

	A	B	C	D	E	F
1	1. FILL IN THESE VALUES FOR YOUR USER.H, MOTOR, and INVERTER HW					
2	USER_SYSTEM_FREQ_MHz	60	MHz		sub-calculations	
3	Maximum Bus Voltage	12	V			
4	Maximum Target RPM	25000	RPM		1250 Hz	
5	USER_MOTOR_NUM_POLE_PAIRS	3	PAIRS		6 POLES	
6	USER_PWM_FREQ_kHz	30	kHz			
7	USER_NUM_PWM_TICKS_PER_ISR_TICK	3	ticks		10 kHz	
8	USER_NUM_ISR_TICKS_PER_CTRL_TICK	1	ticks		10 kHz	
9	USER_NUM_CTRL_TICKS_PER_CURRENT_TICK	1	ticks		10 kHz	
10	USER_NUM_CTRL_TICKS_PER_EST_TICK	1	ticks		10 kHz	
11	USER_ZEROSPEEDLIMIT	0.002				
12	2. The following are set by HW design, use defaults for TI EVM or your own HW					
13	USER_VOLTAGE_FILTER_POLE_Hz	663	Hz		Tune to HW Pole	
14	USER_ADC_FULL_SCALE_CURRENT_A	176.00	A			
15	USER_ADC_FULL_SCALE_VOLTAGE_V	15	V			
16	4. Once Motor ID is attempted, update these as best you can and check IQ_V Scaling					
17	USER_MOTOR_RATED_FLUX	0.0043	V/Hz		0.001714286 V/Hz	
18	USER_MOTOR_Ls_d	4.50E-06	H			
19	5. * Ideal Pole Design when you build wwn HW					
20	Minimum Pole	200	Hz MIN			
21	Ideal* Pole >=	343.75	Hz			
22	Half** Pole >=	171.875	Hz			
23						

E	F	G	H	I
			3. THESE VALUES ARE RECOMMENDED FOR USE once TRUE checks are satisfied	
sub-calculations				
			Ideal USER_IQ_FULL_SCALE_FREQ_Hz for Motor	1375.0
1250 Hz			Maximum allowed USER_IQ_FULL_SCALE_FREQ_HZ for HW	2519.0
6 POLES			USER_IQ_FULL_SCALE_FREQ_Hz	1375.0
			Maximum RPM Supported	54450
10 kHz			USER_MOTOR_FLUX_EST_FREQ_Hz	125.0
10 kHz			USER_MAX_ACCEL_EST_Hzps	17.0
10 kHz			CURRENT_Hz > MAX_Hz * 7	TRUE
10 kHz			EST <= CTRL	TRUE
			EST > 10 * USER_VOLTAGE_FILTER_POLE_Hz (+10% margin)	TRUE
			EST > 8 * TARGET_Hz	FALSE
Tune to HW Pole			FLUX_EST_FREQ > ZEROSPEEDLIMIT * FULL_SCALE_FREQ	TRUE
			USER_IQ_FULL_SCALE_CURRENT_A	89.0
			starting USER_IQ_FULL_SCALE_VOLTAGE_V	12.0
ling			4. Check after valid USER_MOTOR_RATED_FLUX Identification	
0.001714286 V/Hz			= Minimum Flux that can be measured with new USER_IQ_FULL_SCALE_VOLTAGE_V (cell I20)	
			Minimum Flux Measurement < 0.9 * RATED_FLUX	TRUE
			IQ_VOLTAGE < RATED_FLUX * EST Hz	TRUE
			new USER_IQ_FULL_SCALE_VOLTAGE_V	12.0
			a) Bus Voltage	12.0
			b) Bemf Generated @ Target Hz + 10% buffer	5.9

Fig. 6.19 InstaSPIN user variable guide (split in two parts for readability) [2]

shown in the dialog boxes, for the drive. A useful guide for this task is provided by Texas Instruments, in the form of an Excel sheet [2]. In this document the user must enter a set of basic drive settings, on the basis of which a set of guidelines and parameters are provided. An application example of this user guide, for the drive discussed in this case study is shown in Fig. 6.19. Observation of said figure (top section) shows a 'yellow' column, which represents the drive data that must be provided by the user. This includes the CPU frequency, DC bus voltage, motor pole pair number, anticipated maximum shaft speed and the selected PWM frequency for the converter. The latter variable is critical, as it in turn, sets three divisor values (referred to as 'ticks'), which define the ADC sample frequency, current controller and estimator frequency. A 'sub calculation' dialog confirms the frequency used by the ADC, current control and estimator. In this case all are set to 10 kHz.

If as is the case here, the ADC full scale voltage/current and LPF are known then these must be provided under row `12`, which also shows a 'red' dialog box, designed to alert the user to the need to set the low-pass voltage filter to the same pole as used in the hardware. Finally, under row `16`, some data on the motor must be provided, which is the stator flux (in V/Hz) that is equal to $\psi_s\,(Wb).2\,\pi$ and the motor stator inductance (found for example by using an L-C-R meter or identified). On the basis of the above information the Excel sheet provides guidance in terms of variables which should be selected. In addition use is made of boolean indicators ('green boxes') to alert the reader to any changes that may need to be made to parameters. This user guide also provides guidance (not shown in Fig. 6.39) when incompatible parameters are entered. Below the recommendations for the Helicopter drive:

- Ideal corner frequency of the LPF (if the user is still in the converter design stage see data underneath row `19`): greater or equal to 343 Hz is recommended.
- `Ideal_USER_IQ_FULL_SCALE_FREQ_Hz`: recommended value that can be set the 'Operations Variables' dialog box (see Fig. 6.13). Underneath the recommended value of 1375 Hz is the maximum value of 2519 Hz.
- `Maximum RPM supported`: highest shaft speed which InstaSPIN can operate with given the current settings in use, equal to 54450 rpm.
- `USER_MOTOR_FLUX_EST_FREQ_Hz`: recommended frequency used during estimator phases 4, 7 and 10 (see Fig. 6.16) and set in the 'Motor Estimation Variables' dialog box (see Fig. 6.13).
- `USER_MAX_ACCCEL_EST_Hzps`: frequency ramp rate during estimation which can be set in the 'Motor Estimation Variables' dialog box.
- `USER_IQ_FULL_SCALE_CURRENT_A`: full scale current recommended value that can be set in the 'Operations Variables' dialog box (see Fig. 6.13).
- `new USER_IQ_FULL_SCALE_VOLTAGE_V`: recommended value that can be set in the 'Operations Variables' dialog box (see Fig. 6.13).

In addition to the above a series of 'green' boolean indicators are set, as mentioned above, pending the data provided in the yellow boxes. On the basis of these indicators the user can adjust relevant parameters. Note in this example there is a 'red' boolean indicator, which points to the need to increase the estimator frequency. Choosing a PWM frequency of 45 kHz and estimator frequency of 15 kHz (as recommended in Sect. 6.1.1) will flip the 'red' boolean indicator to 'green'. For further details the reader is referred to MotorWare user variable guide [2].

Finally it is noted that this case study has considered a 'low inductance motor'. In some extreme cases the motor time constant $\frac{L_s}{R_s}$ of machines may be almost zero, in which case current measurement becomes problematic due to the rapid current decay. Typically the PWM frequency is then increased to try and compensate for this effect. However, if this does not offer an adequate solution it may be prudent to add additional line inductors between converter and motor to improve current measurement. Note that under these circumstances the voltage measurement for InstaSPIN should be done on the MOTOR side so that the additional inductance is not included in the motor identification process. However, the current controller gain and bandwidth must in this case be adjusted to suit the total inductance seen by the converter (hence the total inductance of motor and added inductance must be used for these calculations).

6.2 Case Study V2: CoMoCo Drive

The purpose of this project is to consider the use of a mains connected 'FAST prepared' converter which will be connected to a three-phase AC induction machine. The drive setup given in Fig. 6.20 shows a 'CoMoCo-1.0-ED04' 1.2 kW converter (with cover removed) [3], which is a highly versatile extremely robust industrial unit that makes use of a Texas Instruments TMS320F28069F MCU (version 'F' implies omission of the SpinTAC library). Consequently, the drive can be used for sensorless control of three-phase AC machines up to 1kW, with corresponding motor identification in line with the laboratory examples discussed in previous two chapters (with exception of the SpinTAC examples). Observation of Fig. 6.20 learns that approximately 35 % of the available converter space has been allocated to a line-filter (metal box 'left side'), two toroidal coils and associated power electronics, which are part of the 'Power Factor Correction' (PFC) boost input circuit. This feature gives this module the ability to generate an input current waveform that is in phase with the single phase mains voltage and almost devoid of higher harmonics (due to the presence of the line filter). Furthermore, the 'boost' feature of this PFC unit can generate a DC bus voltage that exceeds the peak mains voltage level (which is the highest DC bus voltage that can be realized with a diode rectifier). This means that standard three-phase 380 V (which is the line to line voltage in Europe) AC motors can be directly connected to this converter whilst operating from a single phase supply.

In terms of project content, both converter commissioning and field-oriented control with full motor parameter identification using instaSPIN will be discussed in the first two subsections. The third and final subsection of this project is allocated to the commissioning and use of the PFC-boost converter section of the CoMoCo drive.

Central to this project is the Marathon 5k336N2A induction machine with nameplate data as shown in Table 6.3. Further inspection of this machine learns that it has an additional set of windings, which can be added if operation at a higher (than shown above) supply voltage is required. In this case the machine is 'low voltage' star connected, as shown on the nameplate. Measurement of the line to line inductance using a L-C-R meter (set to 100 Hz) and resistance using an ohm meter leads to the results shown in Table 6.4. Accordingly, the stator resistance is equal to $R_s = {}^{R_{sLL}}/_{2} = 10.75\,\Omega$. Similarly, an estimate for the leakage inductance can be found using $L_\sigma = {}^{L_{sLL}}/_{2} = 50$ mH. Furthermore, the number of pole pairs can

Fig. 6.20 CoMoCo [3] converter with Marathon 5k336N2A induction machine setup

Table 6.3 Nameplate data Marathon 5k336N2A induction machine

Parameter	Value	Units
Power	1/4	hp
RMS line to line voltage	208–230	V
RMS phase current	1.3–1.4	A
Mains frequency	60	Hz
Rated speed	1725	rpm

Table 6.4 Marathon 5k336N2A measured motor parameters

Parameter	Value	Units
Inductance: L_{sLL}	100	mH
Resistance: R_{sLL}	21.5	Ω

be inferred from the rated speed and mains frequency, given that the latter would correspond to a synchronous speed of $60 * 60/p = 1800$ rpm , if a pole pair number of $p = 2$ is chosen. The corresponding rated slip speed would then be 1800–1725= 75 rpm, which is a realistic value, hence the pole pair number is 2 in this case. The estimated peak rated stator flux for this machine, based on the lowest line to line voltage of $u_l = 208$ V can be found using

$$\hat{\psi}_s = \frac{u_l\sqrt{2}}{\sqrt{3}\,2\,\pi\, f_s} \tag{6.4}$$

where f_s Hz is the mains frequency. Subsequent evaluation of (6.4) gives a peak stator flux estimate of $\hat{\psi}_S = 45\,\text{mWb}$. The required line voltage of the machine gives a guideline for the required minimum DC bus voltage which can be found using

$$u_{\text{bus}}^{\text{min}} = u_l\sqrt{2} \tag{6.5}$$

which corresponds to a minimum DC bus voltage of $u_{\text{bus}}^{\text{min}} = 294\,\text{V}$. Consequently the Marathon induction machine used in this project can be run directly from a rectified 220 V RMS mains supply input. With some basic engineering mathematics as shown above, the framework has been set to commission the drive.

6.2.1 *Case Study V2a: CoMoCo Drive: Commissioning, Phase C*

Prior to using current control in any application, it is helpful to consider drive operation under voltage control first. Using this approach, the voltage/current measurement setup can be examined. Furthermore, the phase relationship between voltage and current can be established, which is important prior to using current control. The following information is relevant for this case study:

- Reference program: `Case2a_069phCv4.vsm`.
- Description: Commissioning of CoMoCo drive.
- Equipment/Software: CoMoCo-1.0-ED-4 converter [3] with Texas Instruments F28069F MCU, Spectrum Digital XDS510 USB Emulator and VisSim simulation program.
- Outcomes: Generate the `.out` file needed to represent the drive structure and calculate the full scale ADC values as well as LPF frequency of the voltage filter.

The voltage controller as discussed in Sect. 3.1 is used for this analysis. However, emphasis in this section will be on configuring the ADC-PWM module, triggering, allocating ADC channels and confirming that the current/voltage waveforms are correct in terms of amplitude and phase relationships. For safety reasons, a regulated variable 400 V DC supply is connected to the single phase AC input terminals of the converter, to provide over-current protection. The phase C model given in Fig. 6.21 has been adapted for this application. Inputs are the electrical frequency and the direct axis voltage reference variable. Furthermore, a channel select input `chan_sel` has been added to toggle between diagnostic scope channels 1-2 and 3-4 in phase C+. Moving one level down into the 'Open loop voltage controller' reveals a set of modules and dialog boxes as indicated in Fig. 6.22. This figure also shows a dialog box which appears upon a 'right click' on the 'main control module'. The critical entry in this case is the 'execute on interrupt' feature which must be set as shown. This implies that the module in question will be executed

Table 6.5 Allocation of input variables to ADC input channels

Input variable	ADC input channel
Phase current i_U	0
Phase current i_V	1
Phase current i_W	2
Phase voltage u_U	3
Phase voltage u_V	4
Phase voltage u_W	5
Bus voltage u_{DC}	6

on an interrupt signal set by the ADC unit. More specifically, the interrupt is in this case allocated to `ADCINT1:6`, which implies that the signal is generated by ADC input channel 6. The ADC input channels are allocated within the ADC-PWM module, which appears when moving one level lower into the 'main control module' as shown in Fig. 6.23. The modules shown are as discussed in Sect. 3.1. However, in this case configuring the ADC-PWM module is of interest. This type of module is board specific, which is why the board/converter name is included in the title. Data acquisition via VisSim, is done by way of a MCU specific 'input channel' module, such as `F28069M Input Channel` in which the user has the option of selecting an 'analog or digital' channel. Furthermore, the user must allocate a channel number (within the range of analog channels available) as mentioned above. Table 6.5 shows the channel allocation used in this laboratory. Note that the allocation of phase currents and voltages to input channels is up to this stage done arbitrarily. A critical commissioning step is to allocate the channel numbers introduced in Table 6.5 to actual MCU variables. This requires access to the `ADC F28069M Properties` dialog box, found by selecting `Embedded` in the main VisSim menu and then selecting `F280x` and finally `ADCconfig` which brings up the module shown in Fig. 6.24. Also given in this figure is part of the circuit diagram of the board under consideration, which shows the pin allocation of the current/voltage variables. The CoMoCo board makes use of the TMS320F28069M processor shown (partly) in Fig. 6.24. Associated with the current/voltage variables and MCU pin numbers are alphanumeric variables, which need to be assigned in the `ADC F28069M Properties` dialog box. In this case an important decision has been made to systematically allocate ADC inputs in pairs, ie `A0`, `B0` etc, which is particularly effective as a so called 'dual sampling' mode

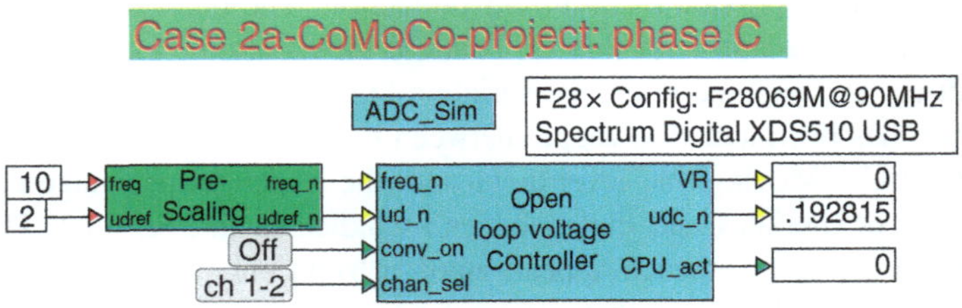

Fig. 6.21 Development Phase C: voltage controller set up for CoMoCo drive

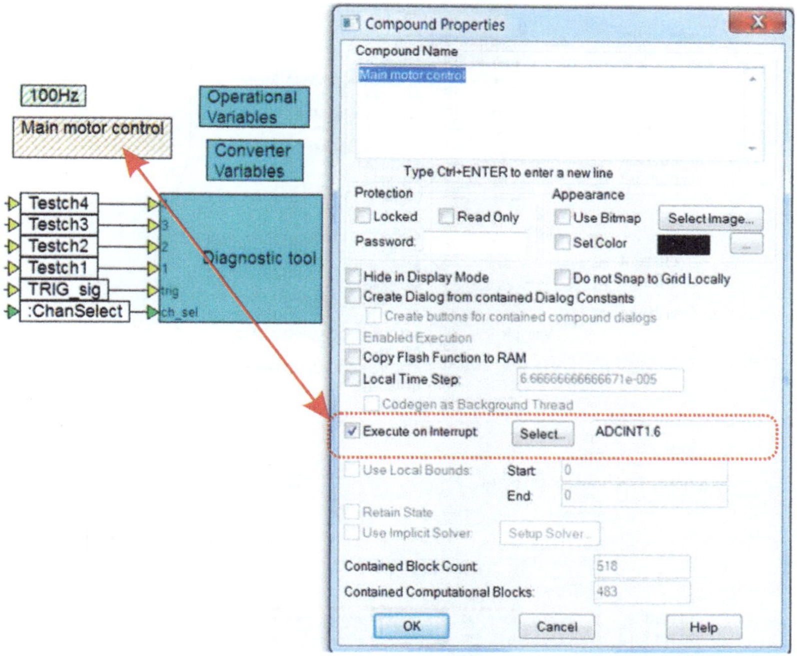

Fig. 6.22 Development Phase C, one level into 'Open loop voltage controller'

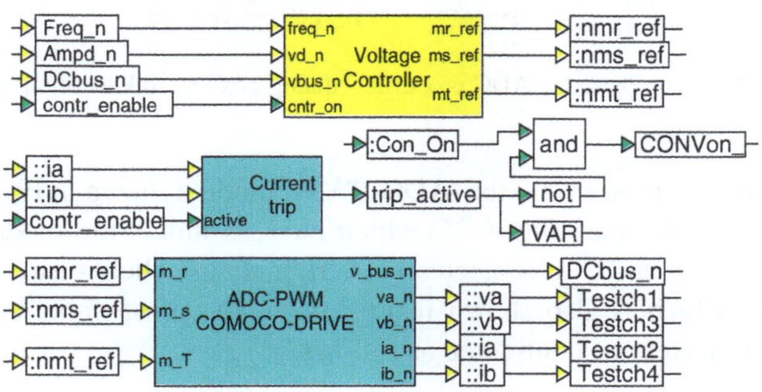

Fig. 6.23 Development Phase C: one level into 'main motor control' module

can then be invoked (as shown by the 'tick') which means two ADC samples are converted simultaneously. This reduces the time required to complete the ADC acquisition task, but it requires specific hardware channel allocation to be done at the design stage of the board.

In addition to assigning the input variables the trigger signal for the input channels need to be assigned. Triggering of the ADC is linked to the PWM module as was discussed in Sect. 2.1.6. For all the input channels the 'Start Of Conversion' signal `ePWM1-SOCA` is used which implies that the pulse width modulation module 1 generates the trigger for the input ADC channels in use.

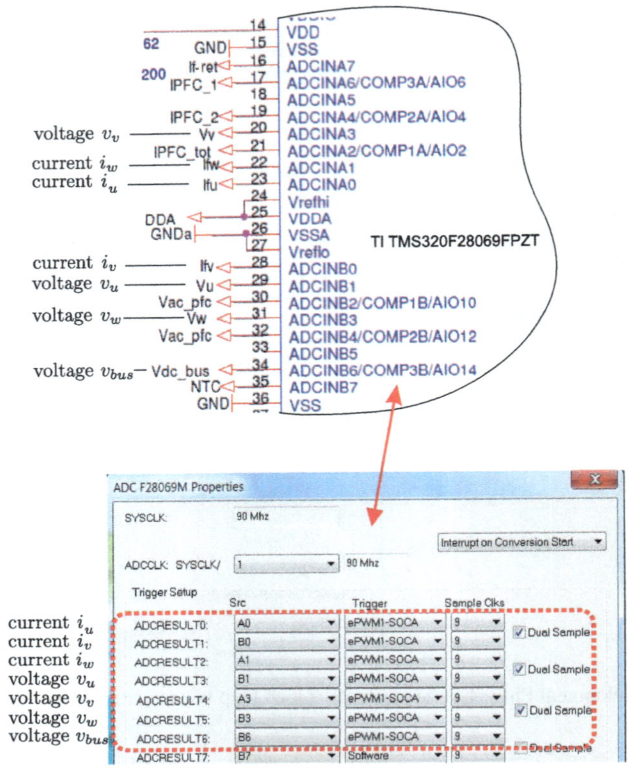

Fig. 6.24 Development Phase C: 'ADC properties' dialog box and link to circuit diagram

Moving one level lower into the 'ADC-PWM' module shown in Fig. 6.23 reveals three PWM modules (see Fig. 6.25),which have as inputs the modulation index variables `R_p`, `S_p` and `T_p` respectively. A 'right mouse click' on a module reveals a dialog box, which is also shown in Fig. 6.25. The most important entries are numbered and need to be configured as follows:

1. `Timer Period`: represent the number of cycles of the PWM staircase as discussed in Sect. 2.1.6. The number to be entered here is defined by the clock frequency divided by the timer period times 2. The factor 'times 2' is due to the selection of count mode `UpDown` in Fig. 6.25. In this case study a 15 kHz PWM frequency is required, which with a system clock frequency of 90 MHz, corresponds to a timer period value of 3000 as shown.
2. `Rising Edge Delay`, `Falling Edge Delay`: these two parameters set the 'dead-time' of the converter, as dictated by the power electronic devices used. For this converter both parameters are set to 135, which relates to the number of MCU clock cycles. Hence the 'dead-time' is equal to $^{135}/_{90MHz} = 1.5\,\mu$ s.
3. `Send start ADC Conversion Pulse A (SOCA)`: this entry is critical, in that it determines when the conversion pulse should be given with respect

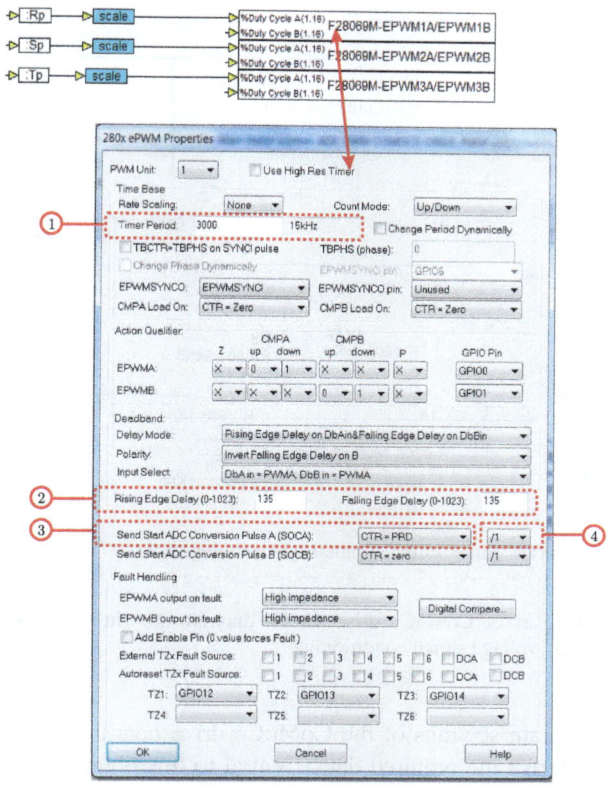

Fig. 6.25 Phase C: 'PWM properties' dialog box

to the PWM cycle, as shown in Fig. 2.23. In this case the option `PRD` is chosen because triggering of the ADC should occur at the top of the PWM staircase, as shown in Fig. 2.22. Note that in Fig. 6.24 ADC triggering refers to `ePWM1-SOCA`, which implies that ePMW module 1 generates the SOC signal for all ADC channels.

4. `1`: this entry specifies if the ADC pulse should be given after every first , second or third PWM clock cycle. In this case a 15 kHz PWM frequency has been selected and the ADC sample frequency in use is 15 kHz (which is also the rate at which the controller module will be executed, with a sampling frequency set in the main VisSim module) hence the PWM to ADC ratio must be set to $1/1$.

Note that only a subset of the ADC and PWM dialog entries have been discussed, given their importance. The reader is however, advised to refer to the relevant Texas Instruments application notes for further details on these modules.

Computation of the ADC full scale voltage/current parameters and the low-pass filter corner frequency are critical to successful sensorless operation. Accordingly these parameters must be either given or derived by careful examination of the schematic diagram of the converter under consideration. Part of the relevant (to this

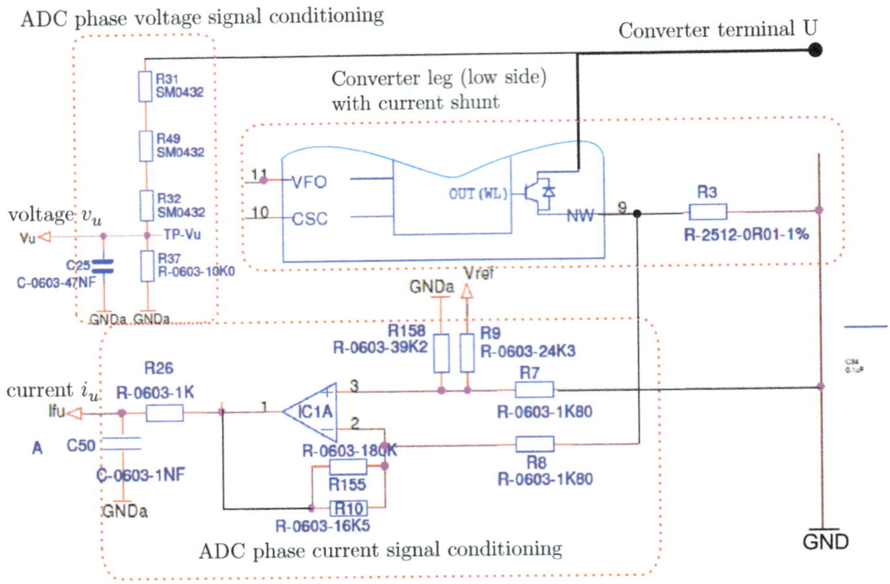

Fig. 6.26 Phase C: Part of CoMoCo drive circuit diagram showing one converter leg, with associated current and voltage signal conditioning

topic) circuit diagram sections of the CoMoCo drive board, as given in Fig. 6.26 will be used to derive the required data. Central to this figure is the lower half of one leg of the three-phase converter, which is connected to a 'Gate driver' circuit that controls all six electronic devices. Inputs to the gate drive module are the signals generated by the PWM modules (see Fig. 6.25). Each of the three lower switches is connected to a shunt resistor which is used to measure the phase current when said device is on. An ADC phase current signal conditioning circuit amplifies the measured shunt voltage shown as `Ifu`, which is used by the ADC converter of the MCU. The ADC full scale current is the peak to peak current value that corresponds to a peak to peak ADC input voltage of 3.3 V. Analysis of the current signal conditioning circuit shows that the ADC full scale current is found using

$$\textit{ADC full scale current}\,(\mathrm{A_{pp}}) = 3.3\left(\frac{R8}{R_p\,R3}\right) \tag{6.6}$$

where $R_p = {R155\,R10}/{(R155+R10)}$, $R3$ and $R8$ are the relevant resistors of the current gain circuit. Substitution of the resistance values (as shown in the diagram) into Eq. (6.6) yields an ADC full scale current value of 39.3 A. Note that the current signal conditioning circuit inverts the shunt resistance voltage, which implies that a positive voltage `Ifu` is represented by an outgoing current from the converter terminal `U`.

Computation of the ADC full scale voltage requires evaluation of the ADC phase and DC bus voltage conditioning circuits. For both circuits, the same voltage attenuation is used, hence only the phase signal conditioning network is considered here. For the phase voltage circuit the attenuation is defined as $R37/(R31+R32+R37+R49)$ and the output `Vu` is connected to the ADC converter of the MCU. The ADC full scale voltage is the peak to peak phase voltage that corresponds to a peak to peak ADC input voltage of 3.3 V. Analysis of the phase voltage conditioning circuit, shows that the ADC full scale voltage is found using

$$\textit{ADC full scale voltage}\,(\mathrm{V_{pp}}) = 3.3\left(\frac{R31 + R32 + R37 + R49}{R37}\right) \tag{6.7}$$

where $R31$, $R32$, $R49$ and $R37$ are the relevant resistors of the voltage gain circuit. Substitution of the resistance values (as shown in the diagram) into Eq. (6.7) yields an ADC full scale voltage value of 430.98 V. In conclusion, the reader is reminded of the fact that the phase voltage signal conditioning circuit is also a low-pass filter, of which the corner frequency may be found using

$$\text{LPF corner freq (Hz)} = \left(\frac{1}{2\pi}\right)\left(\frac{R_v + R37}{R_v\, R37\, C25}\right) \tag{6.8}$$

with $R_v = R31 + R32 + R49$. Substitution of the resistance and capacitance values (as shown in the diagram) into Eq. (6.8) yields a low-pass corner frequency value of 341.24 Hz.

6.2.2 *Case Study V2a: CoMoCo: Commissioning, Phase C+*

Phase C+, is the operational component of the laboratory and is basically a run version of the `.out` file compiled and downloaded to the MCU in phase C (see previous subsection).

The following information is relevant for this laboratory component:

- Reference program : `Case2a_069phCv4_d.vsm`.
- Description: Commissioning of CoMoCo drive.
- Equipment/Software: CoMoCo-1.0-ED-4 converter [3] with Texas Instruments F28069F MCU, Marathon 5k336N2A IM motor, Spectrum Digital XDS510 USB Emulator, DELTA ELEKTRONIKA SM 400-AR-4 Power supply and VisSim simulation program.
- Outcomes: to confirm basic drive operation by observing the voltage and current waveforms.

The run version as shown in Fig. 6.27 uses a VisSim run module, which executes the `.out` file, shown in said module. Two sliders are used, which set the reference frequency and amplitude of the voltage vector. A post-scaling module is used to

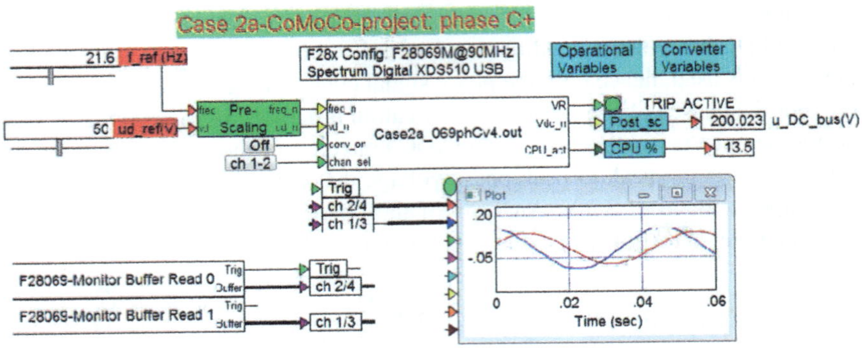

Fig. 6.27 Phase C+: Embedded voltage controller

convert the per unit measured DC bus voltage to actual voltage, as shown with a numeric display. Furthermore, a variable `VR` has been added, which serves to display internal variables, as required. In this case the variable is set to show the 'trip active' signal which is logic '1' when a current trip occurs. The third output `CPU_act` with its post processing module monitors CPU activity. Note that the VisSim 'target interface' module has a dialog box with the option `show CPU utilization` which can also be used to obtain an estimate for CPU activity. However, the approach used here provides a more accurate CPU activity value.

Two 'Monitor Buffer' modules are used to display two selected diagnostic signals. An enlarged view of the VisSim scope module given in Fig. 6.28 shows the per unit current $i_{n_{\alpha}}$ and voltage $u_{n_{\alpha}}$. Multiplication by the full scale current value of 10 A and 400 V for the per unit current and voltage respectively, gives the full scale values. The dialog box entries for this drive are shown for completeness in Fig. 6.29. Note that the ADC full scale shown are those which have been calculated in the previous development phase. If not stop the program and restart. With the correct DC voltage level established by making use of the DC power supply connected to the AC input of the drive, turn on the converter (using the ON button) and monitor the motor shaft and diagnostic scope.

Several important observations can be made on the basis of the information provided in Fig. 6.8 namely:

- The reference frequency has been set to 21.6 Hz, hence the period time of the waveforms should be 46.2 ms.
- A reference voltage amplitude of 50 V has been set by the slider `ud_ref` hence the amplitude of the per unit u_{α} should be approximately (as dead-time effects and the voltage drop across the converter switches will affect the amplitude shown) equal to $^{50}/_{u_{fs}} = 0.125$, where u_{fs} is the full scale voltage value set to 400 V.
- Verification of the current amplitude can be done by making use of a DC-true current probe and oscilloscope, which leads to the result given in Fig. 6.30.

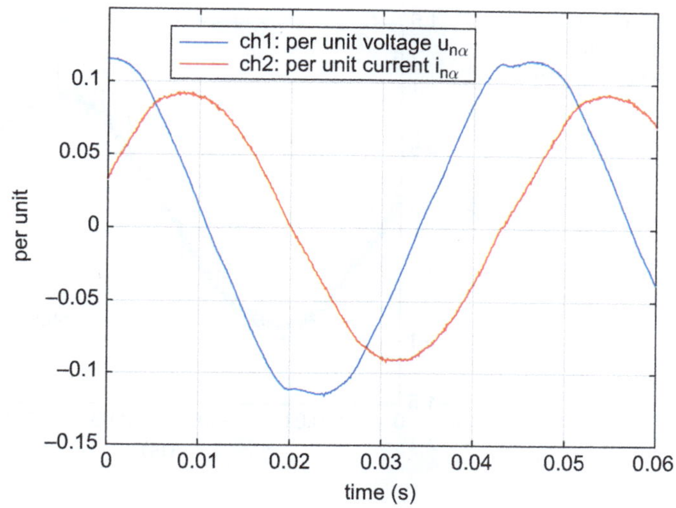

Fig. 6.28 Phase C+ simulation: Scope display showing per unit α current and voltage (CoMoCo drive running with friction load only)

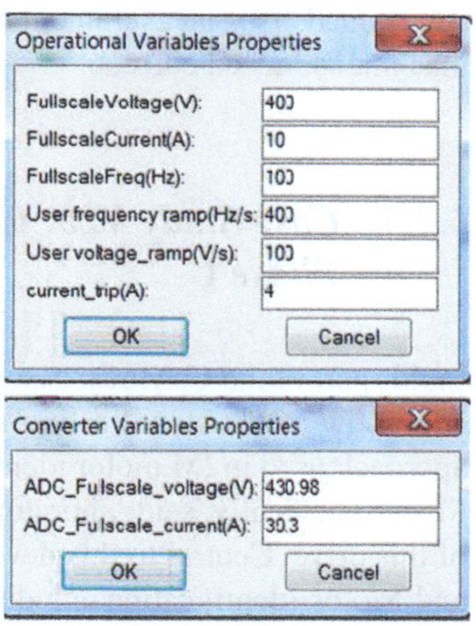

Fig. 6.29 Phase C+: dialog boxes for CoMoCo drive

Observation of the experimental current waveform reveals a current amplitude of 0.9 A, which corresponds to a per unit value of ${}^{0.9}/_{i_{\mathrm{fs}}} = 0.09$, where i_{fs} is the full scale current value set to 10 A. Examination of the VisSim scope result (see Fig. 6.28) confirms that the per unit current is indeed correct.

- The current waveform has a phase lag relative to the voltage waveform which is expected given the presence of an inductive load and (in this case) a rotating machine, which causes a back EMF component.

Finally, observe that the motor is rotating correctly. With a reference frequency of 21.6 Hz the speed of the 4 pole motor will be approximately 648 rpm (no external

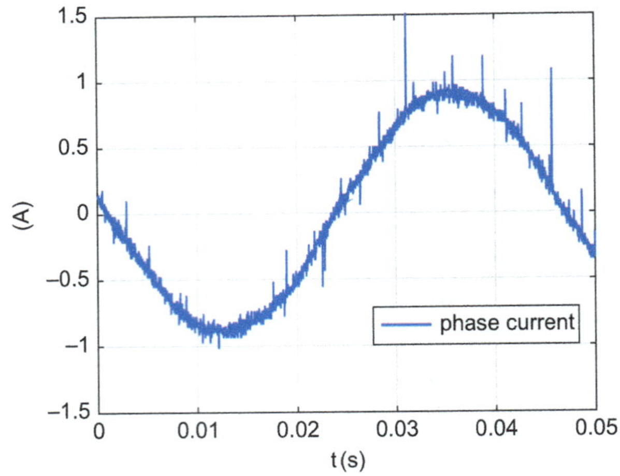

Fig. 6.30 Phase C+: phase current measured with Tektronic current probe and oscilloscope

load, hence speed will be the 'synchronous' speed), which is difficult to measure without suitable equipment. However, if the reference frequency is set to 2 Hz the rotational speed will be 60 rpm, i.e. one revolution per second, which can be visually determined. This measurement will verify that the assumed pole pair number for the machine is indeed correct.

6.2.3 Case Study V2b: CoMoCo Drive: Sensorless Operation, Phase C

In this part of the CoMoCo case study, attention is given to a Phase C module which will allow motor identification and speed control of the drive. For this purpose the approach used in IM motor identification laboratory 3:2 (see Sect. 5.3) will be used. More specifically, said laboratory will be adapted to suit the specific requirements of this drive. Central to this development phase is the 'Drive Controller with FAST and Motor Identification' module, shown in Fig. 6.31, which represents the drive under consideration. Hence details of the drive structure and corresponding dialog box parameter assignments will be discussed in this subsection.

The following information is relevant for this laboratory component:

- Reference program : `Case2b_069phCv6.vsm`.
- Description: Motor Parameter Identification and sensorless control of a CoMoCo drive using the InstaSPIN module.
- CoMoCo-1.0-ED-4 converter [3] with Texas Instruments F28069F MCU, Spectrum Digital XDS510 USB Emulator, DELTA ELEKTRONIKA SM 400-AR-4 Power supply and VisSim simulation program.
- Outcomes: Generate the `.out` file needed to represent the drive structure under investigation.

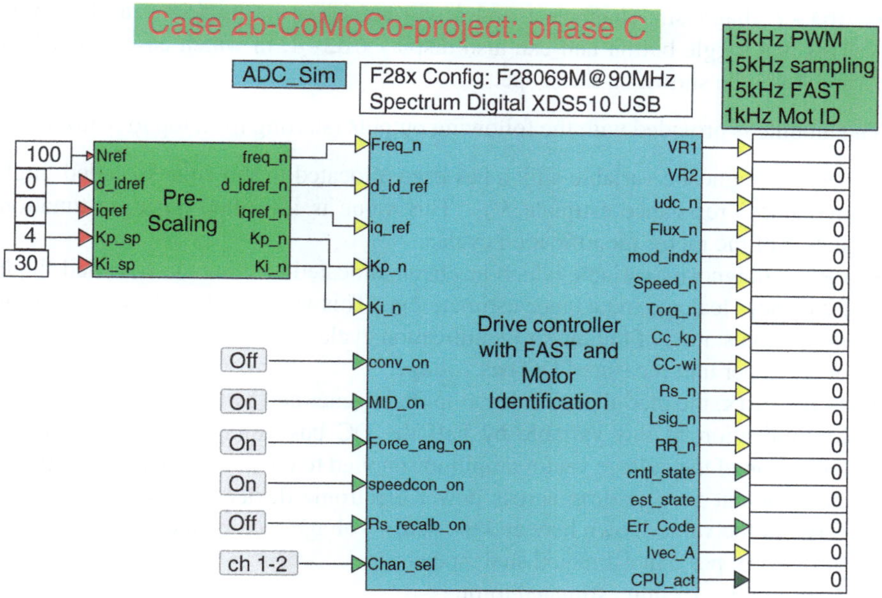

Fig. 6.31 Phase C simulation of a InstaSPIN-FOC CoMoCo drive with motor identification capability

Inputs to the controller are five variables, which set the reference shaft speed, direct/quadrature reference currents and speed controller gains. In this laboratory, six buttons have been allocated to the controller namely:

- `conv_on`: activates the drive.
- `MID_on`:when activated, executes the identification sequence, after the converter has been activated (using the converter ON button) that identifies:
  - `CC_kp/wi`: proportional gain K_p (V/A) and bandwidth ω_i (rad/s) used for the current controller.
  - `Rs`: stator resistance R_s of the machine.
  - `FLux_est`: rotor flux of the machine.
  - `L_sig`: leakage inductance L_σ of the machine.
  - `RR`: rotor resistance of the machine.
- `Force_ang_on`: when activated, forces the FAST algorithm in the chosen (specified by the reference speed) direction. This feature is particular useful for PM with cogging (detent) torque, which would otherwise make it difficult to realize startup under full or partial load conditions. However, it is also advisable to activate this feature during motor identification and full-load torque start of induction machines.
- `Rs_recal_on` :when activated, measures the stator resistance using the 'off-line' identification procedure used during the motor identification sequence, after the converter is enabled and prior to normal drive operation.
- `Speedcon_on`: enables speed or torque control.

- `ch1-2`: diagnostic channels 1 and 2 will be shown in the scope in phase C+. This is a toggle button and can also display `ch3-4`, in which case channels 3 and 4 will be shown on the scope.

The module is provided with the following outputs (starting from top to bottom):

- `VR1`: a diagnostic variable which has been allocated to show the so called 'high frequency' resistance estimate R_{HF}. This value is identified during estimation state 2 of the motor identification cycle.
- `VR2`: a diagnostic variable which has been allocated to show the so called 'high frequency' leakage inductance estimate L_{HF}. This value is also identified during estimation state 2 of the motor identification cycle.
- `udc_n`: per unit measured DC bus voltage.
- `mod_indx`: the modulation index amplitude generated by the current controller. Multiplication of this variable by half the DC bus voltage gives an accurate indication of the voltage vector amplitude applied to the motor. Dead-time effects and inherent voltage drop across power electronic devices will cause an error between the value shown here and the actual voltage seen by the machine.
- `Speed_n`: per unit estimated shaft speed.
- `Torque_n`: per unit estimated torque.
- `CC_kp`: estimate for the proportional current controller gain `K_p`. This value is calculated using $\mathrm{K_p} = {}^{L_{\mathrm{HF}}}/_{\mathrm{T_s}}$, where $\mathrm{T_s}$ is the ADC sampling time, set to 100 μs in this case. L_{HF} is the 'high frequency' inductance value, mentioned above.
- `CC_wi`: estimate for the current controller bandwidth (rad/s). This value is calculated using $\omega_i = {}^{R_{\mathrm{HF}}}/_{L_{\mathrm{HF}}}$, where R_{HF} is the 'high frequency' resistance value, mentioned above.
- `Rs_n`: per unit estimate for the stator resistance. Multiplication of this variable by a gain factor $({}^{u_{\mathrm{fs}}}/_{i_{\mathrm{fs}}})$ gives an estimate for the stator resistance `Rs` (Ω).
- `Lsig_n`: estimate for the leakage inductance (4-parameter model) in (H).
- `RR_n`: per unit estimate for the rotor resistance (4-parameter model). Multiplication of this variable by a gain factor $({}^{u_{\mathrm{fs}}}/_{i_{\mathrm{fs}}})$ gives an estimate for the stator resistance `RR` (Ω).
- `cntrl_state`: status of the controller.
- `est_state` : status estimator.
- `Err_Code`: a diagnostic variable which generates an error code when motor identifications fails during specific critical stages. The following codes are present:

  - Error 0: no errors encountered.
  - Error 1: failure during open-loop flux estimation, state 5 (for PM this error happens in state 6).
  - Error 2 or 3: failure during closed-loop flux estimation, state 7.
  - Error 4: failure during inductance estimation, state 10.

- `Ivec_A`: a diagnostic output which represents the per unit amplitude of the current vector. This variable is used for monitoring the identification process.
- `CPU_act`: a variable used to monitor CPU activity.

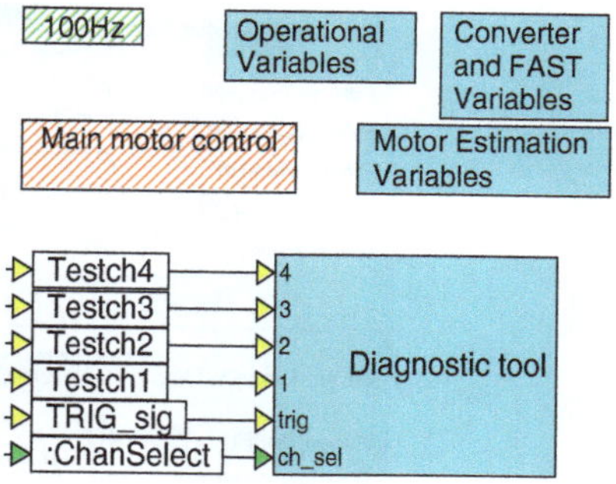

Fig. 6.32 Phase C simulation of a FAST based encoderless FOC drive, with motor identification: one level into the drive controller module

An ADC sampling frequency of $f_{\mathrm{ADC}} = 15\,\mathrm{kHz}$ has been selected for this laboratory, as was mentioned above. The current/speed controllers have also been set to operate at the ADC sampling frequency.

> The FAST and Motor Identification algorithms must be set to operate at sampling frequencies of 15000/i kHz and 15000/j kHz respectively, where i, j are integer values only.

For this lab the values used are $i = 1$ and $j = 15$ which implies that the sampling frequency for the FAST and Motor ID algorithm have been set to 15000 Hz and 1000 Hz respectively. These frequencies are set in the appropriate dialog boxes as will be discussed below.

Moving one level lower into the 'Drive Controller' module reveals the modules shown in Fig. 6.32. Of these shown in the figure, the '100 Hz' and 'Diagnostic tool' modules have been discussed in previous laboratories. The dialog box parameters for this laboratory are given in Fig. 6.33. The parameters associated with the 'Operation variables' and 'Converter and FAST Variables' dialog boxes, have been discussed in previous chapters. However, the readers attention is drawn to several key parameters which should be set prior to motor identification namely:

- `User current_ramp (A/s)`: rate of current change which is also used during motor identification. A too high rate, may cause motor ID to fail because the required current cannot be delivered by the converter. A too low value, may lead to the current not reaching the intended target value for a given motor ID phase.
- `speedcontroller_limit (A)` value sets the limit quadrature current reference value for the speed controller. For induction machines a motor ID sequence occurs where the rotor must be locked, in which case the speed controller will set the quadrature current required to maintain the user set stator frequency. Said current value will depend on the slip frequency in use and will be constrained by the (absolute) limit value set here.
- `Stator_flux_est (Wb)` estimate for the stator flux, this value follows from the analysis given in the previous stage, see Eq. (6.4), which is the value which

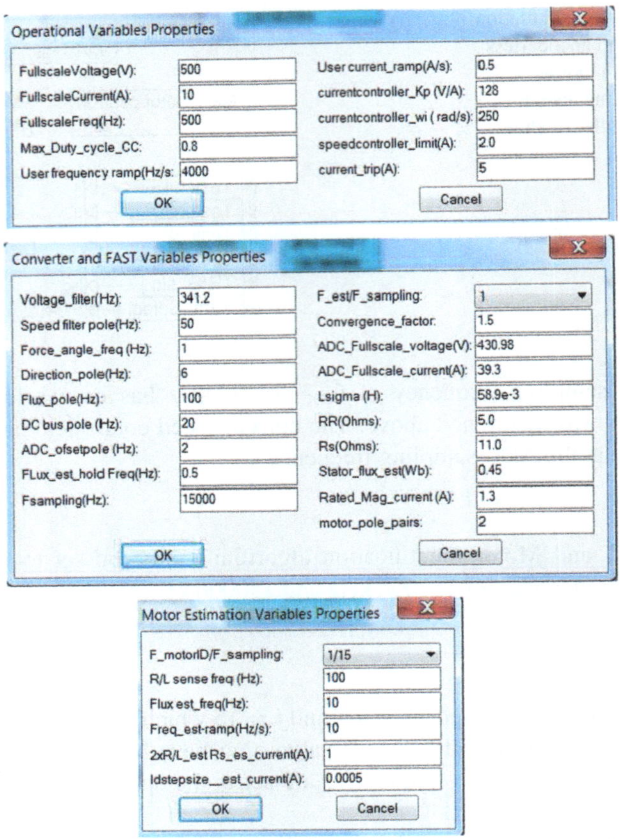

Fig. 6.33 Phase C simulation of a FAST based encoderless FOC drive, with motor identification: Dialog box entries

must be set here. This value is important as the identification routine will adjust the direct axis current in incremental steps until the rated stator flux level is reached.

- `Rated_Mag_current (A)` rated magnetizing current value identified during a motor identification cycle. This value is used when the machine is not identified prior to use. Hence its value can only be set once it has been identified or derived from a machine data sheet.

Also shown in Fig. 6.33 is the dialog box 'Motor Estimation Variables', where the following motor identification parameters must be assigned:

- `F_motorID/F_sampling`: ratio between sampling frequency used by the Identification algorithm and ADC sampling frequency. The ratio is an integer value, hence the user can select the ratio from a dialog box pull-down menu. For example, a choice of `1/15` implies an estimator sampling frequency of 1000 Hz, given that the ADC sampling frequency is set to 15000 Hz.

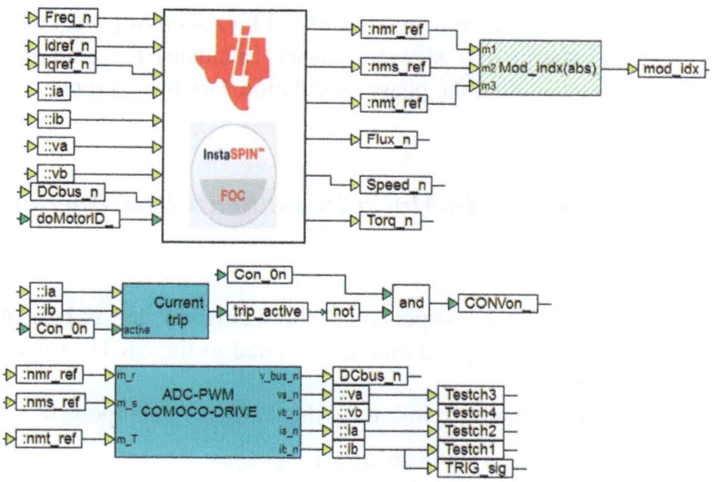

Fig. 6.34 Phase C simulation of a FAST based encoderless field-oriented controlled (FOC) drive, with motor identification: One level into the 'Main Motor Control' module

- `R/L sense freq (Hz)`: frequency used to estimate the current controller gain and bandwidth.
- `Flux est_freq(Hz)`: frequency used to estimate magnetizing current and leakage inductance.
- `Freq_est-ramp(Hz/s)`: rate of frequency change used for motor identification purposes.
- `2xR/L_est,Rs_es_current (A)`: current amplitude used for current controller gain/bandwidth and stator resistance estimation.
- `Idstepsize__est_current(A)`: current amplitude used during the magnetizing current identification of the machine. During this process the direct axis current is changed in increments governed by this parameter, until the stator flux of the machine matches the user set reference flux level.

Moving one level lower into the 'main motor control' module, reveals a set of modules given in Fig. 6.34. Key inputs to the InstaSPIN controller module are the per unit α, β current/voltage variables: `ia`,`ib`, `va`, `vb` and DC bus voltage variable `DCbus_n`, generated by the ADC-PWM module. User reference variables: `Freq_n`, `idref_n`, `iqref_n` represent the users inputs for shaft-speed and direct/quadrature reference currents. Logic input `doMotorID_` is used to activate the motor identification sequence. Outputs of the 'InstaSPIN-FOC module' are (among others) the modulation indices: `nmr_ref`, `nms_ref`, `nmt_ref` generated by the internal current controller. The remaining outputs are the per unit rotor flux `Flux_n`, shaft speed `Speed_n` and shaft torque `Torq_n`. For diagnostic purposes the test variables: `TESTch2`, `TESTch1` have been connected to per unit variables: i_{n_α} and i_{n_β} respectively. Furthermore, test variables: `TESTch3`, `TESTch4` are used to display the filtered per unit converter voltages u_{n_α} and u_{n_β} respectively.

The module `Mod_idex`, which operates with a 1 kHz sampling frequency instead of 15 kHz (which is why it is shown hashed), calculates the amplitude of the modulation index vector using the phase modulation variables as inputs.

6.2.4 *Case Study V2b: CoMoCo Drive: Sensorless Operation, Phase C+*

Phase C+, is the operational component of the laboratory and is basically a run version of the `.out` file compiled and downloaded to the MCU in phase C (see previous subsection).

The following information is relevant for this laboratory component:

- Reference program: `Case2b_069phCv6_d.vsm`.
- Description: FOC sensorless control of an IM machine, with motor parameter identification.
- Equipment/Software: CoMoCo-1.0-ED-4 converter [3] with Texas Instruments F28069F MCU, Marathon 5k336N2A IM motor, Spectrum Digital XDS510 USB Emulator, DELTA ELEKTRONIKA SM 400-AR-4 Power supply and VisSim simulation program.

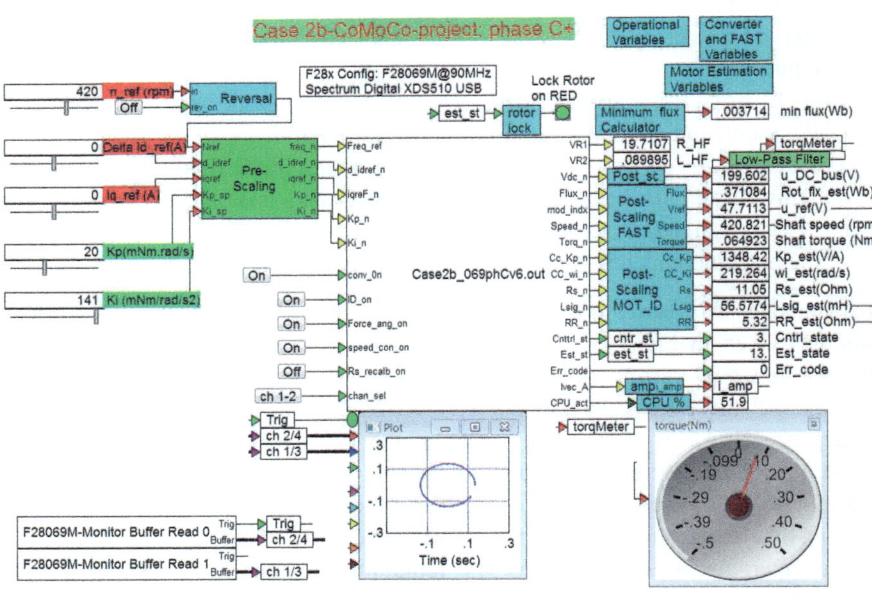

Fig. 6.35 Phase C+ simulation of a FAST based encoderless field-oriented controlled (FOC) drive, with motor identification: example shows CoMoCo drive operation under no load at 420 rpm, after identification has been done

- Outcomes: application of the InstaSPIN algorithm for Marathon IM motor parameter identification purposes and sensorless FOC operation.

The run version shown in Fig. 6.35, uses a VisSim run module, which executes the `.out` file, shown in said module. Prior to running this `.out` file, the following steps should be undertaken:

- Compile the `.out` file in phase C, with 'Flash' option disabled (use of RAM to simplify VisSim operation).
- Prior to start up (by activating the VisSim 'go' button), set the `conv_on` button to OFF and the remaining button as shown in Fig. 6.35.

When the above sequence is completed, hit the go button (VisSim 'green' arrow) and monitor the DC bus voltage. Upon verification that the value is correct the `conv_on` button should be enabled, in which case the motor identification cycle should start as may be verified by monitoring the current vector amplitude, estimated shaft speed, estimator state, estimated flux, leakage inductance and rotor resistance given in Fig. 6.36 for this case study. The scope module given in Fig. 6.35 has been set to show the per unit plot of the current vector trace. In this case its primary diagnostic value is to observe that the circle is centered relative to the origin, which implies that ADC offset correction has been carried out correctly. A brief outline of the experimental identification sequence with corresponding estimator states is given below, where use is made of Fig. 6.36:

- Estimator state 1: estimator idle, ADC offset measurement sequence occurs after the converter has been activated (converter button ON and motor ID on). Controller state cycles from 'idle' via 'offline' to 'online.
- Estimator state 2: identification of the current controller gain and bandwidth. Frequency and amplitude (set in motor ID dialog box) are 100 Hz and 1.0 A (the current value used in this sequence is half the value used in the Rs estimation phase, i.e. 0.5 A during this phase). Motor should NOT rotate whilst a rotating current vector of 0.5 A is applied to the machine. At the end of this sequence the so called 'high frequency' resistance R_{HF} (Ω) and inductance L_{HF} (H) estimates are returned. In this laboratory these variables are shown by diagnostic variables `VR1` and `VR2` respectively. If this stage is executed correctly the readings provide an initial estimate for the motor parameters used to set the current controller gain and bandwidth. The R_{HF} value gives an estimate of the sum of the rotor and stator resistance, while L_{HF} provides an estimate for the leakage inductance.
- Estimator state 3: stator resistance measurement, DC current level (set in motor ID dialog box) is set to 1.0 A. A stationary current vector is applied with an amplitude of 1.0 A and at the end of this sequence the stator resistance value is returned.
- Estimator state 4: open-loop speed control, with a current vector amplitude set to 1.0 A, which is the amplitude used by the resistance measurement. Current ramp up amplitude of 0.5 A/s (set in the 'operational' dialog box). The rotational speed of the vector will ramp up at a rate of 20 Hz/s to 10 Hz. Motor speed should increase to the level that corresponds to the frequency set in the motor

ID dialog box, before reaching the end of this sequence. Hence, in this case, with 10 Hz, being the selected frequency the rotational speed will be approximately $^{5*60}/_{p} = 300$ rpm, where $p = 2$ is the pole pair number of the machine in use.

- Estimator state 5: transitioning from open to closed-loop speed control after which the direct axis current is incremented in steps until the estimated flux (which is the stator flux in this case) matches the reference flux of 0.45 Wb in this case. In this cycle the rated direct axis current is determined.
- Estimator state 7: estimation of the (in this case) stator flux under closed-loop speed control. Current amplitude, should be the direct axis current that corresponds to the selected user stator flux level times a constant $k = 0.4$ set in the 'InstaSPIN Motor control module'. This constant allows the user to choose a magnetizing current during inductance and flux estimation which is a factor k lower than the estimated value. This is useful to be able to operate with a lower flux level and corresponding torque level during locked rotor conditions. Furthermore, a value of $k < 1$ is used (as is the case here) when significant saturation of the machine is present at rated flux. In this case the magnetizing current value used is reduced to 0.55 A as may be observed from the current

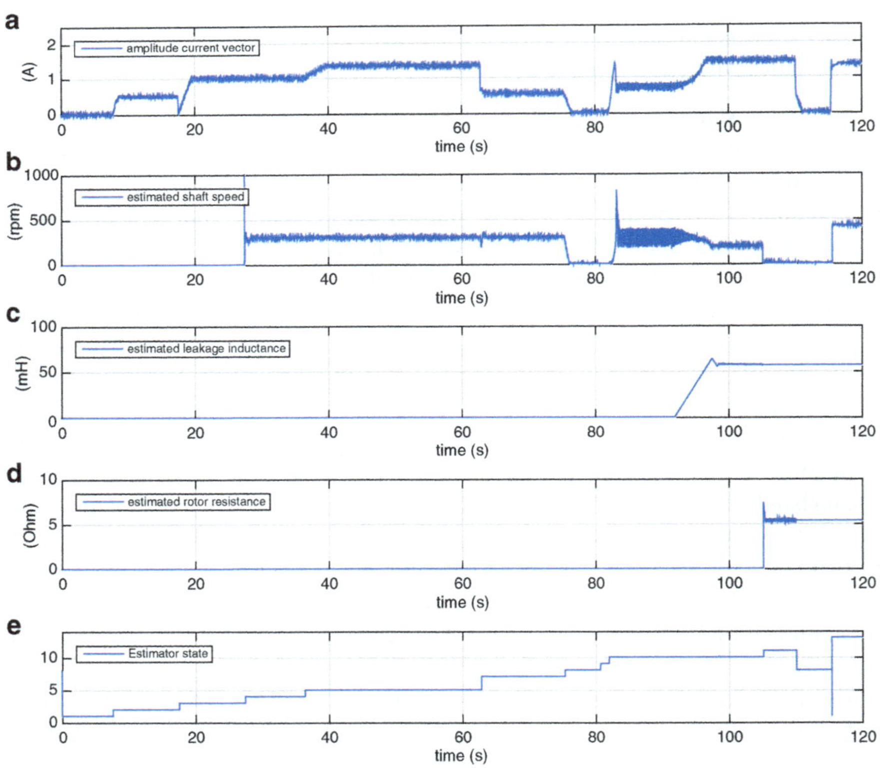

Fig. 6.36 Motor parameter identification sequence for CoMoCo/Marathon drive

vector amplitude (see Fig. 6.36). Speed set by flux estimation frequency of (in this case) 10 Hz.

- Estimator state 8: speed reduced to zero, so that the rotor can be locked by the user.
- Estimator state 9: A time delay which can be set by the user to execute the rotor locking action.
- Estimator state 10: Measurement of the leakage inductance. Current amplitude (and the torque required to hold the rotor) will be determined by the choice of user frequency (set to 10 Hz in this case), magnetizing current used and the speed controller limit value (set to 2 A in this case). Operation under closed-loop speed control. At the end of this cycle the leakage inductance value (4-parameter model representation) is returned to the user. Note that this measurement will only return a meaningful result if the rotor shaft can be locked. The magnetizing current is set by the estimated rated value times $k = 0.4$ (as discussed in state 5).
- Estimator state 11: Measurement of the rotor resistance. Current amplitude (and the torque required to hold the rotor) will be determined by the choice of user frequency (set to 10 Hz in this case) and the speed controller limit value (set to 2 A in this case). Operation under closed-loop speed control. At the end of this cycle the rotor resistance value (4-parameter model representation) is returned to the user. Note that this measurement will only return a meaningful result if the rotor shaft can be locked.
- Estimator state 8: Ramp down motor frequency to zero so that shaft can be unlocked.
- Estimator state 12: completion of motor identification sequence.
- Estimator state 13: Drive online and control returned to the user.

Note that the leakage inductance/rotor estimation states 10, 11 require the rotor to be locked. In industrial applications where this cannot be done, the reader has the following options:

- Redo the motor identification with a motor that can be locked, by using for example a second machine (of the same type).
- Use the high frequency inductance and resistance estimates obtained during estimation state 2 together with the measured stator resistance in state 3 to obtain estimates for the leakage inductance and rotor resistance.

Upon completion of the identification procedure (and control was returned to the user) the reference speed was set to 420 rpm in order to identify the magnetization characteristic of the machine. For this purpose the (unloaded) machine speed was kept constant, whilst the direct axis current was increased in steps of $\pm$ 0.1 A using the `d_Id _ref` slider. With this slider at zero the current vector amplitude (being the direct axis current) is accurately measured from Fig. 6.36 and this value of $i_d = 1.3$ A serves as the reference value. At each change of the slider `d_Id _ref` both current vector amplitude and rotor flux must be recorded. The results of this experiment, given in Fig. 6.37, provide the reader with an indication of the level of saturation present for the chosen operating point (as set by the user

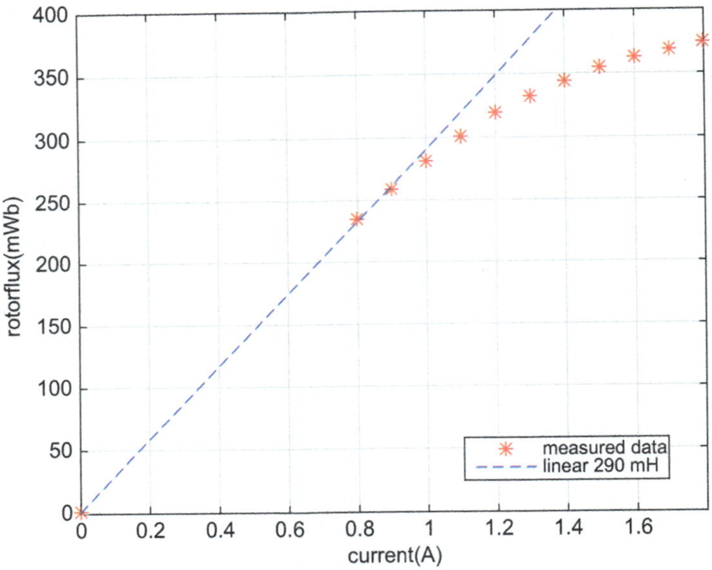

Fig. 6.37 Magnetization curve of the Marathon 5k336N2A IM motor

defined stator flux level). Added to Fig. 6.37 is a linear approximation function, the gradient of which represents the magnetization inductance L_M (4-parameter model representation) of the machine in use. These results confirm that there is significant magnetic saturation present at rated magnetizing current 1.3 A, which is the reason why a lower value was used for the identification of the leakage inductance.

Undertaking identification with a machine operating at an operating point where severe magnetic saturation is present may lead to erroneous estimation of the leakage inductance and rotor resistance.

The motor identification routine, as mentioned above, will provide accurate and meaningful results, provided that prudent choices are made with respect to the ADC full-scale voltage/current values and motor identification dialog box parameters. For this purpose so called 'Motor Identification goodness' plots have been introduced in previous chapters which provide a 'post-processing' analysis of the motor identification process. The 'goodness plots' for this drive shown Fig. 6.38 are briefly discussed below.

The voltage 'ADC transfer' function given in Fig. 6.38 (top plot) shows the 'ADC count', which is the decimal output value that corresponds to the input voltage range $0 \rightarrow \mathit{ADC\,full\,scale\,voltage}$, where the *ADC full scale voltage* (see Eq. (6.2)) value is equal to 430.98 V for this converter. The least significant bit $\mathrm{LSB_v}$ for

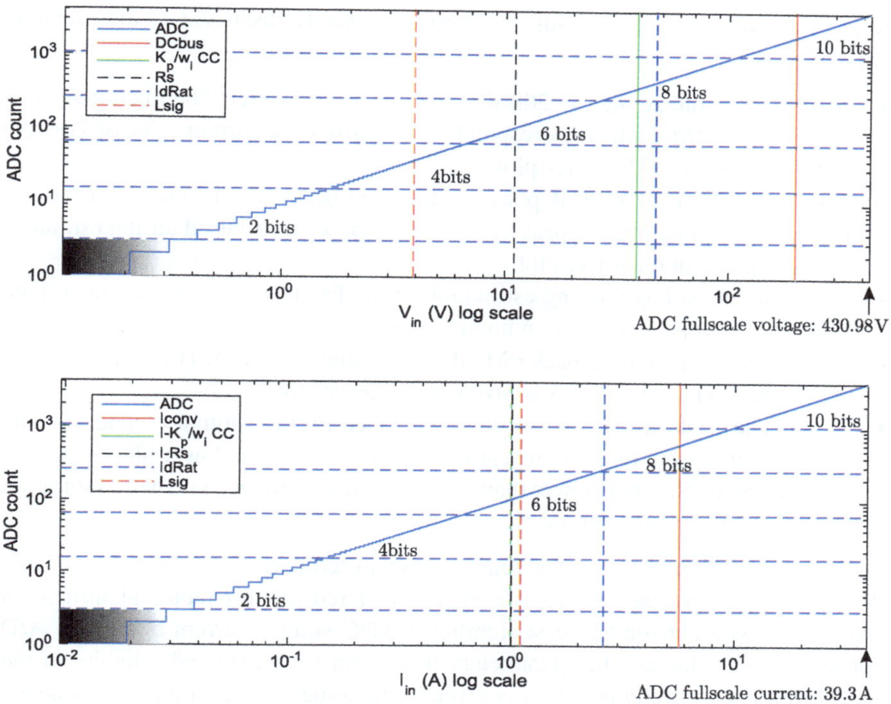

Fig. 6.38 Motor identification 'goodness' plots

the voltage transfer function is equal to $^{ADCfull\,scale\,voltage}/_{4095} = 105.2\,\text{mV}$. This value is used to set the horizontal logarithmic scale, which ranges from $LSB_v \rightarrow$ *ADCfull scale voltage*. The lower plot given in Fig. 6.38 shows the the current ADC transfer function, with input current range of $LSB_i \rightarrow$ *ADCfullscale current*, where the *ADCfullscale current* is equal to 39.3 A (peak to peak current) for this board. Furthermore, the least significant bit LSB_i for the current transfer function is equal to $^{ADCfullscale\,current}/_{4095} = 9.6\,\text{mA}$. The vertical axis of both plots show the ADC count from $1 \rightarrow 4095$. Note that the quantization error becomes significant, when input voltage/current levels are of the same order of magnitude as LSB_v/LSB_i respectively. Consequently, the figures show how well the ADC converters are utilized for a chosen set of motor ID input variables. Note that this 'post-processing' type of analysis, makes use of the measured motor parameters. As such these plots show whether or not the identification procedure has been carried out with appropriate parameter settings (as set in the motor identification dialog box). Furthermore, these plots can provide some guidelines as to what input parameters may need to be changed to improve identification, given the machine under consideration. A brief overview of the voltage/currents levels found during motor identification (see Fig. 6.38) is given below, for each of the legend variables. Note that only the results of the computations is presented here, as the method used to derive this type of result has been extensively discussed in the previous two chapters.

- `DCbus`: DC voltage $u_{DC} = 200$ V in use for the converter. The rated converter current is 2 A RMS, which corresponds to a peak to peak current value of `Iconv`: 5.65 A, as shown in the lower plot.
- `K_p/w_i CC`: peak to peak phase voltage that occurs during estimator state 2. The corresponding peak to peak current `I-K_p/w_i cc` used during estimator state 2 is given in the lower plot.
- `R_s`: The DC voltage during estimation of the DC resistance in estimator state 3, where `I_Rs` is the corresponding DC current value used during identification.
- `Idrat`: The peak to peak back EMF during estimator state: 5. The corresponding peak to peak current of 2.6 A is given in the lower plot.
- `Lsig`: The peak voltage across the inductance during estimator state 10. The corresponding peak to peak current of 1.1 A is given in the lower plot. Note that this current amplitude is lower than the rated value to reduce magnetic saturation effects during this measurement.

Overall observation of the voltage/current levels used during estimation learns that the peak to peak voltage /current levels encountered for this motor identification process are well outside the least significant ADC voltage/current bits of the A/D converter. Consequently, the parameters found can be considered reliable. In the sequel to this case study attention is given to the issue of selecting key parameters such as those shown in the dialog boxes, for the drive. A useful guide for this task is provided by Texas Instruments, in the form of an Excel sheet [2]. In this document the user must enter a set of basic drive settings, on the basis of which a set of guidelines and parameters are provided. An application example of this user guide, for the drive discussed in this case study is shown in Fig. 6.39. Observation of said figure (top section) shows a 'yellow' column, which represents the drive data that must be provided by the user. This includes the CPU frequency, DC bus voltage, motor pole pair number, anticipated maximum shaft speed and the selected PWM frequency for the converter. The latter variable is critical as it in turn sets three divisor values (referred to as 'ticks'), which define the ADC sample frequency, current controller and estimator frequency. A 'sub calculation' dialog confirms the frequency used by the ADC, current control and estimator. In this case all are set to 15 kHz.

If, as is the case here, the ADC full scale voltage/current and LPF are known then these must be provided under row `12`, which also shows a 'red' dialog box designed to alert the user to the need to set the low-pass voltage filter to the same pole as used in the hardware. Finally, under row `16`, some data on the motor must be provided, which is the stator flux (in V/Hz) that is equal to ψ_s (Wb). 2π and the motor leakage inductance (found for example by using an L-C-R meter or identified). On the basis of the above information the Excel sheet provides guidance in terms of variables which should be selected. In addition, use is made of boolean indicators ('green boxes') to alert the reader to any changes that may need to be made to parameters. This user guide also provides guidance (not shown in Fig. 6.39) when incompatible

1. FILL IN THESE VALUES FOR YOUR USER.H, MOTOR, and INVERTER HW			
USER_SYSTEM_FREQ_MHz	90	MHz	sub-calculations
Maximum Bus Voltage	500	V	
Maximum Target RPM	1800	RPM	60 Hz
USER_MOTOR_NUM_POLE_PAIRS	2	PAIRS	4 POLES
USER_PWM_FREQ_kHz	15	kHz	
USER_NUM_PWM_TICKS_PER_ISR_TICK	1	ticks	15 kHz
USER_NUM_ISR_TICKS_PER_CTRL_TICK	1	ticks	15 kHz
USER_NUM_CTRL_TICKS_PER_CURRENT_TICK	1	ticks	15 kHz
USER_NUM_CTRL_TICKS_PER_EST_TICK	1	ticks	15 kHz
USER_ZEROSPEEDLIMIT	0.002		
2. The following are set by HW design, use defaults for TI EVM or your own HW			
USER_VOLTAGE_FILTER_POLE_Hz	341.2	Hz	Tune to HW Pole
USER_ADC_FULL_SCALE_CURRENT_A	39.20	A	
USER_ADC_FULL_SCALE_VOLTAGE_V	430.98	V	
4. Once Motor ID is attempted, update these as best you can and check IQ_V Scaling			
USER_MOTOR_RATED_FLUX	2.83	V/Hz	0.047619048 V/Hz
USER_MOTOR_Ls_d	5.00E-02	H	
5. * Ideal Pole Design when you build wwn HW			
Minimum Pole	200	Hz MIN	
Ideal* Pole >=	16.5	Hz	
Half** Pole >=	8.25	Hz	

	3. THESE VALUES ARE RECOMMENDED FOR USE once TRUE checks are satisfied	
sub-calculations		
	Ideal USER_IQ_FULL_SCALE_FREQ_Hz for Motor	66.0
60 Hz	Maximum allowed USER_IQ_FULL_SCALE_FREQ_HZ for HW	1297.0
4 POLES	**USER_IQ_FULL_SCALE_FREQ_Hz**	66.0
	Maximum RPM Supported	3920
15 kHz	**USER_MOTOR_FLUX_EST_FREQ_Hz**	6.0
15 kHz	**USER_MAX_ACCEL_EST_Hzps**	5.0
15 kHz	CURRENT Hz > MAX_Hz * 7	TRUE
15 kHz	EST <= CTRL	TRUE
	EST > 10 * USER_VOLTAGE_FILTER_POLE_Hz (+10% margin)	TRUE
	EST > 8 * TARGET_Hz	TRUE
Tune to HW Pole	FLUX_EST_FREQ > ZEROSPEEDLIMIT * FULL_SCALE_FREQ	TRUE
	USER_IQ_FULL_SCALE_CURRENT_A	21.0
	starting USER_IQ_FULL_SCALE_VOLTAGE_V	500.0
ling	**4. Check after valid USER_MOTOR_RATED_FLUX Identification**	
0.047619048 V/Hz	= Minimum Flux that can be measured with new USER_IQ_FULL_SCALE_VOLTAGE_V (cell I20)	
	Minimum Flux Measurement < 0.9 * RATED_FLUX	TRUE
	IQ_VOLTAGE < RATED_FLUX * EST Hz	TRUE
	new USER_IQ_FULL_SCALE_VOLTAGE_V	500.0
	a) Bus Voltage	500.0
	b) Bemf Generated @ Target Hz + 10% buffer	186.8

Fig. 6.39 InstaSPIN user variable guide (split in two parts for readability) [9]

parameters are entered. Below the recommendations for the CoMoCo/Marathon drive:

- Minimum corner frequency of the LPF (if the user is still in the converter design stage see data underneath row `19`): 200 Hz is recommended.
- `Ideal_USER_IQ_FULL_SCALE_FREQ_Hz`: recommended value that can be set in the 'Operations Variables' dialog box (see Fig. 6.33). Underneath the recommended value of 66 Hz is the maximum value of 1297 Hz.

- `Maximum RPM supported`: highest shaft speed which InstaSPIN can operate with given the current settings in use, equal to 3920 rpm.
- `USER_MOTOR_FLUX_EST_FREQ_Hz`: recommended frequency used during estimator phases 4, 5 and 10 (see Fig. 6.16) and set in the 'Motor Estimation Variables' dialog box (see Fig. 6.33).
- `USER_MAX_ACCCEL_EST_Hzps`: frequency ramp rate during estimation which can be set in the 'Motor Estimation Variables' dialog box.
- `USER_IQ_FULL_SCALE_CURRENT_A`: full scale current recommended value that can be set in the 'Operations Variables' dialog box (see Fig. 6.33).
- `new USER_IQ_FULL_SCALE_VOLTAGE_V`: recommended value that can be set in the 'Operations Variables' dialog box (see Fig. 6.33).

In addition to the above, a series of 'green' boolean indicators are set, as mentioned above, pending the data provided in the yellow boxes. On the basis of these indicators the user can adjust relevant parameters. For further details the reader is referred to the MotorWare document under consideration [2].

6.2.5 *Case Study V2c: CoMoCo Drive: PFC+boost, Introduction*

Hitherto in this book a DC power supply has been used to meet the power requirement for the converters considered. However, many industrial applications utilize either a single or three-phase mains supply for the drive, hence for completeness a case study involving a 'mains connected' converter is appropriate. In this case study a single phase 'mains' supply source `V_ac` is considered as shown in Fig. 6.40. In this example a full bridge rectifier is used together with DC bus capacitor `C2`, which serves to reduce the AC ripple on the DC bus output `Vdc`. In this part of the case study the load (represented by `RLOAD`) is the CoMoCo drive which operates as a V/f drive. The input power of said drive is held approximately constant and equal to 131 W, while its output is connected to a Vostermans [16], induction machine based fan unit. For this operating scenario the input and current were measured as

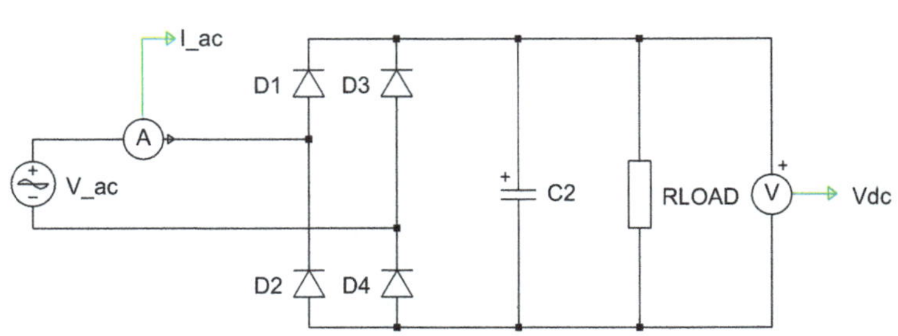

Fig. 6.40 Single phase mains supply with rectifier

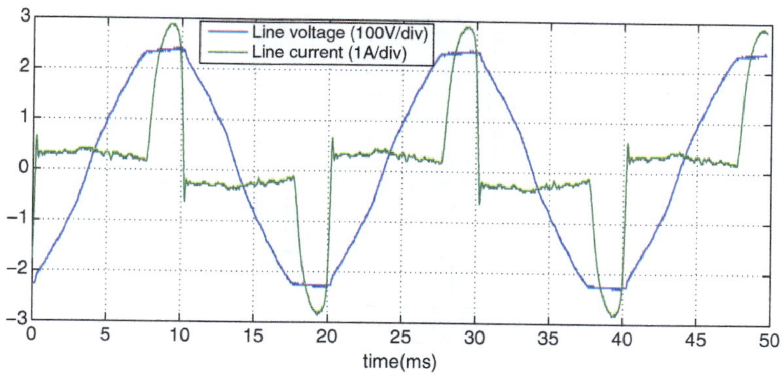

Fig. 6.41 Measured input voltage/current waveforms for operation with a full-bridge rectifier only

may be observed from Fig. 6.41. For safety reasons a variac was connected between the 230 V RMS/50 Hz single phase grid and the input terminals of the CoMoCo drive. Observation of said results shows a 169 V RMS sinusoidal input voltage and a current waveform (which is NOT sinusoidal) with a peak of approximately 3.0 A. The type of current waveform present, is typical of power supplies which utilize a full-bridge rectifier. The pulsed nature of the waveform follows from the need to cyclically charge the typically large DC bus capacitor `C2`. Power supplies which have a pulsed current waveform as shown in Fig. 6.41 are likely to be rejected for consumer use as they usually do not meet IEC standard for mains connected appliances. Typically the pulsed current waveform introduces unwanted higher harmonics in the grid. A second problem of the rectifier circuit of Fig. 6.40 is that the maximum DC bus voltage is inherently limited by the peak of the input voltage waveform. Hence if the drive in its current setup is connected directly to the mains the maximum DC voltage would be $220\sqrt{2} = 311$ V. A standard three-phase motor when directly connected to the grid would then operate with a line voltage of $220\sqrt{3} = 380$ V. If a power electronic converter was to be used to provide the same rated output for the motor the corresponding minimum DC bus voltage would need to be $381\sqrt{2} = 538$ V. Consequently, a simple single phase rectifier network as introduced above will not suffice on the grounds of not being able to deliver the required DC bus voltage level and the presence of a non sinusoidal input current waveform.

A better alternative to the above approach is the input power circuit shown in Fig. 6.42, which is known as a 'two phase interleaved Power Factor Correction Circuit (PFC)' [1]. The term 'two phase' refers to the presence of two 'boost' converters namely:

- Phase 1: Power electronic switch `Q1`, inductance `L3` and diode `D5`. The power electronic device is controlled by PWM input `PFC_PWM1` via a gate driver circuit (represented by the `NOT` module)

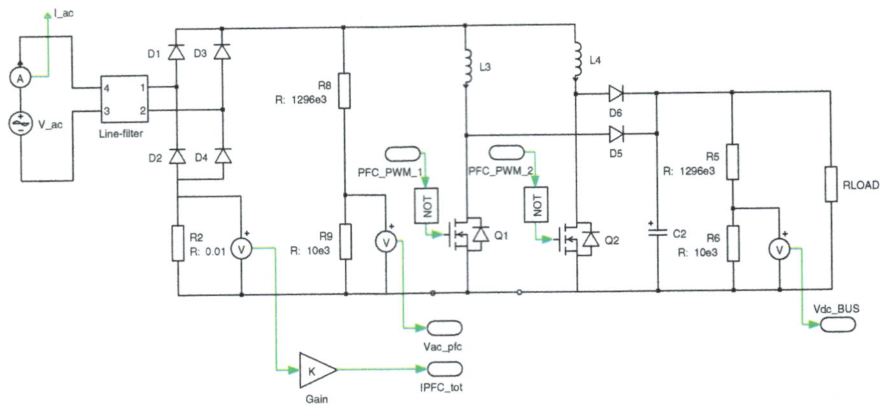

Fig. 6.42 Single phase mains supply with PFC+boost circuit

- Phase 2: Power electronic switch Q2, inductance L4 and diode D6. The power electronic device is controlled by PWM input PFC_PWM2 via a gate driver circuit (represented by the NOT module)

When either device is turned on, magnetic energy is stored in the corresponding coil, which upon turn off is transferred to the bus capacitor C2. This implies that a bus voltage higher than the peak rectified voltage will be present. The term 'PFC' refers to the presence of a controller (to be discussed more extensively in the ensuing subsection) which ensures that the rectified current measured by resistance R2 is a mirror image of the rectified voltage. Furthermore, the amplitude of said current is controlled in such a manner that the DC output voltage equals the user set reference voltage. The CoMoCo drive input circuit is identical in concept to the circuit representation shown in Fig. 6.42, as may be deduced (to some extent) by observation of Fig. 6.20 which shows two toroidal coils which represent circuit elements L3, L4, the metal enclosure of the input line filter and three large capacitors which together represent circuit element C2. The functionality of this PFC module is illustrated with the aid of the measured results given in Fig. 6.43. Note that the same DC bus load conditions as used for the previous measurement are also applied in this case. Observation of the results, shows the presence of the main voltage waveform together with the current, which is now sinusoidal and almost in phase with the latter. A small phase delay is visible, which is due to the presence of the line filter, that is required to ensure that switching harmonics from the two boost converters do not appear on the AC line side. In this example the reference DC bus voltage was set to 350 V. Furthermore, the current amplitude required to deliver the same (as in the previous case, without PFC) input power is considerably lower namely 1.1 A. Hence the use of the PFC+boost approach delivers the required outcomes of boosting the DC bus voltage and generating a sinusoidal AC current waveform that can in principle satisfy IEC requirements. The ensuing two subsections of this case study outline an embedded control approach which can control the PFC-boost section of the CoMoCo drive.

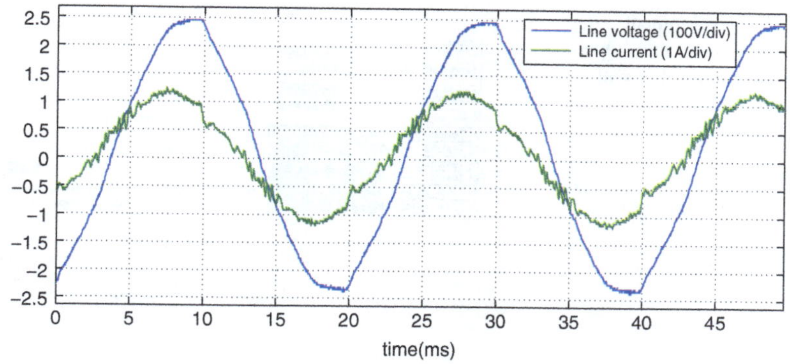

Fig. 6.43 Measured input voltage/current waveforms for operation with PFC+boost

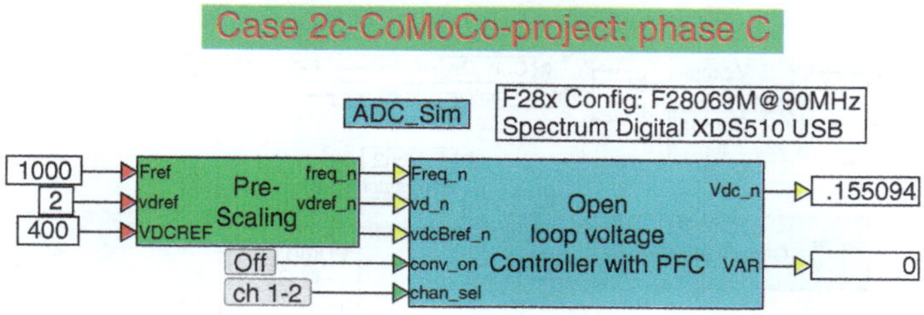

Fig. 6.44 Phase C simulation of a CoMoCo based V/f converter with PFC+boost capability

6.2.6 Case Study V2c: CoMoCo Drive: PFC+boost, Phase C

In this part of the CoMoCo case study, attention is given to the Phase C module which contains a PFC+boost compound module that is of interest for this part of the case study. Central to this development phase is the 'Open loop voltage controller with PFC' module, as shown in Fig. 6.44, which represents the control structure under consideration. The following information is relevant for this laboratory component:

- Reference program : `Case2c_069phCv3.vsm`.
- Description: PFC+boost control of the CoMoCo drive.
- CoMoCo-1.0-ED-4 converter [3] with Texas Instruments. F28069F MCU, Spectrum Digital XDS510 USB Emulator and VisSim simulation program.
- Outcomes: Generate the `.out` file needed to control the PFC structure under investigation.

Inputs to the controller are three variables, which set the reference frequency, direct axis reference voltage for the V/f drive (as introduced in Sect. 3.1) and the reference DC bus voltage for the PFC controller. The logic inputs are `conv_on` which activates the V/f converter and PFC module simultaneously. A channel select

Fig. 6.45 Development Phase C, one level into 'Open loop voltage controller with PFC' module

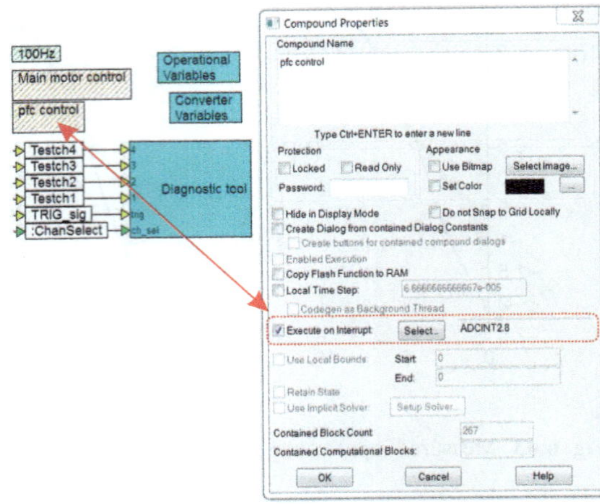

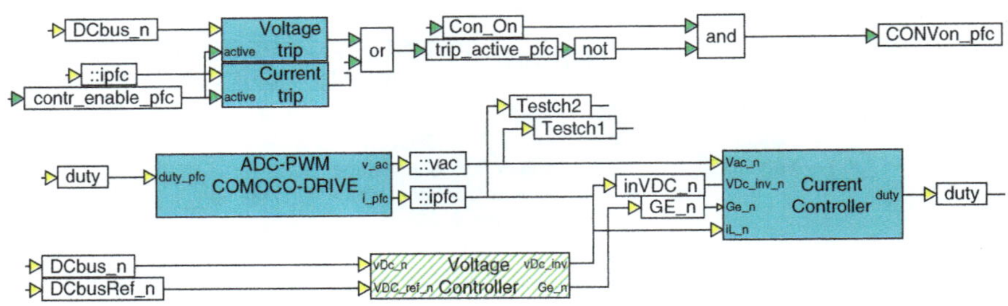

Fig. 6.46 Development Phase C, one level into 'pfc control' module

button `ch1-2` is used to toggle the variables display on the diagnostic scope in phase C+. Two outputs are present which represent the per unit measured DC bus voltage `Vdc_n` and a diagnostic variable `VAR` used for debugging purposes. In this case this variable has been assigned to the duty cycle value generated by the PFC controller, which in turn controls the two boost converters. Moving one level into the 'open loop voltage controller with PFC' module leads to the set of module shown in Fig. 6.45. This figure also shows a dialog box which appears upon a 'right click' on the 'PFC control module'. The critical entry in this case is the 'execute on interrupt' feature which must be set as shown. This implies that the module in question will be executed on an interrupt signal set by the ADC unit. More specifically, the interrupt is in this case to `ADCINT2:8`, which implies that the signal is generated by ADC input channel 8. The ADC PFC input channels are allocated within the ADC-PWM module (Fig. 6.46), which appears when moving one level lower into the 'pfc control' module shown in Fig. 6.45. The two output variables of the ADC module are `vac_pfc`, which is the attenuated rectified AC input voltage and `IPFC_tot`. The latter represents the rectified AC current as may be observed from Fig. 6.42. Both signals are used by a 'current controller' module which generates the duty cycle `duty` for the two boost converters. The 'voltage controller' module controls the magnitude of the peak rectified current

Table 6.6 Allocation of input variables to PFC-ADC input channels

Input variable	ADC input channel
Rectified voltage vac_pfc	7
Rectified current IPFC_tot	8

which is set by comparing the measured DC bus voltage DCbus_n with the user set reference DC bus voltage DCbusRef_n. The hatching of this module indicates that its execution rate is different to that of the surrounding modules. In this example the voltage controller runs at a sampling rate of 15 kHz, whereas the other modules operate with 45 kHz, as will be come apparent shortly. For details on the control approach used here, the reader is referred to De Gusseme, et al. [5]. Also shown in Fig. 6.46 are a voltage and current trip module, which shut down the PFC module (duty to zero) when the DC bus voltage or the rectified current IPFC_tot exceed the limit values set in the 'operational variables' dialog box.

The next step of development is concerned with configuring the PFC-ADC-PWM module. Data acquisition via VisSim, is done by way of a MCU specific 'input channel' module, such as F28069M Input Channel where the user has the option of selecting either an 'analog or digital' channel. Furthermore, the user must allocate a channel number (within the range of analog channels available) as mentioned above. Table 6.6 shows the channel allocation used in this case study. Note that the allocation of the two PFC variables to input channels is up to this stage done arbitrarily. However, in this case it is prudent to simply allocate the successive (as used by the V/f converter) channel numbers. A critical commissioning step is to allocate the channel numbers introduced in Table 6.6 to actual MCU variables. This requires access to the ADC F28069M Properties dialog box, found by selecting Embedded in the main VisSim menu and then selecting F280x and finally ADCconfig which brings up the module shown in Fig. 6.47. Also given in this figure is part of the circuit diagram of the board under consideration, which shows the pin allocation of the current/voltage variables. The CoMoCo board makes use of the TMS320F28069M processor shown (partly) in Fig. 6.47. Associated with the current/voltage variables and MCU pin numbers processor are alphanumeric variables, which need to be assigned in the ADC F28069 Properties dialog box, as shown.

In addition to assigning the input variables, the trigger signal for the input channels needs to be assigned. Triggering of the ADC is linked to the PWM module as was discussed in Sect. 2.1.6. For the PFC input channels the 'Start Of Conversion' signal ePWM4-SOCB is used, which is also the PWM module that controls the two switches Q1, Q2 shown in Fig. 6.42.

Moving one level lower into the 'ADC-PWM' module shown in Fig. 6.46 reveals a PWM module (see Fig. 6.48), which has as input the duty cycle variable d_u generated by the PFC current controller. A 'right mouse click' on this PWM module reveals a dialog box, which is also shown in Fig. 6.48. The most important entries are numbered and need to be configured as follows:

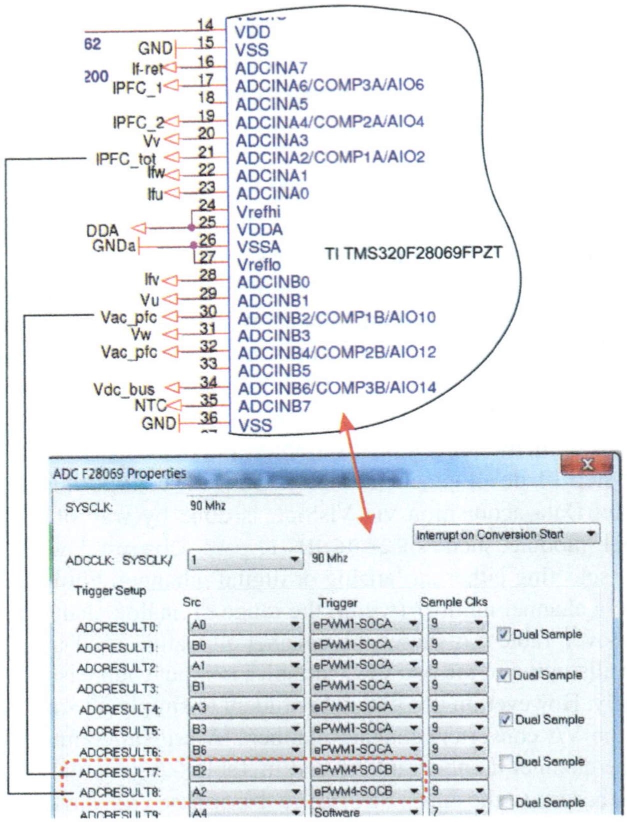

Fig. 6.47 Development Phase C: ' PFC ADC properties' dialog box and link to circuit diagram

1. `Timer Period`: represents the number of cycles of the PWM staircase as discussed in Sect. 2.1.6. The number to be entered here is defined by the clock frequency divided by the timer period times 2. In this case study a 90 kHz PWM frequency is required for the boost converters, which with a system clock frequency of 90 MHz, corresponds to a timer period value of 500 as shown.
2. `EPWMA`, `EPWMB` outputs of this PWM unit are tied to `GPIO` pin variables `GPIO6` and `GPIO7` respectively. Hence PWM signal `EPWMA` is connected to `GPIO6` which in turn activates device `Q1` using gate input `PFC_PWM_1`. Likewise, PWM signal `EPWMB` controls device `Q2` using gate input `PFC_PWM_2`. Note that the gate drivers invert the signals, hence a logic 0 turns ON a device. The `Action Qualifier` settings for both PWM signals correspond to the outputs shown in Fig. 6.49. In this example the duty cycle input for `EPWMA` is set to 0.25, which corresponds to a duty cycle input for `EPWMB` of 0.75 as may be observed from Fig. 6.49. The triangular waveform shows the decision points where the output of the PWM are changed.

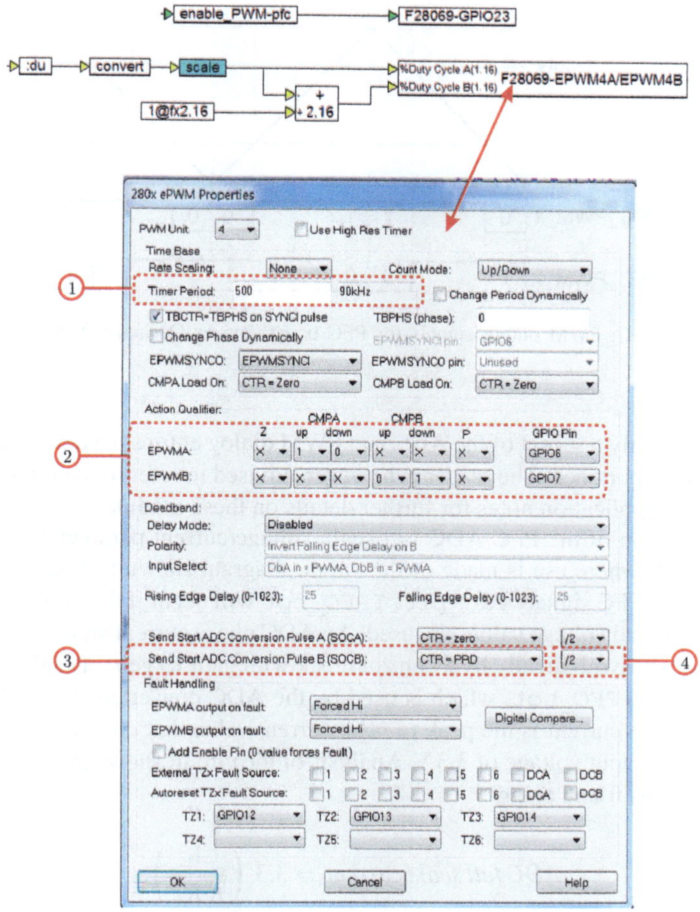

Fig. 6.48 Phase C: 'PWM properties' dialog box

3. `send start ADC Conversion Pulse B (SOCB)`: this entry is critical, in that it determines when the conversion pulse should be given with respect to the PWM cycle. In this case the option `PRD` is chosen because triggering of the ADC should occur at the top of the PWM staircase, as shown in Fig. 6.49. Note that in Fig. 6.48 ADC triggering refers to `ePWM4-SOCB`, which implies that EPMW module 4 generates the SOC signal for both ADC-PFC channels.
4. `2`: this entry specifies if the ADC pulse should be given after every first , second or third PWM clock cycle. In this case a 90 kHz PWM frequency has been selected and the ADC sample frequency in use is 45 kHz (which is also the rate at which the controller module will be executed) hence the PWM to ADC ratio must be set to $1/2$.

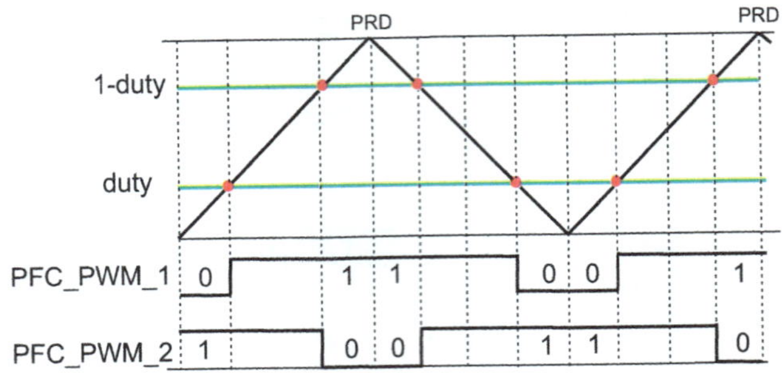

Fig. 6.49 Phase C: PWM output signals for PFC boost devices Q1 and Q2 (device ON when PFC-PWM '0'

Note that only a subset of the ADC and PWM dialog entries have been discussed, given their importance. The reader is however, advised to refer to the relevant Texas Instruments application notes for further details on these modules.

Computation of the PFC ADC full scale voltage/current parameters remains to be discussed, where use is made of the circuit diagram shown in Fig. 6.42. Proper scaling of the PFC signals `vac_pfc`, `IPFC_tot` will occur if the correct ADC full scale current and voltage values are used. An ADC phase current signal conditioning circuit (shown symbolically by a control block with gain `K`) generates the measured shunt voltage `IPFC_tot`, which is used by the ADC converter of the MCU. The ADC full scale current is the peak to peak current value that corresponds to a peak to peak ADC input voltage of 3.3 V. Analysis of the circuit shows that the ADC full scale current is found using

$$ADC\,full\,scale\,current = 3.3\left(\frac{1}{K\,R2}\right) \tag{6.9}$$

where $K = 11.11$ for the CoMoCo circuit. Substitution of the resistance values (as shown in the diagram) into Eq. (6.9) yields an ADC full scale current value of 29.68 A. The current signal conditioning circuit inverts the shunt resistance voltage, which implies that the voltage `IPFC_tot` will be in phase with the attenuated rectifier voltage signal `vac_pfc`. In this case the voltage attenuation network used for the PFC is identical to that used for the drive, hence the ADC full scale voltage level is also equal to 430.98 V.

6.2.7 Case Study V2c: CoMoCo Drive: PFC+boost, Phase C+

Phase C+, is the operational component of this case study and is basically a run version of the `.out` file compiled and downloaded to the MCU in phase C (see previous subsection).

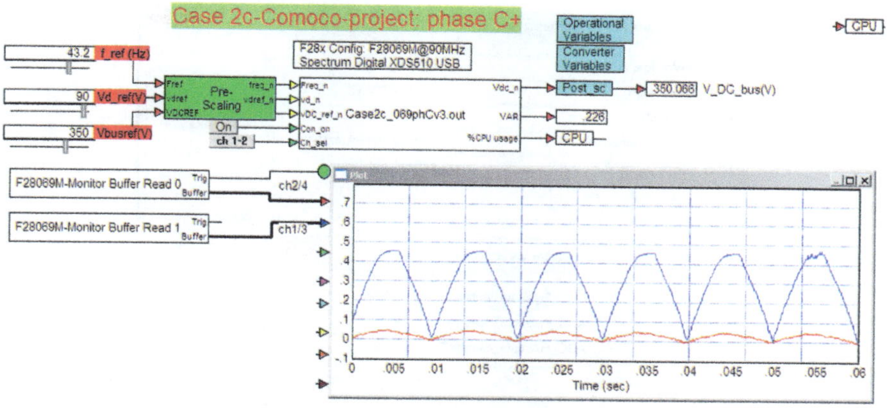

Fig. 6.50 Phase C+: Embedded voltage controller with PFC

The following information is relevant for this laboratory component:

- Reference program : `Case2c_069phCv3_d.vsm`.
- Description: PFC+boost control of the CoMoCo drive.
- Equipment/Software: CoMoCo-1.0-ED-4 converter [3] with Texas Instruments F28069F MCU, Vostermans Fan unit, Spectrum Digital XDS510 USB Emulator, Variac and VisSim simulation program.
- Outcomes: to confirm basic PFC-boost operation by observing the voltage and current waveforms.

The run version as shown in Fig. 6.50 uses a VisSim run module, which executes the `.out` file, shown in said module. Three sliders are used, of which two set the reference frequency and amplitude of the voltage vector for the V/f drive. The third slider is used to set the reference DC bus voltage. A post-scaling module is used to convert the per unit measured DC bus voltage to actual voltage, as shown with a numeric display. Furthermore, a variable `VR` has been added, which shows in this case the instantaneous duty cycle of the PFC current controller.

Two 'Monitor Buffer' modules are used to display two selected diagnostic signals which represent in this case the per unit rectified voltage `vac` (shown in 'blue') and per unit rectified measured current `ipfc` (shown in 'red'). Note that the actual voltage and current values are found by multiplying the per unit values with the full-scale values of 500 V and 25 A respectively. The results shown in the VisSim scope module are consistent with the results (measured using external voltage/current probes) shown in Fig. 6.43. Note that the voltage and current waveforms shown in the VisSim scope are in phase, which confirms the correct functioning of the PFC controller. The current phase lead in the measured results shown in Fig. 6.43 is, as was mentioned earlier, due to the presence of the line-filter. The dialog box entries for this drive are shown for completeness in Fig. 6.51. Note that the ADC full scale shown in the 'Converter Variables' dialog box are those which have been calculated in the previous development phases. The parameters shown in the 'Operation

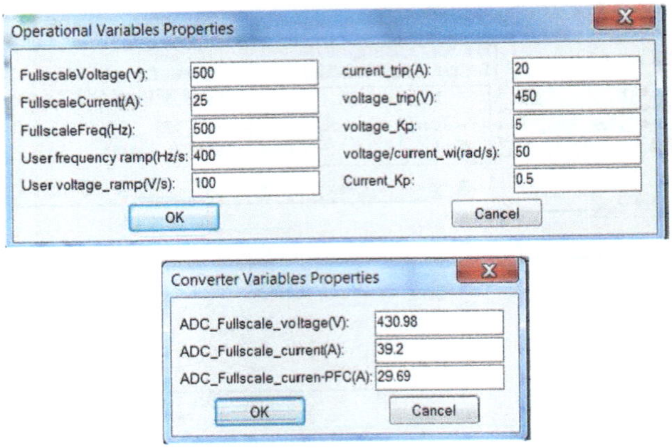

Fig. 6.51 Phase C+: dialog boxes for CoMoCo drive

Variables' dialog are the full-scale voltage/frequency and current values together with the voltage/frequency ramp setting as required for the V/f drive. The current trip setting represents the current limit for PFC and V/f drive controller where a software trip is invoked. Likewise, a voltage trip will occur if the DC bus voltage exceed the selected trip value of 450 V in this case. The three remaining parameters `voltage_Kp`, `current_Kp` and `voltage/current_wi` represent the gains and bandwidth settings of the PFC voltage/current controller.

In conclusion the reader is reminded of the fact that this introduction of a PFC controller as discussed above will inherently add an additional processing load to the MCU, which may be problematic if other computational time intensive tasks must also be executed. Hence this consideration must be weighed up against the option of using a dedicated integrated circuit for realizing PFC control.

6.3 Case Study V3: e-Traction Drive

The purpose of this project is to consider the use of a 12–450 V_{dc} bus connected 'FAST prepared' converter which will be connected to the three-phase AC induction machine used in the previous case study. The drive setup given in Fig. 6.52 shows a 'e-Traction motor controller-Rev.3 (CI.01.1065)' 3 kW converter (with cover removed) [6], which is a highly versatile extremely compact industrial unit that makes use of a Texas Instruments F28069M control card. Other Texas Instruments control cards can also be used, with or without an external JTAG emulator. Consequently, the drive can be used for sensorless control of three-phase AC machines up to 3 kW, with corresponding motor identification in line with the laboratory examples discussed in the previous two chapters. Observation of Fig. 6.52 confirms the compact nature of the unit which is provided with a separate (to the main DC supply) 24 V_{dc} input used to power the logic circuits of the converter. In addition,

Fig. 6.52 e-Traction [6] 3 kW converter with Marathon 5k336N2A induction machine setup

this module has a CAN-BUS interface, hence multiple units can be easily controlled from a central location and also readily connected in parallel to main 12–450 V_{dc} and auxiliary 24 V_{dc} bus supplies. The unit, designed for automotive applications, has three Hall-effect current sensors together with a three phase resistance/capacitive network used to measure the motor voltages. A common mode choke (not shown in Fig. 6.52) is typically connected between the three converter terminals and the motor. For short motor/converter cable distances, as is the case here, the common mode choke is not required and is therefore not used in this case study. In terms of project content, both converter commissioning and field-oriented control using FAST as a software encoder according to the approach shown in laboratory 3:1 (see Sect. 5.2) will be discussed in this study. In addition, a SpinTAC based speed controller is introduced as shown in Fig. 6.53 together with a SpinTAC inertia estimation module used earlier in laboratory 1:4 (see Sect. 3.4). Shown in Fig. 6.53 are four 'green' modules referred to as the 'SpinTAC Motion Control Suite' [8]. These modules, located in the F28069M micro-processor controller unit (MCU), provide the reader with the ability to implement a sophisticated motion control capability for the drive. The 'SpinTAC Control Suite'developed by LineStream Technologies [8], is part of MotorWare [9] and is used for sensorless operation in this project. A brief description of said modules is given here:

- SpinTAC Velocity Control: generates the quadrature reference current value for the current controller when this module is active (switch in position 'C'). Inputs are the reference and actual shaft speed value, shown here as frequencies $f_{\mathrm{m}}, f_{\mathrm{m}}^{\mathrm{ref}}$ respectively. In addition, the reference rate of frequency change $a_{\mathrm{m}}^{\mathrm{ref}} = {}^{df_{\mathrm{m}}^{\mathrm{ref}}}/_{dt}$ is also required.

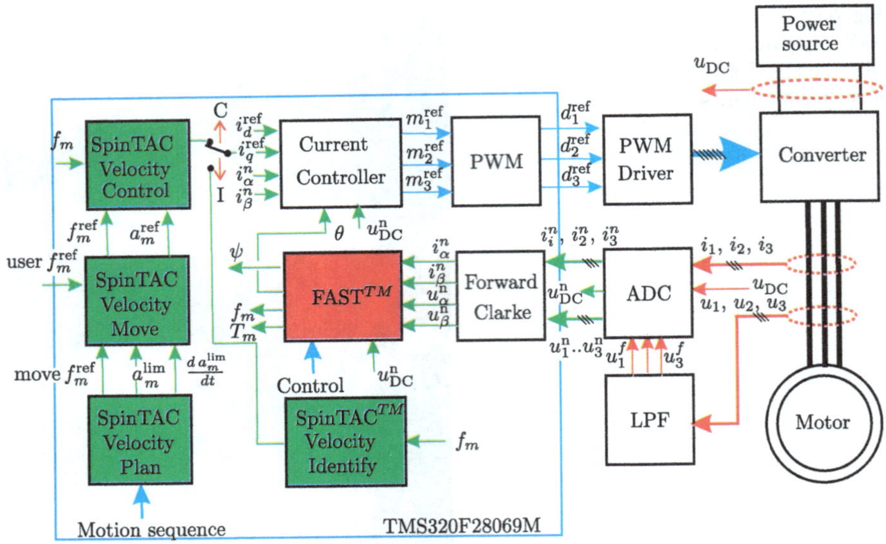

Fig. 6.53 Sensorless drive setup with the SpinTAC motion control suite

- SpinTAC Velocity Move: This module allows the user to execute a complex speed versus time profile [9], which is then translated to a set of outputs $a_{\rm m}^{\rm lim}$, $da_{\rm m}^{\rm lim}/dt$, as discussed above, together with a shaft speed reference move$f_{\rm m}^{\rm ref}$.
- SpinTAC Velocity Identify: Its function is to identify the system inertia, which in this case is the inertia of the Marathon Induction machine.
- SpinTAC Velocity Plan: generates the reference shaft speed/acceleration variables, shown here as $f_{\rm m}^{\rm ref}$, $a_{\rm m}^{\rm ref}$ respectively. Inputs are a user set reference shaft speed user$f_{\rm m}^{\rm ref}$ and limit variables $a_{\rm m}^{\rm lim}$, $da_{\rm m}^{\rm lim}/dt$, which define the shaft acceleration and its derivative (known as 'jerk') respectively. Also shown is a shaft speed reference 'move$f_{\rm m}^{\rm ref}$', which is used instead of the user speed reference 'user$f_{\rm m}^{\rm ref}$' when the 'Velocity Plan' module is used. Note that the 'Velocity Plan' module is not implemented in VisSim, instead it needs to be accessed via the InstaSPIN-Motion package [9].

This case study, as others in this chapter, serves to show how previous (Chaps. 3–5) laboratory examples can be applied to specific industrial converter/machine combinations.

6.3.1 *Case Study 3a: e-Traction Drive: Commissioning, Phase C*

Prior to using current control in any application, it is helpful to consider drive operation under voltage control first. Using this approach, the voltage and current

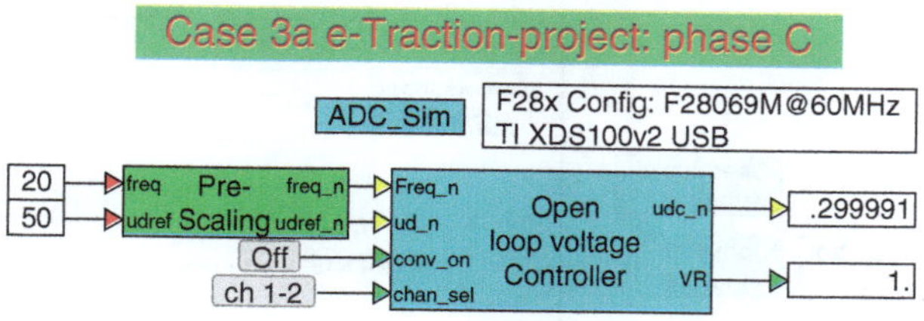

Fig. 6.54 Development Phase C: voltage controller for e-Traction drive

measurements can be examined. Furthermore, the phase relationship between voltage and current can be established, which is important prior to using current control. The following information is relevant for this case study:

- Reference program : `Case3a_069phCv3.vsm`.
- Description: Commissioning of e-Traction drive.
- Equipment/Software: e-Traction motor controller-Rev.3 (CI.01.1065) converter [6] with Texas Instruments F28069M control card, 24 V_{dc} Power supply and VisSim simulation program.
- Outcomes: Generate the `.out` file needed to represent the drive structure and calculate the full scale ADC values as well as LPF frequency of the voltage filter.

The voltage controller as discussed in Sect. 3.1 is used for this analysis. However, emphasis in this section will be on configuring the ADC-PWM module, triggering, allocating ADC channels and confirming that the current/voltage waveforms are correct in terms of amplitude and phase relationships. The phase C, development phase, as given in Fig. 6.54 has been adapted for this application. Inputs are the electrical frequency and direct axis voltage reference variable. Furthermore, a channel select input `chan_sel` has been added to toggle between diagnostic scope channels 1-2 and 3-4 in phase C+. Moving one level down into the 'Open loop voltage controller' reveals a set of modules and dialog boxes as indicated in Fig. 6.55. This figure also shows a dialog box which appears upon a 'right click' on the 'main control module'. The critical entry in this case is the 'execute on interrupt' feature which must be set as shown. This implies that the module in question will be executed on an interrupt signal set by the ADC unit. More specifically, the interrupt is in this case `ADCINT1:8`, which implies that the signal is generated by ADC input channel 8. The ADC input channels are allocated within the ADC-PWM module, which appears when moving one level lower into the 'main control module' as shown in Fig. 6.55. The modules shown are as discussed in Sect. 3.1. However, in this case configuring the ADC-PWM module is of interest, because this module is board specific. Data acquisition via VisSim, is done by way of a MCU specific 'input channel' module, such as `F28069M Input Channel` in which the user has the option of selecting an 'analog or digital' channel. Furthermore, the user must allocate a channel number (within the range of analog channels available) as mentioned above (Fig. 6.56). Table 6.7 shows the channel allocation used in this laboratory. Note that the allocation of phase currents and voltages to input

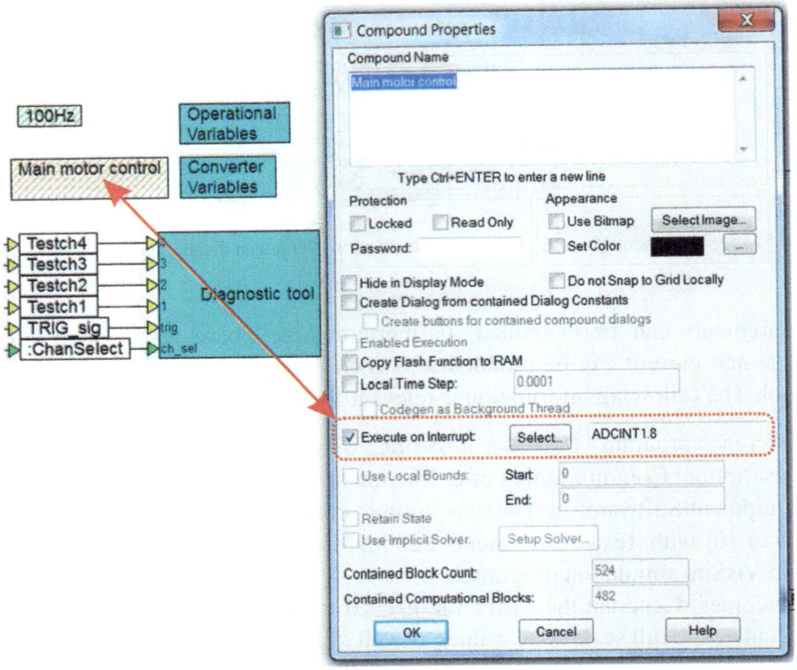

Fig. 6.55 Development Phase C, one level into 'Open loop voltage controller'

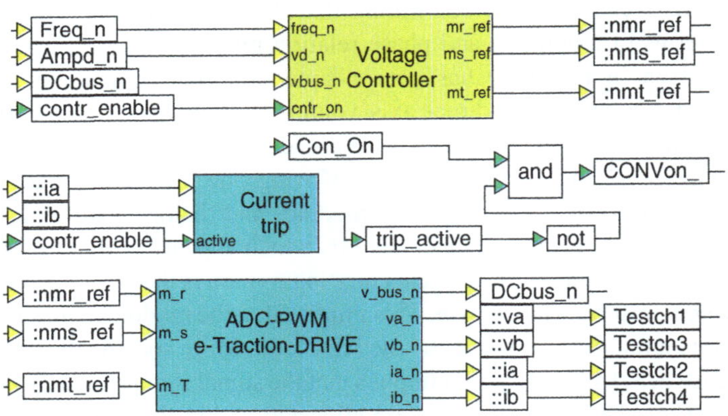

Fig. 6.56 Development Phase C: one level into 'main motor control' module

channels is up to this stage done arbitrarily. A critical commissioning step is to allocate the channel numbers introduced in Table 6.7 to actual MCU variables. This requires access to the `ADC F28069M Properties` dialog box, found by selecting `Embedded` in the main VisSim menu and then selecting `F280x` and finally `ADCconfig`, which brings up the module shown in Fig. 6.57. Also given in this figure is part of the circuit diagram of the board under consideration, which

Table 6.7 Allocation of input variables to ADC input channels

Input variable	ADC input channel
Phase current i_U	2
Phase current i_V	3
Phase current i_W	4
Phase voltage u_U	5
Phase voltage u_V	6
Phase voltage u_W	7
Bus voltage u_{DC}	8

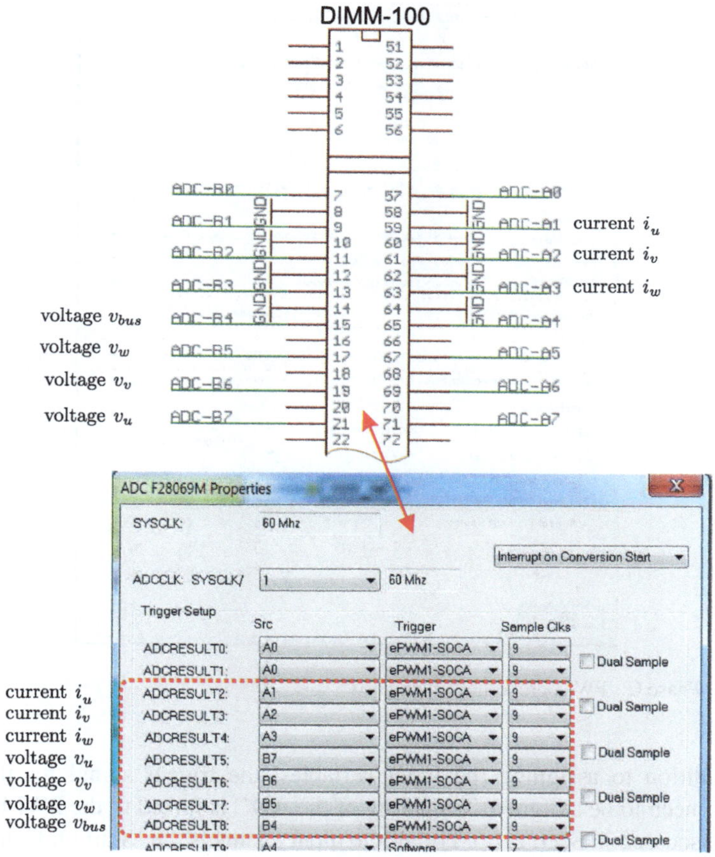

Fig. 6.57 Development Phase C: 'ADC properties' dialog box and link to circuit diagram

shows the pin allocation for the current/voltage variables. The e-Traction board makes use of the TMSF28069M control card, which slots into the `DIMM-100` connector shown in Fig. 6.57. Associated with these current/voltage variables and pin numbers are alphanumeric variables which must be assigned in the `ADC F28069M Properties` dialog box, as shown.

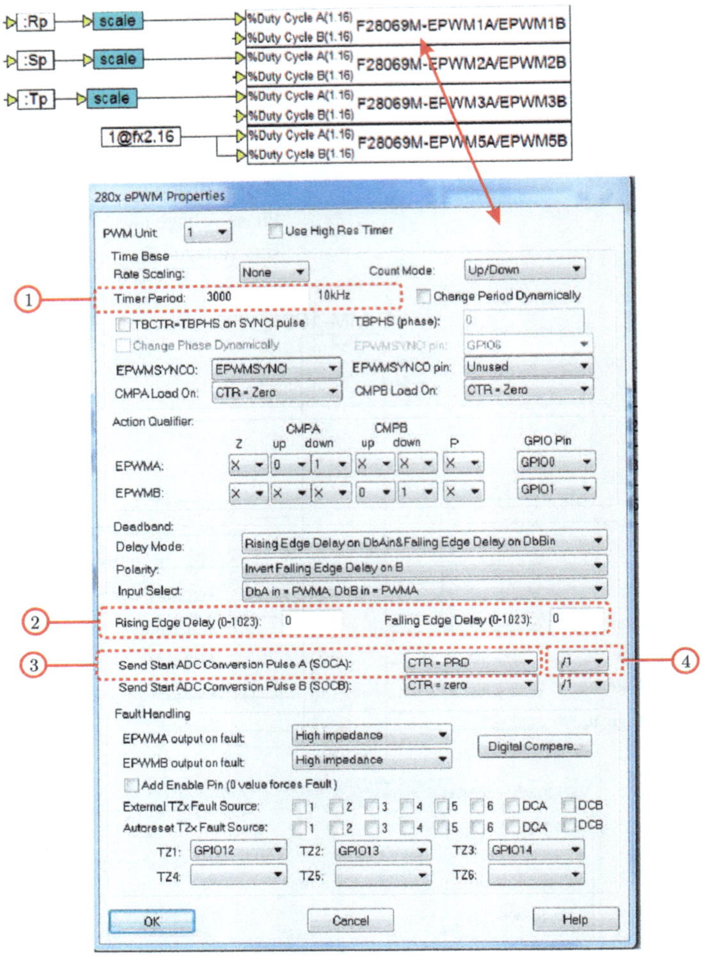

Fig. 6.58 Phase C: 'PWM properties' dialog box

In addition to assigning the input variables, the trigger signal for the input channels need to be assigned. Triggering of the ADC is linked to the PWM module as was discussed in Sect. 2.1.6. For all the input channels the 'Start Of Conversion' signal: `ePWM1-SOCA` is used which implies that pulse width modulation module 1 generates the trigger for the input ADC channels in use.

Moving one level lower into the 'ADC-PWM' module shown in Fig. 6.56 reveals four PWM modules (see Fig. 6.58). Modules 1 to 3 provide the PWM signals for the converter, whilst the fourth module can be used to set the current trip level for the Hall-effect current sensors used for this converter. This feature is not used in this case study. Inputs for PWM modules 1 to 3 are the modulation index variables `Rp`, `Sp` and `Tp` respectively. A 'right mouse click' on ePWM module 1 reveals a dialog box, which is also shown in Fig. 6.58. The most important entries are numbered and need to be configured as follows:

1. `Timer Period`: represents the number of cycles of the PWM staircase as discussed in Sect. 2.1.6. The number to be entered here is defined by the clock frequency divided by the timer period times 2. In this case study a 10 kHz PWM frequency is required, which with a system clock frequency of 60 MHz, corresponds to a timer period value of 3000 as shown.
2. `Rising Edge Delay`: `Falling Edge Delay`: these two parameters set the 'dead-time' of the converter, as dictated by the power electronic devices used. For this converter both parameters are set to 0, given that the gate drivers for the power electronics devices set the dead-time in this case.
3. `send start ADC Conversion Pulse A (SOCA):` this entry is critical, in that it determines when the conversion pulse should be given with respect to the PWM cycle, as shown in Fig. 2.23. In this case the option `PRD` is chosen because triggering of the ADC should occur at the top of the PWM staircase, as shown in Fig. 2.23. Note that in Fig. 6.57 ADC triggering refers to `ePWM1-SOCA`, which implies that ePMW module 1 generates the SOC signal for all ADC channels.
4. `1`: this entry specifies if the ADC pulse should be given after every first, second or third PWM clock cycle. In this case a 10 kHz PWM frequency has been selected and the ADC sample frequency in use is 10 kHz (which is also the rate at which the controller module will be executed, with a sampling frequency set in the main VisSim module) hence the PWM to ADC ratio must be set to $^1/_1$.

Note that only a subset of the ADC and PWM dialog entries have been discussed, given their importance. The reader is however, advised to refer to the relevant Texas Instruments application notes for further details on these modules.

Computation of the ADC full scale voltage/current parameters and the low-pass filter corner frequency are critical to successful sensorless operation. Accordingly, these parameters must be either given or derived by careful examination of the schematic diagram of the converter under consideration. Part of the relevant (to this topic) circuit diagram sections of the e-Traction drive board, given in Fig. 6.59 will be used to derive the required data. Central to this figure is one leg of the three-phase converter, which is part of an integrated 'six-pack' module. An ADC phase current signal conditioning circuit measures the phase current in each converter leg using a Hall-effect sensor `IC9`. The output `ADC_A1` of this module is attenuated prior to use by the ADC converter of the MCU. The ADC full scale current is the peak to peak current value that corresponds to a peak to peak ADC input voltage of 3.3 V. Analysis of the current signal conditioning circuit shows that the ADC full scale current is found using

$$ADC\,full\,scale\,current\,(\mathrm{A_{pp}}) = 3.3\left(\frac{(R37+R38)}{R37\,K_{\mathrm{HeC}}}\right) \tag{6.10}$$

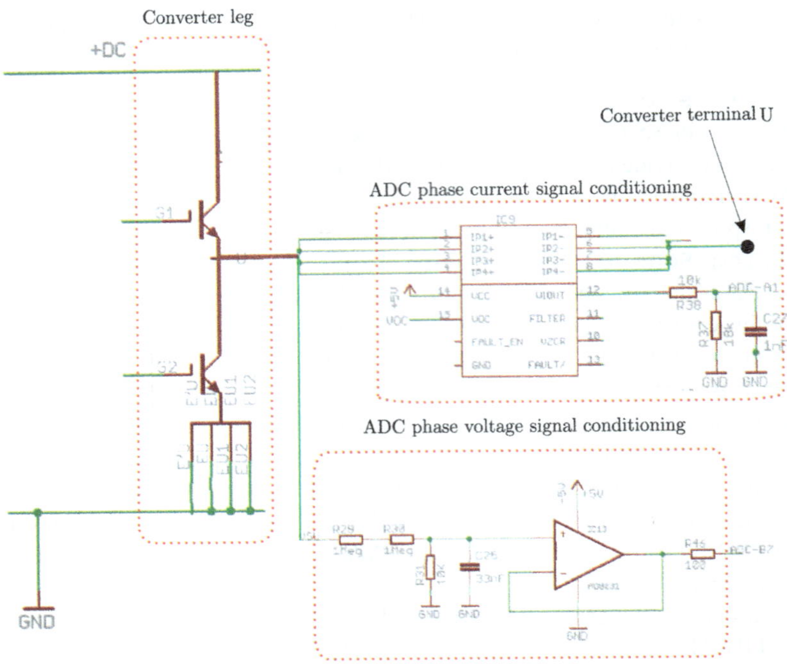

Fig. 6.59 Phase C: Part of e-Traction drive circuit diagram showing one converter leg, with associated current and voltage signal conditioning

where K_{HeC} is the transconductance of the Hall-effect sensor. Substitution of the resistance values (as shown in the diagram) into Eq. (6.10) and a given transconductance value of $K_{HeC} = 28\,\text{mV/A}$ yields an ADC full scale current value of 183.33 A.

Computation of the ADC full scale voltage requires evaluation of the ADC phase and DC bus voltage conditioning circuits. For both circuits, the same voltage attenuation is used, hence only the phase signal conditioning network is considered here. For the phase voltage circuit the attenuation is defined as $R31/(R29+R30+R31)$ and the output `ADC_B7` is connected to the ADC converter of the MCU via an 'emitter follower' configured OP-AMP. The ADC full scale voltage is the peak to peak phase voltage that corresponds to a peak to peak ADC input voltage of 3.3 V. Analysis of the phase voltage conditioning circuit, shows that the ADC full scale voltage is found using

$$ADC\,full\,scale\,voltage\,(\text{V}_{\text{pp}}) = 3.3\left(\frac{R29 + R30 + R31}{R31}\right) \tag{6.11}$$

where $R29$, $R30$ and $R31$ are the relevant resistors of the voltage gain circuit. Substitution of the resistance values (as shown in the diagram) into Eq. (6.11) yields an ADC full scale voltage value of 663.3 V. In conclusion, the reader is reminded of the fact that the phase voltage signal conditioning circuit is also a low-pass filter, of which the corner frequency may be found using

$$LPF\,corner freq\,(\mathrm{Hz}) = \left(\frac{1}{2\,\pi}\right)\left(\frac{R29 + R30 + R31}{(R29 + R30)\;R31\;C25}\right) \tag{6.12}$$

Substitution of the resistance and capacitance values (as shown in the diagram) into Eq. (6.12) yields a low-pass corner frequency value of 341.24 Hz.

6.3.2 Case Study V3a: e-Traction Drive: Commissioning, Phase C+

Phase C+, is the operational component of the case study and is basically a run version of the `.out` file compiled and downloaded to the MCU in phase C (see previous subsection).

The following information is relevant for this laboratory component:

- Reference program : `Case3a_069phCv3_d.vsm`.
- Description: Commissioning of e-Traction drive.
- Equipment/Software: e-Traction motor controller-Rev.3 (CI.01.1065) converter [6] with Texas Instruments F28069M control card, 24 V_{dc} power supply, Marathon 5k336N2A IM motor, DELTA ELEKTRONIKA SM 400-AR-4 Power supply and VisSim simulation program.
- Outcomes: to confirm basic drive operation by observing the voltage and current waveforms.

The run version shown in Fig. 6.60 uses a VisSim run module, which executes the `.out` file, shown in said module. Two sliders are used, which set the reference frequency and amplitude of the voltage vector. A post-scaling module is used to convert the per unit measured DC bus voltage to actual voltage, as shown with a numeric display. Furthermore, a variable `VR` has been added, which serves to display internal variables, as required. In this case the variable is set to show the ‘trip active’ signal. Which is logic ‘1’ when a current trip occurs.

Two ‘Monitor Buffer’ modules are used to display two selected diagnostic signals. The VisSim scope module shows the per unit α voltage/current waveforms and an enlarged view of these results is given in Fig. 6.61. Multiplication of these waveforms by the full scale values of $u_{fs} = 400\,\mathrm{V}$ and $i_{fs} = 10\,\mathrm{A}$ respectively, gives actual current values.

The dialog box entries for this drive are shown for completeness in Fig. 6.62. Note that the ADC full scales shown are these which have been calculated in the previous development phase. When the correct DC voltage level is shown on the numerical display, turn on the converter (using the ON button) and monitor the motor shaft and diagnostic scope. Several important observations can be made on the basis of the information provided in Fig. 6.61 namely:

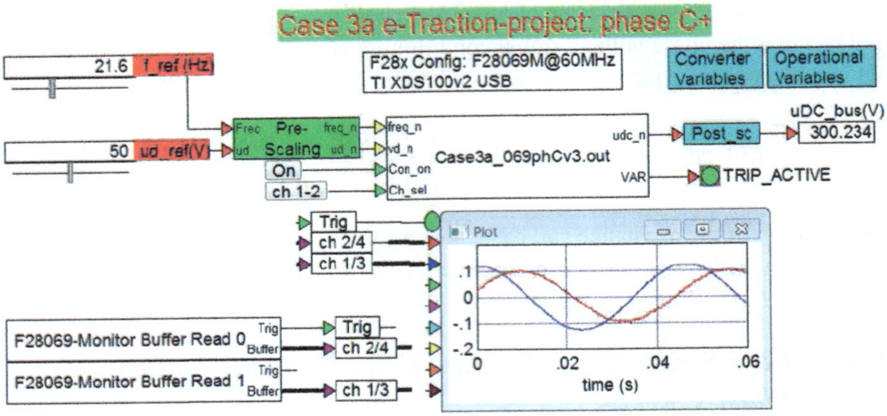

Fig. 6.60 Phase C+: e-Traction based voltage controller

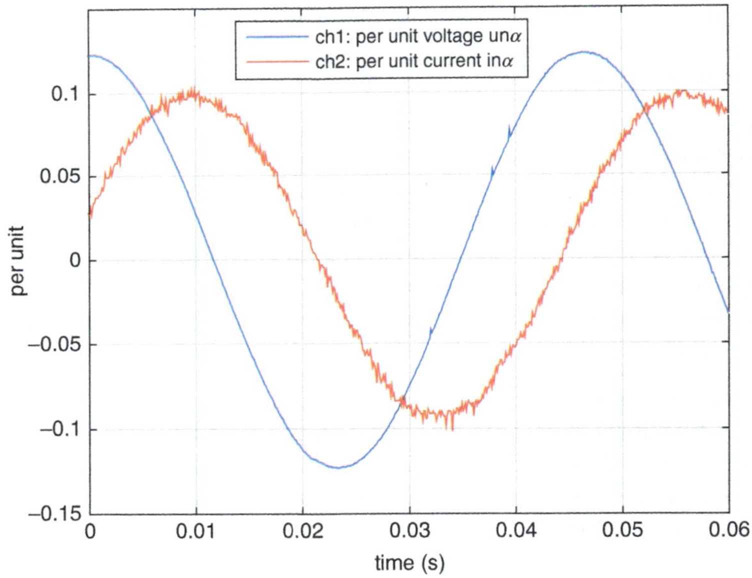

Fig. 6.61 Phase C+ Operation of the drive: scope results with channel select option: 'ch 1-2'

- The reference frequency has been set to 21.6 Hz, hence the period time of the waveforms should be 46.2 ms.
- A reference voltage amplitude of 50 V has been set by the slider `Vd_ref` hence the amplitude of the per unit u_α should be approximately (as dead-time effects and the voltage drop across the converter switches will affect the voltage waveform shown) equal to $^{50}/_{u_{fs}} = 0.125$, where u_{fs} is the full scale voltage value set to 400 V.
- Verification of the current amplitude can be done with a DC-true current probe and an oscilloscope, which leads to the result given in Fig. 6.63. Observation of

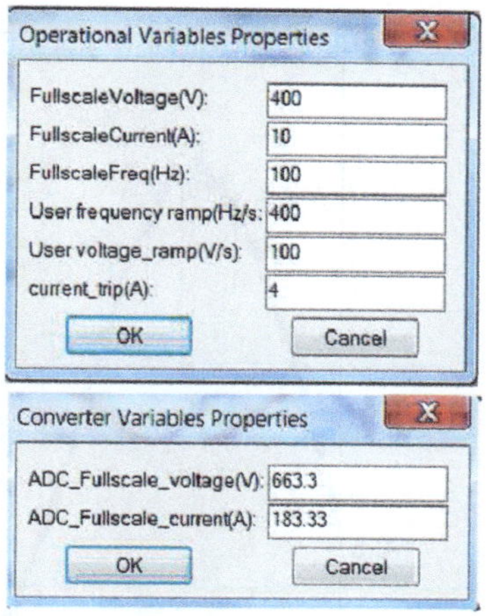

Fig. 6.62 Phase C+: dialog boxes for e-Traction drive

the experimental current waveform reveals a current amplitude of 0.9 A, which corresponds to a per unit value of $^{0.9}/_{i_{\text{fs}}} = 0.09$, where i_{fs} is the full scale current value set to 10 A. Examination of the scope module in Fig. 6.60 confirms that the per unit current is indeed correct.

- The current waveform has a phase lag relative to the voltage waveform which is expected given the presence of an inductive load and (in this case) a rotating machine, which causes a back EMF component. Note that verification of the correct phase relationship should also be undertaken for the β voltage/current waveforms.

Finally, observe that the motor is rotating correctly. With a reference frequency of 21.6 Hz the speed of the 4 pole motor will be approximately 648 rpm (no external load, hence speed will be the 'synchronous' speed), which is difficult to measure without suitable equipment. However, if the reference frequency is set to 2 Hz the rotational speed will be 60 rpm, i.e. one revolution per second, which can be visually determined. This measurement will verify that the assumed pole pair number for the machine is indeed correct.

6.3.3 Case Study V3b: e-Traction Drive: Sensorless Operation, Phase C

In this part of the e-Traction case study, attention is given to the Phase C development stage, which will allow sensorless operation of the drive, using FAST as a software encoder. Furthermore, a SpinTAC based speed controller and inertia estimator will be used for this drive. For this purpose the approach used in laboratory 3:1 (see Sect. 5.2) will be used, which will be adapted to suit the specific requirements of this drive. Specifically, this means the inclusion of the SpinTAC speed control module, hitherto not used in this book and the inertia module

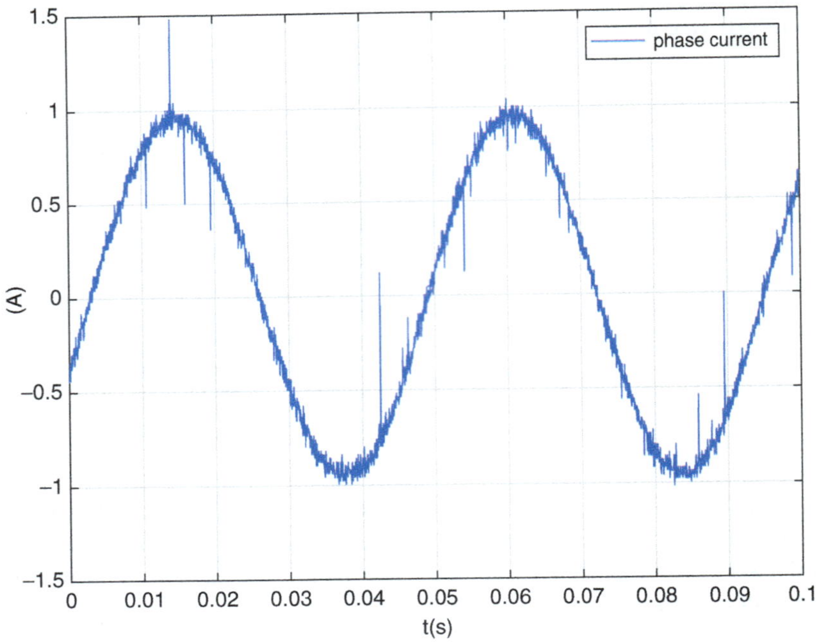

Fig. 6.63 Phase C+: phase current measured with Tektronic DC-true probe and oscilloscope

introduced earlier. Central to this development phase is the 'External FOC FAST Controller with SpinTAC' module, as shown in Fig. 6.64, which represents the drive under consideration. Hence details of the drive structure and corresponding dialog box parameter assignments will be discussed in this subsection.

The following information is relevant for this laboratory component:

- Reference program : `eTRAC-069MphCv1.vsm`.
- Description: Sensorless SpinTAC based speed control of e-Traction drive using the InstaSPIN module.
- Equipment/Software: e-Traction motor controller-Rev.3 (CI.01.1065) converter [6] with Texas Instruments F28069M control card, 24 V DC power supply and VisSim simulation program.
- Outcomes: Generate the `.out` file needed to represent the drive structure under investigation.

Inputs to the Controller are four variables, which set the reference shaft speed, direct/quadrature reference currents and SpinTAC speed controller speed bandwidth. In this laboratory, six buttons have been allocated to the controller namely:

- `Inertia_ID_on`: activates inertia estimation procedure provided the button `speed_loop control` is disabled and button `closed_loop` control is enabled.

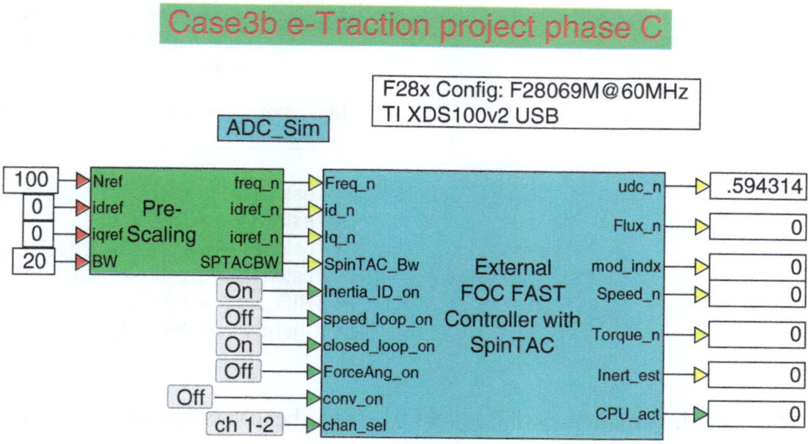

Fig. 6.64 Phase C simulation of a InstaSPIN-FOC e-Traction drive with SpinTAC speed control

- `speed_loop_on`: when activated provides speed control and torque control via the `iqref` user input when disabled.
- `closed_loop_on`: when activated, makes use of the FAST generated rotor angle to achieve field oriented control. When disabled the angle used for the current controller is indirectly set by `Nref`, while the current is set by the user variables `idref`, `iqref`.
- `ForceAng_on`: the force angle function (for induction machine based drives) is used when a high start up torque is required. When speed reversals are required, the function should be turned off.
- `conv_on`: enables converter operation.
- `ch1-2`: when this information is displayed diagnostic channels 1 and 2 will be shown in the scope in phase C+. This is a toggle button and can also display `ch3-4`, in which case channels 3 and 4 will be shown on the scope.

The module is provided with the following (per unit) outputs (starting from top to bottom):

- `udc_n`: measured DC bus voltage.
- `mod_indx`: the modulation index amplitude generated by the current controllers. Its value, when multiplied by half the DC bus voltage, gives an indication of the reference voltage amplitude in use. The actual voltage will differ, due to the voltage drop across the power electronics devices and dead-time effects.
- `Flux_n`: estimated rotor flux.
- `Speed_n`: estimated shaft speed.
- `Torque`: estimated torque.
- `Inert est`: estimate for the inertia.
- `CPU_act`: output which gives an estimate of CPU activity.

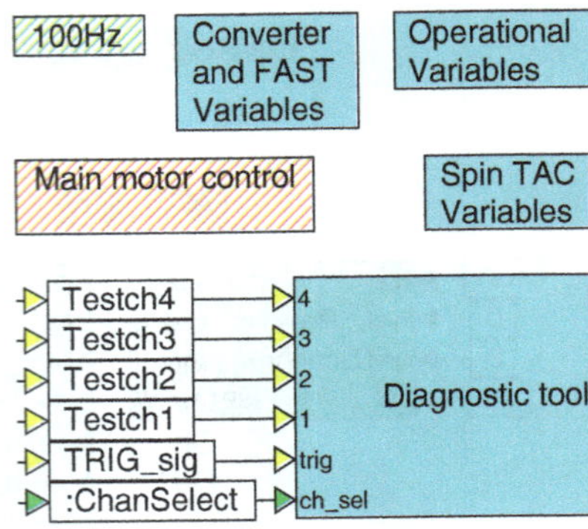

Fig. 6.65 Phase C simulation of a FAST based encoderless FOC drive,with SpinTAC speed control: one level into the drive controller module

An ADC sampling frequency of $f_{ADC} = 10\,\text{kHz}$ has been selected for this laboratory, which is also used for the current controllers.

The FAST Identification algorithm must be set to operate at a sampling frequency of $10000/i\,\text{kHz}$, where i is an integer value.

For this lab the value used is $i = 1$ which implies that the sampling frequency for the FAST algorithm has been set to 10 kHz. Moving one level lower into the 'Drive Controller' module reveals the modules shown in Fig. 6.65. Of these shown in the figure, the '100 Hz' and 'Diagnostic tool' modules have been discussed in previous laboratories. The dialog box parameters for this laboratory are given in Fig. 6.66. The parameters associated with the 'Operation variables' dialog box, have been discussed in the previous section. Of interest here are therefore, the dialog boxes 'Converter and FAST Variables' and 'SpinTAC variables'. The entries for the former dialog box are those required for FAST operation and contain (among others) the motor parameters and full scale ADC voltage/current values. Further details on the entries are given in laboratory 3:1 (see Sect. 5.2). For the SpinTAC dialog box the following entries are required:

- `Velocity limit (inertia est) rpm`: maximum shaft speed allowable for the SpinTAC inertia algorithm.
- `Acceleration limit rpm/s`: maximum acceleration rate of shaft speed magnitude change allowable by the SpinTAC speed controller.
- `Deceleration limit rpm/s`: maximum deceleration rate of shaft speed magnitude change allowable by the SpinTAC speed controller.
- `Jerk n limit rpm/s^2`: maximum absolute rate of shaft acceleration change allowable by the SpinTAC speed controller.

Moving one level lower into the 'main motor control' module, reveals a set of modules given in Fig. 6.67. Key inputs to the InstaSPIN controller module are the per unit alpha/beta current/voltage variables: `ia`, `ib`, `va`, `vb` and DC bus voltage

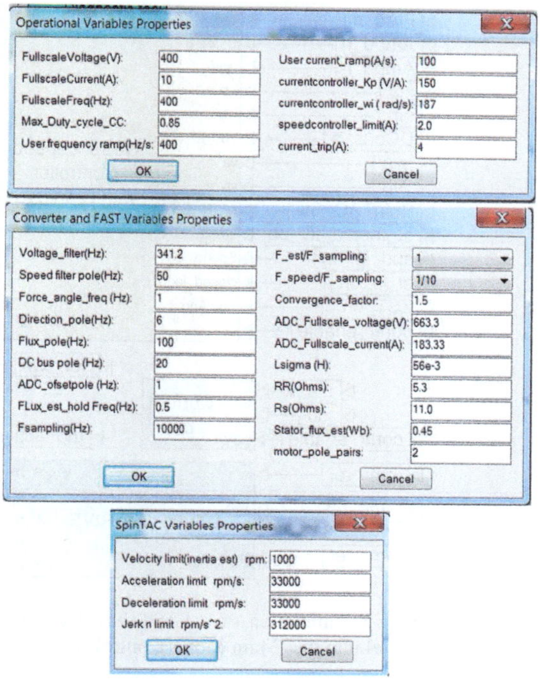

Fig. 6.66 Phase C simulation of a FAST based encoderless FOC drive, with SpinTAC speed control: Dialog box entries

variable `DCbus_n`, generated by the ADC-PWM module. User reference variables: `Freq_n`, `Ampd`, `Ampq_n` and SpinTAC variable `SPinTAC_Bw` represent the scaled user inputs for shaft-speed, direct/quadrature currents and speed bandwidth. Outputs of the 'Current/speed controller with SpinTAC' are (among others) the modulation indices: `nmr_ref`, `nms_ref`, `nmt_ref` generated by the current controller. In addition, this module also provides a variable `mod_ind`, which is the amplitude of the modulation index. The remaining outputs are the flux, speed and torque estimates generated by the FAST algorithm and the per unit inertia estimate `inertia_est`. Moving one level into the 'Current/speed controller with SpinTAC' module (see Fig. 6.68), reveals the FOC current controller structure, exemplified by module `PI-Kp-wi-Lim`. Inputs to the current controller are the reference direct/quadrature variables `idref` and `iqref`. In addition, the measured (from the ADC module) and forward Park converted currents are supplied to this module together with the inverted DC bus voltage `inVdc`. The angle information for the FOC controller in the form of the variables `cos_th`,`sin_th` (which are the instantaneous Sine and Cosine values of the angle), is provided by either an integrator (with the required Sine/Cosine outputs) or the InstaSPIN module. The latter module makes use of the per unit α, β current/voltages and DC bus voltage supplied by the ADC module. Outputs of the FAST module are: flux angle, rotor flux amplitude, electrical shaft frequency, torque and inverse DC bus voltage. Also shown in Fig. 6.68 is the 'SpinTAC SPEED Solution' module based on the SpinTAC modules shown in Fig. 6.53 (without the Velocity Plan module), which executes

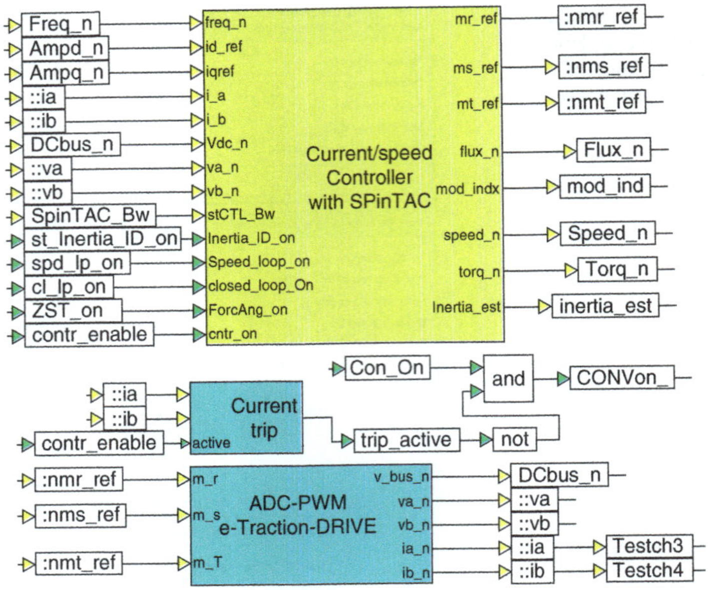

Fig. 6.67 Phase C simulation of a FAST based encoderless FOC drive, with SpinTAC speed control: One level into the 'Main Motor Control' module

the speed control and inertia identification components of the drive. Inputs to this module are the user reference frequency `freq_n`, estimated shaft frequency (from the FAST module) `rotfreq` and the speed bandwidth setting `stCTL_Bw`. Also shown is the variable `st_InerEST`, which is the estimated per unit inertia value that is also used as an input for the SpinTAC speed controller. The output variable `st_iqref` is the per unit quadrature current, which is generated by either the inertia estimator or the speed controller. This value is used by the current controller when either speed control or inertia estimation is active (controlled via the switch in the top section of the figure). A series of logic modules are used to ensure that the conditions for speed and inertia estimation are met. This implies that closed-loop control must be enabled (flag `closloopON` is logic 1), and inertia estimation will only occur if speed control is disabled (flag `spedloop_on` is logic 0).

Moving one level lower into the 'SPinTAC SPEED Solution' module reveals a set of modules shown in Fig. 6.69. Of the three modules shown, the module `ST_VelId` executes the SpinTAC inertia estimation algorithm. Input is the estimated shaft frequency variable `VELFdb` and key output is the per unit quadrature current variable `IDOut` which serves as a reference for the current controller during inertia estimation. The remaining two modules are responsible for drive speed control. The module `ST_VelMove`, with input: reference speed frequency `spedRef`, generates the parameters for the module `ST_VelCtl`, that provides the quadrature current controller reference `CTLOut` when speed control is active. These two modules `ST_VelMove`, `ST_VelCtl` provide the user with the ability to setup different

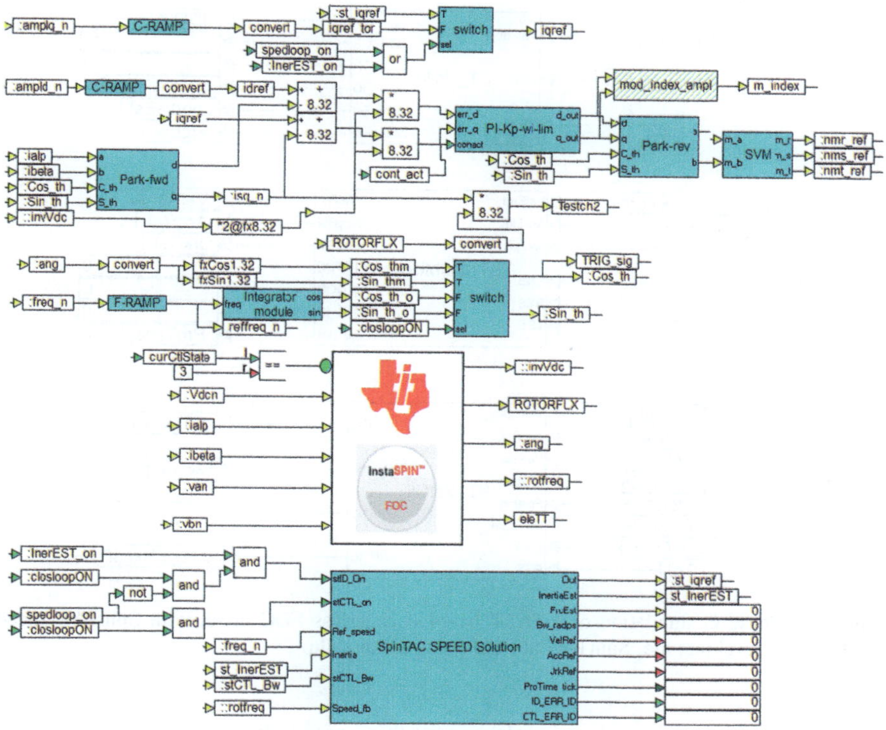

Fig. 6.68 Phase C simulation of a FAST based encoderless FOC drive,with SpinTAC speed control: One level into the 'Current/speed controller with SpinTAC' module

types of speed profiles and also sets the users defined quadrature current limits. In this case study a 'standard S curve' has been selected, which provides the drive with a smooth transition from standstill to motion as will be demonstrated in the next section. Each of the modules shown in Fig. 6.69 has a series of dialog boxes which must be addressed by the user.

6.3.4 Case Study 3b: e-Traction Drive: Sensorless Operation, Phase C+

Phase C+, is the operational component of the laboratory and is basically a run version of the `.out` file compiled and downloaded to the MCU in phase C (see previous subsection).

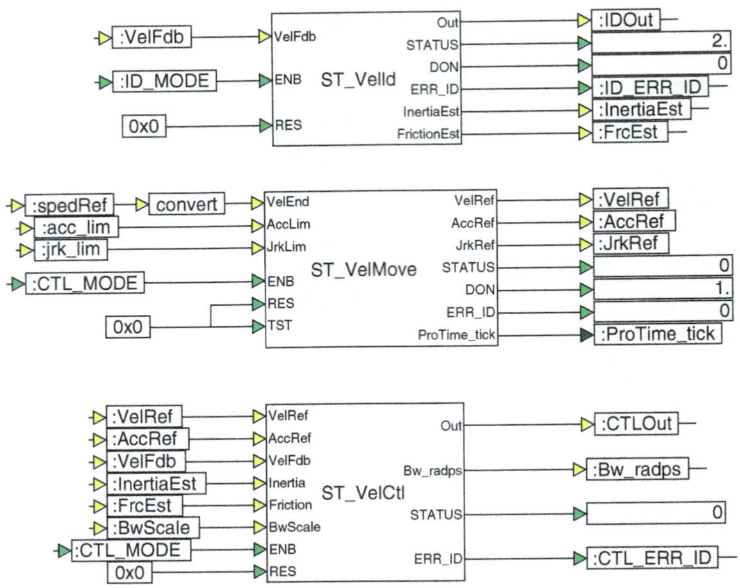

Fig. 6.69 Phase C simulation of a FAST based encoderless FOC drive, with SpinTAC speed control: One level into the 'SpinTAC SPEED Solution' module

The following information is relevant for this case study component:

- Reference program: `eTRAC-069MphCv1_d.vsm`.
- Description: Sensorless SpinTAC based speed control of e-Traction drive using the InstaSPIN module.
- Equipment/Software: e-Traction motor controller-Rev.3 (CI.01.1065) converter [6] with Texas Instruments F28069M control card, 24 V_{dc} supply, Marathon 5k336N2A IM motor, DELTA ELEKTRONIKA SM 400-AR-4 Power supply and VisSim simulation program.
- Outcomes: application of the InstaSPIN algorithm for sensorless FOC operation using SpinTAC technology for inertia estimation and speed control.

The run version shown in Fig. 6.70, uses a VisSim run module, which executes the `.out` file, shown in said module. Prior to running this `.out` file, the following steps should be undertaken:

- Compile the `.out` file in phase C, with 'Flash' option disabled.
- Download the `.out` file via to the MCU during phase C.
- Prior to start up (by activating the VisSim 'go' button), set the `conv_on` button to OFF.

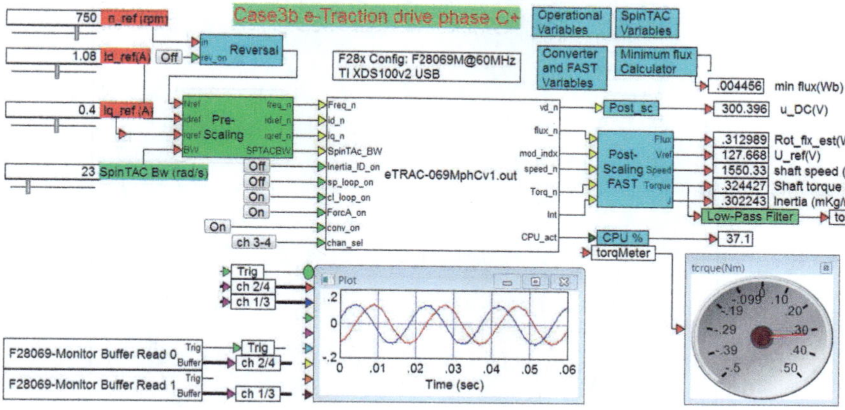

Fig. 6.70 Phase C+ simulation of a FAST based field-oriented controlled (FOC) e-Traction drive, torque control active

Hit the go button and monitor the DC bus voltage. Upon verification that the value is correct, the conv_on button should be enabled, in which case sensorless operation under torque control will occur, given that SpinTAC speed control and inertia estimation have been initially turned off, as may be observed from Fig. 6.70. A provisional load (shaft hand loaded) was used to avoid a motor over speed situation whilst the results shown in the diagnostic scope were taken. An enlarged view of the VisSim scope module given in Fig. 6.71 shows the per unit currents i_{n_α}, i_{n_β}. Multiplication by the full scale current value of 10 A gives the actual value. Some observation of the information provided in Figs. 6.70, 6.71 is briefly summarized below:

- The DC supply level was set to 300 V.
- The results shown are the per unit alpha/beta currents from the ADC module. On the basis of the used reference direct and quadrature currents, the peak current value should be $\sqrt{1.08^2 + 0.4^2} = 1.15$ A. This corresponds to a per unit value of 0.115 (full scale current value of 10 A in use), which is in agreement with the results shown.
- The theoretical shaft torque produced is equal to $T_m = 1.5 p\, i_q\, \psi_R$, where the variables shown are the pole pair number $p = 2$, quadrature current $i_q = 0.4$ A and estimated rotor flux $\psi_R = 0.312$ Wb. This leads numerically to a shaft torque value of $T_m = 0.37$ Nm, which is slightly higher than shown on the torque meter and display. The torque difference is attributed to noise on the estimated torque signal.
- The frequency of the currents shown in the diagnostic scope is 52.63 Hz (derived by measurement of the period time in Fig. 6.71), hence the magnetic field rotates at a speed of approximately 1578 rpm. Calculation of the theoretical shaft slip speed is realized using $60 i_q R_R / (\psi_R 2\pi p)$, which corresponds to $n_{slip} = 32$ rpm hence the average theoretical shaft speed should be $1578 - 32 = 1546$ rpm. The reading

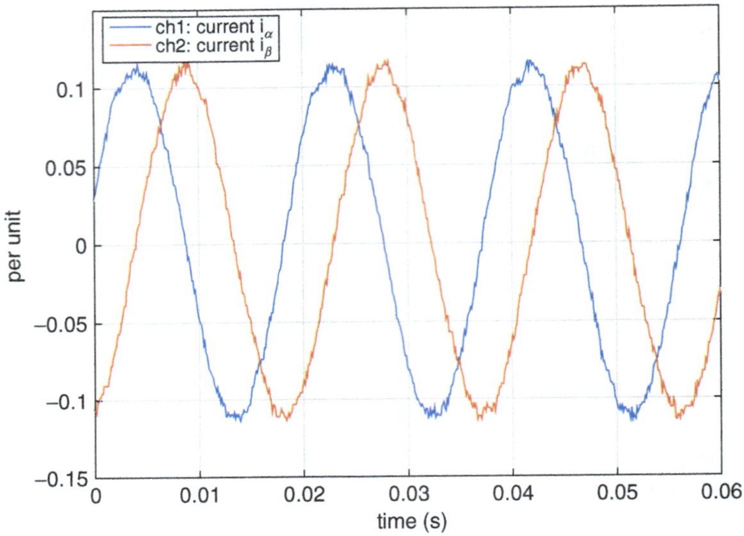

Fig. 6.71 Phase C+ simulation: Scope display showing per unit α, β currents (e-Traction drive running with hand induced load)

shown is slightly higher, and this is attributed to rotor resistance R_{R} inaccuracies, i.e. the value calculated by FAST differs slightly from the actual value given that both stator and rotor resistance are temperature dependent. Note that the speed reference slider reading, shown as 750 is NOT in use given that torque control is operational, hence speed is dictated by the applied load.

Hence some relatively simple computational checks provide insight, with respect to the results shown.

The results shown above were used to verify correct FOC sensorless operation, without SpinTAC technology, hence the use of torque control. In the concluding part of this section the SpinTAC components of the case study will be considered. SpinTAC inertia estimation is activated by the button`inertia _ID_on`, whilst the remaining buttons remain unchanged (speed loop OFF, closed loop ON, force angle OFF, converter ON). The machine (under no-load) will now be exposed to a specific torque curve, which results in an acceleration/deceleration sequence from near standstill to near standstill, as discussed earlier in Chap. 3 (see Sect. 3.4). At the end of this sequence an inertia estimate is generated by the SpinTAC module `ST_VelId`, which in this case yields a value of $J = 300\,\mu\mathrm{kgm}^2$.

SpinTAC speed control is initiated by the button `speed_loop_on`, in which case the machine shaft speed will track the reference shaft speed set by the slider. Furthermore, this speed control case study has been configured to give a smooth speed reversal when the speed reversal button is used, as is apparent from the result

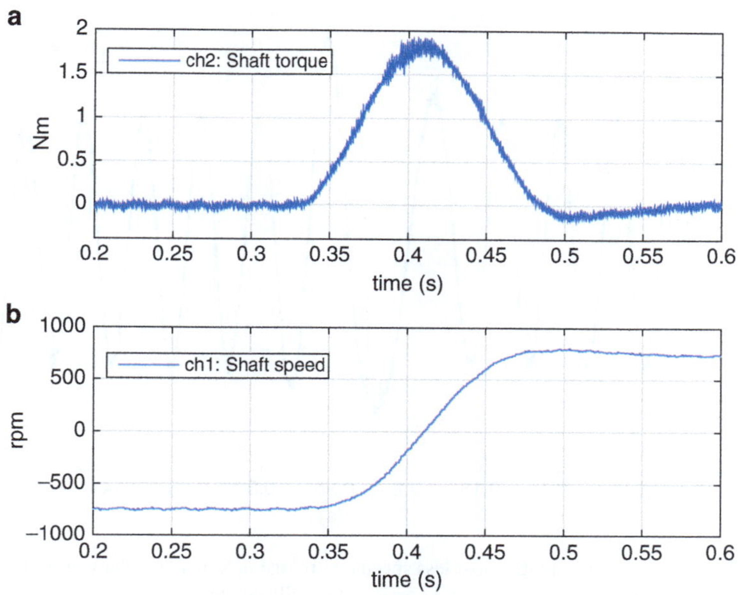

Fig. 6.72 Estimated shaft torque and speed during a speed reversal −750 → 750 rpm, e-Traction drive, FOC sensorless control and SpinTAC active

shown in Fig. 6.72. The results shown in Fig. 6.72 were obtained by allocating diagnostic channels `Ch1`, `Ch2` to the per unit estimated shaft speed and the product of estimated rotor flux and per unit measured quadrature current respectively. These waveforms, obtained using the VisSim scope set to operate with a buffer size of 6001 (which corresponds to a measurement time of 0.6 s), were then rescaled to provide readings in shaft speed and shaft torque.

Observation of Fig. 6.72 learns that the speed reversal is now a smooth 'S' function as programmed in SpinTAC. The corresponding torque waveform is 'bell' shaped and the current limit on the speed controller was purposely set to 2 A to ensure that this bell curve is not clipped by the speed controller. Note that 'force angle' should be turned OFF when undertaking speed reversals. Further evidence of the speed reversal is shown in Fig. 6.73, in which case the phase current was measured with the aid of a DC-true current probe and oscilloscope. Note that the absolute time used in Fig. 6.73 is not the same as used in Fig. 6.72.

It is emphasized that the SpinTAC control approach offers the user a complete library of speed and positioning profiles to suit any industrial requirement [8]. In this case only a relatively simple example was shown.

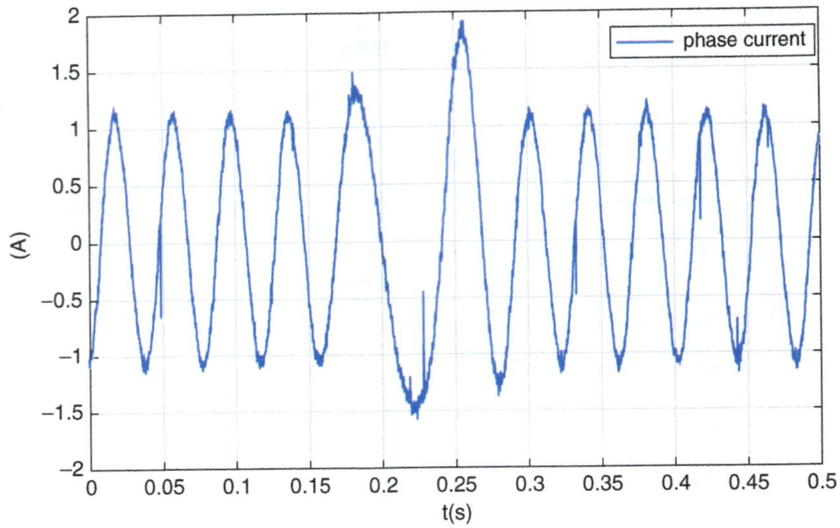

Fig. 6.73 e-Traction drive, FOC sensorless speed control using SpinTAC: phase current measured during speed reversal with DC-true current probe and oscilloscope

Chapter 7
PLECS PIL Based Case Studies

In the previous three chapters extensive use has been made of VisSim embedded control software to realize sensorless control of three-phase AC machines using the InstaSPIN approach. In all cases a hardware set up as shown in Fig. 7.1 (top part) was used to access the LAUNCHXL-F28069M, whereby a PWM/ADC module was provided for data acquisition/control of the converter via a gate driver. In this chapter the PLECS based 'processor in the loop' approach is considered, as shown in Fig. 7.1 (bottom part). This PIL module can replace the ADC input variables (and additional inputs if needed) by those generated by an 'ADC model', which in turn is connected to an electrical model of the drive. In addition, the PWM output signals which are normally used to drive the hardware converter are now redirected to a 'PWM model' that generates the gate signals for the model converter. A high speed data link ensures that the code running on the actual MCU is tied to the PLECS simulation model. Furthermore, the execution of the simulation and embedded code is fully synchronized. In contrast to earlier chapters the Texas Instruments development environment 'Code Composer' is used here, which has access to a series of InstaSPIN MotorWare development laboratories [9]. Specifically two MotorWare development laboratories will be considered in this chapter, given that these are 'PIL' prepared, which implies that the necessary additional PLECS code has been added to enable use of the PIL approach discussed in this chapter. A processor board with a F28069M device must be connected via USB, to enable PLECS PIL operation. The key differences with the VisSim PIL approach discussed previously is the use of a switching converter that may contain ideal or non-ideal devices. With the aid of the approach set out in this chapter a user can fully develop a PLECS PIL based software environment that can be tested with a simulated drive,

---

Electronic supplementary material: The online version of this chapter (doi: 10.1007/978-3-319-20043-9_7) contains supplementary material, which is available to authorized users.

D.W.J. Pulle et al., *Applied Control of Electrical Drives*,
Power Systems, DOI 10.1007/978-3-319-20043-9_7

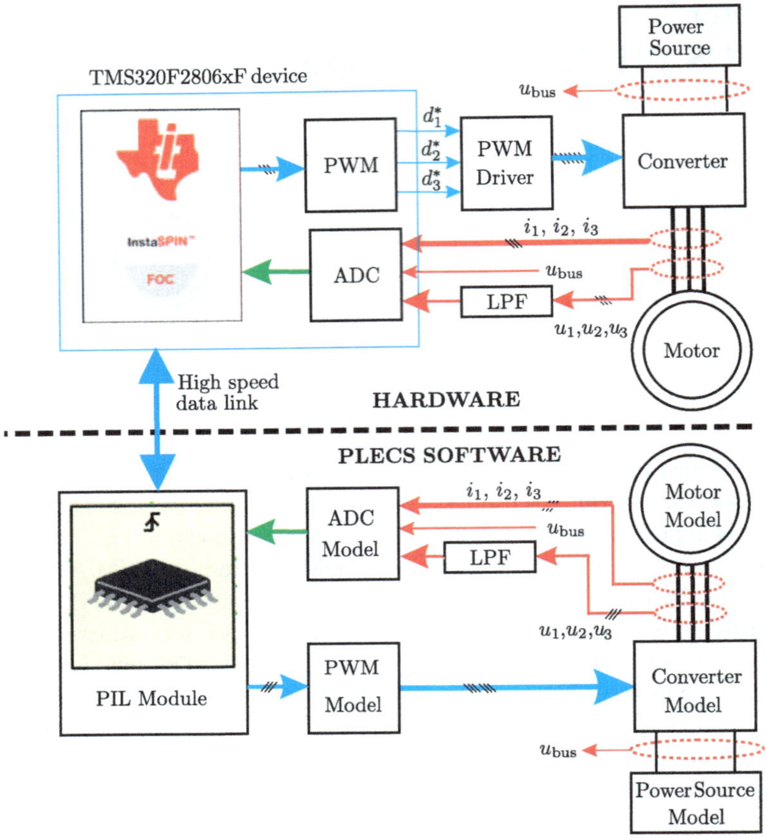

Fig. 7.1 Sensorless drive operation using InstaSPIN with hardware or PLECS processor in the loop (PIL) simulation

prior to use with the actual hardware. Note however, that this type of development requires a certain amount of C-code literacy.

The PLECS PIL approach provides the reader with the ability to evaluate InstaSPIN technology by comparing, for example, the flux vector generated by the estimator with the machine model flux vector whilst using a detailed switching model of the converter. Hence an objective and detailed analysis can be made, which also allows changes to InstaSPIN parameters to be evaluated. Most importantly, this approach is effective for large drive systems, where it is highly desirable (for safety and damage control) to evaluate drive performance under simulated conditions (and in this case with the InstaSPIN MCU in the loop) prior to using the actual hardware. Furthermore, the PLECS approach can be deployed by users which already have Simulink based models of their drive. Hence this allows said users to explore the InstaSPIN sensorless drive approach in combination with their simulated drive setup.

In terms of content this chapter will consider three case studies, where use is made of PIL prepared MotorWare laboratories 2/4, which allow motor identification/ sensorless operation and FAST as software encoder respectively. The first two case studies model the e-Traction converter/Marathon induction machine and LAUNCHXL-F28069M/LVSERVOMTR PM machine combination, which require MotorWare laboratories 2 and 4 respectively. The third and final case study considers an e-Traction converter model with a *single* phase induction machine. For this special case MotorWare laboratory 4 is adapted to allow FOC sensorless control of said machine and explore its use with the InstaSPIN approach. Note that this chapter does not consider the steps needed to 'PIL' prepare MotorWare (or any other executable C-code) laboratories, as this is outside the scope of this book. Instead the reader is referred to appropriate PLECS documentation on this topic [10].

7.1 Case Study P1: e-Traction Converter with Marathon IM Motor

The purpose of this project is to apply the PLECS 'processor in the loop' (PIL) approach to VisSim case study example V3 (see Sect. 6.3) where use was made of an e-Traction converter connected to a Marathon 5k336N2A induction machine. In the latter case, as in the present case, InstaSPIN based FOC torque control is to be deployed, with the important change that the hardware is replaced by a PLECS model, as may be observed from Fig. 7.2 whereby the PIL module communicates with the sensorless ROM based algorithm located on the F28069M MCU hardware. The following information is relevant for this case study:

- Reference program : `Case3C_eTRAC_069v6.plecs`.
- Description: Sensorless control of an e-Traction converter connected to a Marathon Induction machine using the PLECS PIL approach and MotorWare laboratory 4.
- Equipment/Software:Texas Instruments LAUNCHXL-F28069M (no boost packs required) and PLECS Simulation software version 3.6 with PIL capability.
- Outcomes: To demonstrate the tracking capability of InstaSPIN and FOC torque control using vector plots.

As indicated above no boost packs are required for this case study, nor is an external power supply needed, because power for the MCU is directly provided via the USB cable connected to the laptop. In this case the following jumper and dip-switch positions on the LAUNCHXL-F28069M module are required:

- Jumpers `JP1` and `JP2` CLOSED
- Jumpers `JP4` and `JP5` OPEN
- Jumper `JP6` OPEN
- Jumpers `JP3` and `JP7` CLOSED
- Dip-switches `SW1` to `SW3` ON

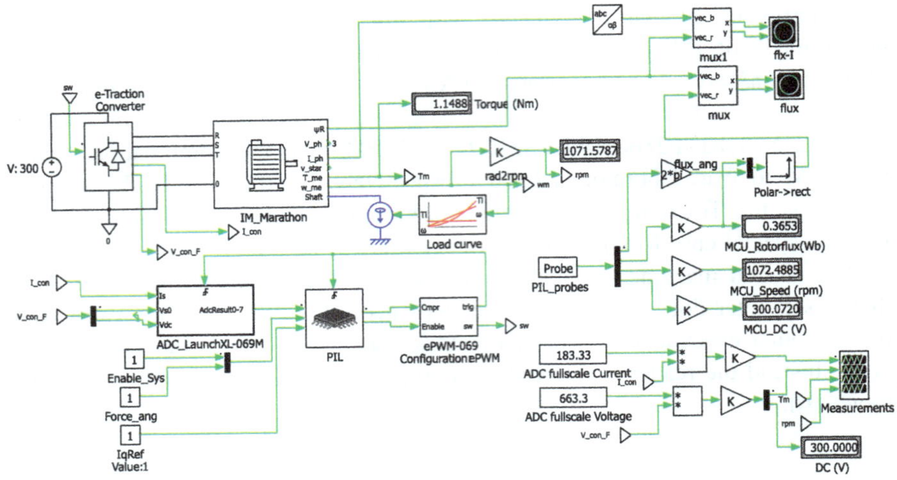

Fig. 7.2 e-Traction converter with Marathon IM motor model and PIL operation using InstaSPIN

Central to the PLECS model shown in Fig. 7.2 is the drive, which is formed by the converter and machine models. Outputs of the converter are the low-pass filtered voltage vector `V_con_F` and measured current vector `I_con` both of which are connected to the 12 bit ADC model `ADC_LAUNCH-069M`. The ADC output of this model is supplied to the PIL module together with (in this case) three additional inputs namely:

- `Enable_Sys`: flag that must be set to activate the InstaSPIN controller.
- `Force_ang`: be set by the user to enable Force Angle operation.
- `IqRef`: the reference quadrature current value(A) to be used by the FOC controller.

During PIL operation input data is accessed (via the PIL module) by the InstaSPIN algorithm and then returns (among others) the modulation indices for the PWM model `ePWM_069M`, which in turn generates the control signals `sw` for the converter and most importantly the trigger signal for the ADC and the PIL module itself. The latter is critical in ensuring that the correct operation of the control algorithm is maintained using pseudo real-time operation. Consequently, any issues related to the execution of the control algorithm, resource corruption as well as processor utilization can be identified during PIL operation. A `PIL_probes` module provides access to a set of (user selectable) variables from the InstaSPIN module and (in this case) the estimated rotor flux amplitude/angle, shaft speed and DC bus voltage are shown on the numerical displays. For comparison purposes, the estimated rotor flux vector is shown together with the rotor flux vector generated by the machine model on a vector scope `flux`. Furthermore, FOC operation can be verified by vector scope module `fx-I`, which shows the stator current vector together with the rotor flux vector generated by the machine model. For this case

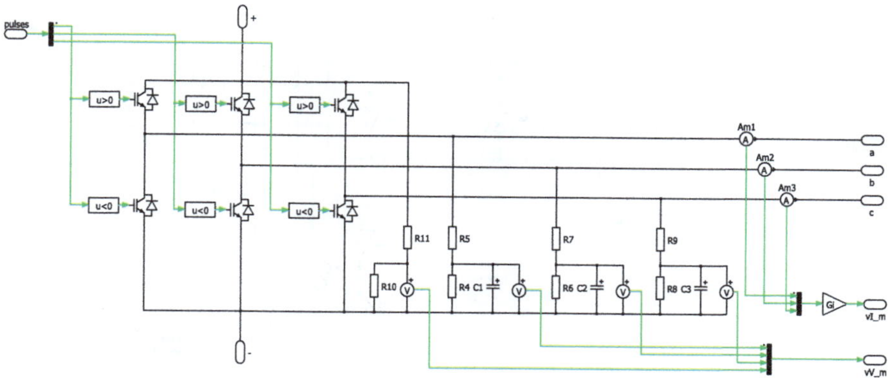

Fig. 7.3 e-Traction converter model with voltage/current measurement

study FOC torque control is deployed, hence it is prudent to include a load model, which is represented by the module `load_curve`, where a quadratic load curve has been selected that generates a torque of 1.0 Nm at 1000 rpm. The output of this module is connected to a mechanical torque block, which in turn is connected to the mechanical output of the machine shaft.

For measurement purposes a four channel `measurements` scope is used to display the actual converter voltages/currents (after multiplication by the relevant `ADC_full scale value`/3.3), shaft torque and shaft speed. The DC bus voltage is also shown on a numerical display so that its value can be compared with the InstaSPIN generated DC bus voltage level `MCU_DC(V)`.

Moving one level into the `e-Traction Converter` module (see Fig. 7.3) reveals the converter structure and voltage/current signal conditioning circuits used to generate the input vectors `V_con_F`, `I_con` for the ADC module. The vector `pulses` generated by the PWM model, controls the six ideal (in this case) power switches. Voltage attenuation and filtering of the converter voltages makes use of a resistance/capacitor network as discussed in Fig. 6.59. Said figure also shows a Hall-effect sensor gain `G`, that has been introduced in Fig. 7.3 to achieve the correct ADC current scaling.

Moving one level lower into the `IM_Marathon` machine model given in Fig. 7.2 reveals a set of PLECS electrical and control blocks as shown in Fig. 7.4. Central to this model is a so called 'IRTF module', which is common to all AC machine models, discussed in our previous books [4, 14]. In this case a four parameter squirrel cage amplitude invariant space vector induction machine model is shown, with parameters:

- `Rs`: stator resistance.
- `Lsi`: stator referred leakage inductance.
- `LM`: magnetizing inductance.
- `RR`: rotor resistance.

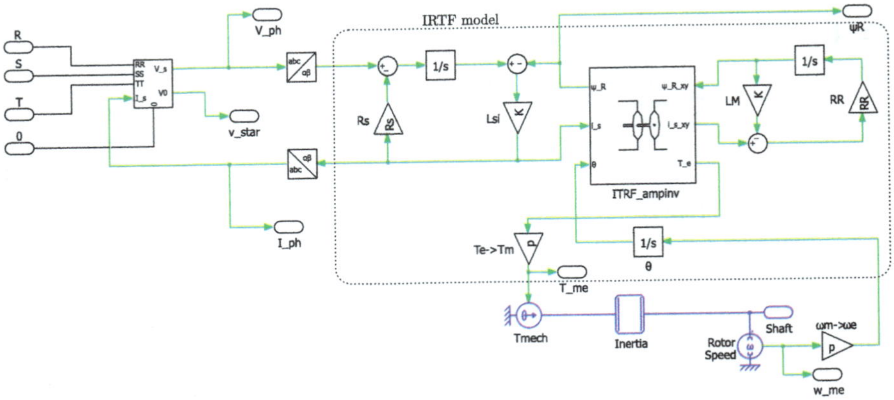

Fig. 7.4 IRTF based space vector model of Marathon induction machine

For this case study the parameters identified in the CoMoCo case study (see Sect. 6.2.4) are used together with the inertia estimate obtained during the e-Traction case study (see Sect. 6.3.4). Input to the space vector 'IRTF model' is the voltage vector `V_ph`, which is generated by an electrical to control block interface module and Forward Clarke conversion module. The interface module has as inputs the three electrical phase terminals `R`, `S`, `T` and `0` (which must be connected to the negative side of the DC bus). A star connected winding is assumed and the star point of the winding is also accessible via control output `v_star`. Input to the interface module is the Reverse Clarke transformed stator current vector `I_ph`. The space vector based rotor flux variable `psi_R` is provided as an output, given that the InstaSPIN generated flux vector must (when operating correctly) be aligned with said vector. The two remaining input/output variables are the electrical shaft angle `theta` and electrical torque `T_e`. A gain `p` (pole pair number of the machine) is used for the conversion from electrical torque `T_e` to shaft torque `T_me`, which is also used as input to a mechanical block `Tmech`. A set of mechanical block modules is used to derive the shaft speed variable `w_me`, for a given inertia `Inertia` value.

7.1.1 Case Study P1a: e-Traction Converter with Marathon IM Motor: PIL Drive Commissioning

With the VisSim PIL approach considered earlier in this book, a 'phase B' development phase was introduced to compile a drive laboratory example which led to the creation of an `.out` file (see Sect. 5.2.1). Prior to using the PLECS PIL approach a similar set of development steps must be undertaken. However, in this chapter use is made of the Texas Instruments MotorWare software directory structure, which contains a wide range of C-code modules that cover all aspects of drive operation for different MCU's and hardware platforms. Included in this

1. CPU and Inverter Set-up and introduction to interfacing to the ROM library
2a. Using InstaSPIN™-FOC for the first time, Motor ID, full control system from ROM [F2806x only]
2b. Run InstaSPIN™-FOC from user memory (RAM/FLASH), only FAST™ in ROM
2c. Advanced Motor ID for low inductance motors
2d. FPU32 enabled [F2806x only]
3a. Saving your motor parameters and loading from user.h
3b. Saving your motor and hardware parameters and loading from user.h
3c. Using parameters from user.h with FPU32 [F2806x only]
4. Running InstaSPIN™-FOC only as a Torque controller
4a. Torque Mode with FPU32 [F2806x only]
5a. Adjusting the supplied current controllers
5b. Tuning the supplied speed controller
5g. Tuning the speed controller with FPU32 [F2806x only]
7. Using Rs Online Recalibration
7a. Rs Online Recalibration with FPU32 [F2806x only]
9. Experimentation with Field Weakening
9a. Field Weakening with FPU32 [F2806x only]
10a. Experimentation with Space Vector Over-Modulation
10c. Over-Modulation with FPU32 [F2806x only]

Fig. 7.5 MotorWare laboratory examples [9]

structure is a set of MotorWare laboratories, as given in Fig. 7.5, which provides the reader with a set of InstaSPIN based examples that can be directly adapted to a specific application. Consequently, the user does not need to design a C-code based drive algorithm from the ground up, instead the task is reduced to adapting the code to suit the application at hand. The laboratories shown in Fig. 7.5 are run with the aid of the Texas Instruments 'Code Composer' (CCS) environment, as will be shown shortly. This approach can in fact be used independently of either the VisSim or PLECS. However, when using the PLECS PIL approach additional C-code must be added to the above laboratory examples prior to use [10]. In this chapter MotorWare laboratories 2a and 4 will be used, given that these have been ' PLECS PIL prepared'. For this case study, PIL commissioning using MotorWare laboratory 4, renamed as `instaspin_foc_28069F` (given that it is PIL prepared) will be considered. This CCS laboratory example is very similar to VisSim case study V3b (see Sect. 6.3.3) in terms of functionality.

A set of detailed steps will be shown below, a process which is not repeated for the subsequent two cases studies, given that the approach is essentially identical. Firstly, CCS should be started, after which lab 4 is loaded by selecting `Project`, then option `Import existing CCS project`. Finally, identify the location of the PIL prepared lab `instaspin_foc_28069F`. It is prudent to keep this CCS lab together with the corresponding PLECS file in the same folder. After import of said laboratory the CCS screen shot shown in Fig. 7.6 should appear. Note that this file has two options namely use of DRV8312 or LaunchXL-069 (selectable, right side, next to the 'hammer' symbol). In this case the option `BOOSTXL-DRV8301` must be selected, which effectively means using the LaunchXL-069 option (in this case WITHOUT the BOOST pack present). Apparent on the top right side of Fig. 7.6 is the so called 'watch window' `expressions` which contains variables and flags that may be either set by the

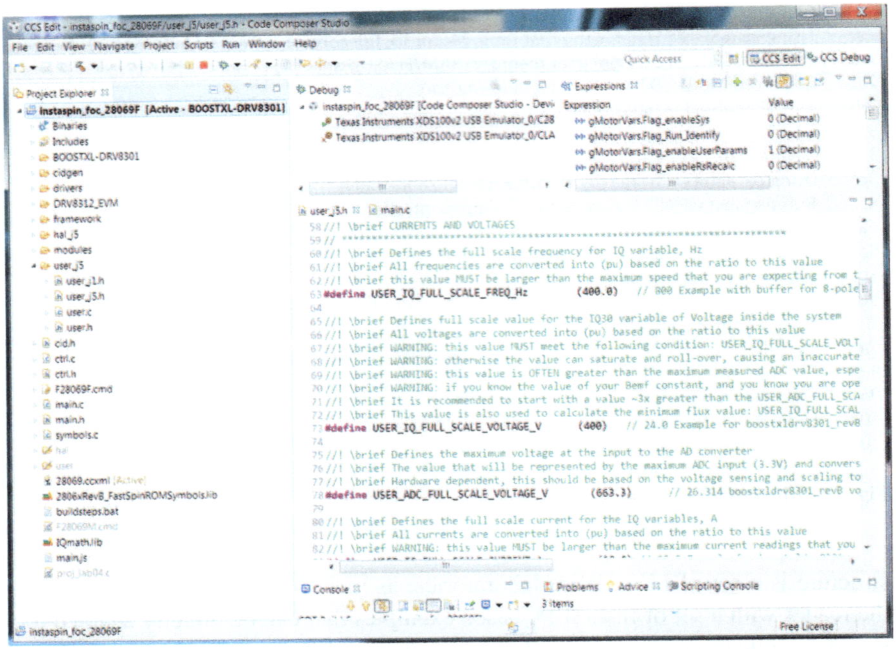

Fig. 7.6 Code Composer (CCS) Edit Screen shot: after import of MotorWare project 4

user during 'debug operation', to be discussed shortly. When said operation is active, the watch window also shows the variables, such as estimated shaft speed and flux (in V/Hz) generated by InstaSPIN. The variables which appear in the watch window can either be entered manually or inserted via the `scripting console` window in the CCS edit window. If the latter is not visible select `view` and then `scripting window\index{scripting window}`, which brings up the required window. Select `open command file` to locate the `main.js` file (located in the same folder as the file `instaspin_foc_28069F`). Note that the watch window is normally used to control the drive and read back key data. With PIL active, partial drive control may be allocated to the PLECS program (as is the case here).

Shown in the window on the left side of Fig. 7.6 is the directory associated with laboratory 4: `instaspin_foc_28069F`, which shows a key file `user_j5.h` that is also expanded in the center window. Note that this file resides under directory `user_j5`, which signifies that it is associated with the 'aft' (furthest away from USB) boost pack (if it is installed) location on the LAUNCHXL-F28069M module. This `user_j5.h` file contains all the data entries, which hitherto have been made in the VisSim labs via dialog box entries. Note that the term `_j5` of the `user_j5.h` file refers to the parameters which are linked with the 'aft' boost pack. After compilation in 'VisSim phase C' of, for example, case study V3b, a C-code file: `eTRAC-069MphCv1.c` is automatically generated which contains all the code required to generate the `.out` file. Note that the `user_j5.h` file

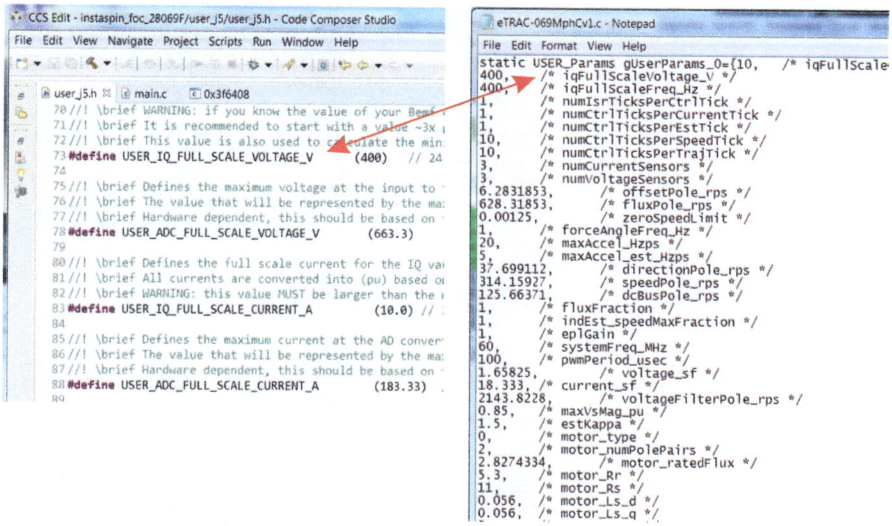

Fig. 7.7 Part of Code Composer (CCS) Screen shot 'user-j5.h' and corresponding part of VisSim C-code file

is also used by the PLECS file `Case3C_eTRAC_069v6.plecs` (within the `Initialization` section of the `Simulation` pull-down menu). Part of the VisSim generated `eTRAC-069MphCv1.c` file is shown together with part of the CCS `user_j5.h` file in Fig. 7.7, in order to demonstrate the commonality of these two files in terms of user entries. Apparent from Fig. 7.7 is the need for the user to enter all the relevant (as discussed earlier via dialog box entries) manually. For this case study, the motor type and motor parameters, as used in case study V3b (see Sect. 6.3.3) also need to be entered prior to undertaking a `Rebuild`. Note that in this case study a clock frequency of 60 MHz is used, which implies the need to also edit the `hal.c` file (under the `hal_j5` directory) and the `user.h` file (under the `user_j5` directory) in order to set the appropriate clock frequency.

Subsequent compilation is achieved by a right click on the file name: `instaspin_foc_28069F[Active-BOOST-DRV8301]` and selecting `Rebuild Project`, after which a file `.out` is generated. A 'debug' session must then be activated in order to run the compiled out file `.out`. Note that 'running' the file, implies in this case that the MCU will be placed under the control of the PLECS program where the triggering of the MCU is carried out by the PWM model `ePWM_069M` (see Fig. 7.2). A 'debug' session is activated by hitting the 'green bug', after which the compiled 'out' file is downloaded to the MCU and the CCS debug screen (shown partly) should appear as given in Fig. 7.8.

Prior to starting debug operation, enable the `continuous refresh` feature (yellow double arrow button top right of screen), which will ensure that the watch window variables are updated during PIL operation. Start debug by hitting the `Resume` button (green arrow), in which case a 'red' LED on the LAUNCHXL-F28069M module should start to flash.

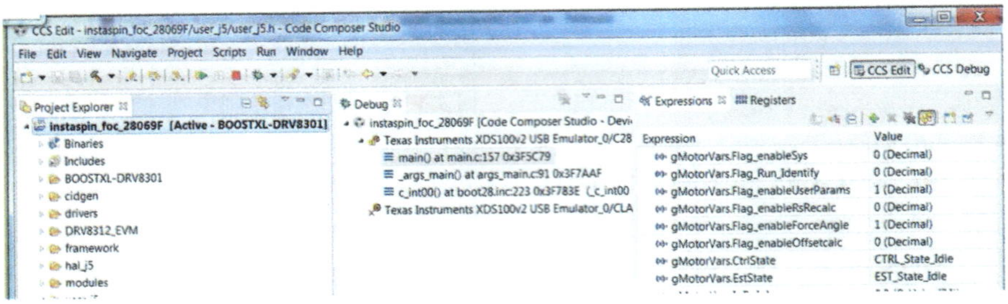

Fig. 7.8 Code Composer (CCS) Debug Screen shot: after ‘debug’ (green ‘bug’) start

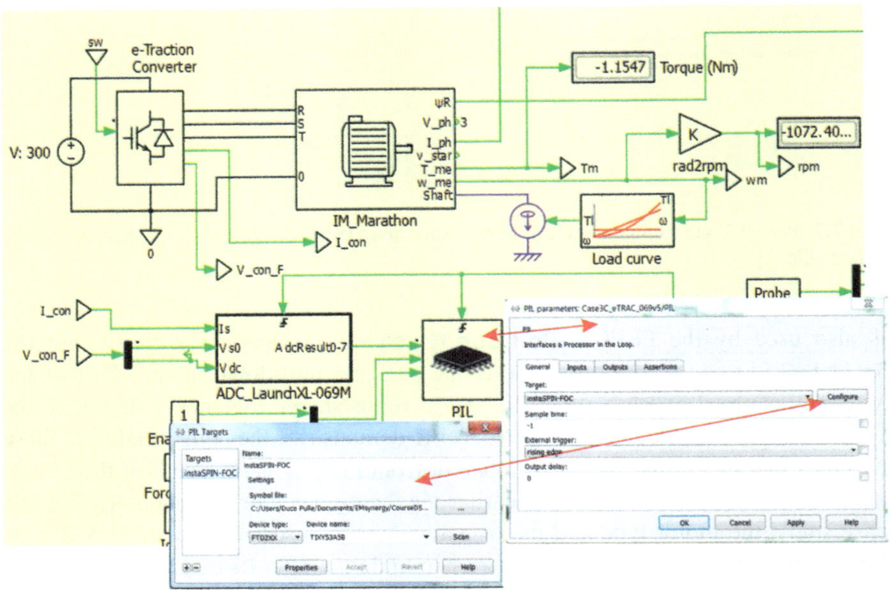

Fig. 7.9 Part of PLECS Screen shot and PIL module related dialog boxes

The second phase of PIL commissioning requires a revisit to the PLECS program `Case3C_eTRAC_069v6.plecs`, where the PIL must be prepared for operation. The composite set of screen shots given in Fig. 7.9 shows the steps required to activate the PIL module. A left double click on the PIL module leads to the window `PIL parameters`, which shows (among others) the `input`, `output` buttons, which provide detailed information on the inputs and outputs of the PIL module. A left click on `configure` leads to dialog box `PIL Targets`, in which the `symbols file` entry must reflect the path needed to access the file `.out` generated in the previous step via CCS (this file will be in the folder `instaspin_foc_28069F`). In addition, this dialog box has a `Scan` button which is used to identify the attached device and the corresponding name should appear in the dialog box.

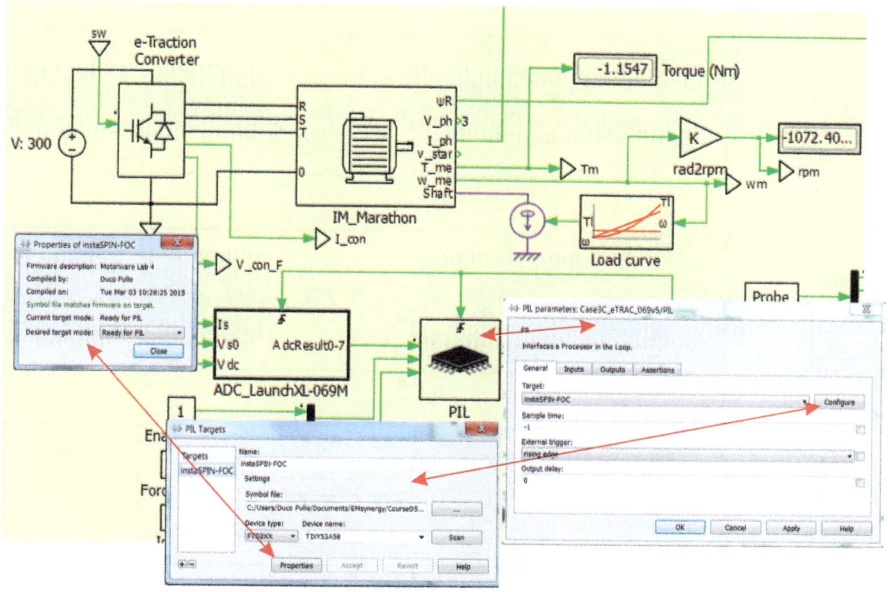

Fig. 7.10 Part of PLECS Screen shot and PIL module related dialog boxes, final phase of PIL commissioning

The final PIL commissioning step requires a return to the PIL window `PIL targets` as shown in Fig. 7.10. A double left click on the PIL module is required to reach the `PIL parameters` window, followed by a left click on `Configure`, which brings up the required window. Then click on `Properties` to bring up a window which shows, among others, the line `Symbol file matches firmware on target`, which signifies that PIL commissioning has been successful and that the PLECS run phase can start provided the option `Desired target mode:Ready for PIL` has been selected. Note that the alternative selection mode is `Normal Operation` which means running drive operation directly from the CCS watch window, i.e. without the use PLECS and WITH the hardware (actual drive) active, as will be discussed at a later stage.

7.1.2 Case Study P1b: e-Traction Converter with Marathon IM Motor: PIL Drive Operation

With PIL commissioning completed and the MCU online, attention is directed towards running the PLECS program `Case3C_eTRAC_069v6.plecs` given in Fig. 7.2. However prior to starting the program, check that the simulation run is set to 3 s and activate the variables `Enable_sys` and `Force_ang`. Furthermore, set the quadrature current variable `IqRef` to 1.0 A.

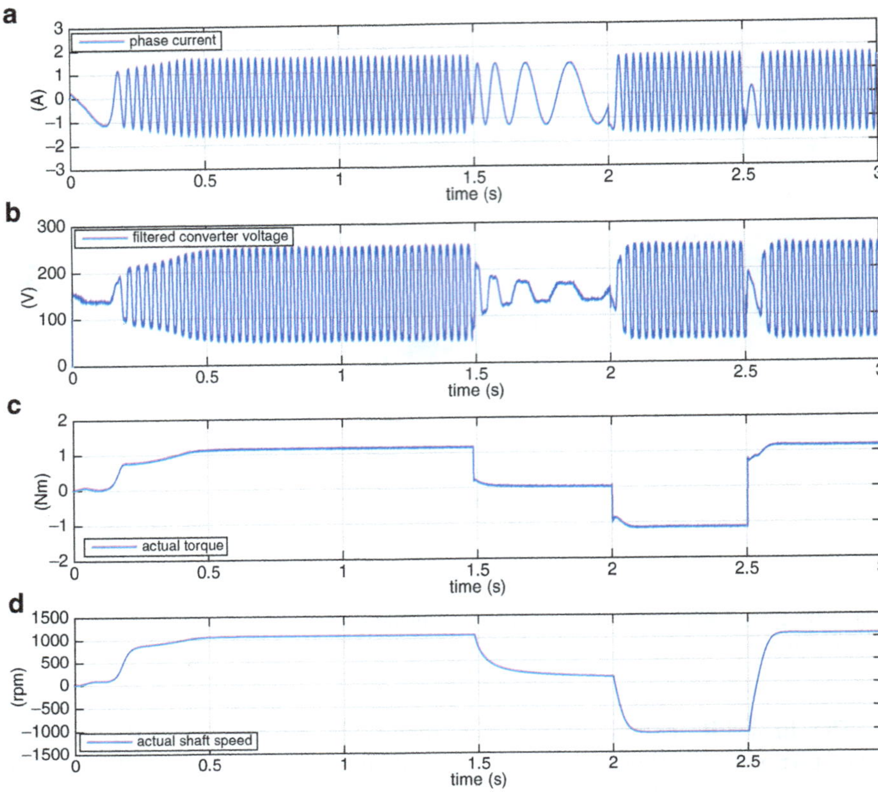

Fig. 7.11 Results shown on the measurement scope at completion of a 3.0 s simulation cycle

Operation can be monitored by observation of the vector plots `flux`, `flx-I` and the `Measurements` scope. On the basis of the results shown in Fig. 7.11, the operation sequence and observed phenomena are itemized after completion of the chosen simulation interval.

- At the start of operation the machine is full de-fluxed, hence during the initial interval $0 \rightarrow\approx 0.5$ s the estimator flux and actual flux come into alignment and at the same time rotor flux builds to its steady-state value, which is governed by the direct axis current of 1.3 A. After rotor flux alignment is achieved, 'force angle' is turned off. Note that the rate of current change (which affects the torque build up in the initial stage in this case) is set by the parameter `USER_MAX_CURRENT_SLOPE` in the `user_j5.h` file. The quadrature reference current has been set to 1.0 A.
- At $t \approx 1.5$ s the quadrature current `IqRef` is set to zero. This causes the speed to reduce (given the presence of a quadratic load torque/speed characteristic) and the phase amplitude must then match the set direct axis current value of 1.3 A. Torque will be zero, while the filtered voltage amplitude will also reduce because the EMF amplitude reduces with speed.

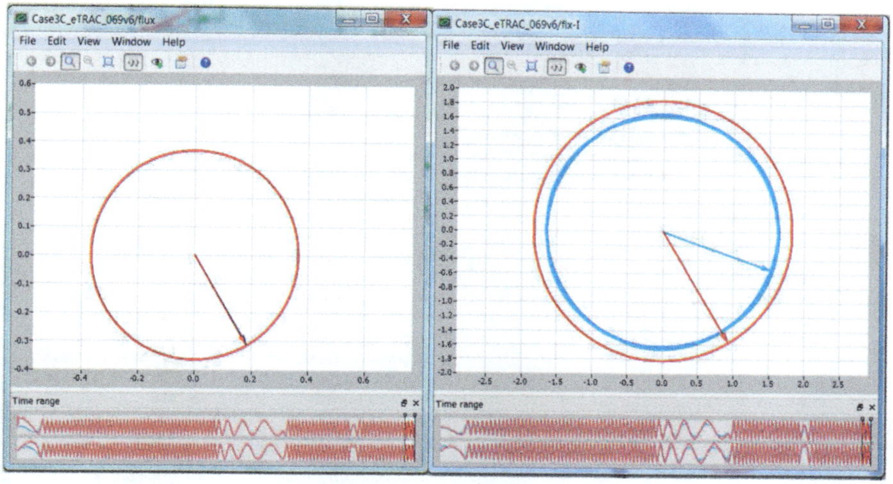

Fig. 7.12 Vector plot Screen shots: LEFT, estimated ('red') and actual rotor flux, RIGHT, actual rotor flux (times 5) and current vector ('blue')

- At $t \approx 2.0$ s the quadrature current `IqRef` is set to -1.0 A, which causes a torque step in the negative direction, hence speed will become negative and stabilize to its steady-state value (torque generated by the machine matches the load torque). Phase current amplitude will increase as its value is determined by the direct and quadrature current components.
- At $t \approx 2.5$ s the quadrature current `IqRef` is set to 1.0 A, which leads to a torque reversal followed by a speed change and constant steady-state speed value at completion of the designated time sequence.

As mentioned earlier the vector plots provide valuable information on drive behavior. For example, the vector plot screen shots given in Fig. 7.12 show operation during a time interval in the region $t \approx 2.9$ s (as indicated by the highlighted time sequence underneath the plots) The vector plot on the left hand side of Fig. 7.12 shows the estimated ('red') flux together with the flux vector generated by the machine model. The two are, as may be expected, in complete alignment and remain so after the initialization time interval has passed. The vector plot on the right hand side of Fig. 7.12 shows the machine rotor flux vector (multiplied by a factor 5) and the current vector ('blue'). In this instance the current vector leads the flux vector (as expected, given that `IqRef` equals 1.0 A during this time instance), hence a positive torque is produced. Note that the current vector amplitude is defined by the 1.3 A magnetizing current AND the 1.0 A (for this operating point) quadrature current.

It is emphasized that an extensive range of operating conditions under different motor parameter conditions, load conditions etc. can be examined, where most importantly, operation of InstaSPIN can be evaluated objectively against well defined models.

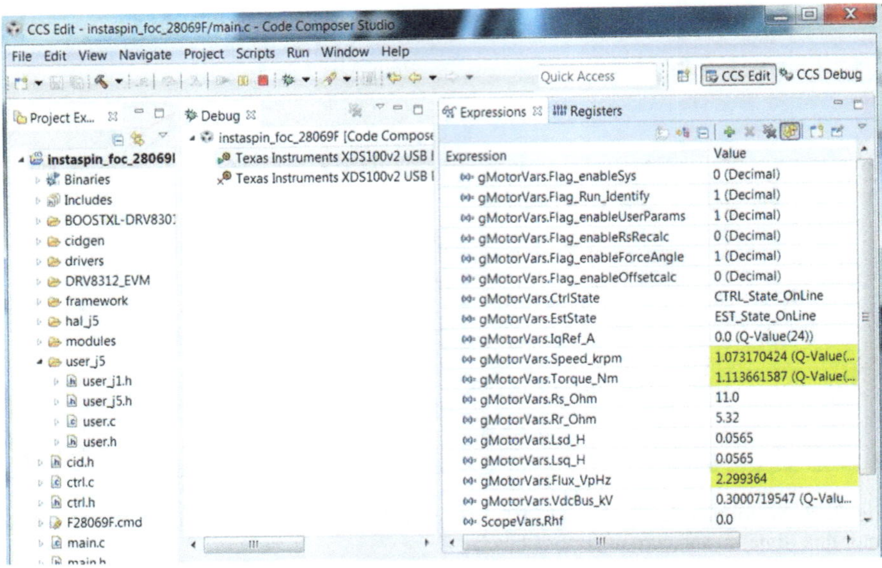

Fig. 7.13 CCS screen shot: watch window during PIL operation

Apart for the results shown in the PLECS program, the user can also observe the Watch window parameters during operation. A screen shot example, given in Fig. 7.13 shows highlighted in yellow the estimated rotor flux (Voltage per Hz notation), shaft torque and shaft speed. In this screen shot example the drive is operating under steady-state motoring conditions with a reference quadrature current setting of `IqRef`=1.0 A and external load. Also indicated are the (user set) machine parameters and measured DC bus voltage. As mentioned earlier, a number of the flags are set via PLECS directly and cannot therefore be changed in the watch window. However, other variables such as the `enablePowerWarp` flag (used to optimize the flux level for a given torque) can be activated from the watch window.

7.2 Case Study P2: Motor Identification Using InstaSPIN and PIL Technology

The purpose of this case study is to apply the PLECS 'processor in the loop' (PIL) approach to VisSim laboratory 2:2 (see Sect. 4.3), where use was made of the LAUNCHXL-F28069M module, with BOOSTXL-DRV8301 boost pack connected to a LVSERVOMTR PM machine. In the latter case as in the present case, InstaSPIN based Motor identification is to be used, with the important change that the hardware is replaced by a PLECS model, as may be observed from Fig. 7.14 whereby the PIL module communicates with the sensorless ROM based algorithm located on the F28069M MCU hardware. In addition, standalone motor identification using a so called 'normal operation' model of the PIL module will also be discussed. The following information is relevant for this case study:

- Reference program : `LaunchXL_Teknic_IDv5.plecs`.
- Description: Motor parameter identification of a PM motor using the PLECS PIL approach and MotorWare laboratory 2a.
- Equipment/Software: Texas Instruments LAUNCHXL-F28069M, with BOOSTXL-DRV8301 module ('aft' position), LVSERVOMTR PM machine and PLECS Simulation software version 3.6 with PIL capability.
- Outcomes: To demonstrate motor parameter identification using the PLECS PIL approach and with actual attached machine.

As indicated above a boost pack is used in this laboratory, which is located in the 'aft'(furthest away from the USB port). In this case the following jumper and dip-switch positions on the LAUNCHXL-F28069M module are required:

- Jumpers `JP1` and `JP2` OPEN
- Jumpers `JP4` and `JP5` CLOSED
- Jumper `JP6` OPEN
- Jumpers `JP3` and `JP7` CLOSED
- Dip-switches `SW1` to `SW3` ON

Central to the PLECS model shown in Fig. 7.14 is the electrical drive which consists of the `BOOST-8301` converter and `PM_Teknic` machine model. Outputs of the converter are the low-pass filtered voltage vector `V_con_F` and measured current vector `I_con` which are connected to the 12 bit ADC model `ADC_LAUNCHXL_069M`. The ADC output of this model is supplied to the PIL module. During PIL operation (PLECS simulation synchronized to the MCU via the data link, 'blue' LED on), input data for the PIL is used by the InstaSPIN algorithm and then returns (among others) the modulation indices for the PWM model `ePWM_069M`, which in turn generates the control signals `sw` for the converter and most importantly the trigger signals for the ADC and the PIL module itself. The latter is critical in ensuring that correct operation of the control algorithm is maintained using pseudo real-time operation. Consequently, any issues related to the execution of the control algorithm, resource corruption (interrupt timing issues) as well as processor utilization can be identified during PIL operation. A `PIL_probes` module provides access to a set of (user selectable) variables from the InstaSPIN module and in this case the estimated rotor flux angle (not used in this study), rotor flux amplitude, DC bus voltage, estimator state and estimated stator inductance/resistance are available.

For this case study a mechanical brake model and 10 mNm friction have been added, in line with the friction torque level of the actual machine. The purpose of the brake module is to ensure that the rotor remains locked during the `RoverL`, phase, which undertakes an initial coarse identification of the motor, in order to set the current controller gains for the drive.

For measurement purposes a six channel `Measurements` scope is used to show the key probe variables (rotor flux amplitude, estimated R_s, L_s), actual shaft speed and motor phase current.

The `BOOST-8301` module contains the converter and voltage/current signal conditioning circuits used to generate the inputs vectors `V_con_F`, `I_con` for the ADC module as discussed earlier (see Fig. 3.2). The vector `sw` generated by the PWM model, controls the six ideal power switches. Voltage attenuation and filtering of the converter voltages makes use of a resistance/capacitor network as discussed in Fig. 3.5.

Moving one level lower into the `PM_Teknic` (LVSERVOMTR) machine model given in Fig. 7.14 reveals a set of PLECS electrical and control blocks as shown in Fig. 7.15. Central to this model is a so called 'ideal rotating transformer' (IRTF) control block which is central to all AC machine models, as discussed in our previous books [4, 14]. In this case an amplitude invariant space vector PM machine model is shown, with parameters:

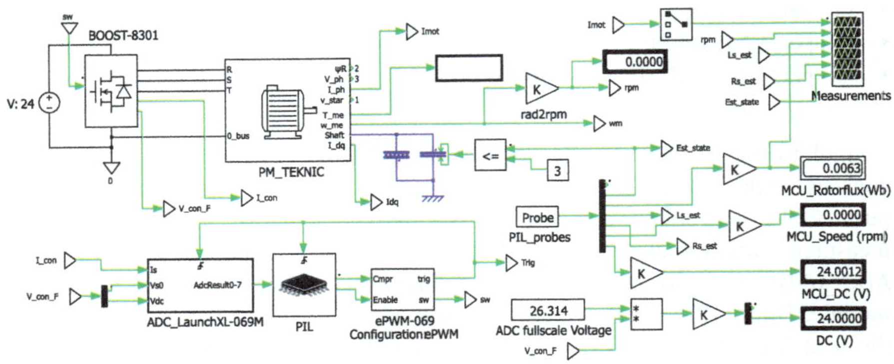

Fig. 7.14 LAUNCHXL-F28069M, LVSERVOMTR PM motor and PIL operation using InstaSPIN

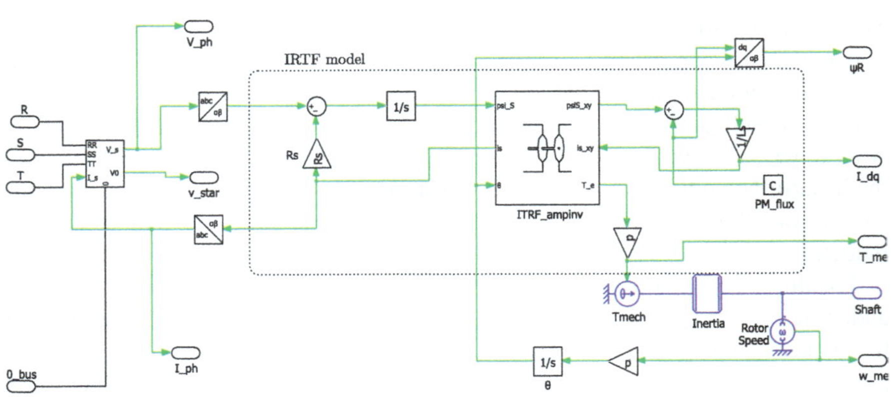

Fig. 7.15 IRTF based space vector model of LVSERVOMTR PM machine

- `Rs`: stator resistance.
- `Ls`: stator inductance.
- `PM_flux`: PM flux (Wb).

For this case study the parameters used are those identified earlier in this book, given that the drive setup modeled here matches the actual drive present. Furthermore, the MotorWare lab used for the PIL approach is also used at a later stage to identify the parameters of the machine. Inputs to the space vector IRTF model are the Forward Clark transformed phase voltages, which are generated by an electrical to control block interface module. This module has as inputs the three electrical phase terminals `R`,`S`,`T` and `0` (which is connected to the negative side of the DC bus). A star connected winding is assumed and the star point of the winding is also accessible via control output `v_star`. Input to the interface module is the Reverse Clarke transformed stator current vector `I_ph`. The space vector based PM rotor flux variable `psi_R` is provided as an output, given that the InstaSPIN generated flux vector must be aligned with said vector. The two remaining input/output variables are the electrical shaft angle `theta` and electrical torque `T_e`. A gain `p` (pole pair number of the machine) is used for the conversion from electrical to shaft torque, where the latter is used as input to a mechanical block `Tmech`. A set of mechanical block modules is used to derive the shaft speed variable `w_me`, for a given inertia.

7.2.1 *Case Study P2a: Motor Identification Using InstaSPIN and PIL Technology: PIL Drive Commissioning*

The commissioning of the PIL requires a virtually identical set of steps as described for the previous case study (see Sect. 7.1.1). Consequently, only the basic steps are discussed in this section. After start-up of code composer (CCS), load the PIL prepared `instaspin_id_28069F` folder, which is the MotorWare motor identification laboratory 2 that is adapted for the drive in use. After import of said laboratory the CCS screen shot shown in Fig. 7.16 should appear. Readily observable in Fig. 7.16 is the directory associated with `instaspin_id_28069F`, which shows the additive: `Active-BOOSTXL-DRV8301`, The latter option must be set by the user (in for example the 'properties' box of this file) given that this application can be used with the DRV8312 or BoostXL module. A sub folder of this directory is the `user_j5.h` file, that is also expanded in the center window, which contains all data entries that hitherto have been made in the equivalent VisSim lab 2.2 via dialog box entries. This file is linked to the 'aft' boost converter, whereas file `user_j1.h` (not used in this lab) is tied to the 'forward' boost. Apparent from Fig. 7.16 is the need for the user to enter all the relevant entries (undertaken previously via dialog boxes) manually. For this laboratory the motor type and known motor parameters, as derived in Sect. 7.2.3 (in this case) need to be entered (see center window Fig. 7.16) prior to undertaking a `Rebuild`. A rebuild is activated by a right click on the file name: `instaspin_id_28069F` and selecting `Rebuild`, after which a file `.out` should be generated (located in the same folder that contains

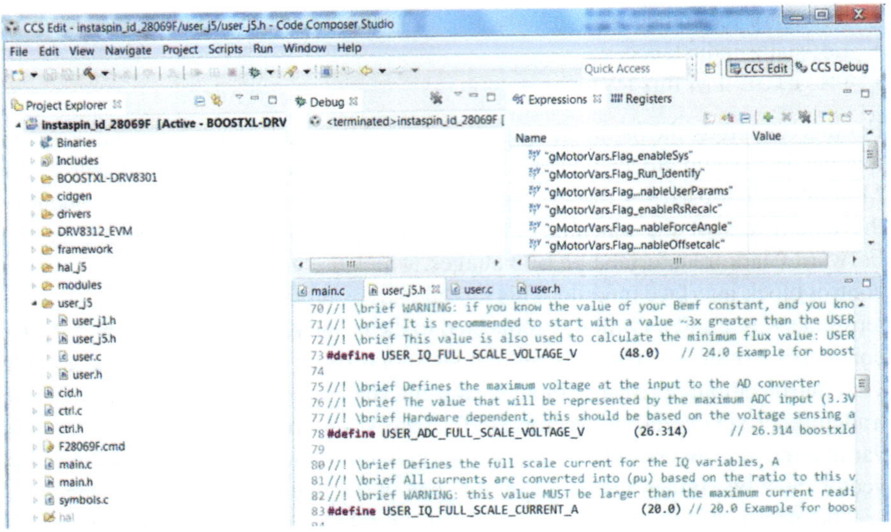

Fig. 7.16 Code Composer (CCS) Edit Screen shot: after import of MotorWare project 2a

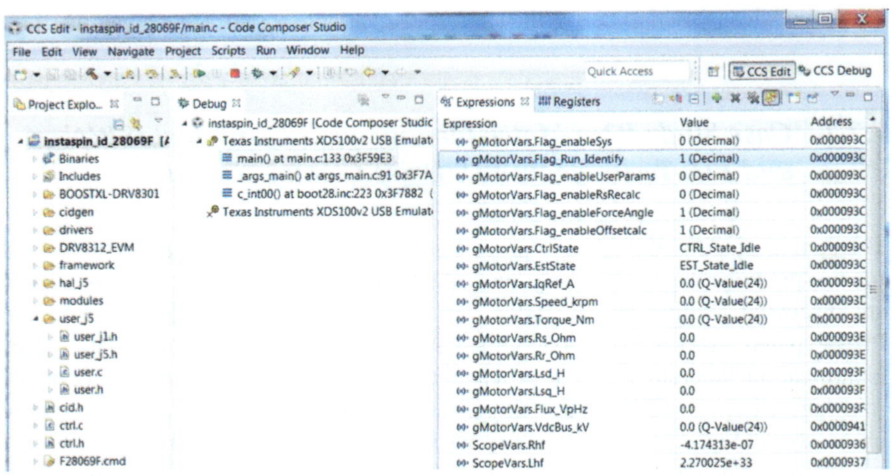

Fig. 7.17 Code Composer (CCS) Edit Screen shot: start of motor ID debug session

the PLECS simulation and folder `instaspin_id_28069F`). A 'debug' session must now be activated by hitting the 'green bug', after which the compiled out file is downloaded to the MCU and the CCS debug screen shown in Fig. 7.17 should appear. Apparent on the top right side of Fig. 7.17 is the 'watch window' `expressions` which contains variables and flags that may be either set by the user during 'debug operation'. When said operation is active, the watch window also shows a range of variables, such as estimated shaft speed and flux (in V/Hz) generated by InstaSPIN.

Note that 'running' the file, implies in this case that the MCU will be placed under the control of the PLECS program where the triggering of the MCU is carried out by the PWM model `ePWM_069M`. **Prior** to activating the debug session enable the `continuous refresh` feature (yellow doubled arrow button top right of screen), which will ensure that the watch window variables are updated during PIL operation. Furthermore, check that the flag `Flag_Run_Identify` is set to 1, to ensure that motor identification will start after PLEC PIL operation is activated. Also confirm that the flag `Flag_enableSys` is still 0, as this variable must be set to 1 **after** the PLECS program has been activated. Start debug by hitting the `Resume` button (green arrow), in which case a 'red' LED on the LAUNCHXL-F28069M module should start to flash.

The second phase of PIL commissioning requires access to the PLECS program `LaunchXL_Teknic_IDv5.plecs`, where the PIL (processor in the loop) module must be prepared for operation (see Fig. 7.14). A left double click on the PIL module leads to the window `PIL parameters`, which shows (among others) the name of the target. A left click on `configure` leads to dialog box `PIL Targets`, in which the `symbol file` entry must reflect the path needed to access the file `.out` generated in the previous step via CCS (this file will be in the folder `BOOSTXL-DRV8301`) which is inside folder `instaspin_id_28069F`.

The final PIL commissioning step requires a click on `Properties` to bring up a window which shows among others the line `Symbol file matches, firmware on target`, which signifies that PIL commissioning has been successful and that the PLECS run phase can start provided the option `Desired target mode:Ready for PIL` has been selected. Note that the alternative selection mode is `Normal operation` which means running drive operation directly from the CCS watch window, i.e. without the use of PLECS and WITH the hardware (actual drive) active, which will be discussed in the last section of the case study.

7.2.2 Case Study P2b: Motor Identification Using InstaSPIN and PIL Technology: PIL Drive Operation

With PIL commissioning completed and the MCU online, (flashing 'red' LED on the LAUNCHXL-F28069M module) attention is directed towards running the PLECS program `LaunchXL_Teknic_IDv3.plecs`. Prior to starting the program, select items `Simulation` and `Simulation Parameters` of the respective pull-down menus and check that the simulation run time is set to 70 s. Note that the simulation initialization parameters are set indirectly from the `user_j5.h` file.

After starting this program return to the 'watch window' and set the flag `gMotorVars.Flag_enableSys` to 1, in order to start actual operation of the algorithm, which in turn provides the triggering for the PLECS PIL module.

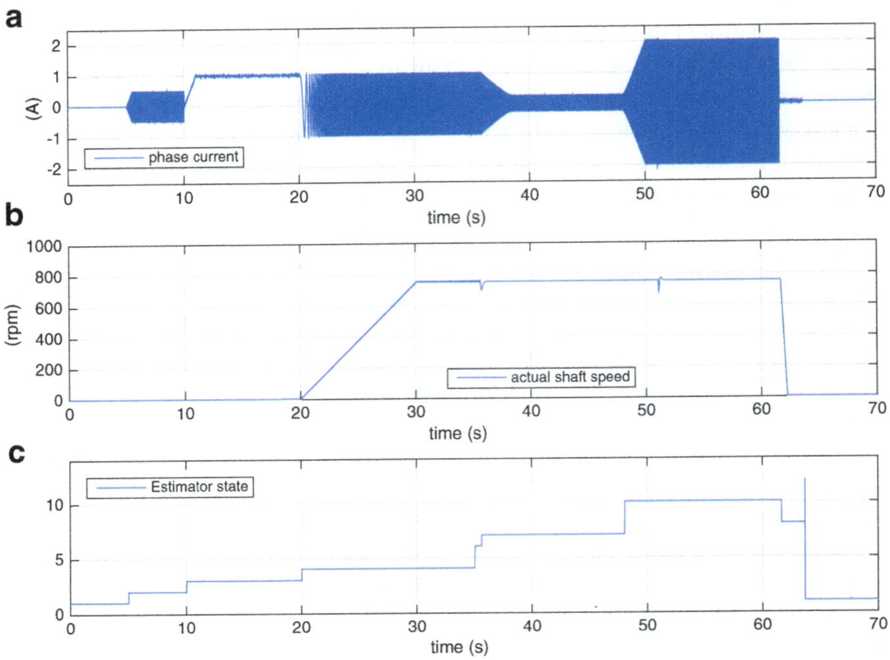

Fig. 7.18 Use of PLECS PIL approach for PM drive with motor identification capability: operational data

Operation can be monitored by observation of the `Measurements` scope data during the actual run sequence, which in this case can take three hours for a 70 s time period. On the basis of the processed scope results shown in Figs. 7.18, 7.19 the operation sequence and observed phenomena are itemized after completion of the chosen simulation interval. Figure 7.18 shows the phase current, actual shaft speed and estimator states present during identification. The second Fig. 7.19 shows the estimated parameters: rotor (PM in this case) flux, stator inductance and resistance. These results are shown together with the actual (reference) parameters as used in the model of the LVSERVMTR PM machine.

- At the start of operation ADC offset correction (Estimator state 1) takes place, which eliminates any hardware drift errors in the voltage/current measurement circuits.
- At $t \approx 5$ s the `EST_State_RoverL` (Estimator state 2) sequence starts, where a 300 Hz rotating current vector of 0.5 A is applied to the stationary motor (rotor locked during this phase). This measurement generates outputs `Lhf`, `Rhf`, which are the so called 'high frequency' estimates for the stator inductance and stator resistance respectively. These 'HF' values are used to dimension the current controllers in subsequent motor ID phases, hence this stage must yield credible and consistent results, prior to proceeding to the next phase.

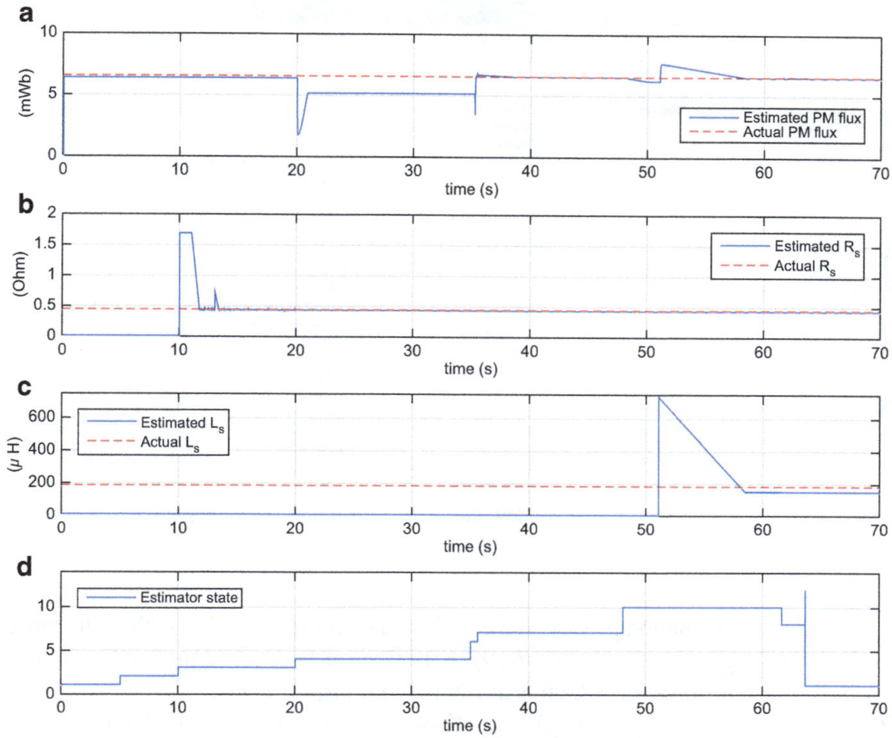

Fig. 7.19 Use of PLECS PIL approach for PM drive with motor identification capability: parameter data

- At $t \approx 10$ s the `EST_State_Rs` (Estimator state 3) sequence starts (see Fig. 7.19) which estimates the stator resistance. In the simulation the rotor remains locked during this phase. Current vector amplitude is 1 A, at 0 Hz.
- At $t \approx 20$ s the `EST_RampUp` (Estimator state 4) sequence starts where a rotating current vector with an amplitude of 1 A is applied with a frequency ramp from $0 \rightarrow 50$ Hz, the rotor lock is removed at this stage. Operation occurs under 'open-loop' speed control takes place where the shaft speed ramps up from zero to a steady-state speed of (in this case) 750 rpm) as may be observed from Fig. 7.18. At the end of this cycle the `EST_State_RateFlux_Ol`(Estimator state 6) is entered which estimates the 'open-loop' rotor flux.
- At $t \approx 37$ s the `EST_State_RateFlux` (Estimator state 7) sequence starts where the speed loop is closed. The current vector amplitude is ramped down and the resultant current level is now dictated by the load, which is typically due to friction (no external load should be connected).
- At $t \approx 48$ s the `EST_State_Ls`(Estimator state 10) sequence starts where a direct axis current is gradually applied $0 \rightarrow -2$ A (negative on purpose), whilst the machine remains under closed-loop speed control. At $t \approx 52$ s the actual

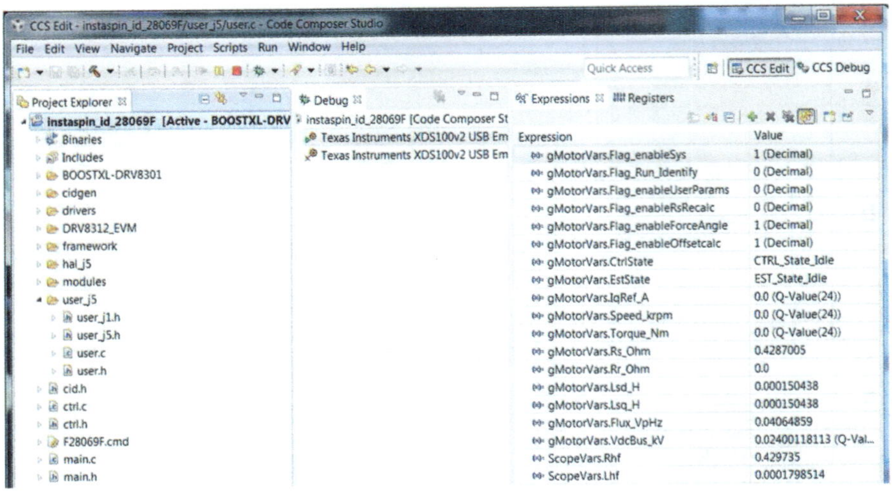

Fig. 7.20 CCS screen shot: watch window during PIL operation after completion of estimation cycle

inductance estimation process starts and should continue until a stable inductance estimate is generated (see Fig. 7.19). A 17 % error between actual and estimated inductance is apparent, which is attributed to transport delays.

- At $t \approx 62$ s the `EST_Ramp_down` (Estimator state 8) starts, which ramps down the current and concludes the identification sequence. In this case the estimator state is set to 'idle' (Estimator state 1) at $t \approx 64$ s.

Note that the ramp-up settings used for the frequency and current during the identification process are the 'default' values set in MotorWare.

Apart for the results shown in the PLECS program, the user can also observe the Watch window parameters during operation. A screen shot example, as given in Fig. 7.20 shows the estimated machine parameters: `Lsd`, `Rs` the PM flux level `Flux_VpHz` (flux in V/Hz) and HF inductance/resistance parameters `Lhf` and `Rhf` after completion of the identification sequence. Also indicated are the measured DC bus voltage, status of the controller and estimator. In this case all the flags are controlled via the watch window.

7.2.3 *Case Study P2c: Motor Identification Using InstaSPIN and PIL Technology: 'Normal Drive' Operation*

So called 'normal operation' refers to operation of the actual drive, i.e. with the actual converter and motor attached. The commissioning of 'normal operation' requires a virtually identical set of steps as described for the previous case study. Consequently, only the basic steps are discussed in this section. After startup of code composer (CCS), load the PIL prepared `FOC_Sensorless_28069F_MotorID` file, which is the MotorWare motor identification laboratory 2a that must be re-configured for the drive in use. After import of said laboratory the CCS screen shot shown in Fig. 7.16 should appear. A rebuild is activated by a right click on the file name: `FOC_Sensorless_28069F_MotorId` and selecting `Rebuild`, after which a file `.out` should be generated (located in the same folder that contains the PLECS simulation and folder `instaspin_id_28069F`). A 'debug' session is activated by hitting the 'green bug', after which the compile out file `.out` is downloaded to the MCU and the CCS debug screen will appear. Start debug by hitting the `Resume` button (green arrow), in which case a 'red' LED on the LAUNCHXL-F28069M module should start to flash.

The second phase of PIL commissioning requires access to the PLECS program `LaunchXL_Teknic_IDv3.plecs`, where the PIL module must be prepared for operation. A left double click on the PIL module leads to the window `PIL parameters` which shows the PIL target name. A left click on `Configure` leads to dialog box `PIL Targets`, in which the `symbols file` entry must reflect the path needed to access the file `xxx.out` generated in the previous step via CCS (this file will be in the folder `FOC_Sensorless_28069F_MotorId`).

The final PIL commissioning step requires a left click on the `Properties` dialog box which opens the `Properties of InstaSPIN-FOC` window. This window should show, among others, the line `Symbol file matches firmware on target` as shown in Fig. 7.21, which confirms that

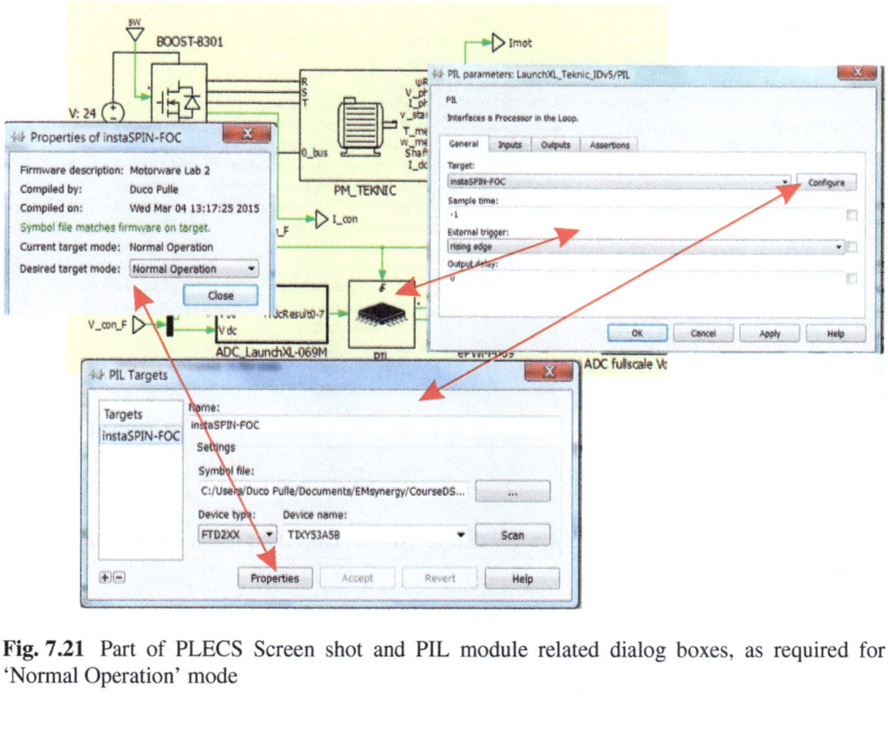

Fig. 7.21 Part of PLECS Screen shot and PIL module related dialog boxes, as required for 'Normal Operation' mode

communication with the PIL module is correct. To allow operation with the actual hardware select for the `Desired target mode` dialog, the option `Normal Operation`, which means running drive operation directly from the CCS watch window.

Actual drive operation, which implies motor identification must now be initialized by the user via the watch window, by setting the following flags:

- `Flag_Run_Identify` to 1: which ensures that the motor identification sequence starts after the flag `Flag_enableSys` is set.
- `Flag_enableSys` to 1: control initialized for operation.
- `Flag_enable ForceAngle` to 1: Activates the 'force angle' option, which is helpful for PM machines with cogging torque.

Upon completion of the identification sequence the watch window, as shown in Fig. 7.22 should appear. The motor ID sequence is carried out in real time and the motor ID stages are identical to those discussed in the previous section, i.e. with PIL active, as may be observed from Fig. 7.23, which shows the measured phase current. Results shown were obtained with a DC-true current probe and oscilloscope.

Readily apparent from this figure are the key motor identification phases as identified earlier in Fig. 7.18 using the same time sequence.

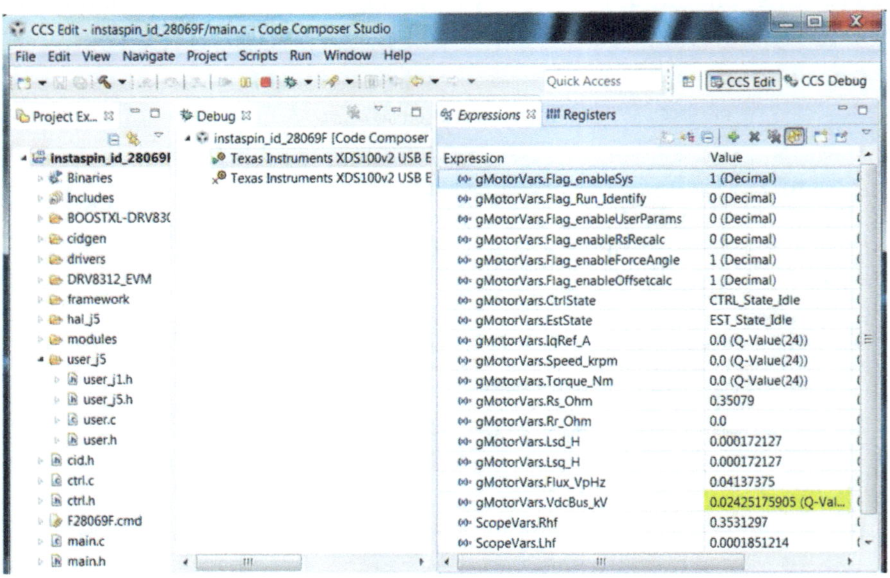

Fig. 7.22 CCS screen shot: watch window during 'Normal Operation' upon completion of motor identification

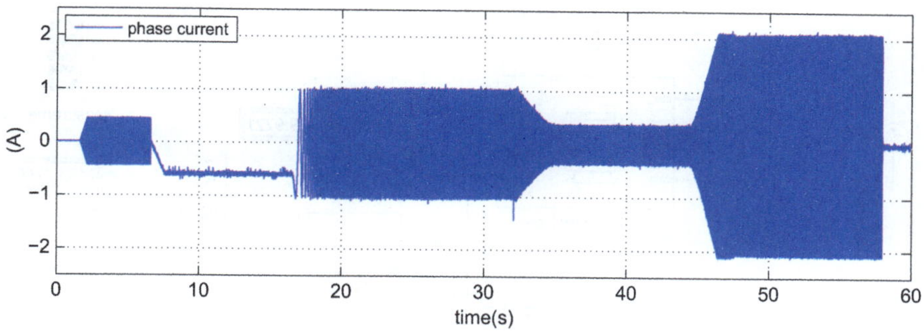

Fig. 7.23 Measured phase current during 'Normal Operation' after completion of motor identification: oscilloscope data displayed with Matlab

7.3 Case Study P3: e-Traction Converter with Single-Phase IM Motor

The purpose of this case study is to apply the PLECS 'processor in the loop' (PIL) approach to a unique drive setup which consists of an (unnamed) single phase induction machine connected to the e-Traction converter discussed earlier. In the latter case, as in the present case, InstaSPIN based FOC torque control is to be used, with the important change that the hardware is replaced by a PLECS model of the converter and single phase induction machine. The following information is relevant for this case study:

- Reference program : `IMsinphPIL_eTRAC_069v4.plecs`.
- Description: Sensorless control of a single phase induction machine connected to an e-Traction converter. Use of the PLECS PIL approach to model the drive, using MotorWare laboratory 4e software.
- Equipment/Software: Texas Instruments LAUNCHXL-F28069M (no boost packs required) and PLECS Simulation software version 3.6 with PIL capability.
- Outcomes: To demonstrate the possibility of sensorless InstaSPIN-FOC based control for a *single phase* induction machine.

As indicated above no boost packs are used in the laboratory, nor is an external power supply required, given that power to the MCU is directly provided via the USB cable connected to the laptop. In this case the following jumper and dip-switch positions on the LAUNCHXL-F28069M module are required:

- Jumpers `JP1` and `JP2` CLOSED
- Jumpers `JP4` and `JP5` OPEN
- Jumper `JP6` OPEN
- Jumpers `JP3` and `JP7` CLOSED
- Dip-switches `SW1` to `SW3` ON

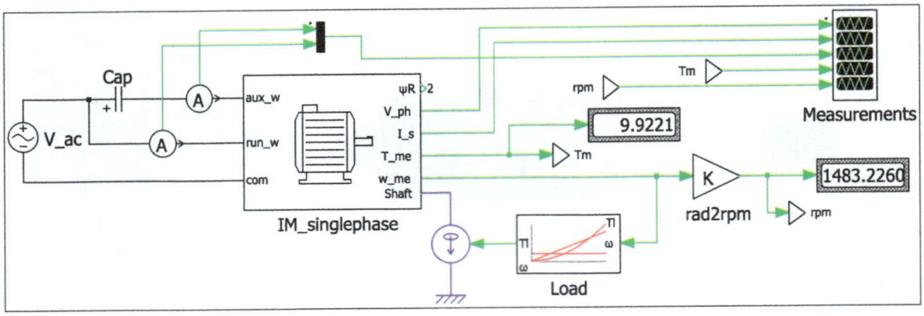

Fig. 7.24 PLECS model: Mains connected single phase capacitor run IM machine

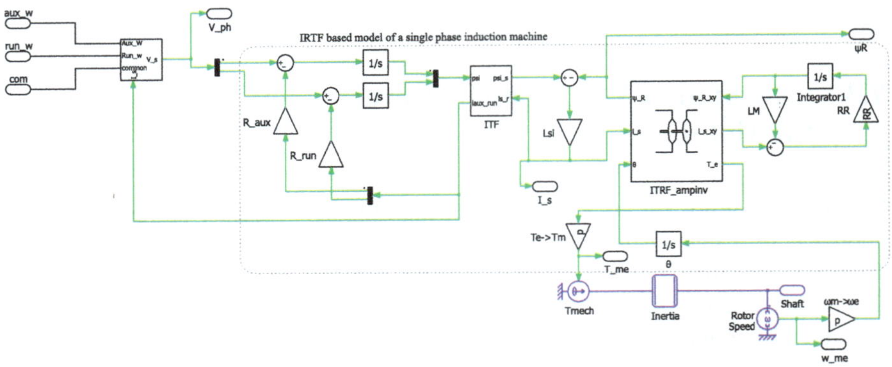

Fig. 7.25 IRTF based space vector model of single phase induction machine

Prior to discussing the sensorless approach it is instructive to consider the single phase machine characteristics under line start conditions. The PLECS model given in Fig. 7.24 shows a model of the single phase IM machine, of which the so called 'auxiliary' (aux) winding is connected to the 230 V_{rms}/50 Hz single phase supply via a 25 μF capacitor `cap`. The 'run' winding is connected directly to the mains supply. A quadratic load is assumed, which ensures a load torque of 2 Nm at 1400 rpm. Moving one level lower into the `IM_singlephase` machine model given in Fig. 7.24 reveals a set of PLECS electrical and control blocks as shown in Fig. 7.25. Central to this model is a so called 'ideal rotating transformer' (IRTF) control block as discussed in our previous books [4, 14]. The space vector component of this model is similar to that found in a 'standard' four parameter squirrel cage induction machine. The 'single' phase IM machine [4] is in fact a two phase induction machine, with two unbalanced (unequal number of turns) windings and unequal phase resistances. Hence the parameter set of this machine is somewhat different to the standard three-phase machine. The machine in question is defined by the following set of parameters:

- `R_run`: run winding resistance.
- `R_aux`: auxiliary winding resistance.

- `Lsi`: stator referred combined leakage inductance.
- `LM`: magnetizing inductance.
- `RR`: rotor resistance.
- `Aux-RunWindRat`: winding ratio (k) between the auxiliary and run winding.

The winding ratio parameter (conveniently referred to as k) is used in the 'ideal transformer' (ITF)[14] model shown in Fig. 7.25. Inputs to this ITF model are a flux vector $[\psi_{aux}\,\psi_{run}]^T$ and current vector `I_s`= $\left[i^r_{aux}\,i_{run}\right]^T$, where i^r_{aux} is known as the 'referred' auxiliary current. The ITF output vectors are defined by the current vector $[i_{aux}\,i_{run}]^T$ and flux vector $\left[\psi^r_{aux}\,\psi_{run}\right]^T$, where ψ^r_{aux} represent the 'referred' auxiliary flux of the machine. The relationship between the referred and actual auxiliary winding variables are defined as:

$$\psi^r_{aux} = \frac{\psi_{aux}}{k} \tag{7.1a}$$

$$i_{aux} = \frac{i^r_{aux}}{k} \tag{7.1b}$$

where k is the aux/run winding ratio typically in the range $1.1 \rightarrow 1.6$ for most single phase machines. Note that the 'run' winding flux and current variables are not transformed by the ITF module. Inputs to the electrical/control block interface module are the (electrical) `aux_w`, `run_w` terminals and the common (to both) connection `com`. Output of this module is the voltage vector `V_ph` defined as $[u_{aux}\,u_{run}]^T$ and these represent the voltage across the auxiliary and run winding respectively. Input to the interface module is the current vector $[i_{aux}\,i_{run}]^T$, which represents the current in the auxiliary and run winding respectively. Note that the machine in question would be a 'conventional' two phase machine in the event that the phase resistance would be equal and the winding ratio was set to $k = 1$. However, this is not the case, and the machine parameters used in this case study reflect an actual (unmarked) single phase four pole machine. A set of mechanical block modules is used to derive the shaft speed variable `w_me`, for a given inertia, which is similar to the approach used for standard AC machines.

A `Measurement` module is used to display (among others) the voltages/currents of the 'run' and 'aux' winding together with the shaft speed and shaft torque. An example of the results generated by this model, as given in Fig. 7.26 reflect the typical behavior of this type of drive. Clearly visible are the large transient currents and torque ripple of the machine, even under steady-state conditions, at a speed of 1483 rpm. Furthermore, these results show the auxiliary winding voltage and run winding voltage and corresponding auxiliary and run currents.

The challenge in this particular PLECS case study is to connect this machine to a three-phase converter and apply the required transformations to the measured voltages and current so that InstaSPIN perceives the inputs to be those of a 'standard' three-phase machine. Central to the PLECS model shown in Fig. 7.27

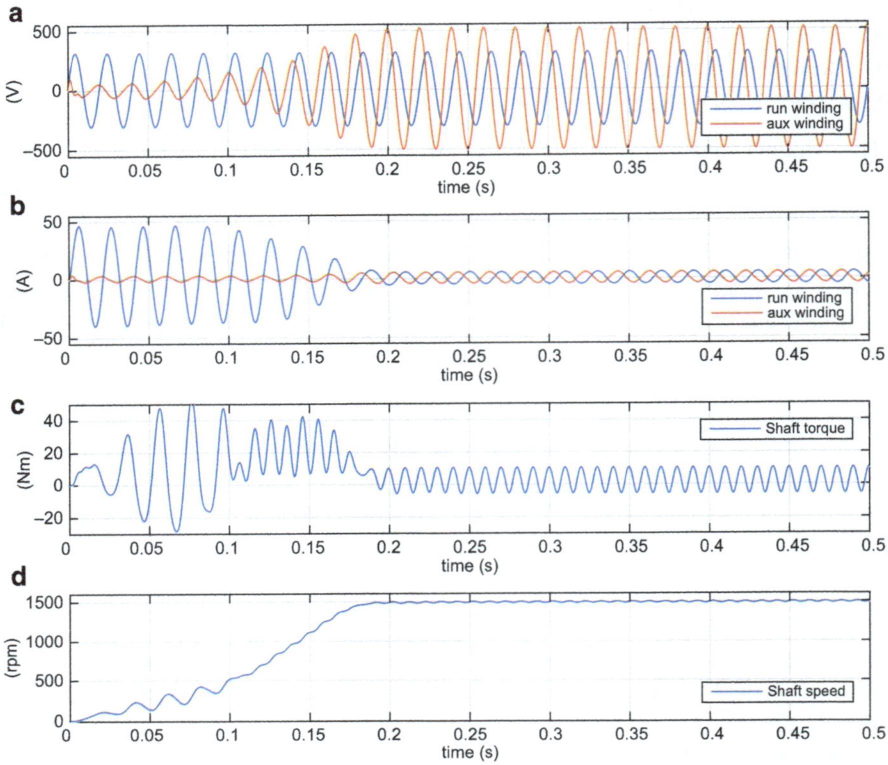

Fig. 7.26 Results mains connected single phase induction machine at completion of a 0.5 s simulation cycle

is the drive, which consists of the converter and machine model. Outputs of the converter are the low-pass filtered voltage vector `V_con_F` and measured current vector `I_con`, which, are connected to module `IM1ph>IM3ph` (see bottom right section in Fig. 7.27), that undertakes the transformation of the currents/voltage variables to pseudo three-phase outputs `v_ADC,I_ADC`. These are in turn connected to the 12 bit ADC model `ADC_LaunchXL-069M`. The ADC output of this model is supplied to the PIL module together with (in this case) three additional inputs namely:

- `Enable_Sys`: flag that is set by the user to enable operation.
- `Force_ang`: flag used to enable the user to enable Force Angle operation.
- `Iq`: the reference quadrature current value to be used by the FOC controller. This variable is controlled by a `PI speed controller`, which has been added for convenience to ensure that the steady-state speed matches the value used during mains connected operation (see Fig. 7.26).

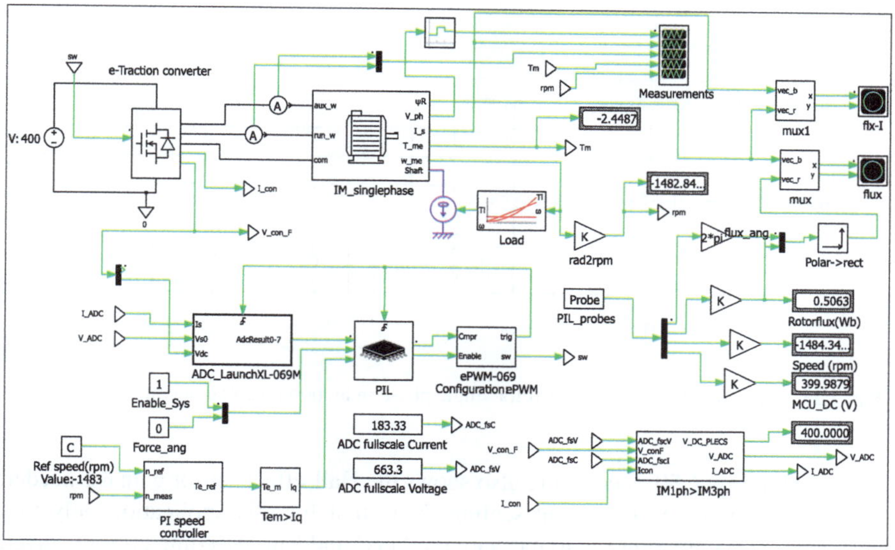

Fig. 7.27 e-Traction converter with single phase IM motor model and PIL operation using InstASPIN

During PIL operation (PLECS simulation synchronized to the MCU via the data link, 'blue' LED on), input data for the PIL is used by the InstaSPIN algorithm and then returns the modulation indices for the PWM model `ePWM_069M`, which in turn generates the control signals `sw` for the converter and most importantly the trigger signal for the ADC and the PIL module itself. A `PIL_probes` module provides access to a set of (user selectable) variables from the InstaSPIN module and (in this case) the estimator rotor flux amplitude/angle, shaft speed and DC bus voltage are shown on the numerical displays. For comparison purposes, the estimated rotor flux vector is shown together with the rotor flux vector generated by the machine model on a vector scope `flux`. Furthermore, FOC operation can be verified by vector scope module `flx-I`, which shows the stator current vector generated by the machine model together with the actual rotor flux vector. The load model which is generated by the module `Load`, is identical to that used for mains connected operation to ensure that the encountered load conditions are identical.

For measurement purposes a five channel `Measurements` scope is used to display the machine voltages/currents, shaft torque and shaft speed as undertaken for the main connected approach.

Moving one level into the `e-Traction Converter` module (see Fig. 7.28) reveals the converter structure and voltage/current signal conditioning circuits used to generate the inputs vectors `V_con_F`, `I_con` for the ADC module. The vector `sw` generated by the PWM model, controls the six ideal (in this case) power switches. For the single-phase application, the space vector algorithm needs to be adapted, as will be discussed in the following section. Voltage attenuation and filtering of the converter voltages makes use of a resistance/capacitor network as

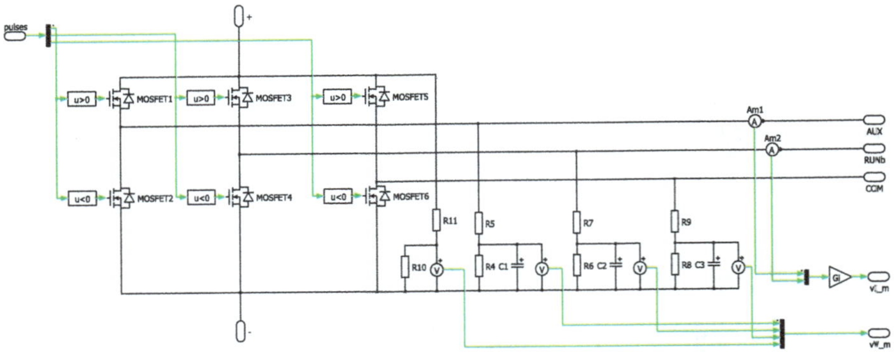

Fig. 7.28 e-Traction converter adapted for single phase induction machine operation

discussed in Fig. 6.59. Said figure also shows the Hall-effect sensor gain `Gi` needed to achieve correct ADC current scaling. Note that for this application only two currents sensors are required to measure the 'aux' and 'run' winding currents given that the 'common' `COM` phase carries the sum current of said two windings.

7.3.1 *Case Study P3a: e-Traction Converter with Single Phase IM Motor: PIL Drive Commissioning*

The commissioning of the PIL requires a virtually identical set of steps as described for the previous two case studies. Consequently, only the basic steps are discussed in this section. The PWM frequency, clock frequency and ADC fullscale current/voltage settings used for the e-Traction drive example (see Sect. 7.1) are also used here. After start-up of code composer (CCS), load the PIL prepared `instaspin_foc_28069F` folder which is the MotorWare motor identification laboratory 4 within the folder `IMsingPH`. After import of said laboratory the CCS screen shot shown in Fig. 7.29 should appear. Shown in this window is the directory associated with `instaspin_foc_28069F`, which includes a `user_j5.h` file, that is also expanded in the center window. This `user_j5.h` file contains all the parameter settings needed for sensorless operation using InstaSPIN. Note that the `_j5` extension signifies operation associated with the 'aft' boost pack. For the single phase case the 'run' winding is taken as the reference winding, hence the stator resistance value used here, is equal to R_{run} set in the motor model. The rated flux level used is calculated on the basis of the run winding rated RMS voltage and frequency values.

For single phase operation the space vector modulation module `svgen.h` must be adapted as shown in Fig. 7.30. The module changes are based on the case $k = 1.5$, which is the aux/run winding ratio of the motor in use. A rebuild is activated by a right click on the file name: `instaspin_foc_28069F` and selecting `Rebuild`,

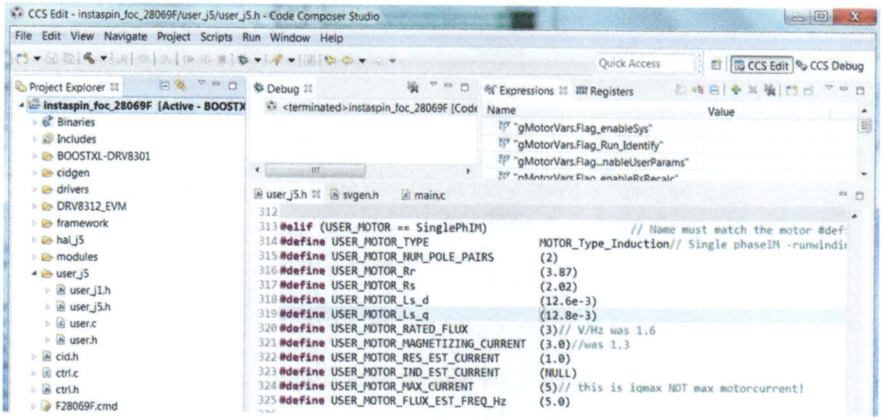

Fig. 7.29 Code Composer (CCS) Edit Screen shot: after import of adapted MotorWare project 4

after which a file `.out` should be generated (located in the same folder `IMsingPH` that contains the PLECS simulation and folder `instaspin_foc_28069F`). A 'debug' session must now be activated in order to run the compiled out file `.out`. Note that 'running' the file, implies in this case that the MCU will be placed under the control of the PLECS program, where the triggering of the MCU is carried out by the PWM model `ePWM_069M`. A 'debug' session is activated by hitting the 'green bug', after which the compiled `.out` file is downloaded to the MCU and the CCS debug screen will appear. Enable the `continuous refresh` feature ('yellow' double arrow button top right of screen), which will ensure that the watch window variables are updated during PIL operation. Start debug operation by hitting the `Resume` button ('green' arrow), in which case a 'red' LED on the LaunchXL module should start to flash.

The second phase of PIL commissioning requires access to the PLECS program `IMsinphPIL_eTRAC_069v4.plecs`, where the PIL must be prepared for operation. A left double click on the PIL module leads to the window `PIL parameters`, which shows the dialog box `Configure`. A left click on said dialog box leads to dialog box `PIL Targets`, in which the `symbol file` entry must reflect the path needed to access the file `xxx.out` generated in the previous step via CCS (this file will be in the folder `instaspin_foc_28069F`).

The final PIL commissioning step requires a left click on the `Properties` dialog box, which brings up the required window (provided that the 'red' LED on the LaunchXL is flashing). Then confirm that the 'green' line `Symbol file matches firmware on target` is shown, which signifies that PIL commissioning has been successful and that the PLECS run phase can start provided the option `Desired target Mode:Ready for PIL` has been selected.

```
{
    // modulator for  running two phase IM machine of three-p
     // _iq Vmax,Vmin,Vcom;
     // _iq Va,Vb,Vc;
    //   _iq maxModulation = SVGEN_getMaxModulation(handle);
    //   _iq maxModulationNeg = -maxModulation;
      _iq Va_tmp = pVab->value[0] + (pVab->value[0] >> 1); //
          _iq Vb_tmp = pVab->value[1]; // Vb

          _iq Vmax, Vmin;

          if(Va_tmp > Vb_tmp)
          {
                  Vmax = Va_tmp;
                  Vmin = Vb_tmp;
          }
          else
          {
                  Vmax = Vb_tmp;
                  Vmin = Va_tmp;
          }

          _iq Vcom = (Vmax+Vmin)>>1;

          pT->value[0] = Va_tmp-Vcom;
          pT->value[1] = Vb_tmp-Vcom;
          pT->value[2] = -Vcom;
  return;
} // end of SVGEN_run() function
```

Fig. 7.30 Code Composer (CCS) Screen shot of module 'svgen.h'

7.3.2 Case Study P1b: e-Traction Converter with Single Phase Induction Motor: PIL Drive Operation

With PIL commissioning completed and the MCU 'on-line' attention is directed towards running the PLECS program `IMsinphPIL_eTRAC_069v4.plecs`. However, prior to starting the program, select items; `Simulation`, `Simulation Parameters` of the respective pull-down menus and check that the simulation run is set to 4.0 s. Note that all the relevant drive parameters such as PWM and ADC sampling frequency are directly linked to the `user.h` file (via the `Simulation, Simulation Parameters, Initialization` window). Prior to running, set the `Force_ang` and `Enable_Sys` flags to 1. In addition, set the reference shaft speed variable `Ref speed` to 1483 rpm (which is the steady-state operating speed achieved with the mains connected machine and identical load). The screen shot example shown earlier (see in Fig. 7.27) was taken during running operation and shows

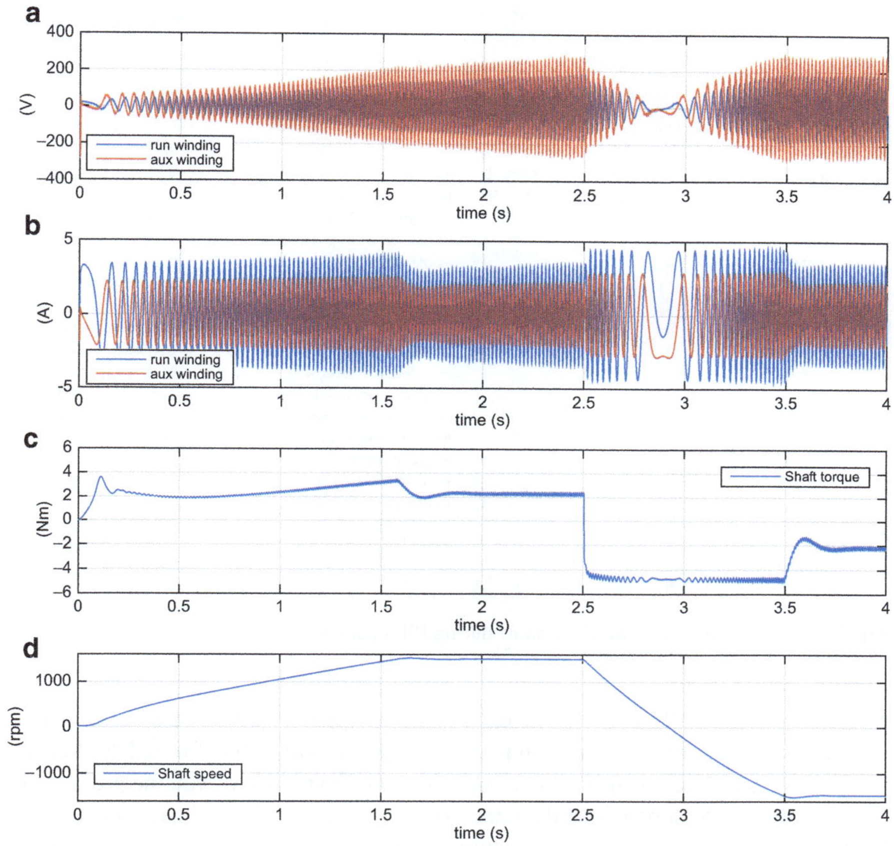

Fig. 7.31 Results sensorless FOC single phase induction drive at completion of a 4 s simulation cycle

(among others) the `PIL_probes`, which have been configured to show the MCU generated shaft speed, DC bus voltage and rotor flux. If operation is correct the DC voltage and speed must match those generated by the PLECS models directly (as shown by numerical displays linked to `rpm` and the bus voltage `V_DC_PLECS` (output from module `IM1ph>IM3ph`). Operation can be monitored by observation of the vector plots `flux`, `flx-I` and the `Measurements` scope. On the basis of the results shown in Fig. 7.31, the operation sequence and observed phenomena are itemized after completion of the chosen simulation interval.

- At the start of operation the machine is full de-fluxed, hence during the initial interval 0 $\rightarrow\approx$ 1.7 s the estimator flux and actual flux come into alignment and at the same time rotor flux builds to its steady-state value, which corresponds to a direct axis current of 3.0 A. At the end of this period the reference speed value of 1483 rpm has been reached and force angle is turned off (this is done because a speed reversal is required).

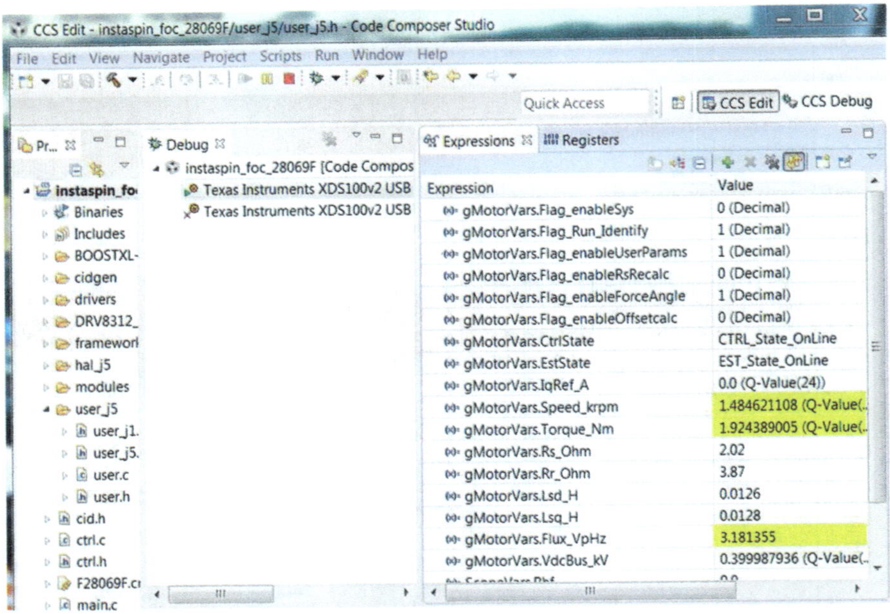

Fig. 7.32 CCS screen shot: watch window during PIL operation

- At $t \approx 2.5$ s a reference speed reversal $1483 \rightarrow -1483$ rpm is invoked which causes a quadrature current step change (maximum value limited to 5 A via `user_j5.h` file setting). This causes the speed to reduce (given the presence of a quadratic load torque/speed characteristic). Torque will also be reversed (given that it is directly linked to the quadrature current change, while the rotor flux is kept constant).
- At $t \approx 3.6$ s the shaft speed stabilizes to its steady-state value (torque generated by the machine matches the load torque) and will be equal to -1483 rpm.

Apart for the results shown in the PLECS program, the user can also observe the Watch window parameters during operation. A screen shot example, as given in Fig. 7.32 shows (marked in 'yellow') the estimated rotor flux. Also indicated are the machine parameters (which show the stator resistance and leakage inductance of the 'run' winding), measured DC bus voltage and status of the controller/estimator both of which are `_OnLine` as expected. As mentioned earlier, a number of the flags are set via PLECS directly and cannot therefore be changed in the watch window.

A comparison between the measured results obtained with the mains connected machine (Fig. 7.26) and the FOC sensorless controlled machine (Fig. 7.31) shows a substantial improvement in drive performance and controllability. Furthermore, this case study highlights the effectiveness of the PIL approach for (among others) developing a drive concept for a non-standard InstaSPIN (as is the case here) sensorless application.

Erratum:

Applied Control of Electrical Drives

Real Time Embedded and Sensorless Control using VisSim™ and PLECS™

Duco W.J. Pulle, Pete Darnell, and André Veltman

D.W.J. Pulle et al., *Applied Control of Electrical Drives*,
Power Systems, DOI 10.1007/978-3-319-20043-9

DOI 10.1007/978-3-319-20043-9_8

The print and online versions of the book contain some errors, and the corrections to these versions are given below:

Chapter 3
The following text is missing in the first page of Chapter 3

Electronic supplementary material: The online version of this chapter (doi: 10.1007/978-3-319-20043-9_3) contains supplementary material, which is available to authorized users.

Chapter 4
The following text is missing in the first page of Chapter 4

Electronic supplementary material: The online version of this chapter (doi: 10.1007/978-3-319-20043-9_4) contains supplementary material, which is available to authorized users.

Chapter 5
The following text is missing in the first page of Chapter 5

Electronic supplementary material: The online version of this chapter (doi: 10.1007/978-3-319-20043-9_5) contains supplementary material, which is available to authorized users.

D.W.J. Pulle et al., *Applied Control of Electrical Drives*,
Power Systems, DOI 10.1007/978-3-319-20043-9_8

Chapter 6
The following text is missing in the first page of Chapter 6

Electronic supplementary material: The online version of this chapter (doi: 10.1007/978-3-319-20043-9_6) contains supplementary material, which is available to authorized users.

Chapter 7
The following text is missing in the first page of Chapter 7

Electronic supplementary material: The online version of this chapter (doi: 10.1007/978-3-319-20043-9_7) contains supplementary material, which is available to authorized users.

The online version of the original book can be found at
http://dx.doi.org/10.1007/978-3-319-20043-9

References

1. C2000 Systems and Applications Team (2010) Implementing Sensorless Field Oriented Control of AC Induction Motor with PFC front end running on CLA using C2000 Microcontrollers. Tech. rep., Texas Instruments, location: ControlSUITE installation directory\development_kits\HVMotorCtrl+PfcKit_v1.7\PFC2PhiLCLA+ACI\~Docs\PFC+ACI_Integration.pdf
2. Clearman C (2014) motorware_selecting_user_variables.xlsx. Location: motorware installation directory/docs/labs/
3. Contronics (2015) Comoco intelligent motor control. URL www.comoco.nl
4. De Doncker R, Pulle DWJ, Veltman A (2011) Advanced Electrical Drives. Power Systems, Springer Netherlands, Dordrecht, URL link.springer.com/10.1007/978-94-007-0181-6
5. De Gusseme K, de Sype DMV, Van Den Bossche AP, Melkebeek JA (2003) Sample correction for digitally controlled boost PFC converters operating in both CCM and DCM. In: Eighteenth Annual IEEE Applied Power Electronics Conference and Exposition, 2003. APEC '03, IEEE, Miami Beach, Florida, vol 1, pp 389–395, DOI 10.1109/APEC.2003.1179243, URL ieeexplore.ieee.org/xpl/articleDetails.jsp?arnumber=1179243
6. e-Traction (2015) e-Traction. URL www.e-traction.eu
7. LaunchXL (2015) LAUNCHXL-F28069m. URL www.ti.com/tool/launchxl-f28069M
8. LineStream (2015) LineStream Technologies. URL linestream.com
9. MotorWare (2015) MotorWare Software. URL www.ti.com/tool/motorware
10. Plecs (2015) Plecs Simulation Software for Power Electronics. URL www.plexim.com
11. Pulle DWJ (2015) Laboratory files for Applied Control of Electrical Drives. URL extras.springer.com
12. Thomas E (2015) A2spin LLC. URL www.a2spin.com
13. TMS320F28027 errata (2008) TMS320f28027, TMS320f28026, TMS320f28023, TMS320f28022, TMS320f28021, TMS320f28020, TMS320f280200, TMS320f280270, TMS320f280260, TMS320f280230, TMS320f280220 Piccolo MCU Silicon Errata. URL https://www.ti.com/product/tms320f28027
14. Veltman A, Pulle DWJ, De Doncker RW (2007) Fundamentals of electrical drives. Power systems. Springer, Berlin. http://dx.doi.org/10.1007/978-1-4020-5504-1; http://www.springerlink.com/openurl.asp?genre=book&isbn=978-1-4020-5503-4
15. VisSim (2015) VisSim Embedded™ Suite. URL www.vissim.com/products/vissim/embedded.html
16. Vostermans (2015) Vostermans Ventilation. URL ventilation.vostermans.com
17. Walkera (2015) Walkera. URL www.walkera.com

D.W.J. Pulle et al., *Applied Control of Electrical Drives*,
Power Systems, DOI 10.1007/978-3-319-20043-9

Abbreviations

AC	Alternate Current
A/D	Analog-Digital
ADC	Analog to Digital Converter
CCS	Code Composer Studio
CPU	Central Processing Unit
DC	Direct Current
DFO	Direct Field Orientation
EMF	Electro Motive Force
EOC	End Of Conversion
FOC	Field Oriented Control
ID	(motor) IDentification
IM	Induction Machine
IRTF	Ideal Rotating TransFormer
ITF	Ideal Transformer
JTAG	Joint Test Action Group
LSB	Least Significant Bit
LED	Light Emitting Diode
LPF	Low Pass Filter
MCU	Micro Controller Unit
OP-AMP	Operational Amplifier
PFC	Power Factor Correction
PI	Proportional Integral
PIL	Processor In the Loop
PM	Permanent Magnet
PWM	Pulse Width Modulation
RAM	Random Access Memory
RHS	Right Hand Side
RMS	Root Mean Square

D.W.J. Pulle et al., *Applied Control of Electrical Drives*,
Power Systems, DOI 10.1007/978-3-319-20043-9

ROM	Read Only Memory
S/H	Sample and Hold
SOC	Start Of Conversion
SVM	Space Vector Module (pulse centering)
TBCTR	Time Base CounTeR
TBPRD	Time Base PeRioD
TI	Texas Instruments
V/f	Voltage frequency controller

List of symbols

a	acceleration
θ	angle in electrical degrees
C	capacitance
I	current
i	current
D	diode
e	back e.m.f.
ψ	flux linkage
f	frequency
Z	impedance
L	inductance
J	inertia
K	parameter
k	factor (e.g. winding factor)
m	modulation index
T	period
p	power
p	pole pair number
r	radius
X	reactance
R	resistance
T	simulation or sampling time
s	slip
ω	angular speed
n	speed
S	switch
t	time
T	torque

D.W.J. Pulle et al., *Applied Control of Electrical Drives*,
Power Systems, DOI 10.1007/978-3-319-20043-9

U	voltage
u	voltage
x	auxiliary variable
$\mathrm{G_i}$	Gain current
$\mathrm{G_R}$	Gain
$\mathrm{G_v}$	Gain voltage
$\mathrm{K_e}$	EMF constant
$\mathrm{K_i}$	integral gain (e.g. of current controller)
$\mathrm{K_p}$	proportional gain (e.g. of current controller)
$\mathrm{K_{pi}}$	effective Gain of PI controller (3.6)
a	transformation factor
τ	time constant
τ_{R}	Time constant (rotor)
Sw	switching signal

List of indices

$\hat{X}$	amplitude
$X^{\alpha\beta}$	fix stator coordinates
X^{dc}	Direct Current
X^{dq}	field oriented coordinates
X^{fs}	full scale value)
X^{HF}	High Frequency
X^{IM}	Induction mot
X^{lim}	limit
X^{LL}	line to line
X^{max}	maximum (e.g. maximum current)
X^{nom}	nominal or rated (e.g. rated stator current)
X^{n}	normalized (e.g. normalized stator current)
X^{PM}	PMmot
X^{ref}	reference (e.g. reference current)
X^{*}	reference
X^{rms}	Root Mean Square
X^{r}	in rotor flux oriented coordinate system
X^{s}	in stator flux oriented coordinate system
X^{xy}	rotor-oriented coordinates
X^{i}	input
X^{o}	output
X_{α}	real component of quantity in stator coordinates
X_{A}	amplitude
X_{aux}	auxiliary (e.g. auxiliary winding of single-phase induction machine)
X_{b}	bandwidth
X_{β}	imaginary component of quantity in stator coordinates
X_{cc}	current controller
X_{conv}	converter

D.W.J. Pulle et al., *Applied Control of Electrical Drives*,
Power Systems, DOI 10.1007/978-3-319-20043-9

X_d	real component of quantity in field oriented coordinates
X_{DC}	DC (e.g. DC link voltage)
X_e	electrical
X_f	excitation (e.g. excitation flux)
X_i	integral (e.g. integral component)
$X_{\{101\}}$	switching states for given voltage vector
X_k	integer variable used as counter (e.g. time step in discrete system)
X_l	load (e.g. load torque)
X_M	transformed main quantity
X_m	main quantity (e.g. main inductance)
X_m	mechanical
X_n	normalized; per unit
X_A	phase A
X_B	phase B
X_C	phase C
X_i	phase index
X_{PM}	Permanent Magnet
X_p	proportional (e.g. proportional component)
X_q	imaginary component of quantity in field oriented coordinates
X_R	transformed rotor quantity
X_r	rotor quantity (e.g. rotor resistance)
X_{run}	run (e.g. run winding of single-phase induction machine)
X_s	sampling (e.g. sampling time)
X_{sc}	short circuit (e.g. short circuit current)
X_{slip}	slip (e.g. slip frequency)
X_{sp}	sampling (e.g. sampling time of speed controller)
X_s	stator quantity (e.g. stator resistance)
X_σ	stray or leakage (e.g. stray or leakage flux)
X_i	indicates a certain instant in time
X_x	real component of quantity in rotor-oriented coordinates
X_y	imaginary component of quantity in rotor-oriented coordinates
X_0	zero
X_{en}	per unit electrodynamic torque
X_{ein}	integral component of per unit torque
X_{pp}	peak to peak value
$\vec{X}$	space vector

Index

D.W.J. Pulle et al., *Applied Control of Electrical Drives*,
Power Systems, DOI 10.1007/978-3-319-20043-9

Lightning Source UK Ltd.
Milton Keynes UK
UKOW06n1237120916

282798UK00001B/17/P